Resolving Erroneous Reports in Toxicology and Therapeutic Drug Monitoring

Resolving Erroneous Reports in Toxicology and Therapeutic Drug Monitoring

A Comprehensive Guide

Amitava Dasgupta, PhD, DABCC
Professor of Pathology and Laboratory Medicine
University of Texas Health Sciences Center at Houston

WILEY
A JOHN WILEY & SONS, INC., PUBLICATION

Published by John Wiley & Sons, Inc., Hoboken, New Jersey.
Published simultaneously in Canada.

Library of Congress Cataloging-in-Publication Data:

Dasgupta, Amitava, 1958–
Resolving erroneous reports in toxicology and therapeutic drug monitoring : a comprehensive guide / Amitava Dasgupta.
p. ; cm.
Includes bibliographical references.
ISBN 978-1-118-14965-2 (cloth)
I. Title.
[DNLM: 1. Drug Monitoring–methods. 2. False Positive Reactions. 3. Pharmaceutical Preparations–analysis. 4. Substance Abuse Detection–methods. WB 330]

615.7'040724–dc23

2012002022

10 9 8 7 6 5 4 3 2 1

Contents

Preface

Therapeutic drug monitoring and testing for drugs of abuse are important clinical laboratory tests that have a significant impact on patient safety and patient management. Physicians rely on serum or whole blood levels of a therapeutic drug for dosage adjustment and also to interpret significant drug–drug interactions. Therefore a falsely elevated or falsely lowered drug concentration due to the presence of an interfering substance in the specimen has a very serious impact on patient safety. The interference can be a false positive or a false negative. A clinician usually questions the validity of a test result if the concentration of a therapeutic drug is unexpectedly high. However, negative interference has more serious clinical consequences because it occurs infrequently compared with positive interference, and a clinician may simply increase the dosage of the medication without recognizing that the drug concentration may be falsely low due to the presence of an interfering substance in the specimen. There are reports in the literature of severe digoxin toxicity due to increased digoxin dosage based on reports of low digoxin concentration due to negative interference (see Chapter 12). Drugs of abuse testing using immunoassays is subjected to interference. Many over-the-counter cold and cough medications containing ephedrine or pseudoephedrine may cause a false-positive amphetamine immunoassay test result due to cross-reactivity with antibodies used in these immunoassays. A false-positive phencyclidine test result due to the presence of dextromethorphan, a common ingredient in many over-the-counter cold medications, is well documented in the literature. Such false-positive test results are not of concern for workplace drug testing because positive immunoassay test results are always confirmed by a chromatographic method, most commonly gas chromatography combined with mass spectrometry (GC/MS). However, for medical drug testing where

GC/MS confirmation is not available, a physician may falsely accuse a patient regarding his or her drug abuse based on a false-positive immunoassay result, although in reality the patient is not a drug abuser. This may hamper physician–patient trust or may cause mental agony to the patient. A false-positive blood alcohol result using an enzymatic alcohol assay has a similar consequence. A false-positive blood alcohol level measured by breath analyzer may have a serious legal impact because a driver may be falsely accused of driving under the influence of alcohol (see Chapter 7 for a detail discussion on this topic).

This book provides a comprehensive guide for laboratory professionals and clinicians regarding the sources of errors in therapeutic drug monitoring and drugs of abuse testing and how to resolve such errors and identify discordant specimens. Error-free laboratory results are essential for patient safety. Because herbal medicines are widely used by the general population, drug–herb interactions are discussed. For example, warfarin is known to interact pharmacokinetically and pharmacodynamically with many drugs; Chapter 9 discusses many clinically significant interactions of warfarin with herbal supplements. Chapter 16 is devoted to other important drug–herb interactions where an apparent erroneous result in therapeutic drug monitoring is due not to the presence of an interfering substance but to clinically important drug–herb interactions. Clinical laboratory testing is also helpful in the diagnosis of certain plant poisoning and toxicities from the use of certain herbal supplements (Chapter 10).

This book is intended as a practical guide for laboratory professionals and clinicians who deal regularly with erroneous results in therapeutic drug monitoring and drugs of abuse testing. I hope this book will help them become more aware of such sources of errors and empower them to eliminate such errors when feasible.

I would like to thank Robert L. Hunter, chairman of the Pathology and Laboratory Medicine Department at the University of Texas–Houston Medical School, for his support when I worked on the project. I also thank Alice Wells for critically reading the entire manuscript and making helpful suggestions. Last but not least I thank my wife, Alice, for tolerating my long hours spent on writing the book on weekdays and weekends. Finally, readers will be the judge of the final success of this book. If they find this book useful, that will be my best reward for writing it.

Amitava Dasgupta
Houston, Texas

Chapter 1

An Introduction to Tests Performed in Toxicology Laboratories

1.1. INTRODUCTION

In general therapeutic drug monitoring, urine toxicology drug screens, analysis of blood alcohol and volatiles as well as emergency toxicology drug screenings are commonly offered tests in a toxicology laboratory. Certain drugs with a narrow therapeutic range require routine monitoring, and in general serum or plasma is the preferred specimen. However, certain immunosuppressant drugs such as cyclosporine, tacrolimus, sirolimus, and everolimus are monitored in whole blood, although another immunosuppressant, mycophenolic acid, is monitored in serum or plasma. Drug screening for a patient with a suspected drug overdose is more commonly performed using urine specimens, but blood and gastric fluid specimens are also analyzed for the screening of drugs in case of a suspected recent overdose. In addition, blood alcohol analysis is also commonly conducted in toxicology laboratories because alcohol use alone may cause life-threatening intoxication. In addition, many abusers of illicit drugs also consume alcohol at the same time to achieve euphoria. Bogstrand et al reported that psychoactive substances were found in approximately 50% of the patients admitted to the hospital within 12 hours of injury. Of a total of 1272 patients studied (510 women and 762 men), 38% of the women and 48% of the men had a positive blood sample of a psychoactive drug on admission. Alcohol was the most prevalent substance; 27% of patients had a positive blood alcohol test. Cannabis was the most prevalent illicit drug (6.2%); diazepam was the

Resolving Erroneous Reports in Toxicology and Therapeutic Drug Monitoring: A Comprehensive Guide, First Edition. Amitava Dasgupta.

most common drug, detected in 7.4% of patients. The authors concluded that alcohol was the most common substance found in these patients and was particularly related to violence, whereas medicinal drugs were most prevalent in accidents at home (1). Alcohol is also a risk factor for injury in adolescents. Injured adolescents are more likely to visit the emergency department with an alcohol-related event during the early hours of the morning (2). Multiple abused drugs are also encountered in severely intoxicated patients and individuals who die from a drug overdose. Dickson et al reported a case of a 22-year-old white man who died from a drug overdose. Routine toxicological analysis detected morphine in the decedent's blood (0.06 mg/mL). In his urine specimen, 6-monoacetyl morphine (a marker compound for heroin abuse), morphine, codeine, doxylamine, and mephedrone were confirmed (3). In addition to poisoning due to alcohol, an overdose with various drugs may provoke a visit to the emergency department. Both salicylate and acetaminophen are commonly encountered drug in poisoned patients, and such drug levels are often screened in a toxicology laboratory using serum or plasma specimens.

1.2. ACETAMINOPHEN AND SALICYLATE ASSAYS

Acetaminophen (paracetamol) overdose, both intentional and accidental, remains a significant public health concern. In one report, the authors calculated that from 2000 to 2006, an age-adjusted rate of hospitalization related to acetaminophen was 13.9 per 100,000 population in the United States. Most acetaminophen overdoses were intentional (4). Acetaminophen can also cause liver toxicity. Because acetaminophen is a component of many medications, both prescription and over the counter, unintentional overdose can occur. Concurrent use of alcohol may also potentiate hepatoxicity of acetaminophen (5). Chronic alcohol abusers are also at an increased risk of acetaminophen-induced hepatotoxicity even after therapeutic use (6).

Salicylate poisoning is also common, and an adult can die from it. In 2005, according to the Toxic Exposure Survey from the American Association of Poison Control Center's National Poisoning and Exposure Database, there were more than 20,000 reported exposures from salicylate, and 64% of these patients were treated in a health care facility. It was considered that 50% of all exposures were intentional, and 60 patients died from a salicylate overdose (7). Galbois et al reported the case of a 74-year-old schizophrenic patient who died of salicylate poisoning; his blood salicylate level was 876 mg/L (87.6 mg/dL, a very toxic level) (8). Salsalate is a nonsteroidal anti-inflammatory drug that is mostly metabolized to two molecules of salicylic acid. However, approximately 7–10% of the drug is not hydrolyzed to salicylic acid and can be recovered in the urine either as the unchanged drug or as glucuronide conjugate (9). Delayed salicylate toxicity without early manifestation may occur

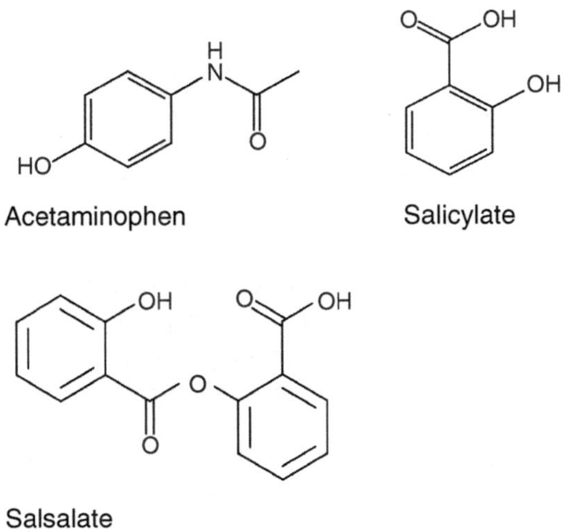

Figure 1.1. Chemical structures of acetaminophen, salicylate, and salsalate.

after overdose with both salicylate and salsalate. Chemical structures of salicylate, salsalate, and acetaminophen are given in Figure 1.1.

Case Study

A 14-year-old girl ingested 120 tablets of 81 mg aspirin extended release and 6 tablets of ciprofloxacin 2 hours prior to arrival at the emergency department. Upon arrival she denied nausea, shortness of breath, diaphoresis, or abdominal pain. Activated charcoal 50 g with sorbitol was administered orally for decontamination. No acetaminophen or ethanol was detected in her blood. In addition, a urine drug of abuse screen was also negative. The first salicylate blood level was 1 mg/dL (therapeutic: 10–20 mg/dL) drawn 4 hours after ingestion, but the salicylate level was elevated to 13 mg/dL 6 hours after ingestion, and the patient remained asymptomatic. The patient remained asymptomatic until 35 hours after exposure when she developed dizziness, tinnitus, and epigastric discomfort, and her blood salicylate concentration was elevated to 46 mg/dL. A second dose of 50 g of activated charcoal was administered along with bicarbonate infusion. She did not develop any renal failure, and after an observation period she was discharged to a psychiatric facility. The cause of delayed salicylate toxicity was unclear. Possibilities include delayed absorption due to enteric-coated or extended-release formulation, or the formation of bezoars (aggregates of drug that form a soft mass with limited surface area exposed to gastric fluid). The interior portion of the drug mass has mostly undissolved drugs. Other than salicylate, a variety of medications may form such bezoars (10).

Case Study

A 31-year-old man with a history of depression, posttraumatic stress disorder, and prior attempted suicide was discovered by his neighbor in the morning with greatly reduced consciousness. He was transferred to a hospital by the emergency medical team, and he admitted that he had attempted suicide the previous night by overdose but did not disclose the medication taken. Thirty-six new and old pill bottles were found on the scene that included acetaminophen, hydrocodone with acetaminophen, hydroxyzine, ibuprofen, lorazepam, magnesium oxide, morphine, oxycodone, paroxetine, ranitidine, salsalate, temazepam, tramadol, venlafaxine, and zolpidem. On arrival at the hospital, his blood pressure was 162/92 mm Hg, pulse 100 bpm, respiratory rate 14/min, and 98% oxygen saturation at room air. Toxicological investigation revealed a serum salicylate level of 29.2 mg/dL, and a urine drug screen was positive for benzodiazepines and cannabinoids. After 3 hours, the patient's level of consciousness and respiratory rate both decreased, and an arterial blood gas showed a pH of 7.31, a pCO_2 of 48, and pO_2 of 111. He was intubated 5.5 hours after admission due to apnea, and sodium bicarbonate was administered intravenously. In addition, two doses of activated charcoal (50-g dose) were administered by nasogastric tube. Salicylate concentration then peaked to 55 mg/dL just over 8 hours after presentation to the emergency department. Later his salicylate blood level declined, and sodium bicarbonate therapy was discontinued. Unfortunately, his blood salicylate level increased again later, peaking at 61.7 mg/dL 67 hours after presentation. The patient was later extubated and kept on a psychiatric hold with a one-to-one sitter. Salicylate ingestions are known to demonstrate unusual toxicokinetics and absorption patterns during overdose, and in this case a return to a toxic salicylate level was observed after apparent resolution of toxicity (11).

Acetaminophen and salicylate in serum, plasma, or urine can be measured by commercially available assays that may be either based on colorimetric principle or are immunoassays. These assays can be run on various automated analyzers. In addition, chromatographic methods such as high-performance liquid chromatography or gas chromatography can also be used for the determination of both acetaminophen and salicylate in various biological matrixes. Gaspari and Locatelli described a simple high-performance liquid chromatographic determination of both salicylate and acetaminophen in plasma after liquid-liquid extraction with hexane and ultraviolet detection at 228 nm (12). Miceli et al also described a liquid chromatographic method for the determination of salicylate and acetaminophen in human plasma using 8-chlorotheophylline as the internal standard (13). However, chromatographic procedures are labor intensive, and in toxicology laboratories, various automated assays are commonly used for routine determination of both salicylate and acetaminophen.

Unfortunately, these automated assays are subjected to interferences, and the presence of high bilirubin in serum or plasma may affect both the acetaminophen and salicylate assays. Stewart and Watson reviewed various methods available for the estimation of salicylate and acetaminophen in serum, plasma, and urine (14).

A false-positive acetaminophen level due to hyperbilirubinemia has been reported. In one report, the authors observed false-positive acetaminophen levels in two patients who had high bilirubin concentrations (25.5 mg/dL and 40.1 mg/dL, respectively) in their sera using the GDS Diagnostics enzymatic acetaminophen assay (GDS Diagnostics, Elkhart, IN). However, enzyme-multiplied immunoassay technique (EMIT) (Syva, Palo Alto, CA), acetaminophen assay, and gas chromatography/ mass spectrometric (GC/MS) assay did not reveal the presence of acetaminophen. The GDS assay utilizes an enzyme (n-arylacylamidase) to convert acetaminophen into para-aminophenol and acetate. Then p-aminophenol reacts with ortho-cresol in the presence of periodate to form the chromophore indophenol, which has a strong absorption spectra at 615 nm. The EMIT assay utilizes an antibody that recognizes acetaminophen. Although the mechanism of interference with the GDS enzymatic assay is unknown, the authors speculated that bilirubin may form a complex with ortho-cresol (15). Polson et al concluded that false-positive acetaminophen test results may occur when bilirubin concentration is above 10 mg/dL, leading to potential clinical errors especially with enzymatic-colorimetric assays (16). Significant positive bias of bilirubin in the Trinder reaction-based salicylate methods (color complex formed due to reaction of salicylate with ferric ions) on automated analyzers has been reported. However, such interference can be eliminated by using the fluorescence polarization immunoassay for salicylate using the AxSYM analyzer (Abbott Laboratories, Abbott Park, IL) (17). Broughton et al also described interference of bilirubin on a salicylate assay performed using the Olympus automated analyzer (18). Mitochondrial acetoacetyl-CoA thiolase deficiency is a rare metabolic disorder causing acute episodes of severe ketosis and acidosis. Tilbrook reported false-positive salicylate in an 18-month-old boy who presented to the hospital with severe acidosis. The authors concluded that false-positive salicylate using the Trinder reagent was due to the interference of a high level of acetoacetate in the specimen that interfered with the assay (19). However, immunoassays for salicylate manufactured by various diagnostic companies are free from such interferences.

Case Study

A 31-year-old woman was admitted to the hospital for abdominal pain, decreased appetite, malaise, confusion, and tea-colored urine. Investigation showed acute liver failure characterized by high bilirubin (70.7 mg/dL), alanine aminotransferase 6170 U/L, aspartate aminotransferase 5080 U/L, lactate dehydrogenase 6830 U/L, and alkaline phosphatase 150 U/L. Plasma acetaminophen concentration of 121 μmol/L (therapeutic up to 100 μmol/L) resulted in suspicion of an acetaminophen overdose as the probable cause of liver failure because serological tests for hepatitis A and B were negative. However, her plasma acetaminophen level remained elevated even on day 3 (104 μmol/L) raising the suspicion of bilirubin interference in acetaminophen measurement because the acetaminophen assay on the Vitros analyzer (Johnson & Johnson, Rochester, NY) is based on the enzymatic conversion of acetaminophen to para-aminophenol and subsequent reaction with ortho-cresol to form the blue-colored complex indophenol, which is measured by change in absorption at 600 nm. High bilirubin interferes with the assay. When the authors measured acetaminophen concentration using protein-free ultrafiltrate, which is free from protein-bound bilirubin, the acetaminophen concentration was below the detection limit of the assay, indicating no acetaminophen was present in the plasma. When acetaminophen was remeasured using a chromatographic method, no acetaminophen level was detected, further establishing that the initial high acetaminophen result was a false-positive result due to the interference of bilirubin with the acetaminophen assay (20).

1.3. ANALYSIS OF ALCOHOL

Alcohol is a major cause of motor vehicle accidents, and such victims are treated in the emergency department of hospitals. Blood alcohol testing is a routine and widely ordered test in a toxicology laboratory. Blood alcohol can be measured by either an enzymatic method or by gas chromatography. Although enzymatic methods can be automated and are often applied for measuring blood alcohol in busy toxicology laboratories, these methods also suffer from interferences, especially if both high lactate and lactate dehydrogenase are present in the specimen. However, gas chromatographic methods are free from such interferences, and such methods should be used for legal alcohol determination. This important topic is discussed in detail in Chapter 7. In addition, gas chromatographic methods are capable of analyzing other volatile compounds such as methanol, isopropyl alcohol, acetone, ethylene glycol, and related volatile compounds along with alcohol (ethyl alcohol) simultaneously.

1.4. THERAPEUTIC DRUG MONITORING

The International Association for Therapeutic Drug Monitoring and Clinical Toxicology adopted the following statement to describe therapeutic drug monitoring, "Therapeutic drug monitoring is defined as the measurement made in the laboratory of a parameter that, with appropriate interpretation, will directly influence prescribing procedures. Commonly, the measurement is in a biological matrix of a prescribed xenobiotic, but it may also be of an endogenous compound prescribed as a replacement therapy in an individual who is physiologically or pathologically deficient in that compound" (21). Therapeutic drug monitoring has been used in clinical practice since the 1970s with the goal of personalizing the dosage of a drug for maximum efficacy and minimal toxicity. Usually therapeutic drug monitoring is necessary for a drug with a narrow therapeutic window, and only a small fraction of all drugs available require therapeutic drug monitoring. In general, therapeutic drug monitoring is not needed for any over-the-counter drug because these drugs usually have a wider margin of safety. However, the intentional or accidental overdose of over-the counter medications such as salicylate or acetaminophen is troublesome because such an overdose may even be fatal.

A drug may be administered to a patient via various routes including oral, rectal, intravenous, intramuscular, transdermal, or through sublingual application. Each route of administration has its advantages and disadvantages. For example, the oral route of administration is easiest for a patient, but the drug may suffer low bioavailability due to first-pass metabolism or intake of food or the bioavailability may be higher if the patient consumes alcohol. Moreover, a peak drug level may be achieved after a long delay. In contrast, peak concentration can be achieved rapidly if the drug is administered intravenously or intramuscularly, but that route of administration may result in patient discomfort. Rapid absorption of a drug can be achieved by sublingual application, but the drug may undergo first-pass metabolism thus reducing the efficacy of the drug. Usually a drug is poorly absorbed after transdermal application, and absorption may also be low after rectal application of a drug. In addition, most drugs that require therapeutic drug monitoring are delivered orally except for vancomycin and aminoglycoside. Criteria for drugs to be a candidate for therapeutic drug monitoring are the following:

1. Narrow therapeutic range where the dose of a drug that produces the desired therapeutic concentrations is also closer to the dose that may also cause toxic serum concentration. Serious toxicity may be encountered if the drug is not monitored.
2. There is an unpredictable relationship between dose and clinical outcome but a predictable relation between serum or whole blood drug level and clinical efficacy as well as toxicity. Significant changes in metabolism due to genetic makeup, age, sex, or disease for these drugs are responsible for the poor relation between the dosage and the drug level in the blood.
3. Drugs that demonstrate nonlinear pharmacokinetic parameters are also candidates for therapeutic drug monitoring.
4. Toxicity of a drug may lead to hospitalization, irreversible organ damage, and even death; for example, vancomycin may cause irreversible ototoxicity.

1.4.1. Drugs Requiring Therapeutic Drug Monitoring

Most drugs monitored in clinical laboratories are administered to patients with chronic diseases. These drugs are often used as a prophylactic agent to prevent reoccurrence of symptoms. For example, phenytoin is used to prevent certain types of convulsions in patients. Patient compliance is a major issue for successful drug therapy, and often patients do not take drugs as recommended, especially when they are dealing with a chronic illness. Gillisen reported that in patients with asthma, the adherence rates to medications are sometimes below 50% (22). Patsalos et al concluded that therapeutic drug monitoring of anticonvulsant drugs is beneficial to assess compliance especially in patients with uncontrolled seizures and breakthrough seizures (23). The cure rate for acute lymphoblastic leukemia (ALL) may exceed 85%, but up to 3 years of maintenance therapy with weekly methotrexate and daily 6-mercaptopurine is needed. Therefore, compliance with therapy is essential for the cure of ALL. In one report, the authors compared direct structured interview, the search of lack of compliance documented in the clinical record, and therapeutic drug monitoring of methotrexate to investigate compliance with therapy among children receiving such treatment. In 5 of 49 interviews, at least an episode of noncompliance was observed; searching clinical records revealed 8 of 49 patients skipped taking medication, and therapeutic drug monitoring revealed that 14 of 49 children had no detectable level of methotrexate in serum, indicating noncompliance. The authors concluded that face-to-face interview and clinical file notes are unreliable to determine patient noncompliance, and therapeutic drug monitoring is the best way to identify noncompliant patients (24). To attain optimal efficacy, a drug must reach the optimal level in the blood. Other than noncompliance, the patient may experience a lower level of drug than expected based on dosage due to reduced absorption or ultrarapid metabolism of the drug due to genetic makeup. Therefore, when a person reaches a steady state, routine therapeutic drug monitoring is also helpful to assess the reasons for an altered drug response (25). In addition to genetic factors, clinically significant drug–drug as well as drug–herb interactions may also significantly alter the drug level in the blood. For example, if quinidine is added to the drug regime of a patient taking digoxin, reduction in digoxin dosage is necessary because an increase in serum digoxin concentration occurs in 90% of patients after initiation of quinidine therapy, which causes a reduction in the renal clearance of digoxin (26).

Usually anticonvulsants, cardioactive drugs, immunosuppressants, antiasthmatic drugs, antidepressants, antiretroviral drugs, antineoplastic drugs, and antibiotics with narrow therapeutic windows such as vancomycin and aminoglycosides are monitored. In most instances, trough blood level (15–30 minutes prior to the next dose) is the preferred specimen for therapeutic drug monitoring except for certain antibiotics (vancomycin and aminoglycosides) where both peak and trough drug levels are monitored. Vancomycin and aminoglycoside can produce serious nephrotoxicity and ototoxicity. Peak serum concentrations for aminoglycosides such as amikacin and kanamycin above

32–34 μg/mL are associated with a higher risk of nephrotoxicity and ototoxicity (27). Vancomycin also has a low therapeutic index, and both nephrotoxicity and ototoxicity can be encountered in patients undergoing vancomycin therapy (28). Therefore, it is also necessary to monitor both peak and trough concentration of vancomycin. Ranges for peak concentrations of 20–40 μg/mL have been widely reported (29). The recommended trough concentration is 5–10 μg/mL because trough concentrations above 10 μg/mL are associated with an increased risk of nephrotoxicity (30,31).

Commonly monitored drugs include classical anticonvulsants such as phenytoin, valproic acid, carbamazepine, and phenobarbital, although certain newer anticonvulsant such as lamotrigine may benefit from routine therapeutic drug monitoring. In addition, certain cardioactive drugs such as digoxin, procainamide, quinidine, lidocaine, and so on, require routine monitoring. All immunosuppressants also need therapeutic drug monitoring. Commonly monitored drugs are listed in Table 1.1. There are also less frequently monitored drugs that may benefit from therapeutic drug monitoring, for example various antiretroviral drugs including protease inhibitors that are used in treating patients with acquired immunodeficiency virus (AIDS). These less frequently monitored drugs are listed in Table 1.2. However, the list of drugs that may benefit from therapeutic drug monitoring is increasing steadily as more publications indicate the benefits of therapeutic drug monitoring in certain drugs including common drugs in a selected patient population. Gerin et al concluded that routine therapeutic drug monitoring of voriconazole is potentially helpful in infants and children even if voriconazole is administered intravenously (32). Pea et al demonstrated that therapeutic drug monitoring in optimizing drug exposure with high-dose daptomycin plus continuous infusion of meropenem in patients with severe cellulitis, morbid obesity, and changing renal function is highly beneficial (33).

1.4.2. Effect of Gender and Pathophysiology on Drugs Disposition and Utility of Therapeutic Drug Monitoring

There are some differences between males and females for the metabolism of certain drugs. In addition, pregnancy can also significantly alter metabolism as well as the protein binding of certain strongly protein-bound drugs. Many disease conditions such as uremia, liver disease, cardiovascular disease, thyroid dysfunction, and other pathological conditions may also affect drug disposition. Elderly people are more susceptible to drug toxicity than younger adult, whereas neonates may also have a reduced capacity to metabolize drugs. Therapeutic drug monitoring is very helpful for the proper dosage adjustments for these patients.

1.4.2.1. Effect of Sex on Drug Disposition. Men and women may show both pharmacokinetic and pharmacodynamic differences in response to a particular drug therapy. There are anatomical differences between male and female. In general, males have a higher bodyweight, greater body surface area,

TABLE 1.1. Commonly Monitored Therapeutic Drugs

Drug	Specimen Requirement	Therapeutic Range (Trough)
Anticonvulsants		
Carbamazepine	Serum or plasma	4–12 μg/mL
Ethosuximide	Serum or plasma	40–75 μg/mL
Methsuximide	Serum or plasma	10–40 μg/mL
Phenytoin	Serum or plasma	10–20 μg/mL
Phenobarbital	Serum or plasma	15–40 μg/mL
Primidone	Serum or plasma	5–12 μg/mL
Valproic acid	Serum or plasma	50–100 μg/mL
Cardioactive Drugs		
Digoxin	Serum or plasma	0.8–2.0 ng/mL
Lidocaine	Serum or plasma	1.5–5.0 μg/mL
Procainamide and	Serum or plasma	4–10 μg/mL
N-acetylprocainamide		4–8 μg/mL
Quinidine	Serum or plasma	2–5 μg/mL
Antiasthmatics		
Theophylline	Serum or plasma	10–20 μg/mL
Caffeine	Serum or plasma	5–15 μg/mL
Antidepressants		
Amitriptyline and Nortriptyline	Serum or plasma	120–250 ng/mL (Amitriptyline + Nortriptyline) 50–150 ng/mL (Nortriptyline alone)
Clomipramine	Serum or plasma	150–450 ng/mL
Doxepin and Nordoxepin	Serum or plasma	150–250 ng/mL (Doxepin and Nordoxepin)
Imipramine and Desipramine	Serum or plasma	150–250 ng/mL (Imipramine and Desipramine)\
Lithium	Serum or plasma	0.8–1.2 mEq/L
Antibiotics		
Amikacin	Serum or plasma	<5 μg/mL (trough) 15–25 μg/mL (peak)
Gentamicin	Serum or plasma	1–2 μg/mL (trough) 4–8 μg/mL (peak)
Tobramycin	Serum or plasma	1–2 μg/mL (trough) 4–8 μg/mL (peak)
Vancomycin	Serum or plasma	5–15 μg/mL (trough) 30–40 μg (peak)
Immunosuppressants		
Cyclosporine	Whole blood (EDTA)	>150–350 ng/mL
Tacrolimus	Whole blood (EDTA)	5–15 ng/mL
Mycophenolic acid	Serum or plasma	1–3.5 μg/mL
Everolimus	Whole blood (EDTA)	3–8 ng/mL
Sirolimus	Whole blood (EDTA)	4–12 ng/mL

TABLE 1.2. Less Commonly Monitored Drugs

Drug	Specimen Requirement	Therapeutic Range (Trough)
Anticonvulsants		
Carbamazepine, 10, 11-epoxide (active metabolite of carbamazepine)	Serum or plasma	0.4–4 μg/mL
Clonazepam	Serum or plasma	10–50 ng/mL
Gabapentin	Serum or plasma	2–12 μg/mL
Lamotrigine	Serum or plasma	1–4 μg/mL
Zonisamide	Serum or plasma	10–40 μg/mL
Cardioactive Drugs		
Amiodarone	Serum or plasma	1.0–2.5 μg/mL
Flecainide	Serum or plasma	0.2–1.0 μg/mL
Mexiletine	Serum or plasma	0.5–2.0 μg/mL
Tocainide	Serum or plasma	5–12 μg/mL
Verapamil	Serum or plasma	50–200 ng/mL
Antidepressants		
Fluoxetine and Norfluoxetine	Serum or plasma	300–1000 ng/mL
Paroxetine	Serum or plasma	20–200 ng/mL
Sertraline	Serum or plasma	30–200 ng/mL
Haloperidol	Serum or plasma	2–15 ng/mL
Antiretroviral Agents		
Amprenavir	Serum	150–400 ng/mL
Atazanavir	Serum	100 ng/mL
Indinavir	Serum	80–120 ng/mL
Lopinavir	Serum	700 ng/mL
Nelfinavir	Serum	700–1000 ng/mL
Saquinavir	Serum	100–250 ng/mL
Nevirapine	Serum	150–400 ng/mL
Efavirenz	Serum	100 ng/mL

and total water content (both extracellular and intracellular) than females, causing difference between volumes of distribution of certain drugs, especially lipophilic drugs, between males and females. Although absorption of a drug is not different between men and women, the absorption rate may be slightly slower in females. Hepatic metabolism of drugs by phase I (via CYP1A2, CYP2D6, and CYP2E1) and phase II (by glucuronyl transferase, methyltransferases, and dehydrogenases) reactions appear to be faster in males than females, although metabolisms of drugs by CYP2C9, CYP2C19, and N-acetyltransferase or clearance of drugs that are substrates for P-glycoprotein appear to be similar in both males and females (34). Women also have high levels of sex hormones and may also take oral contraceptives. Managing women with antiepileptic drugs is a challenge because in general estrogen is a proconvulsant and progesterone is an anticonvulsant. Hormonal contraceptives usually contain progesterone alone or in combination with estrogen (natural or synthetic). Both natural and synthetic estrogens and progesterones (including common ethinyl estradiol) are metabolized by CYP3A4. The antiepileptic drugs such as phenobarbital, primidone, carbamazepine, oxcarbazepine, and phenytoin induce the cytochrome P450 system and may cause higher clearance of oral contraceptives, resulting in lack of contraception. Hormonal contraceptives can also interact with antiepileptic drugs. Ethinyl estradiol in combination with other components of oral contraceptive preparations can reduce the serum lamotrigine level by 50% (35). Soldin and Mattison reviewed sex differences in pharmacokinetics and pharmacodynamics. In general there are differences in pharmacokinetic parameters between men and women for many drugs including acebutolol, cefotaxime, clozapine, metronidazole, ofloxacin, valproic acid, and verapamil. In addition, pregnancy also significantly affect disposition of certain drugs (36). Gender differences in the pharmacokinetics of common drugs are listed in Table 1.3. Women are also more susceptible to adverse effects of certain drugs than men. Women are at increased risk of QT prolongation with certain antiarrhythmic drugs compared with men even at the same levels of serum drug concentrations. In contrast, certain psychotropic drugs such as chlorpromazine, fluspirilene, and various antipsychotic drugs appear to be more effective in women than men for the same dosage (37).

Epidemiologic surveys have indicated that between one third and two thirds of all pregnant women take at least one medication during pregnancy. Drug therapy in pregnant women usually focuses on safety of the drug on the fetus (tetragonic effect of drug), and therapeutic drug monitoring during pregnancy aims to improve individual dosage improvement, taking into account pregnancy-related changes in drug disposition (38). In general, the bioavailability of a drug is not significantly altered during pregnancy, but increased plasma volume and changes in protein binding may alter the volume of distribution of many drugs. The renal excretion of unchanged drugs is increased

TABLE 1.3. Gender Differences in Pharmacokinetics of Some Representative Drugs

Drug	Comments
Atenolol	Higher clearance in men than women, so greater reduction in blood in women than men
Ciprofloxacin	Clearance is lower in women
Clozapine	For similar dose, serum concentration higher in women than men
Diazepam	Plasma protein binding decreases during pregnancy increasing pharmacologically active free fraction
Erythromycin	Oral bioavailability may decrease in pregnant women
Lithium	Clearance increases during pregnancy
Ofloxacin	Clearance in lower in women compared with men
Phenobarbital	Clearance increases during pregnancy
Quinidine	Reduced plasma protein binding during pregnancy
Valproic acid	Plasma protein binding decreases during pregnancy

in pregnancy. In addition, metabolism of drugs catalyzed by isoenzymes of cytochrome P450 (CYP3A4, CYP2D6, and CYP2C9) and uridine diphosphate glucuronosyltransferase (UGT1A4 and UFT2B7) are increased in pregnancy. Therefore dosages of drugs that are metabolized by these routes may need to be increased during pregnancy to avoid loss of efficacy. In contrast, activities of some isoenzymes (CYP1A2 and CYP2C19) are reduced in pregnancy. Therefore, dosage reduction may be needed for drugs that are metabolized via these isoenzymes (39).

Significant increases in clearance of lamotrigine have been reported in pregnancy. Apparent clearance seems to increase steadily during pregnancy until it peaks approximately at week 32 when 330% increases in clearance from baseline values can be observed (40). Lower serum concentrations of lithium have been reported in pregnancy that may be related to an increase in the glomerular filtration rate (GFR) in pregnancy. Altered pharmacokinetics of ampicillin can be observed in pregnancy where serum concentrations may be lower by 50% in pregnant women compared with nonpregnant women due to altered pharmacokinetics. Faster elimination of phenoxymethylpenicillin (Penicillin V) in pregnant women has also been demonstrated (38). In general, dosage adjustments are required for anticonvulsants, lithium, digoxin, certain β-blockers, ampicillin, cefuroxime, and certain antidepressants in pregnant women. In addition, for certain drugs such as tetracycline, antithyroid medications, coumarin anticoagulants, aspirin, indomethacin, opioids, barbiturates, and phenothiazine unwanted effects in the fetus may occur despite careful adjustment of maternal dosage (41,42).

1.4.2.2. Renal Disease and Drug Disposition. Renal disease causes impairment in the clearance of many drugs by the kidney. Correlations have been established between creatinine clearance and the clearance of digoxin, lithium, procainamide, aminoglycoside, and several other drugs, but creatinine clearance does not always predict the renal clearance of a particular drug. In addition, elderly patients may have unrecognized renal impairment, and caution should be exercised when medications are prescribed for elderly patients. Serum creatinine remains normal until the GFR has fallen by at least 50%. Nearly half of the older patients have normal serum creatinine but reduced creatinine clearance. Dose adjustments based on renal function is recommended for many medications in elderly patients even for medications that exhibit large therapeutic windows (43). Renal disease also causes impairment of drug protein binding because uremic toxins compete with drugs for binding to albumin. Such interaction leads to increases in the concentration of pharmacologically active free drug concentration, which is clinically more important for strongly protein-bound drugs. Measuring free drug concentrations of strongly protein-bound anticonvulsant drugs such as phenytoin, carbamazepine, and valproic acid is recommended in uremic patients to avoid drug toxicity (44). Clinical utility of monitoring free anticonvulsants in uremic patients is discussed in detail in Chapter 13.

1.4.2.3. Liver Disease and Drug Disposition. Liver dysfunctions not only reduces clearance of a drug metabolized through hepatic enzymes or biliary mechanism, but they also affect plasm protein binding due to reduced synthesis of albumin and other drug-binding proteins. Mild to moderate hepatic disease may cause an unpredictable effect on drug clearance. Portal-systemic shunting present in patients with advanced liver cirrhosis may significantly reduce first-pass metabolism of high extraction drugs, thus increasing concentrations of such drugs in the blood and increasing the chance of drug overdose and toxicity (45). In addition, activities of several isoenzymes of cytochrome P450 enzymes (CYP1A1, CYP2C19, and CYP3A4/5) are reduced in liver dysfunction while activities of other isoenzymes such as CYP2D6, CYP2C9, and CYP2E1 may not be affected significantly. Therefore, drugs that are metabolized by CYP1A1, CYP3A4/5, and CYP2C19 may show increased blood levels in patients with hepatic dysfunction requiring dosage adjustment to avoid toxicity (46). Although the phase I reaction involving cytochrome P450 enzymes may be impaired in liver disease, the phase II reaction (glucuronidation) seems to be affected to a lesser extent, although both phase I and phase II reactions in drug metabolism are substantially impaired in patients with advanced cirrhosis. At this point there is no universally accepted endogenous marker to access hepatic impairment, and the semiquantitative Child-Pugh score is frequently used to determine the severity of hepatic dysfunction and thus dosage adjustments, although there are limitations to this approach (45). Nonalcoholic fatty liver disease is the most common chronic liver disease. This type of liver disease also affects activities of drug-metabolizing enzymes in the liver with the potential to produce adverse drug reactions from the standard dosage (47).

Mild to moderate hepatitis infection may also alter the clearance of drugs. Trotter et al reported that total mean tacrolimus dose in year 1 after transplant was lower by 39% in patients with hepatitis C compared with patients with no hepatitis C infection. The most likely explanation for these findings is decreased hepatic clearance of tacrolimus caused by mild hepatic injury from recurrent hepatitis C virus (48). Zimmermann et al reported that oral dose clearance of sirolimus (rapamycin) was significantly decreased in subjects with mild to moderate hepatic impairment compared with controls, and the authors stressed the need for careful monitoring of trough whole blood sirolimus concentrations in renal transplant recipients exhibiting mild to moderate hepatic impairment (49). Wyles and Gerber reviewed the effect of hepatitis with hepatic dysfunction on antiretroviral therapy especially highly active antiretroviral therapy (HAART) in patients with AIDS and commented that the dosage of the protease inhibitors indinavir, lopinavir, ritonavir, amprenavir, and atazanavir may require reduction in patients with liver disease, although hepatic dysfunction does not affect pharmacokinetics of nucleoside reverse transcriptase inhibitors because these drugs are not metabolized by liver enzymes (50).

Hypoalbuminemia is often observed in patients with hepatic dysfunction, thus impairing protein binding of many drugs. Because free (unbound) drugs are responsible for pharmacological action, careful monitoring of free concentrations of strongly albumin-bound antiepileptic drugs such as phenytoin, carbamazepine, and valproic acid is recommended in patients with hepatic dysfunction to avoid drug toxicity. See Chapter 13 for a more in-depth discussion on this topic.

1.4.2.4. Effect of Cardiovascular Disease on Drug Disposition. Cardiac failure is often associated with disturbances in cardiac output, influencing the extent and pattern of tissue perfusion, sodium and water metabolism, and gastrointestinal motility that may affect absorption and disposition of many drugs. Hepatic elimination of drugs via oxidative phase I metabolism is also impaired in patients with congestive heart failure due to decreased blood supply in the liver (51). Theophylline metabolism, which is largely independent of hepatic blood flow, is reduced in patients with severe cardiac failure and dose reduction is needed. Digoxin clearance is also decreased. The quinidine plasma level may also be high in these patients due to the lower volume of distribution (52). Therefore therapeutic drug monitoring is crucial in avoiding drug toxicity in these patients. Physiological changes in critically ill patients can significantly affect the pharmacokinetics of many drugs. These changes include absorption, distribution, metabolism, and excretion of drugs in critically ill patients. Understanding these changes in pharmacokinetic parameters is essential for optimizing drug therapy in critically ill patients. Moreover, usually free fractions of strongly protein-bound drugs are elevated in critically ill patients due to low serum albumin concentrations, which are addressed in Chapter 13.

1.4.2.5. Altered Pharmacokinetics in Neonates, Children, and the Elderly. In the fetus, CYP3A7 is the major hepatic cytochrome P450 enzyme with CYP3A5 also present in significant levels in half of the children. However, in adults, CYP3A4 is the major functional cytochrome P450 enzyme responsible for metabolism of many drugs. CYP1A1 is also present during organogenesis, and CYP2E1 may be present in some second trimester fetuses. After birth, hepatic CYP2D6, CYP2C8/9, and CYP2C18/19 are activated. CYP1A2 becomes active during the fourth to fifth months (53). In general, the decreased capacity of the neonatal liver to metabolize drugs may prolong action of drugs such as phenobarbital, theophylline, and phenytoin. However, age is not considered to have a major influence on the absorption of drugs from the gut except for the first few weeks of life when absorption steps may be less efficient. Neonates and infants have an increased total body water to body fat ratio compared with adults, whereas the reverse is observed in the elderly. These factors may affect the volume of distribution of drugs depending on their lipophilic character. Moreover, altered plasma binding of drugs may be observed in both neonates and some elderly due to low albumin, thus increasing the fraction of free drug. In general, drug metabolizing capacity by the liver enzymes is reduced in newborns, particularly in premature infants, but increases rapidly during the first few weeks and months of life to reach values that are generally higher than adult metabolizing rates. In contrast, the efficiency of cytochrome P450 enzymes declines with old age. Renal function at the time of birth is reduced by more than 50% of adult value but then increases rapidly in the first 2–3 years of life. Renal function then starts declining with old age. Oral clearance of lamotrigine, topiramate, levetiracetam, oxcarbazepine, gabapentin, tiagabine, zonisamide, vigabatrin, and felbamate is significantly higher (20–120%) in children compared with adults depending on the drug and the age distribution of the population. However, clearance of these drugs is reduced (10–50%) in the elderly population compared with middle-aged adults (54). Therefore, physiological differences between children and adult result in altered pharmacokinetics and drug effects. In neonates and infants decreased weight-adjusted doses are recommended because of reduced protein binding, renal excretion, and metabolism. However, children older than 1 year require significantly higher weight-corrected doses on many drugs compared with adults that are eliminated by CYP1A2, CYP2C9, and CYP3A4, although weight-corrected doses for drugs eliminated by renal excretion or metabolism via CYP2C19, CYP2D6, and N-acetyltransferase and UDP glucuronosyltransferase in children are similar to adults (55). Premature infants may also metabolize a drug different than adult. For example, only unchanged theophylline and caffeine are found in the urine of premature neonates, indicating the oxidative pathways for theophylline metabolism. In contrast, in children and adults, 3-methylxanthine and 1,3-dimethyl uric acid are the major metabolites of theophylline recovered in the urine (56).

1.4.2.6. Thyroid Dysfunction and Drug Disposition. Patients with thyroid disease may have an altered drug disposition because thyroxine is a potent activator of the cytochrome P450 enzyme system. Therefore, lower levels of drugs may result from high thyroxine levels due to the induction of the hepatic oxidative metabolism pathway. In contrast, hypothyroidism is associated with the inhibition of hepatic oxidative metabolism of many drugs. Hypothyroidism also affects the metabolism of immunosuppressants. Haas et al reported a case where a patient developed hypothyroidism 6 months after single lung transplant and was admitted to the hospital for anuric renal failure. The patient showed a toxic blood level of tacrolimus that was resolved with the initiation of thyroxine replacement therapy and a dose reduction of tacrolimus (57). Therefore, therapeutic drug monitoring of immunosuppressants can aid in avoiding such drug toxicity. Amiodarone is a potent antiarrhythmic drug associated with thyroid dysfunction because due to high iodine content amiodarone inhibits 5-deiodinase activity. Although most patients treated with amiodarone remain euthyroid, amiodarone-induced thyrotoxicosis or amiodarone-induced hypothyroidism may occur depending on the iodine status of the patient as well as the history of prior thyroid disease. Screening of thyroid disease before amiodarone therapy and periodic monitoring of thyroid functions are recommended for patients treated with amiodarone (58).

1.5. ANALYTICAL METHODS USED FOR THERAPEUTIC DRUG MONITORING

Immunoassays are commonly used for routine therapeutic drug monitoring in clinical laboratories. However, commercially available assays are available for roughly 25 different drugs that are commonly monitored in clinical laboratories. For less commonly monitored drugs, chromatographic methods such as gas chromatography, GC/MS, liquid chromatography combined with ultraviolet or fluorescence detection, or liquid chromatography combined with tandem mass spectrometry can be used. Application of gas chromatography for therapeutic drug monitoring is only applicable for drugs that are relatively volatile. For nonvolatile drugs, for example immunosuppressants, only liquid chromatographic techniques can be used. In general, liquid chromatography combined with tandem mass spectrometry can offer best specificity and lower limits of detection for a drug than other methods.

1.5.1. Various Formats of Immunoassays Used for Therapeutic Drug Monitoring

Many therapeutic drugs are measured in blood by immunoassay methods using automated analyzers, and in most immunoassay methods serum or plasma can be used directly without any pretreatment. The assays require very small amounts of sample (mostly less than 100 μL). With respect to assay design, there are two formats of immunoassays: competition and immunometric (commonly referred as "sandwich"). Competition immunoassays are desirable for analytes with small molecular weight such as most therapeutic drugs. In contrast, sandwich immunoassays are mostly used for analytes with larger molecular weight, such as proteins where two different specific antibodies are used. In the competitive immunoassay format, analyte molecules in the specimen, for example a particular drug, compete with analyte (or its analogs), labeled with a suitable tag provided in the reagent, for a limited number of binding sites in the analyte-specific antibody (also provided in the reagent). Thus, in these types of assays, the higher the analyte concentration in the sample, the less of the labeled antigen can bind to the antibody to form the conjugate, or if small amounts of drugs molecules are present, then more labeled antigens can bind to the antibody. If the bound label provides the signal, which in turn is used to calculate the analyte concentration in the sample, the analyte concentration in the specimen is inversely proportional to the signal produced. If the free label provides the signal then the signal produced is proportional to the analyte concentration. The signal is mostly optical: absorbance, fluorescence, or chemiluminescence.

There are several variations in this basic format of a competitive immunoassay such as homogeneous or heterogeneous format. In the homogeneous format, the bound label has different properties than the free-labeled antigen, and no separation of bound versus free labeled antigen is needed. For example, in the fluorescent polarization immunoassay, the free label has different Brownian motion than when the relatively small molecular weight labeled antigen is complexed with a large antibody. Therefore, only when a labeled antigen is bound to the antibody, signal is generated. However, in the EMIT method, the enzyme used for labeling the antigen is glucose 6-phosphate-dehydrogenase, which is active in the free form but inactive when bound to the antibody. The active enzyme during reaction with the substrate also reduces the cofactor NAD to NADH, and the absorbance is monitored at 340 nm. Therefore, only free-labeled antigen can generate the signal. In the cloned enzyme donor immunoassay method, two genetically engineered inactive fragments of the enzyme β-galactosidase are coupled to the antigen and the antibody reagents. When they combine, the active enzyme is produced and the substrate, a chromogenic galactoside derivative, produces the assay signal.

In heterogeneous immunoassays, format bound label is separated from the unbound labels before measuring the signal. The separation is often done magnetically, where the reagent analyte (or its analog) is provided as coupled to paramagnetic particles (PMPs), and the antibody is labeled. Conversely, the antibody may be also provided as conjugated to the PMP, and the reagent analyte may carry the label. After separation and wash, the bound label is reacted with other reagents to generate the signal. This is the mechanism in many chemiluminescent immunoassays, where the label may be a small molecule that generates chemiluminescent signal (59). The label also may be an enzyme (enzyme-linked immunosorbent assay) that generates a chemiluminescent, fluorometric, or colorimetric signal. In older immunoassay formats, the labels used to be radioactive (radioimmunoassay, or RIA). But because of safety and waste disposal issues, RIA is rarely used today. Another type of heterogeneous immunoassay uses polystyrene particles. If these particles are micro-sizes, that type of assay is called microparticle-enhanced immunoassay. Microparticle enzyme immunoassay is also used for the analysis of drugs in biological matrix.

Antibody used in the immunoassay may be polyclonal or monoclonal. Usually polyclonal antibodies are generated using an animal, and for a small molecular weight analyte such as a therapeutic drug, it is most commonly injected as a conjugate of a large protein in the animal to generate the antibody. Since many clones of the antibodies specific for the analyte are generated, these antibodies are called polyclonal and are less specific than newer techniques that generate monoclonal antibody. To generate monoclonal antibody, a mast cell of the animal can be selected as producing the optimum antibody, and then it can be fused to an immortal cell. The resulting tumor cell grows uncontrollably, producing only the single clone of the desired antibody. Such antibodies may be grown in live animals or cell culture. Sometimes, instead of using the whole antibody, fragments of the antibody, generated by digestion of the antibody with peptidases, for example, Fab, Fab′ (or their dimeric complexes) are used as a reagent. Even though the immunoassay methods are now widely used, there are few limitations of this technique. Antibody specificity is the major concern of an immunoassay, which is discussed in various chapters throughout the book.

1.5.2. Chromatographic Methods for Therapeutic Drug Monitoring

Chromatographic techniques used for therapeutic drug monitoring can be broadly classified as gas chromatography and liquid chromatography. Gas-liquid chromatography, also commonly referred to as gas chromatography (GC), is a separation technique first described in 1952 by James and Martin. In most GC columns, the stationary phase is a liquid and the mobile phase is an inert gas. In general the stationary phase has a low vapor pressure so that at column temperature it can be considered nonvolatile. Introduction of the capillary column significantly improves resolution of peaks in GC analysis. Depending on the stationary phase composition, a GC column may have low, intermediate, or high polarity. The sensitivity and specificity of GC analysis depends on the choice of detector. Mass spectrometry (MS) can be used in combination with a gas chromatograph where MS is capable of producing a mass spectrum of any compound coming out of the column of gas chromatograph. The nitrogen phosphorus detector is specific for nitrogen- and phosphorus-containing compounds and is very sensitive. An electron capture detector can detect any halogen-containing compounds. Flame ionization and thermal conductivity detectors are also used in gas chromatography. However, GC combined with MS has the best sensitivity and specificity over other detectors used along with gas chromatography. Although only nonpolar volatile compounds with low molecular weights can be analyzed by GC, if a relatively polar compound with small molecular weight (usually less than 500) can be chemically converted to a nonpolar relatively volatile compound through a process called derivatization, it can also be analyzed by GC.

Application of GC as a separation technique is limited to volatile molecules, but high performance liquid chromatography (HPLC) can be used for separation of both polar and nonpolar molecules. Usually derivatization is not necessary for HPLC analysis. HPLC is based on the principle of liquid–liquid chromatography where both the mobile phase and stationary phase are liquid. In normal liquid chromatography, the stationary phase is polar such as a silica gel column and the mobile phase (eluting solvent) is nonpolar. In reverse phase chromatography, the stationary phase is nonpolar, most commonly derivatized silica, and the mobile phase is polar. Detectors used in the HPLC method include ultraviolet detector, fluorescence detector, conductivity detector, refractive index detector as well as MS. Ultraviolet detection is commonly used in clinical laboratories, although other detection techniques such as the fluorescence technique and electrochemical detection technique may also be used. Mass spectrum is usually considered a molecular fingerprint of a compound. Mass spectrometer used in gas chromatography is usually an electron ionization or chemical ionization mass spectrometric detector. Electron ionization (EI) at 70 eV produces reproducible mass spectrum, which is a common ion source used in GC/MS analysis of several therapeutic drugs with relatively low molecular weight (usually less than 500). It is relatively easy to combine a mass spectrometer with a gas chromatograph because a high-efficiency pump can remove the carrier gas in the gas chromatography, which is an inert gas such as helium. However, combining HPLC with a mass spectrometer is more complex because the liquid mobile phase is not compatible with a mass spectrometer, which usually works in the vacuum. However, there are technologies available for combining liquid chromatography with mass spectrometry but only moving belt and particle beam interfaces are compatible with EI (60). The electrospray interface is very common in the HPLC/MS analyzer used in clinical laboratories. The electrospray interface produces singly or multiple charged ions directly from a solvent system by creating a fine spray of highly charged droplets in the presence of a strong electric field with assistance from heat or from pneumatics. In this process nonvolatile and polar compound can be ionized. The atmospheric pressure chemical ionization interface produces sample ions by charge transfer from reagent ions. The reagent ions are produced from the solvent vapor of the mobile phase.

Chromatographic methods are used for the therapeutic drug monitoring of drugs where there is no commercially available immunoassay, for example antiretroviral and other drugs used in HAART for AIDS patients, some newer anticonvulsants, several anticancer drugs other than methotrexate, and various antidepressants. Yadav et al described a liquid chromatography combined with tandem mass spectrometric method for determining concentrations of the antiretroviral drugs tenofovir, emtricitabine, and lamivudine in human plasma (61). D'Avolio et al described a liquid chromatography combined with mass spectrometric method simultaneous quantification of 14 antiretroviral agents in the peripheral blood mononuclear cell of patients infected with the human immunodeficiency virus optimized using corpuscular volume evaluation (62). Although immunoassays are available for the routine monitoring of certain immunosuppressants such as cyclosporine, tacrolimus, sirolimus, everolimus, and mycophenolic acid, these assays suffer from many limitations, and HPLC combined with tandem mass spectrometry can be used for therapeutic monitoring of these drugs. See Chapter 15 for a more detailed discussion on this topic. Taylor et al commented that liquid chromatography combined with tandem mass spectrometry is the preferable method for therapeutic drug monitoring of immunosuppressants and antiretroviral drugs (63). Although there are immunoassays for determining total tricyclic antidepressant concentration in serum or plasma, chromatographic methods are preferred because commercially available immunoassay cross-reacts with all tricyclic antidepressants and their metabolites as well as other drugs, for example carbamazepine, hydroxyzine, and cetirizine (64,65). See Chapter 14 for a more detailed discussion on this topic.

1.6. TESTING OF DRUGS OF ABUSE

Drugs of abuse including prescription drug abuse are a serious public health and safety concern. According to the survey of health and human services of the U.S. government, marijuana remains the most commonly abused drug in the United States (66). Abuse of prescription medications for nonmedical purpose is also a serious public health issue. Cocaine abuse is also widespread among the general population. In general drugs that are commonly abused by individuals in the United States can be classified under different categories:

Cannabinoids: Marijuana, hashish

Stimulants: Amphetamine, methamphetamine, MDMA, cocaine

Narcotic analgesics and opiates: Opium, heroin, codeine, morphine, oxycodone, hydrocodone, meperidine, methadone, fentanyl and its analogs

Central nervous system depressants and tranquilizers: Barbiturates, various benzodiazepines, methaqualone, γ-hydroxybutyrate (GHB)

Anesthetics: Ketamine, phencyclidine

Hallucinogens: LSD (lysergic acid diethylamide), peyote cactus, magic mushroom

In most toxicology laboratories, drugs of abuse are tested for medical purpose only; forensic drug testing and workplace drug testings are conducted in laboratories specifically certified for this purpose. Workplace drug testings are conducted in toxicology laboratories certified by SAMHSA (Substance Abuse and Mental Health Services Administration, an agency under the Department of Health and Human Services of the U.S. government) or by the College of American Pathologists, which also certifies forensic laboratories. Although SAMHSA-mandated drug testings include only amphetamines (including 3, 4-methylenedioxymethamphetamine; MDMA), cocaine (as benzoylecgonine), opiates, marijuana (as marijuana metabolite), and phencyclidine (PCP), private employers often also test additional drugs such as barbiturates, benzodiazepines, methadone, oxycodone, propoxyphene, and methaqualone in the workplace drug testing panel. For medical drug testing, both SAMHSA and non-SAMHSA drugs are included. Most commonly tested drugs in medical drug testing panel include amphetamine (along with MDMA and 3,4-methylenedioxyamphetamine [MDA]), cocaine metabolite, opiates including oxycodone, propoxyphene, methadone, marijuana metabolites, benzodiazepines, barbiturates, and possibly methaqualone. Although fentanyl is also abused, it is not usually included in the panel for routine drugs of abuse. In addition, less commonly abused drugs such as ketamine, LSD, and some designer drugs such as GHB are not always routinely tested in most toxicology laboratories.

As mentioned earlier, medical drugs of abuse testings are commonly conducted using urine specimens, although blood or gastric content can be also used for drug screening in case of recent overdose. Over 90% of all workplace drug testings are conducted using urine specimens, although workplace drug testings can also be conducted using hair specimens, oral fluid, or sweat. Because urine is the most common specimen for both medical and workplace drug testing, in this chapter drug of abuse tests using urine specimens are discussed. An abused drug or its metabolite can be detected in the urine for few days to weeks depending on the particular drug. The window of detection of short-acting barbiturates such as pentobarbital in urine is 1 day, whereas a long-acting barbiturate such as phenobarbital has a detection window of 1 month. Oxycodone may be detected in urine 2–4 days after use; methaqualone can be detected up to 3 days. The windows of detection of common abused drugs are discussed in Chapter 2.

For both medical drug testing and workplace drug testing, immunoassays approved by the Food and Drug Administration (FDA) are used for the initial screening test. If the screening test is positive for any drug, then the confirmation is usually conducted using GC/MS, although liquid chromatography combined with mass spectrometry is gaining popularity in drugs of abuse testing. Although confirmation is mandatory for workplace drug testing, confirmation is not mandatory for medical drug testing because results cannot be used for legal purposes. Therefore, if the clinician request confirmation, the toxicology laboratory usually confirms the presence of the drug using GC/MS or another analytical technique, but not all hospital toxicology laboratories have the capability of drug confirmation. In addition, chain of custody where a detailed record of the person who has the possession of the specimen from the collection until analysis is maintained; for medical drug testing, chain of custody is not usually required.

Urine drug testings using immunoassays have certain cut-off concentrations. The drug testing is considered negative if the concentration of the drug or the metabolite is below the cut-off concentration. Cut-off concentrations of immunoassays used for drugs of abuse screening are discussed in Chapter 2.

1.6.1. Testing of Amphetamines

After clinical use or abuse of amphetamine, amphetamine is excreted unchanged in the urine along with other metabolites. After ingestion of methamphetamine, a part of methamphetamine is metabolized to amphetamine, and both methamphetamine and amphetamine are detected in urine. Although the cut-off concentration was originally considered 1000 ng/mL, it has been lowered to 500 ng/mL. Moreover, if methamphetamine is confirmed by GC/MS, amphetamine must be present in the specimen because from a physiological point of view, after methamphetamine use, both amphetamine and methamphetamine must be present in the urine. It was reported in 1990s that at a high injector port temperature (during GC/MS confirmation step), pseudoephedrine, a common active ingredient of many over-the-counter cold medications, can be thermally degraded to methamphetamine and may cause methamphetamine-positive confirmatory results as an artifact. However, no amphetamine can be generated in this process. Therefore, detection of amphetamine in a urine specimen ensures that methamphetamine indeed is the source of amphetamine. Amphetamine immunoassays are affected by various over-the-counter cold medications containing ephedrine, pseudoephedrine, and related compounds.

1.6.2. Testing for Cocaine Metabolite Benzoylecgonine

Cocaine is metabolized to benzoylecgonine and ecgonine methyl ester. The major urinary metabolite is benzoylecgonine, an inactive metabolite. In urine drugs of abuse testing, cocaine is detected and confirmed as benzoylecgonine. Although positive confirmation of benzoylecgonine may occur even after use of cocaine as a local anesthetic in ear, nose, or throat surgery as expected, passive inhalation of cocaine smoke results in urinary benzoylecgonine concentration well below the screening cut-off concentration of 300 ng/mL. In one study, when six male volunteers were exposed to cocaine smoke for 1 hour in an unventilated room (cocaine smoke produced by 100 or 200 mg freebase cocaine heated at 200°C) , peak urinary concentration of benzoylecgonine ranged from 2 to 123 ng/mL (67).

1.6.3. Testing for Marijuana Metabolite

The major active component of cannabinoid (marijuana, hashish) is Δ-9-tetrahydrocannabinol (THC), which is metabolized to 11-nor-9-carboxy Δ-9-tetrahydrocannabinol (THC-COOH). The cut-off concentration of this major metabolite is 50 ng/mL; for confirmatory purposes the corresponding cut-off concentration is 15 ng/mL. Bioavailability of THC is approximately 30% after smoking, but 4–12% after oral use. The major metabolite of marijuana (THC-COOH) may stay in urine up to several weeks (68). The urinary concentration of THC-COOH in marijuana abusers varies widely but is usually from 78.7 ng/mL to 2634 ng/mL (average: 1153 ng/mL) according to one published report (69). However, passive inhalation of marijuana should not produce a positive immunoassay screening test.

1.6.4. Testing for Opiates

The original cut-off of concentration for opiates for the screening was 300 ng/mL. However, the cut-off concentration was increased to 2000 ng/mL in 1998 by the National Institute of Drugs of Abuse to avoid false-positive test results due to ingestion of foods containing poppy seeds. Most immunoassays for opiates utilize antibody that recognizes morphine in urine because morphine-3-glucuronide is the major metabolite after the use of both codeine and morphine as well as abuse of heroin, which is first metabolized to 6-monoacetyl morphine and then morphine. The presence of 6-acetylmorphine can only be detected up to 24 hours after heroin abuse, but morphine is present in urine for up to 2–3 days. The mandatory guideline requires testing for 6-acetylmorphine by GC/MS in urine using a cut-off concentration of 10 ng/mL.

Morphine-3- glucuronide (conjugated form of morphine, the major metabolite in urine), codeine, and 6-acetylmorphine have high cross-reactivity with morphine antibody and all if present in urine produces positive opiate screening results. Most opiate immunoassays have poor cross-reactivity with oxycodone and relatively low cross-reactivity with hydromorphone and hydrocodone. If these compounds are present in high amounts in urine, the specimen may test positive for opiates. There are specific immunoassays for detecting the presence of oxycodone in urine because pain medicine oxycodone is also widely abused. There are several synthetic opioid that are not metabolized to morphine. These drugs include fentanyl, buprenorphine, hydrocodone, oxycodone, meperidine, methadone, and oxymorphone. Therefore abuse of fentanyl, meperidine (Demerol), or oxycodone cannot be detected by opiate screening assays, and specific immunoassays capable of detecting oxycodone or methadone must be used for detecting such opioids in urine specimens. For example, the DRI oxycodone assay (Microgenics Corporation, Fremont, CA) is a homogenous immunoassay that can selectively identify urine specimens containing oxycodone at a concentration of 100 ng/mL or higher. This assay has no cross-reactivity with codeine, morphine, hydrocodone, and hydromorphone. Backer et al reported that of 1523 urine specimens collected from pain management patients, 435 specimens tested positive for oxycodone. The GC/MS analysis confirmed the presence of oxycodone or oxymorphone in 433 specimens at a concentration of 100 ng/mL and higher. In addition to oxycodone and oxymorphone, 189 specimens also contained morphine, hydromorphone, hydrocodone, and codeine (70). Methadone is metabolized to two pharmacologically inactive urinary metabolites: 2-ethylidene-1,5-dimethyl-3,3-diphenylpyrrolidine (EDDP) and 2-ethyl-5-methyl-3,3-diphenylpyrrolidine (EMDP). Usually both methadone and EDDP are found in urine in significant amounts, whereas the concentration of EDMP is relatively low. Commercially available immunoassays for screening of methadone in urine have antibodies either directed toward methadone or EDDP. However, an immunoassay designed for detecting the presence of methadone in urine may have a low cross-reactivity with EDDP. Preston et al demonstrated that 2% of urine specimens collected from patients who had taken methadone tested negative for methadone while all specimens tested positive for EDDP (71).

1.6.5. Testing for Phencyclidine

PCP is abused by snorting, smoking, intravenous injection, and also taken orally. PCP is absorbed into the circulation after intake by any one of these routes. PCP is metabolized by the liver to various hydroxylated metabolites, but a portion of PCP is also excreted unchanged in urine. PCP tests targets the unchanged drug for detection both in immunoassay screens as well as in GC/MS confirmatory tests.

1.6.6. Testing for Barbiturates and Benzodiazepines

Barbiturates can be short acting or long acting. Commonly abused barbiturates are secobarbital, pentobarbital, and amobarbital. These are the common barbitals targeted for confirmation if the barbiturate screening assay is positive. Many commercially available immunoassays for barbiturates utilize antibody against secobarbital.

Many benzodiazepines are used in clinical practice, and not all are approved for medical use in the United States. The benzodiazepines commonly abused include diazepam, lorazepam, oxazepam, clonazepam, and triazolam. In addition flunitrazepam (Rohypnol) is a benzodiazepine that is not legally available in the United States. However, this drug is found on the clandestine market and widely abused in the United States, and it is also associated with date rape situations. One of the major metabolite of flunitrazepam is 7-aminoflunitrazepam, which cross-reacts with antibodies used in the benzodiazepine screening assays. Most toxicology laboratories performing workplace drug testing also have the capability of confirming the presence of flunitrazepam and its metabolite in urine using GC/MS.

1.7. METHODS USED FOR TESTING OF DRUGS OF ABUSE

Usually FDA-approved immunoassays are used for initial screening for the presence of an abused drug in urine. Then the drug detected during the screening assay is confirmed using an alternative analytical technique such as GC/MS, which is the gold standard in confirming drugs of abuse. There are many published methods for various classes of drugs for analysis by GC/MS. Immunoassays usually follow the same format as immunoassays for therapeutic drug monitoring, and many commercial diagnostic companies that sell immunoassays for therapeutic drug monitoring also sell immunoassays for drugs of abuse screening.

Although drug testing in urine is most common, drug screening in whole blood is useful to determine the agent responsible for drug overdose. Oiestad et al described an ultra-performance liquid chromatography combined with tandem mass spectrometry for drug screening in whole blood after liquid–liquid extraction on ChemElute columns using ethyl acetate/heptane (4:1). Drugs were confirmed using tandem mass spectrometry using positive ion electrospray mode. Drug or metabolites of codeine, morphine, ethyl morphine, oxycodone, buprenorphine, methadone, cocaine, methylphenidate, amphetamine, methamphetamine, 3,4-methylenedioxymethamphetamine, δ-9-tetrahydrocannabinol, fentanyl, alprazolam, bromazepam, clonazepam, diazepam, nordiazepam, 4-hydroxy diazepam, fenazepam, flunitrazepam, lorazepam, nitrazepam, oxazepam, zopiclone, zolpidem, carisoprodol, and meprobamate can be detected by this method (72).

1.8. CONCLUSIONS

Immunoassays are widely used in both therapeutic drug monitoring and initial screening of drugs of abuse in urine. Immunoassays are subjected to interference causing both false-positive and false-negative results. The challenge for laboratory professionals is to identify such interferences and use alternative methods for eliminating such interference if possible and to alert the clinician regarding a toxicology report that may not be accurate due to an interfering substance present in the specimen. The focus of this book is how to identify such erroneous results. In most cases by using chromatographic methods such interferences can be resolved, but chromatographic methods may not be readily available for commonly monitored drugs in the toxicology laboratories where commercial immunoassays are available for that particular drug. However, interference may be resolved by using an alternative immunoassay for the drug where the other assay is not affected by the interfering substance.

In addition, certain designer drugs may not be detected by routine immunoassays commercially available for the screening of drugs of abuse in the toxicology laboratory. A negative toxicology report may occur in a patient overdosed with a designer drug or date rape drug such as gamma-hydroxy butyric acid that is not usually screened in a toxicology laboratory. This important topic is discussed in Chapter 5. In addition, positive tests for drugs of abuse may be due to the use of a prescription drug or other factors (see Chapter 4 for more detail). A trained and well-informed laboratory professional can help the clinician a great deal in the proper management of the patient by alerting him or her regarding the limitations of therapeutic drug monitoring or drugs of abuse testing and suggesting to the clinician when a specimen must be send to a reference laboratory for further testing.

REFERENCES

1. Bogstrand ST, Normann PT, Rossow I, Larsen M et al. Prevalence of alcohol and other substances of abuse among injured patients in a Norwegian emergency department. Drug Alcohol Depend 2011; 117: 132–138.
2. Linakis JG, Chun TH, Mello MJ, Baird J. Alcohol related visits to the emergency department by injured adolescents: a national perspective. J Adolesc Health 2009; 45: 84–90.
3. Dickson AJ, Vorce SP, Levine B, Past MR. Multiple-drug toxicity caused by the coadministration of 4-methylmethcathinone (mephedrone) and heroin. J Anal Toxicol 2010; 34: 152–158.

4. Manthripragada AD, Zhou EH, Budnitz DS, Lovegrove MC et al. Characterization of acetaminophen overdose related emergency department visits and hospitalization in the United States. Pharmacoepidemiol Drug Saf 2011; 20: 819–826.
5. Guggenheimer J, Moore PA. The therapeutic applications of and risks associated with acetaminophen use: a review and update. J Am Dent Assoc 2011; 142: 38–44.
6. Prescott LF. Paracetamol, alcohol and liver. Br J Clin Pharamacol 2000; 49: 291–301.
7. Herres J, Ryan D, Salzman M. Delayed salicylate toxicity with undetectable initial levels after large-dose aspirin ingestion. Am J Med Toxicol 2009; 27: 1173 e1–3.
8. Galbois A, Ait-Oufella H, Baudel JL, Maury E et al. An adult can still die of salicylate poisoning in France in 2008. Intensive Care Med 2009; 35: 1999.
9. Dromgoole SH, Furst DE, Paulus HE. Metabolism of salsalate in normal subjects. J Pharm Sci 1984; 73: 1657–1659.
10. Rivera W, Kleinschmidt KC, Velez LI, Shepherd G et al. Delayed salicylate toxicity at 35 hours without early manifestations following a single salicylate ingestion. Ann Pharamcother 2004; 38: 1186–1188.
11. West PL, Horowitz BZ. Delayed recrudescence to toxic salicylate concentration after salsalate overdose. J Med Toxicol 2010; 6: 150–154.
12. Gaspari F, Locatelli M. Determination of aspirin and salicylic acid in uremic patient's plasma using reversed phase high performance liquid chromatography. Ther Drug Monit 1987; 9: 242–247.
13. Miceli JN, Aravind MK, Cohen SN, Done AK. Simultaneous measurements of acetaminophen and salicylate in plasma by liquid chromatography. Clin Chem 1979; 25: 1002–1004.
14. Stewart MJ, Watson ID. Analytical reviews in clinical chemistry: methods for the estimation of salicylate and paracetamol in serum, plasma and urine. Anal Clin Biochem 1987; 24: 552–565.
15. Beuhler MC, Curry SC. False positive acetaminophen levels associated with hyperbilirubinemia. Clin Toxicol (Phila) 2005; 43: 167–170.
16. Polson J, Wians FH, Orsulak P, Fuller D et al. False positive acetaminophen concentrations in patients with liver injury. Clin Chim Acta 2008; 391: 24–30.
17. Dasgupta A, Zaidi S, Johnson M, Chow L et al. Use of fluorescence polarization immunoassay for salicylate to avoid positive/negative interference by bilirubin in the Trinder salicylate assay. Ann Clin Biochem 2003; 40: 684–688.
18. Broughton A, Marenah C, Lawson N. Bilirubin interference with a salicylate assay performed on an Olympus analyzer. Ann Clin Biochem 2000; 37: 408–410.
19. Tilbrook LK, Slater J, Agarwal A, Cyriac J. An unusual cause of interference in a salicylate assay caused by mitochondrial acetoacetyl-CoA thiolase deficiency. Ann Clin Biocehm 2008; 45: 524–526.
20. Fong BM, Siu TS, Tam S. Persistently increased acetaminophen concentrations in a patient with acute liver failure. Clin Chem 2011; 57: 9–11.
21. Watson I, Potter J, Yatscoff R, Fraser A et al. Therapeutic Drug Monit [editorial] 1997; 19: 125.

22. Gillisen A. Patient's adherence in asthma. J Physiol Pharmacol 2007; 58 (Suppl 5, Pt 1): 205–222.
23. Patsalos PN, Berry DJ, Bourgeois BF, Cloyd JC et al. Antiepileptic drugs—best practice guidelines for therapeutic drug monitoring: a position paper by the sub commission on therapeutic drug monitoring, ILAE commission on therapeutic strategies. Epilepsia 2008; 49: 1239–1276.
24. Jaime-Perez JC, Gomez-Almaguer D, Sandoval-Gonzalez A, Chapa-Rodriguez A et al. random serum methotrexate determinations for assessing compliance with maintenance therapy for childhood acute lymphoblastic leukemia. Leuk Lymphoma 2009; 50: 1843–1847.
25. Stieffenhofer V, Hiemke C. Pharmacogenetics, therapeutic drug monitoring and non compliance [in German]. Ther Umsch 2010; 67: 309–315.
26. Bigger JT, Leahey EB Jr. Quinidine and digoxin: an important interaction. Drugs 1982; 24: 229–239.
27. Black RE, Lau WK, Weinstein RJ, Young LS et al. Ototoxicity of amikacin. Antimicrob Ag Chemother 1976; 9: 956–961.
28. Duffull SB, Begg EJ. Vancomycin toxicity: what is the evidence for dose dependence? Adverse Drug React Toxicol Rev 1994; 13: 103–114.
29. Begg EJ, Barclay ML, Kirkpatrick C. The therapeutic monitoring of antimicrobial agents [review]. J Clin Pharmacol 1999; 47: 23–30.
30. Ryback MJ, Albrecht LM, Boike SC, Chandrasekar PH. Nephrotoxicity of vancomycin: alone or with an aminoglycoside. J Antimicrob Chemother 1990; 25: 679–687.
31. Cimino MA, Rotstein C, Slaughter RL, Emrich LJ. Relationship of serum antibiotic concentrations to nephrotoxicity in cancer patients receiving concurrent aminoglycoside and vancomycin therapy. Am J Med 1987; 83: 1091–1096.
32. Gerin M, Mahlaoui N, Elie C, Lanternier F et al. Therapeutic drug monitoring of voriconazole after intravenous administration in infants and children with primary immunodeficiency. Ther Drug Monit 2011; 45: e37.
33. Pea F, Cojutti P, Sbrojavacca R, Cadeo B et al. TDM guided therapy in daptomycin and meropenem in a morbidly obese critically ill patient. Ann Pharmacother 2011; July 12 (Epub ahead of print).
34. Schwartz JB. The influence of sex on pharmacokinetics. Clin Pharmacokinet 2003; 42; 107–121.
35. Noe KH. 2007. Gender specific challenges in the management of epilepsy in women. Semin Neurol 27: 331–339.
36. Soldin OP, Mattison DR. Sex differences in pharmacokinetics and pharmacodynamics. Clin Pharamacokinetic 2009; 48: 143–157.
37. Rademaker M. Do women have more adverse drug reactions? Am J Clin Dermatol 2001; 2: 349–351.
38. Loebstein R, Koren G. Clinical relevance of therapeutic drug monitoring during pregnancy. Ther Drug Monit 2002; 24: 15–22.
39. Anderson GD. Pregnancy induced changes in pharmacokinetics: a mechanistic based approach. Clin Pharmacokinet 2005; 44: 989–1008.
40. Pennell PB, Newport DJ, Stowe ZN, Helmers SL et al. The impact of pregnancy and childbirth on the metabolism of lamotrigine. Neurology 2004; 27: 292–295.

41. Hodge LS, Tracey TS. Alterations in drug disposition during pregnancy: implications for drug therapy. Expert Opin Drug Metab Toxicol 2007; 3: 557–571.
42. Mucklow JC. The fate of drugs in pregnancy. Clin Obstet Gynecol 1986; 13: 161–175.
43. Terrell KM, Heard K, Miller DK. Prescribing to older ED patients. Ann J Emerg Med 2006; 24: 468–478.
44. Dasgupta A. Usefulness of monitoring free (unbound) concentrations of therapeutic drugs in patient management [review]. Clin Chim Acta 2007; 377: 1–13.
45. Verbeeck RK. Pharmacokinetics and dosage adjustment in patients with hepatic dysfunction. Eur J Clin Pharmacol 2008; 64: 1147–1161.
46. Villeneuve JP, Pichette V. Cytochrome P450 and liver disease. Curr Drug Metab 2004; 5: 273–282.
47. Merrell MD, Cherrington NJ. Drug metabolism alterations in nonalcoholic fatty liver disease. Drug Metab Rev 2011; 43: 317–344.
48. Trotter JF, Osborne JC, Heller N, Christians U. Effect of hepatitis C infection on tacrolimus does and blood levels in liver transplant recipients. Aliment Pharmacol Ther 2005; 22: 37–44.
49. Zimmermann JJ, Lasseter KC, Lim HK, Harper D et al. Pharmacokinetics of sirolimus (rapamycin) in subjects with mild to moderate hepatic impairment. J Clin Pharmacol 2005; 45: 1368–1372.
50. Wyles DL, Gerber JG. Antiretroviral drug pharmacokinetics in hepatitis with hepatic dysfunction. Clin Infect Dis 2005; 40: 174–181.
51. Ng CY, Ghabrial H, Morgan DJ, Ching MS et al. Impaired elimination of propranolol due to right heart failure: drug clearance in the isolated liver and its relationship to intrinsic metabolic capacity. Drug Metab Dispo 2000; 28: 1217–1221.
52. Benowitz NL, Meister W. Pharmacokinetics in patients with cardiac failure. Clin Pharmacokinet 1976; 1: 389–405.
53. Oesterheld JR. A review of developmental aspects of cytochrome P 450. J Child Adolesc Psychopharmacol 1998; 8: 161–174.
54. Perucca E. Pharmacokinetics variability of new antiepileptic drugs at different age. Ther Drug Monit 2005; 27: 714–717.
55. Anderson GD. Developmental pharmacokinetics Semin Pediatr Neurol 2010; 17: 208–213.
56. Grygiel JJ, Birkett DJ. Effect of age on patterns of theophylline metabolism. Clin Pharamcol Ther 1980; 28: 456–462.
57. Haas M, Kletzmayer J, Staudinger T, Bohmig G et al. Hypothyroidism as a cause of tacrolimus intoxication and acute renal failure: a case report. Wien Klin Wochenschr 2000; 112: 939–941.
58. Padmanabhan H. Amiodarone and thyroid dysfunction. South Med J 2010; 103: 922–930.
59. Dai JL, Sokoll LJ, Chan DW. Automated chemiluminescent immunoassay analyzers. J Clin Ligand Assay 1998; 21: 377–385.
60. Marquet P, Lachatre G. Liquid chromatography-mass spectrometry in forensic and clinical toxicology. J Chromatogr B 1999; 7333: 93–118.

61. Yadav M, Singhal P, Goswami S, Pande UC et al. Selective determination of antiretroviral agents tenofovir, emtricitabine and lamivudine in human plasma by LC-MS-MS method for a bioequivalence study in healthy human subjects. J Chromatogr Sci 2010; 48: 704–713.
62. D'Avolio A, Simiele M, Siccardi M, Baietto L et al. A HPLC-MS method for simultaneous quantification of fourteen antiretroviral agents in peripheral blood mononuclear cell of HIV infected patients optimized using corpuscular volume evaluation. J Pharm Biomed Anal 2011; 54: 779–788.
63. Taylor PJ, Tai CH, Franklin ME, Pillans PI. The current role of liquid chromatography-tandem mass spectrometry in therapeutic drug monitoring of immunosuppressants and antiretroviral drugs. Clin Biochem 2011; 44: 14–20.
64. Saidinejad M, Law T, Ewald MB. Interference by carbamazepine and oxcarbazepine with serum and urine screening of tricyclic antidepressants. Pediatrics 2007; 120: e504–509.
65. Dasgupta A, Wells A, Datta P. False positive serum tricyclic antidepressant concentration using fluorescence polarization immunoassay due to the presence of hydroxyzine and cetirizine. Ther Drug Monit 2007; 29: 134–139.
66. US Department of Health and Human Services. National survey on drug use and health, Washington, DC: US Department of Health and Human Services; 2006 (Office of Applied Studies).
67. Cone EJ, Yousefnejad D, Hillsgrove MJ, Holicky B et al. Passive inhalation of cocaine. J Anal Toxicol 1995; 19: 399–411.
68. McGilveray IJ. Pharmacokinetics of cannabinoids. Pain Res Manag 2005; 10 (A): 15A–22A.
69. Fraser AD, Worth D. Urinary excretion of 11-nor-9-carboxy-delta9-tetrahydrocannaninol and 11-hydroxy-delta9-THC: cannabinoid metabolites to creatinine ratio stuffy IV. Forensic Sci Int 2004; 143: 147–152.
70. Backer RC, Monforte JR, Poklis A. Evaluation of the DRI oxycodone immunoassay for the detection of oxycodone in urine. J Anal Toxicol 2005; 29: 675–677.
71. Preston LP, Epstein DH, Davoudzadeh D, Husetis MA. Methadone and metabolite concentrations in patients maintained on methadone. J Anal Toxicol 2003; 24: 530–535.
72. Oiestad EL, Johansen U, Oiestad AM, Christophersen AS. Drug screening of whole blood by ultra-performance liquid chromatography-tandem mass spectrometry. J Anal Toxicol 2011; 35: 280–293.

Chapter 2

Challenges in Drugs of Abuse Testing

2.1. INTRODUCTION

Drugs of abuse testing can be classified under two broad categories: medical drug testing and workplace drug testing. Medical drug testing is conducted in most hospital laboratories as well as in a doctor's office setting using readily available immunoassays or point-of-care testing devices. A positive test result may or may not be confirmed by an alternative analytical method, preferably gas chromatography/mass spectrometry (GC/MS) because there is no legal consequence of a positive test result. Because medical records are confidential, a clinician by law cannot contact a patient's employer that the person is abusing drugs. In contrast, workplace drug testings have legal consequences and a job can be denied if a drug test is positive for a potential new hire. Therefore, more rigorous procedures are followed to ensure that the test result is indeed positive beyond a reasonable doubt. After collection, the specimen is sealed and a chain of custody form is maintained to ensure who has the possession of the specimen from the time of collection to the time of reporting of the result. Although similar immunoassays may be used for medical drug testing and the initial screening phase of workplace drug testing, every positive specimen must be confirmed by GC/MS or a similar chromatographic technique with mass spectral identification of the abused drugs. Moreover, a medical review officer who is independent of the laboratory that tested the

Resolving Erroneous Reports in Toxicology and Therapeutic Drug Monitoring: A Comprehensive Guide, First Edition. Amitava Dasgupta.

specimen must review the result and if necessary interview the donor to ensure there is no other alternative medical explanation for the positive drug testing result (e.g., positive morphine because the patient was taking a prescription pain medication containing codeine). Both medical and workplace drug testings have their challenges. These challenges include the following:

1. Both medical and workplace drugs testing include commonly abused drugs, and a less common drug, for example the abuse of lysergic acid diethylamide (LSD), may be undetected.
2. Many designer drugs may escape detection (see Chapter 5).
3. A person may attempt to beat the drug test by drinking a large amount of fluid to dilute urine because drug tests have a certain cut-off.
4. In workplace drug testing a person may try to adulterate the specimen; certain adulterants are capable of causing false-negative results with initial immunoassay screening.
5. A urine drug screen has a small window of detection and can only detect a recent abuse. However, hair drug testing can circumvent such limitations.

2.2. WORKPLACE DRUG TESTING

On September 15, 1986, President Reagan issued Executive Order No. 12564 directing all federal employees involved in law enforcement, national security, protection of life and property, public health and safety as well as other functions requiring a high degree of public trust to be subjected to mandatory drug testing. This initial order was intended for drug testing in a limited number of federal employees and at that time had little significance for private employers. A research facility started in 1935 in Lexington, Kentucky, became the Addiction Research Center in 1948. This organization finally became the National Institute of Drug Abuse (NIDA) in 1974 under the U.S. Department of Health and Human Services. NIDA became the lead agency to implement the drug-free federal workplace program as mandated by President Reagan. Later NIDA was transferred under the National Institutes of Health and became more involved in research while a new agency called the Substance Abuse and Mental Health Services Administration (SAMHSA), under the Department of Health and Human Services, became responsible for federal drug testing program and issues the "Mandatory Guidelines for Federal Workplace Drug Testing Programs." The overall testing process under mandatory drug testing guidelines consists of the proper collection of specimen, initiation of the chain of custody, and finally analysis of specimen (screening and confirmation by GC/MS) by SAMHSA-certified laboratory. The screening by immunoassay should be performed using a method approved by the Food and Drug Administration (FDA). The confirmation should be performed by a second technique, preferably by GC/MS. Although started at the federal level, currently workplace drug testing is common among private employers because such practice can prevent workplace accidents as well as improved the morale of people in the workplace. Implementation of a drug screening program in General Motors resulted in a 50% reduction of workplace injury (1). Gerber and Yacoubian assessed the impact of drug testing within the construction industry and concluded that companies with drug testing programs experienced a 51% reduction in injury incident rates within 2 years of implementing drug-testing programs (2).

Although the urine specimen is the most common workplace drug test, on April 13, 2004, a proposal was published in the Federal Register by SAMHSA to establish drug testings in alternative matrices. In SAMHSA drug-testing programs, five drugs are tested (amphetamine, cocaine as benzoylecgonine, opiates, marijuana, and phencyclidine). In addition, a requirement to test each specimen for 3,4-methylenedioxy-methamphetamine (MDMA), 3,4-methylenedioxyethylamphetamine (MDEA), and 3,4-methylenedioxyamphetamine (MDA) was also added later (3). Various cut-off concentrations of SAMHSA drugs are listed in Table 2.1. In addition to these five drugs, many other drugs are abused. Therefore private employers in addition to these five drugs may also test for barbiturates, benzodiazepines, methadone, propoxyphene, oxycodone, and other drugs. Various cut-off concentrations for testing of non-SAHMSA drugs are listed in Table 2.1.

TABLE 2.1. Cut-Off Concentrations of SAMHSA and Non-SAMHSA Drugs

	Cut-Off Concentration, ng/mL		
Drug/Drug Class	Immunoassay	GC-MS Confirmation	
SAMHSA Drugs			
Amphetamines	500	Amphetamine	250
		Methamphetamine	250
		MDMA	250
		MDEA	250
		MDA	250
Marijuana Metabolite	50		15
Cocaine metabolites (Benzoylecgonine)	150		100
Opiates	2000	Morphine	2000
		Codeine	2000
		6-Acetylmorphine	10
Phencyclidine	25		25
Non-SAMHSA Drugs			
Barbiturates	200		200
Benzodiazepines	200		200
Methadone	300		300
Methaqualone	300		300
Propoxyphene	300		300
Oxycodone	100 or 300		100

2.3. CHALLENGES OF MEDICAL VERSUS WORKPLACE DRUG TESTING

Both medical and workplace drug testings have certain limitations. The urine specimen is the most common in both medical and workplace drug testing. Although a drug or drug metabolite can be detected longer in urine than oral fluid (saliva) or blood, positive urine drug testing only indicates a recent abuse of an illicit drug but cannot provide any information whether the donor is an occasional recreational user of a drug or a chronic abuser of a drug. Using a cut-off value of 1000 ng/mL, amphetamines can be detected up to 5 days after ingestion. After smoking a marijuana cigarette, the marijuana metabolite can be detected for 2–4 days in urine. More frequent use may lead to a detection window of 1 month (4). The detection windows of various SAMHSA and non-SAMHSA drugs are listed in Table 2.2.

The small detection window of urine specimens for detecting drugs of abuse can be extended to weeks, months, or even years by using hair as an alternative specimen for drug testing (5). A drug abuser needs to demonstrate that he or she is drug free before receiving an organ transplant. Haller et al studied 42 patients who were denied organ transplants due to recent drug abuse and observed that of positive hair tests, only half were corroborated through conventional laboratory tests such as urinalysis. The authors concluded that based

TABLE 2.2. Detection Window for SAMHSA and non-SAMHSA Drugs

Drug	Detection Window
SAMHSA Drugs	
Amphetamine	2–3 days
Methamphetamine	2 days
Cocaine (as benzoylecgonine)	2–3 days
Morphine	2–3 days
Codeine	2 days
Heroin (as morphine)	2 days
Phencyclidine	8–14 days
Marijuana (as 11-nor-Δ-9-tetrahydrocannabinol-9-carboxylic acid)	2–4 days after single use 30 days in chronic abuser
Non-SAMHSA Drugs	
BARBITURATES	
Short acting (pentobarbital, secobarbital, etc.)	1 day
Long acting (phenobarbital)	21 days
BENZODIAZEPINES	
Short acting (alprazolam, lorazepam, etc.)	3 days
Long acting (diazepam, etc.)	30 days
Methadone	3 days
Methaqualone	3 days
Oxycodone	2–4 days
Propoxyphene	1–2 days

on a 90-day detection window for an illicit drug using hair analysis, two negative tests indicate 6 months of abstinence from any illicit drug (6). Polla et al reported that when a subject underwent a drug testing for regranting of a driver's license, urine drug analysis may miss some cases due to the short elimination half-life of the drug. The authors observed that even with one or two weekly negative drug screen using urine specimens, all subjects were positive using hair testing of cocaine. The authors concluded that hair testing is superior to urine testing for identifying occasional abuser of drugs (7). Although hair testing is superior to urine drug testing, hair testing cannot identify a recent drug user. Oral fluid testing is capable of identifying recent abuse of an illicit drug.

Major limitation of SAMHSA-mandated drug testing is that commonly abuse drugs such as benzodiazepines are not listed. Therefore a person abusing any common benzodiazepine drug such as alprazolam would not be tested positive in a SAMHSA-mandated drug testing program. Like alprazolam, oxycodone is also widely abused. In one report, the authors investigated 2112 fatalities due to unintentional medication overdose and observed that drugs most commonly associated with fatality include methadone (31%), hydrocodone (19%), alprazolam (15%), and oxycodone (15%) (8). An oxycodone or methadone abuser cannot be identified using the SAMHSA drug testing program because these opioids do not cross-react with opiate immunoassays. However, private employers often include benzodiazepines, oxycodone, and methadone in the drugs of abuse testing panel. In this chapter various limitations of medical and workplace drug testings are addressed.

2.4. CUT-OFF CONCENTRATIONS

Cut-off concentrations are a challenge in both medical and workplace drug testing because a drug may be present at a concentration just below the cut-off concentration of the immunoassay. In this case the drug testing is considered negative despite the presence of a low concentration of drug in the specimen. In general, chromatographic methods such as GC/MS and liquid chromatography combined with mass spectrometry are more sensitive and capable of confirming a drug at a much lower concentration than an immunoassay. However, recently Wang et al described a homogeneous immunoassay for the detection of fentanyl in urine at a cut-off concentration of 2 ng/mL (9).

Physicians sometimes determine patient compliance with their medications by using urine drug testing. However, measurement of a benzodiazepine drug in urine to ensure compliance is limited due to specificity and cut-off concentration limits of various benzodiazepine immunoassays. A chromatographic method such as liquid chromatography combined with tandem mass spectrometry (LC-MS/MS) is more suited for determining such compliance. In one report, the authors tested urine specimens from 180 patients receiving clonazepam using a benzodiazepine immunoassay at a cut-off concentration of 200 ng/mL as well as using a LC/MS-MS with a cut-off of 40 ng/mL (for

7-aminoclonazepam, the major metabolite of clonazepam). The positivity rate was only 21% (38 specimens) by immunoassay at 200 ng/mL while the positive rate was 87% (157 specimens) using a 40 ng/mL cut-off of 7-aminoclonazepam when analyses were performed by LC-MS/MS. A significant number of specimens had a drug concentration between 40 and 100 ng/mL. These urine specimens tested negative by the immunoassay. The authors concluded that the 200 ng/mL cut-off concentration for monitoring compliance with clonazepam has limited value and suggest a lower cut-off of 40 ng/mL (10).

Usually to monitor patient compliance of opioid drugs for pain management, a cut-off concentration of 300 ng/mL is used. However, in workplace drug testing the cut-off is 2000 ng/mL (the cut-off value was increased from 300 to 2000 ng/mL in 1998) to avoid false-positive test results after eating food containing poppy seeds. Nevertheless, in monitoring pain management patients, a 300 ng/mL cut-off may not be adequate for monitoring compliance with opioid therapy, and lower cut-off values may be more suitable (11).

A lower cut-off concentration may also be useful to identify illicit drug users because they often drink a large amount of fluid prior to drug testing to avoid a positive test (12). Usually a creatinine concentration below 20 mg/dL or a specific gravity below 1.003 should be considered an indication of diluted urine (13). Beck et al reported that 11% of all urine specimens submitted to their laboratory for drugs of abuse testing were diluted (14). SAMHSA does not currently allow analysis of dilute urine specimens at lower screening and confirmation cut-offs than the recommended guideline, but in Canada, the Correctional Services of Canada (CSC) guidelines incorporates lower screening and confirmation cut-off for drug/metabolites in urine specimens. The amphetamine screening and confirmation cut-off are both set at100 ng/mL, benzoylecgonine screening and confirmation cut-offs are both set at 15 ng/mL, opiates screening and confirmation cut-offs are both set at 120 ng/mL, phencyclidine screening and confirmation cut-offs are both set at 5 ng/mL, and finally for marijuana the screening cut-off is set at 20 ng/mL while the suggested confirmation cut-off is 3 ng/mL for diluted urine specimens. Fraser and Zamecnik reported that 7912 urine specimens collected and analyzed between 2000 and 2002 by the CSC were diluted and of those, only 26% screened positive using the SAMSHA cut-off values. In contrast, when lower values for cut-off and confirmation were adopted, 1100 specimens tested positive for one or more illicit drugs. The drug most often confirmed positive in a diluted specimen is marijuana. Codeine and/or morphine were also commonly confirmed in these urine specimens and ranked second after marijuana in prevalence (15). Soldin reported that there was more than a 100% increase in cocaine-positive specimens when the cut-off was lowered to 8 0 ng/mL from the recommended 300 ng/mL in a pediatric population because neonates cannot concentrate urine like adults (16). Luzzi et al investigated the analytic performance criteria of three immunoassay systems (enzyme-multiplied immunoassay technique [EMIT], Beckman enzyme-linked immunoassay [EIA], and Abbott fluorescent polarization immunoassay [FPIA]) for detecting abused drugs below established cut-off values. The authors concluded that drugs can be screened at concentrations much lower than that established by SAMSHA cut-off values. For example, the authors proposed a THC–COOH cut-off value of 35 ng/mL using EMIT and 14 ng/mL for the Beckman EIA, whereas SAMSHA guidelines stated a cut-off value of 50 ng/mL. The proposed cut-off values were based on the studies of precision of the assays at proposed lower detection limit where the concentrated volume was less than 20%. This lowering of the cut-off values increased the number of positive specimens in the screening tests to 15.6%. A 7.8% increase was also observed in the confirmation stage of drugs of abuse testing (17).

2.5. DRUGS NOT DETECTED IN DRUGS OF ABUSE TESTING PANELS

Both medical and workplace drug testing target commonly abused drugs. However, several abused drugs including designer drugs cannot be detected by such tests. See Chapter 5 for a discussion of designer drugs. Abuse of magic mushroom, peyote cactus, and solvent cannot be detected by routine drugs of abuse programs (see Chapter 6). Several drugs are abused at rave parties that are also known as club drugs. Usually young adults attend such parties or clubs and abuse MDMA (3, 4-methayledioxymethamphetamine), flunitrazepam, ketamine, and γ-hydroxybutyric acid (GHB). Desired effects of these club drugs are euphoria, feelings of intimacy, altered visual perception, increased libido, and increased energy. All of these drugs have serious adverse effects such as agitation, tachycardia, hypertension, arrhythmia, hyperthermia, and rhabdomyolysis (18). Flunitrazepam (Rohypnol), ketamine, and GHB are also used in acquaintance rape (19).

The club drug MDMA can be detected by routine amphetamine immunoassays and can be confirmed by GC/MS. Flunitrazepam (Rohypnol) is a benzodiazepine drug not approved for use in the United States. Immunoassays for benzodiazepines do not discriminate between individual drugs, and the metabolite of flunitrazepam (7-aminoflunitrazepam) may react poorly with certain immunoassay reagents causing false-negative result (20). Forsman et al reported that none of the subjects ingesting 0.5 mg of flunitrazepam (a low dose) tested positive in urinary drug screening of benzodiazepine using cloned-enzyme donor immunoassay (CEDIA), and only 23 of 102 urine specimens collected from subjects taking higher dosage of flunitrazepam (2 mg) tested positive by the benzodiazepine screening assay. In contrast, a liquid chromatography combined with the tandem MS method was able to detect the presence of major metabolites of flunitrazepam in urine using a lower cut-off of 0.5 ng/mL (21). However, ketamine and GHB are not tested in routine urine drug testing. Therefore, an individual exposed to these drugs may show a negative toxicology screen. Kronz described the case of a 30-year-old woman who presented to the emergency department with possible unknown ingestion of date rape drugs. However, the urine toxicology report was negative (22). Both GHB and ketamine can be confirmed in urine or blood using chromatographic techniques. Lee et at described a GC/MS protocol for the simultaneous determination of ketamine and amphetamine and related designer drugs in urine after derivatization using heptafluorobutyric anhydride (23). GHB and its analogs in blood and urine can be determined by LC/MS (24).

Case Study

A 23-year-old woman presented at the emergency department with paranoid delusion and hallucinations. Her previous history included two episodes of psychosis due to abuse of amphetamine and marijuana. On admission, the patient was perplexed and experiencing a hallucination that she had seen a tiger in her apartment. The toxicology screen was negative, however. Because the toxicology urine screen was negative, a schizophrenic disorder was suspected, and the patient was treated with lorazepam and olanzapine. After 24 hours, olanzapine was discontinued and the patient was receiving only lorazepam. Her confusion resolved in less than 72 hours after admission, but she experienced complete amnesia of the episode. At that point the patient revealed that she was using GHB daily for the last 6 months for insomnia and anxiety. One day prior to her admission, she decided to discontinue GHB and experienced severe withdrawal symptoms requiring hospitalization (25).

In addition to natural occurring morphine and codeine, oxycodone, hydromorphone, hydrocodone, oxymorphone, meperidine hydrochloride, fentanyl, buprenorphine, and methadone are also opioid drugs. In general, opiate immunoassays designed to detect morphine show good cross-reactivity with codeine, hydrocodone, and hydromorphone. In contrast, other opioids such as oxycodone, oxymorphone, meperidine, and methadone have very poor cross-reactivity and may not be detected in urine using an opiate screening test. In addition, fentanyl and buprenorphine cannot be detected by commercially available immunoassays. In Table 2.3 cross-reactivity of oxycodone with some commercial opiate immunoassays are listed. In SAMHSA-mandated drug testing, a person abusing a modest amount of oxycodone will most likely pass

TABLE 2.3. Cross-Reactivity of Various Opiate Assays with Oxycodone

Assay	Manufacturer	Oxycodone Cross-Reactivity
Abuscreen ONLINE	Roche, Indianapolis, IN	<0.3%
CEDIA Opiate Assay	Microgenics, Freemont, CA	3.1%
EMIT d.a.u. Opiate	Syva, Palo Alto, CA	6.7%
EMIT II Plus Opiate	Syva, Palo Alto	5.6%
Neogen Opiate Group ELISA	Lexington, KY	10%
AxSYM Opiate	Abbott Park, IL	<1%

the drug test. If an opiate immunoassay is used to monitor compliance of a patient with oxycodone, a false-negative result may result. Due to the high abuse of oxycodone and methadone, many private employers screen for the presence of drugs in urine because immunoassays are commercially available.

Case Study

A 40-year-old man initially diagnosed with migraine over several years developed a chronic daily headache. After several years of using short-acting opioid analgesics, the clinician determined that oxycodone 20 mg dosage twice a day was needed for his headache management. After 1 month the patient complained that the medication was not effective, and over the next 3 months he was advised to take the medication every 6 hours. Although his pain was under control, he routinely called the clinic stating that he finished his medication faster and need a refill. Potential abuse of oxycodone was suspected, and a urine drug test was ordered. The urine drug screen was negative, indicating no opiates in the urine specimen. The clinician suspected that the patient was selling oxycodone and not taking it. Therefore, the patient was dismissed from the clinic. On his behalf, a family member contacted a toxicologist who informed her that oxycodone may show up as false negative in a urine opiate drug screen. An aliquot of the original urine specimen was retested using GC/MS, and the presence of oxycodone at 1124 ng/mL was confirmed. This value confirmed that the patient was compliant with oxycodone. Unfortunately, due to the low cross-reactivity of oxycodone with the opiate immunoassay originally used by the laboratory for screening his urine specimen, the minimum threshold for positive was 2000 ng/mL of oxycodone (26).

LSD is also abused, but this drug is not tested in the SAMHSA drug-testing program. Some private employers test for LSD because commercially available immunoassays can be used for screening of LSD in urine. However, unlike oxycodone, methadone, or benzodiazepines, not all private employers screen their employees for LSD. Although LSD is a synthetic drug, Hawaiian baby woodrose seeds contain a natural hallucinogen lysergic acid. An 18-year-old man ingested Hawaiian baby woodrose and required hospitalization for psychotic behavior (27).

2.6. ADULTERATION OF SPECIMENS AND NEGATIVE DRUGS OF ABUSE SCREEN

People try to beat drug tests by various ways. These include substituting one's own urine with a drug-free urine specimen obtained from a friend, buying synthetic urine from an Internet site, or taking flushing or detoxifying

TABLE 2.4. Various Ways People Try to Beat Drug Tests

Approach to Beat a Drug Test	Comments
Drinking lots of fluid	Diluted urine may aid in beating drug tests by lowering concentration of an illicit drug.
Flushing/Detoxifying agents	These agents often contain diuretics such as caffeine or hydrochlorothiazide that aid in producing diluted urine
Synthetic/Substituted urine	Synthetic urine specimens are commercially available from Internet sites
Drinking goldenseal tea	Produces dark urine that may cause false-negative test results in immunoassay screening by interfering with signal detection.
In vitro *urinary adulterant*	
Various household chemicals	Interfere with various immunoassays of urinary drug screen. EMIT assays are most adversely affected, although other assays may also produce false-negative results.
Potassium nitrite	May cause false negatives by both immunoassay and GC/MS, especially with opiates and marijuana.
Pyridinium chlorochromate	May cause false negatives by both immunoassay and GC/MS, especially with marijuana.
Glutaraldehyde	False-negative immunoassay screen.
Stealth	False- negative immunoassay screen especially for opiates and marijuana.

agents or *in vitro* urinary adulterant again bought from an Internet site. For emergency department drug testing, adulteration of a urine specimen is not an issue because most likely the specimen was collected by a nurse or health care provider. However, for workplace drug testing, tampering of urine specimens is a significant problem because other than the military, the donor has privacy during collection of the specimen. A person might want to cheat on a drug test to get hired or avoid disciplinary action following a positive drug test. In this section various ways by which people try to beat a drug test are discussed along with how to identify such tampered specimens. Adulteration is the most problematic with urine specimens. With hair or oral fluid analysis, the chances of tampering a specimen by the donor are minimal. Table 2.4 summarizes the various ways a person tries to beat drugs tests.

2.6.1. Flushing and Detoxifying Agents: Impact on Drug Testing

Many flushing and detoxifying agents are available through the Internet to beat drug testing. Manufacturers often claim that taking such products along with plenty of fluid can flush illicit drugs out of the body and an individual can pass the drug test. Currently at least 400 different products are available intended for beating drug tests (4). Common examples of flushing and detoxifying agents are Absolute Detox XXL drink, Absolute Carbo Drinks, Ready Clean Drug Detox Drink, Fast Flush Capsules, and Ready Clean Gel Capsules.

These products may produce diluted urine and may contain diuretics such as hydrochlorothiazide. Cone et al studied the ability of Naturally Clean Herbal tea, goldenseal root, and hydrochlorothiazide to cause false- negative results in urine drugs of abuse testing. Volunteers drank 1 gallon of water (divided in four doses over a 4-hour period) or herbal tea or hydrochlorothiazide 22 hours after smoking marijuana cigarettes or intranasal administration of cocaine. Marijuana and cocaine metabolite levels (as measured by both EMIT and FPIA assays) decreased significantly and frequently switched from positive to negative in subjects after consuming 2 quarts of fluid. Even excess water was effective in diluting a urine specimen to cause false-negative results. Consumption of herbal tea produced dilute urine faster compared with drinking water alone (28). Consumption of goldenseal tea produces dark urine, however, which can be identified by visual inspection (29).

Wagner et al reported a decreased signal in the EMIT assays for drugs of abuse testing in urine following the ingestion of aspirin. During routine drug testing for benzoylecgonine, the authors observed several urine specimens showing negative response rates in comparison with that of drug-free urine calibrators. On further examination, salicylate was identified in all these specimens. When six subjects ingested four 325-mg aspirin tablets, negative absorbance was observed in the EMIT assay and the value correlated with peak salicylate concentration. Reduced signals in urine containing salicylate (concentration ranging from 0 to 100 mg/dL) were also observed with EMIT assays for amphetamine, barbiturates, opiate, benzodiazepines, phencyclidine, propoxyphene, and tetrahydrocannabinol. The authors concluded that ingestion of a therapeutic dosage of aspirin may cause false-negative results with EMIT-d.a.u. testing (30).

2.6.2. *In Vitro* Urinary Adulterants

In vitro adulterants can be classified under two broad categories: household chemicals and adulterants available through various Internet sites.

2.6.2.1. House Chemicals as Adulterants. Several household chemicals are used to invalidate drug of abuse testing in urine. The most commonly used chemicals include table salts (sodium chloride), vinegar, Drano, Joy, hand soap, liquid laundry bleach, denture cleansing tablet, lemon juice, ascorbic acid, hydrogen peroxide, and rubbing alcohol (isopropyl alcohol).

Sodium chloride caused negative interference with several drugs tested by EMIT and caused a slight decrease in measured concentrations of benzodiazepines by FPIA. Interestingly, sodium chloride has no effect on the screening of amphetamines, barbiturates, benzodiazepines, phencyclidine (PCP), benzoylecgonine, opiates, and cannabinoid as the major metabolite 11-nor-9-carboxy-Δ–9 tetrahydrocannabinol (THC-COOH) using CEDIA (Microgenics Corporation, Concord, CA). Sodium bicarbonate in sufficient amounts invalidates the EMIT opiate assay and the FPIA PCP assay. Bicarbonate does not interfere with CEDIA assays for drugs of abuse testing. Hydrogen peroxide caused false-positive benzodiazepine results by FPIA (31,32). Uebel and Wium studied the effect of household chemicals sodium hypochlorite, Dettol (chloroxylenol, a topical antiseptic popular in many countries outside the United States), glutaraldehyde, Pearl hand soap, ethanol, isopropanol, and peroxide on cannabis and methaqualone tests using EMIT assays. Most of the agents tested interfered with the tests, and the greatest effect was observed with glutaraldehyde and Pearl hand soap for methaqualone (false negative). Dettol and Pearl hand soap also caused false-negative results in cannabis tests. Addition of isopropanol, ethanol, and peroxide invalidated a methaqualone test (33). Schwarzhoff and Cody studied the effect of 16 different adulterating agents: ammonia-based cleaner, L-ascorbic acid (vitamin C), Visine eye drops, Drano, goldenseal root, lemon juice, lime solvent, Clorox, liquid hand soap, methanol, sodium chloride, tribasic potassium phosphate, toilet bowl cleaner (Vanish, Drackett Products), white vinegar, and ionic detergent (Multi-Terge) for various abused drugs in urine specimens. The authors tested these adulterating agents at 10% by volume concentration of urine with the exception of goldenseal tea because of the insolubility. For goldenseal tea, one capsule was suspended in 60 mL urine. Of six drugs tested (benzoylecgonine, amphetamines, opiates, phencyclidine: PCP, THC metabolite, and barbiturates), the cannabinoid test was the most susceptible to adulteration. Approximately half of the agents (ascorbic acid, vinegar, bleach, lime solvent, Visine eye drops, and goldenseal) tested caused false negatives. Both cannabinoid and opiate assays were susceptible to bleach, and actual degradation of the THC metabolite (11-nor-9-carboxy-Δ–9 tetrahydrocannabinol) was confirmed by GC/MS analysis. The PCP and benzoylecgonine analyses were affected by alkaline agents. The barbiturate assay was particularly sensitive to detergents and the only assay where authors observed false-positive tests due to the presence of adulterants. Goldenseal roots with its high content of insoluble plant material would be difficult to add to urine as an adulterant because it can be detected easily by visual observation. Nevertheless goldenseal produces a significant effect on both the marijuana metabolite and barbiturate assays at a concentration less than 1% due to probable binding of drugs to insoluble plant material (34). Baiker et al reported that hypochlorite (a common ingredient of household bleach) adulteration of urine caused a decreased concentration of THC-COOH as measured by GC/MS. A false-negative result was also observed with the FPIA screen as well as the Roche Abuscreen (35). Another report described adulteration of urine specimens with denture cleaning tablets containing 1.5% sodium perborate, and various salts interfere with FPIA assays for MDMA, benzoylecgonine, and cannabinoids. The possible mechanism of interference is oxidation of the drug/metabolite by sodium perborate. Salts also alter the ionic strength of urine and may interfere with an immunoassay (36). Components of Visine eye drops in urine may be detected by using high-performance liquid chromatography combined with ultraviolet (UV) detection at 262 nm, a method originally developed for the analysis of ophthalmic formulations (37).

2.6.2.2. Specimen Integrity Test. Specimen integrity tests are performed prior to analysis of all urine specimens submitted for workplace drug testing to ensure that specimens are not tampered. The temperature of the urine should be within 32–38°C (89.6–100.4°F) and must be checked within 4 minutes of collection, although urine specimens may retain temperature for 15 minutes and may be warmer than 33°C. The specific gravity should be between 1.005 and 1.030, and the pH should be between 4.5 and 8.0. The creatinine concentration should be 20–400 mg/dL. A specimen is considered diluted if the creatinine is less than 20 mg/dL and the specific gravity is less than 1.003. The laboratory should perform creatinine and pH analysis of all specimens submitted for drugs of abuse testing. Additional tests are also recommended to detect the presence of other adulterants. Determination of specific gravity is mandatory for any specimen with a creatinine concentration less than 20 mg/dL. Substituted urine specimens have creatinine concentrations less than 5 mg/dL and a specific gravity 1.001 or over 1.020. The urine is adulterated if pH is less than 3 or more than 11. Adulteration with sodium chloride at a concentration necessary to produce a false-negative result always produces a specific gravity over 1.035. Use of household chemicals such as bleach, acid, soap, detergent, as well as adulteration with vinegar, alter the pH of urine to a value outside the physiological range and can be easily detected by specimen integrity tests (Table 2.5). However, the presence of Visine eye drops in adulterated urine cannot be detected by specimen integrity testing. Moreover, the presence of newer urine adulterants such as Urine Luck, UrinAid, Klear, and Whizzies cannot be detected by routine specimen integrity tests.

2.6.2.3. Adulterant Urine Luck. Wu et al reported that the active ingredient of Urine Luck is 200 mmol/L of pyridinium chlorochromate (PCC). The

TABLE 2.5. Common Household Adulterants and Specimen Integrity Tests

	Specimen Integrity Tests*			
Household Chemicals	pH	Creatinine	Temperature	Specific Gravity
Diluted urine		X	X	
Sodium chloride				X
Vinegar	X			
Laundry bleach	X			
Liquid soap	X			Cloudy
Drano	X			
Toilet bowl cleaner	X			
Sodium bicarbonate	X			
Sodium hypochlorite	X			
Lemon juice	X			
Goldenseal				Dark urine
Ascorbic acid (vitamin C)	X			
Visine Eye Drops*				

X = Abnormal test indicative of the adulterant.
*Cannot be detected by routine urine specimen integrity testing.

authors reported a decrease in the response rate for all EMIT II drug screens and for the Abuscreen (Roche Diagnostics) morphine and THC assays. This adulteration also caused reduced concentrations of opiates, and THC-COOH, as determined by GC/MS, was reduced. Wu et al also described the protocol for the detection of PCC in urine using spot tests. The indicator solution contains 10 g/L of 1,5-diphenylcarbazide in methanol. The indicator detects the presence of the chromium ion and is colorless when prepared. Two drops of indicator solution is added to 1.0 mL of urine. If a reddish purple color develops, the test is positive (38). Addition of 4–5 drops of 3% hydrogen peroxide in approximately 200 μL of urine adulterated with PCC (approximately 6–7 drops from a transfer pipette) caused rapid formation of a dark brown color (due to reduction of heptavalent chromium by hydrogen peroxide) and a dark brown precipitate appeared on standing. In contrast, unadulterated urine turned colorless after the addition of hydrogen peroxide (39). The capillary ion electrophoresis technique can also be used for detecting chromate ion, as well as nitrite ion, in urine specimens suspected of adulteration (40).

2.6.2.4. Nitrite as Urine Adulterant. The products Klear and Whizzies contain potassium nitrite as first identified by ElSohly et al. This urinary adulterant affects both immunoassay and GC/MS conformation of marijuana metabolite (THC-COOH) but by using a bisulfite step at the beginning of the sample preparation, such interference can be eliminated (41). Both duration of nitrite exposure and the urine matrix also affect the THC-COOH assay. The pH of the urine, and the original drug concentrations, play major roles in dictating the effectiveness of nitrite in causing false-negative THC-COOH tests. With acidic pH, significant decreases in the immunoassay screening results can be observed in all urine specimens within 4 hours of adulteration with nitrite regardless of original concentrations of THC-COOH (range of concentrations 33–488 ng/mL as determined by GC/MS). All specimens were negative for THC-COOH after 1 day. In contrast, the immunoassay results of urine specimens with basic or neutral pH were less affected by nitrite exposure. Approximately two thirds of the samples with pH values greater than 7.0 were immunoassay positive even 3 days after supplementing with nitrite (42).

The presence of nitrite in urine can be easily identified by a simple spot test. Addition of a few drops of a nitrite-adulterated urine specimen to 0.5 mL of 1% potassium permanganate solution followed by addition of a few drops of 2N hydrochloric acid turns a pink permanganate solution colorless with effervescence (39). Nitrite can also be detected by diazotizing sulfanilamide and coupling the product with N-(1-napthyl) ethylenediamine. The presence of nitrite in urine can also be confirmed by analysis using high-performance liquid chromatography as well as capillary electrophoresis (43,44).

2.6.2.5. Stealth as Urine Adulterant. Stealth consists of two vials, one containing a powder (peroxidase) and another vial containing a liquid (hydrogen peroxide). Both products are added to the urine specimen. Stealth is capable of producing false-negative results using Roche ONLINE and Microgenics CEDIA methods when marijuana metabolites, LSD, and opiates (morphine) are present in the urine at 125–150% of cut-off values. Adulteration of an authentic positive sample provided by a marijuana user caused that sample to screen negative using these immunoassay reagents (45). A low concentration of morphine (2500 ng/mL) could be effectively masked by Stealth, but higher concentrations (6000 ng/mL) tested positive by immunoassay (Roche ONLINE and Microgenics CEDIA immunoassay). GC/MS confirmation can be affected if Stealth is present in the urine. Cody et al reported that GC/MS analysis of Stealth-adulterated urine using standard procedures proved unsuccessful in several cases, and in 4 of 12 cases neither the drug nor the internal standard was recovered (46). Valtier and Cody described a rapid color test to detect the presence of Stealth in urine. Addition of 10 μL of urine to 50 μL of tetramethylbenzidine (TMB) working solution followed by addition of 0.5 mL of 0.1 M phosphate buffer solution caused a dramatic color change of the specimen to dark brown. Peroxidase activity could also be monitored by using a spectrophotometer. Routine specimen integrity check using pH, creatinine, specific gravity, and temperature did not detect the presence of Stealth in urine (47).

2.6.2.6. Glutaraldehyde as Urine Adulterant. Glutaraldehyde is sold as UrinAid, intended for beating a drug test. A 10% solution of glutaraldehyde is available from pharmacies as over-the-counter medication for the treatment of warts. Glutaraldehyde at a concentration of 0.75% volume can lead to false-negative screening results for a cannabinoid test using the EMIT II drugs of abuse screen. Amphetamine, methadone, benzodiazepine, opiate, and cocaine metabolite tests can be affected at a glutaraldehyde concentration between 1% and 2% using EMIT immunoassays. At a concentration of 2% by volume, the assay of cocaine metabolite is significantly affected (apparent loss of 90% sensitivity). A loss of 80% sensitivity was also observed with the benzodiazepine assay. Glutaraldehyde also interfered with the CEDIA immunoassay for the screening of abused drugs (48). Goldberger and Caplan reported that glutaraldehyde caused false-negative results with EMIT but also caused false-positive phencyclidine results with the FPIA (Abbott Laboratories) and Kinetic Interaction of Microparticle in a Solution Immunoassay (KIMS, Roche Diagnostics) (49). Wu et al described a simple fluorometric method for the detection of glutaraldehyde in urine. When 0.5 mL of urine was heated with 1.0 mL of 7.7 mmol/L potassium dihydrogen phosphate (pH 3.0) saturated with diethylthiobarbituric acid for 1 hour at 96–98°C in a heating block, a yellow green fluorophore developed if glutaraldehyde was present. Shaking the specimen with n-butanol resulted in the transfer of this adduct to the organic layer, which can be viewed under long wavelength UV light. Glutaraldehyde in urine can also be estimated using a fluorometer (50).

2.6.2.7. Guidelines for Detecting Adulterants and Dipstick Testing. SAMSHA guidelines require additional tests for urine specimens with abnormal physical characteristics or ones that show characteristics of an adulterated specimen during initial screening or confirmatory tests (nonrecovery of internal standard, unusual response, etc.). A pH less than 3 or more than 11, and nitrite concentrations greater than 500 μg/mL indicate the presence of adulterants. A nitrite colorimetric test or a general oxidant colorimetric test can be performed to identify nitrite. Similarly, the presence of chromium can be confirmed by a chromium colorimetric test or a general test for the presence of oxidant. A confirmatory test can be performed using multi-wavelength spectrophotometry, ion chromatography, atomic absorption spectrophotometry, capillary electrophoresis, or inductively coupled plasma mass spectrometry. The presence of halogen (chloride, fluoride, or bleach) should be confirmed by a halogen colorimetric test or a general test for the presence of oxidants. Confirmatory tests may use multi-wavelength spectrophotometry, ion chromatography, atomic absorption spectrophotometer, capillary electrophoresis, or inductively coupled plasma mass spectrometry. The presence of glutaraldehyde should be detected by a general aldehyde test or the characteristic immunoassay response in one or more drug immunoassay tests for initial screening. The presence of pyridinium chlorochromate should be confirmed by using a general test for the presence of oxidant and a GC/MS confirmatory test.

In addition to these tests, on-site adulterant detection devices have become commercially available. These dipstick devices offer an advantage over other tests because an adulteration check can also be performed at the collection site. Peace and Tarnai evaluated the performance of three on-site devices: Intect 7 (Branan Medical Corporation), MASK Ultrascreen (Kacey Inc.), and AdultaCheck 4 (Sciteck Diagnostics). Intect 7 can simultaneously test for creatinine, nitrite, glutaraldehyde, pH, specific gravity, PCC, and bleach. Ultrascreen tests for creatinine, nitrite, pH, specific gravity, and oxidants. AdultaCheck 4 tests creatinine, nitrite, glutaraldehyde, and pH. The authors adulterated urine specimens with Stealth, Urine Luck, Instant Clean ADD-IT-ive, and Klear at their optimum usage concentration and concluded that Intect 7 was the most sensitive and correctly identified adulterants. AdultaCheck 4 did not detect Stealth, Urine Luck, or Instant Clean ADD-it-ive. Ultra Screen detected a broader range of adulterants than AdultaCheck 4. However, in practice, it only detected these adulterants at levels well above their optimum usage, making it less effective than Intect 7 (51). However, King reported that AdultaCheck 4 is an excellent way to detect contamination in urine specimens (52). AdultaCheck 6 or Intect 7 test strips can determine a range of creatinine values, although the precise concentration of creatinine cannot be determined. Similarly, neither test strip can determine precisely the pH of a urine specimen but can only show the range. However, AdultaCheck 6 and Intect 7 test strips are capable of successfully identifying urine specimens with abnormally low creatinine and/or pH (53).

2.7. DRUG TESTING IN ALTERNATIVE SPECIMENS

Almost all medical drug testing and most workplace drug testing is conducted using urine specimens, whereas hair drug testing, testing in meconium, and drug testings in other alternative matrixes are offered in only few reference laboratories and large academic medical centers. A major advantage of testing drugs in alternative specimens is the avoidance of adulteration. Moreover, hair has a much longer window of detection for a drug compared with urine. Although there are products intended for beating drug testing involving hair specimens, such approaches are usually ineffective. Rohrich et al studied the effect of Ultra Clean Shampoo on drug concentrations in human hair and observed that all drugs that were originally present in hair specimens were also present after a single application of shampoo, although a slight decrease was observed in cocaine, 6-monoacetylmorphine, and THC concentrations. The authors conclude that Ultra Clean Shampoo cannot remove drugs in sufficient amounts to cause a positive hair specimen to be tested negative (54). Hair grows more or less in a constant rate that averages approximately 1 cm per month. Usually it takes about 5–6 days for hair to grow from the root to appear above the scalp so that the specimen is available for detection. A hair specimen of 3 cm indicates whether a person abused drugs or not in a 3-month window period. However, concentration of a drug is much lower than urine concentration, which is an analytical challenge. Drugs are found in hair from picogram to nanogram level per milligram of hair. In one study, the maximum amounts of cocaine and benzoylecgonine detected in hair specimens were 1093.8 ng/mg and 163.7 ng/mg, respectively; the maximum amount of cocaethylene was 13.9 ng/mg in hair. In the same study, the maximum amounts of morphine, codeine, heroin, and 6-acetylmorphine (a metabolite of heroin) detected in subjects were 291.3 ng/mg, 68.4 ng/mg, 146.4 ng/mg, and 220.8 ng/mg, respectively. Similarly, the minimum amounts of cocaine and benzoylecgonine detected in hair specimens were 200 pg/mg and 100 pg/mg, respectively. The minimum amount of cocaethylene detected was also 100 pg/mg. For marijuana, the minimum level detected was 3 pg/mg; the minimum THC-COOH detected was 1 pg/mg (55). Recently Barroso et al summarized the application of hair drug testing in forensic toxicology (56).

Oral fluid is easy to collect and is noninvasive. Oral fluid collection does not require the presence of any medically trained personnel, and oral fluid can be tested at the site of collection by point-of-care devices providing a quick answer whether a drug is present or not in the oral fluid specimen. Drug testing in oral fluid is useful in assessing any impairment in people from drug abuse who are involved in driving, operating machinery, and other tasks requiring high-level psychomotor functioning (57). Unlike urine drug level, oral fluid drug level can be correlated with degree of impairment. There are commercially available devices for oral fluid testing, but a positive test result must be confirmed in a laboratory using GC/MS or a similar analytical technique.

Recently Langel et al described a GC/MS protocol for the analysis of 50 drugs of abuse and medical drugs in oral fluid (58).

2.8. CONCLUSIONS

Urine drug tests have certain limitations including a narrow detection window, sensitivity, and specificity of various immunoassays utilized in the screening of urine specimens. Nevertheless, the urine specimen is the most commonly used specimen for drug testing because the collection process is noninvasive and the urine specimen can be subjected to analysis directly without any specimen pretreatment. Urine specimens are also subjected to easy adulteration, but if attention is paid in the preanalytical step prior to analysis and proper tests are conducted to detect any such adulterant, this major problem can be easily circumvented. Hair analysis has a major advantage of offering a longer window of detection for illicit drugs, but it requires extensive preanalytic manipulation prior to analysis.

REFERENCES

1. MacDonald S, Wells S. The impact and effectiveness of drug testing programs in the workplace. In: S. Macdonald and P. Roman, eds., *Research Advances in Alcohol and Drug Problems. Vol. 11, Drug Testing in the Workplace*. New York: Plenum Press, 1994.
2. Gerber JK, Yacoubian GS Jr. An assessment of drug testing within the construction industry. J Drug Educ 2002; 32: 53–68.
3. Vandevenne M, Vandenbussche H, Verstraete A. Detection window of drugs of abuse in urine. Acta Clin Belg 2000; 55: 323–333.
4. Bush D. The U.S. Mandatory Guidelines for Federal Workplace Drug Testing Programs: current status and future considerations. Forensic Sci Int 2008; 174: 111–119.
5. Kintz P, Villain M, Cirimele V. Hair analysis for drug detection. Ther Drug Monit 2006; 28: 442–446.
6. Haller DL, Acosta MC, Lewis D, Miles DR et al. Hair analysis versus conventional methods of drug testing in substance abusers seeking organ transplantation. Am J Transplant 2010; 10: 1305–1311.
7. Polla M, Stramesi C, Pichini S, Vignali C et al. Hair testing is superior to urine to disclose cocaine consumption in driver's license re-granting. Forensic Sci Int 2009; 189: e41–43.
8. Piercefield E, Archer P, Kemp P, Mallonee S. Increased in unintentional medication overdose deaths: Oklahoma, 1994–2006. Am J Prev Med 2010; 39: 357–363.
9. Wang G, Huynh K, Barhate R, Rodrigues W et al. Development of a homogenous immunoassay for the detection of fentanyl in urine. Forensic Sci Int 2011; 206: 127–131.

10. West R, Pesce A, West C, Crews B et al. Comparison of clonazepam compliance by measurement of urinary concentration by immunoassay and LC-MS/MS in pain management population. Pain Physician 2010; 13: 71–78.
11. Pesce A, West C, West R, Crews B et al. Determination of medication cutoff values in a pain patient population. J Opioid Manag 2011; 7: 117–122.
12. George S, Braithwaite RA. An investigation into the extent of possible dilution of specimens received for urinary drugs of abuse screening. Addiction 1995; 90, 967–970.
13. Fraser AD, Zamecnik J, Keravel J, McGrath L, Wells J. Experience with urine drug testing by the correctional service of Canada. Forensic Sci Int 2001; 121: 16–22.
14. Beck O, Bohlin M, Bragd F, Bragd J et al. Adulteration of urine drug testing—an exaggerated cause of concern [in Swedish]. Lakartidningen 2000; 97, 703–706.
15. Fraser AD, Zamecnik J. Impact of lowering the screening and confirmation cutoff values for urine drug testing based on dilution indicators. Ther Drug Monit 2003; 25: 723–727.
16. Soldin SJ, Morales AJ, D'Angelo LJ, Bogema SC, et al. The importance of lowering the cut-off concentrations of urine screening and confirmatory tests for benzoylecgonine/cocaine [abstract]. Clin Chem 1991; 37: 993.
17. Luzzi VI, Saunders AN, Koenig JW, Turk J et al. Analytical performance of immunoassays for drugs of abuse below established cutoff values. Clin Chem 2004; 50: 717–722.
18. Smith KM, Larive LL, Romanelli F. Club drugs: methylenedioxymethamphetamine, flunitrazepam, ketamine hydrochloride and gamma-hydroxybutyrate. Am J Health Syst Pharm 2002; 59: 1067–1076.
19. Smith KM. Drugs used in acquaintance rape. J Am Pharm Assoc (Wash) 1999; 39: 519–525.
20. Raymon LP, Steele BW, Walls HC. Benzodiazepines in Miami-Dade County, Florida driving under the influence (DUI) cases (1995–1998) with emphasis on Rohypnol: GC-MS confirmation, pattern of use, psychomotor impairment and results of Florida legislation. J Anal Toxicol 1999; 23: 490–496.
21. Forsman M, Nystrom I, Roman M, Berglund L et al. Urinary detection times and excretion patterns of flunitrazepam and its metabolites after a single oral dose. J Anal Toxicol 2009; 33: 491–501.
22. Kronz CS. A 30 year old woman with possible unknown ingestion of date rape drugs. J Emerg Nurs 2000; 26: 544–548.
23. Lee HH, Lee JF, Lin SY, Chen PH et al. Simultaneous determination of HFN-derivatized amphetamines and ketamine in urine by gas chromatography-mass spectrometry. J Anal Toxicol 2011; 35: 162–169.
24. Johansen SS, Windberg CN. Simultaneous determination of g-hydroxybutyrate (GHB) and its analogs (GBL, 1,4-BD, GVL) in whole blood and urine by liquid chromatography coupled to tandem mass spectrometry. J Anal Toxicol 2011; 35: 8–14.
25. Constantinides P, Vincent P. Chronic gamma-hydroxybutyric acid use followed by gamma-hydroxybutyric acid withdrawal mimic schizophrenia: a case report. Cases J 2009; 2: 7520.

26. Von Seggern RL, Fitzgerald CP, Adelman LC, Adelman JU. Laboratory monitoring of OxyContin (oxycodone): clinical pitfalls. Headache 2004; 44: 44–47.
27. Gertsch JH, Wood C. Case report: an ingestion of Hawaiian baby woodrose seeds associated with acute psychosis. Hawaiian Med J 2003; 62: 127–129.
28. Cone EJ, Lange R, Darwin WD. In vivo adulteration: excess fluid ingestion cause false negative marijuana and cocaine urine test results. J Anal Toxicol 1998; 22: 460–473.
29. Mikkelsen SL, Ash KO. Adulterants causing false negative in illicit drug testing. Clin Chem 1988; 34: 2333–2336.
30. Wagner RE, Linder MW, Valdes R. Decreased signal in EMIT assays of drugs of abuse in urine after ingestion of aspirin: potential for false negative results. Clin Chem 1994; 40: 609–612.
31. Warner A. Interference of household chemicals in immunoassay methods for drugs of abuse. Clin Chem 1989; 35: 648–651.
32. Wu AH, Forte E, Casella G, Sun K, et al. CEDIA for screening drugs of abuse in urine and effect of adulterants. J Forensic Sci 1995; 40: 614–618.
33. Uebel RA, Wium CA. Toxicological screening for drugs of abuse in samples adulterated with household chemicals. S Afr Med J 2002; 92: 547–549.
34. Schwarzhoff R, Cody JT. The effects of adulterating agents on FPIA analysis of urine for drugs of abuse. J Anal Toxicol 1993; 17: 14–17.
35. Baiker C, Serrano L, Lindner B. Hypochlorite adulteration of urine causing decreased concentration of delta-9-THC-COOH by GC/MS. J Anal Toxicol 1994; 18: 101–103.
36. Stolk LM, Scheijen JL. Urine adulteration with denture cleaning tablets [letter]. J Anal Toxicol 1997; 21: 403.
37. Rojsitthisak P, Wichitnithad W, Pipitharome O, Sanphanya K, Thanawattanawanich P. Simple HPLC determination of benzalkonium chloride in ophthalmic formulations containing antazoline and tetrahydrozoline. PDA J Pharm Sci Technol 2005; 59: 323–327.
38. Wu A, Bristol B, Sexton K, Cassella-McLane G et al. Adulteration of urine by Urine Luck. Clin Chem 1999; 45: 1051–1057.
39. Dasgupta A, Wahed A, Wells A. Rapid spot tests for detecting the presence of adulterants in urine specimens submitted for drug testing. Am J Clin Pathol 2002; 117: 325–329.
40. Freslew KE, Hagardorn AN, Roberts TA. Capillary ion electrophoresis of endogenous anions and anionic adulterants in human urine. J Forensic Sci 2001; 46: 615–626.
41. ElSohly MA, Feng S, Kopycki WJ, Murphy TP et al. A procedure to overcome interferences caused by adulterant "Klear" in the GC-MS analysis of 11-nor-Δ9-THC-9-COOH. J Anal Toxicol 1997; 20: 240–242.
42. Tsai LS, ElSohly MA, Tsai SF, Murphy TO et al. Investigation of nitrite adulteration on the immunoassay and GC-MS analysis of cannabinoids in urine specimens J Anal Toxicol 2000; 24: 708–714.
43. Singh J, Elberlind JA, Hemphill DG, Holmstrom J. The measurement of nitrite in adulterated urine samples by high performance ion chromatography. J Anal Toxicol 1999; 23: 137–140.

44. Kinkennon AE, Black DL, Robert TA, Stout PR. Analysis of nitrite in adulterated urine samples by capillary electrophoresis. J Forensic Sci 2004; 49: 1094–1100.
45. Cody JT, Valtier S. Effects of Stealth adulteration on immunoassay testing for drugs of abuse. J Anal Toxicol 2001; 25: 466–470.
46. Cody JT, Valtier S, Kuhlman J. Analysis of morphine and codeine in samples adulterated with Stealth. J Anal Toxicol 2001; 25: 572–575.
47. Valtier S, Cody JT. A procedure for the detection of Stealth adulterant in urine samples. Clin Lab Sci 2002; 15: 111–115.
48. Wu AH, Forte E, Casella G, Sun K et al. CEDIA for screening drugs of abuse in urine and the effect of adulterants. J Forensci Sci 1995; 40: 614–618.
49. Goldberger BA, Caplan YH. Effect of glutaraldehyde (UrinAid) on detection of abused drugs in urine by immunoassay [letter]. Clin Chem 1994; 40: 1605–1606.
50. Wu A, Schmalz J, Bennett W. Identification of Urin-Aid adulterated urine specimens by fluorometric analysis [letter]. Clin Chem 1994; 40: 845–846.
51. Peace MR, Tarnai LD. Performance evaluation of three on-site adulteration detection devices for urine specimens. J Anal Toxicol 2002; 26: 464–470.
52. King EJ. Performance of AdultaCheck 4 test stripes for the detection of adulteration at the point of collection of urine specimens used for drugs of abuse testing. J Anal Toxicol 1999; 23: 72.
53. Dasgupta A, Chughtai O, Hannah C, Davis B et al. Comparison of spot tests with AdultaCheck 6 and Intect 7 urine test strips for detecting the presence of adulterants in urine specimens. Clin Chem Acta 2004; 34: 19–25.
54. Rohrich J, Zorntlein S, Potsch L, Skopp G et al. Effect of shampoo Ultra Clean on drug concentrations in human hair. Int J Legal Med 2000; 2: 102–106.
55. Tsanaclis L, Wicks JF. Patterns in drug use in the United Kingdom as revealed through analysis of hair in a large population sample. Forensic Sci Int 2007; 170: 121–128.
56. Barroso M, Gallardo E, Vieira DN, Lopez-Rivadulla M et al. Hair: a complementary source of bioanalytical information in forensic toxicology. Bioanalysis 2011; 3: 67–79.
57. Dryer KR, Wilkinson C. The detection of illicit drugs in oral fluid: another potential strategy to reduce illicit drug related harm. Drug Alcohol Rev 2008; 27: 99–107.
58. Langel K, Gunner T, Ariniemi K, Rajamaki O et al. A validated method for the detection and quantitation of 50 drugs of abuse and medical drugs in oral fluid by gas chromatography-mass spectrometry. J Chromatogr B Analyt Technol Biomed Life Sci 2011; 879: 859–870.

Chapter 3

False-Positive Results Using Immunoassays for Drugs of Abuse Testing

3.1. INTRODUCTION

Drugs of abuse testing are important both in workplace and in medical settings. Workplace drug testing requires confirmation of a positive specimen by an alternative analytical technique. preferably gas chromatography/mass spectrometry (GC/MS). but such confirmations may not be requested by a clinician during medical drug testing. Drug testing is important in the medical setting because Chen et al reported that 66.6% of patients who tested positive for amphetamine did not report their amphetamine abuse to the clinician in the emergency department (1). Similarly, 70% of patients did not report their opiate abuse.

Urine specimens are the most common for drugs of abuse testing. Immunoassays are available for drug of abuse testing in clinical laboratories for application on various automated analyzers as well as point-of-care devices for analysis in a doctor's office. Although in general immunoassays have good sensitivity and specificity, immunoassays are also subjected to interferences. Standridge et al commented that immunoassays are the preferred test for initial screening for drugs of abuse, but positive results should be confirmed by GC/MS or high-performance liquid chromatography. In addition, an extended panel of immunoassays is needed for opiate testing to avoid false negatives, and appropriate collection techniques are essential for preserving

Resolving Erroneous Reports in Toxicology and Therapeutic Drug Monitoring: A Comprehensive Guide, First Edition. Amitava Dasgupta.

specimen integrity and avoiding specimen tampering (2). Amphetamines immunoassays are subjected to many interferences and are more affected compared with other drugs of abuse assays. Nevertheless other immunoassays are also affected by interfering substances. Dietzen et al reviewed confirmatory results of drugs of abuse testing that were positive by immunoassays and reported that there were 175 instances of false-positive screening during a 1-year period. Positive predictive values of various immunoassays varied from 0% (amphetamine) to 100% (tetrahydrocannabinol) (3).

3.2. INTERFERENCES IN AMPHETAMINE IMMUNOASSAY

Other than amphetamine and methamphetamine, two similar synthetic designer drugs, 3,4-methylenedioxymethamphetamine (MDMA, commonly called Ecstasy) and 3,4-methylenedioxyamphetamine (MDA) are also widely abused. Immunoassays are the primary screening methods for the detection of amphetamines, and in most of these assays the antibodies appear to be directed toward the amino group of amphetamine, a primary amine, and methamphetamine, a secondary amine. Both monoclonal and polyclonal antibodies have been used in these assays. Although MDMA and MDA both show significant cross-reactivities with commercially available amphetamine immunoassays, a separate immunoassay is also available for MDMA. In one report, the authors evaluated eight commercially available immunoassays for their effectiveness for detecting MDMA during immunoassay screening test and observed that cross-reactivities were lower at higher concentrations of these drugs (Table 3.1) (4). Kunsman et al also reported that the cross-reactivity of MDMA with an amphetamine immunoassay depends on concentration and varied from 118% at 150 ng/mL to 18% at 10,000 ng/mL (5).

Because of the structural similarity with both amphetamine and methamphetamine, many sympathomimetic amines cross-react with immunoassays designed for amphetamines/methamphetamine screening. Immunoassays designed for the detection of amphetamines can be classified into three general types based on antibody specificity:

1. Assays highly selective for either amphetamine and MDA or methamphetamine and MDMA.
2. Assays that are able to detect both amphetamine and methamphetamine to varying extents but that also exhibit higher levels of cross-reactivity to various over-the-counter (OTC) cold medications containing sympathomimetic amines.
3. Assays for amphetamine and methamphetamine with lower cross-reactivity to OTC medications (6).

The choice of which amphetamine immunoassay to use depends on several factors including the availability of automated analyzers and intended use. Laboratories performing workplace drug testing prefer amphetamines immunoassays specific for only amphetamine and methamphetamine, but hospital

TABLE 3.1. Cross-reactivities of MDMA and MDA with Various Commercial Amphetamine and Amphetamine/MDMA Immunoassays

Immunoassay	Manufacturer	%Cross-reactivity*	
		MDMA	MDA
Amphetamine Immunoassays			
DRI-Amphetamine	Diagnostics Reagents (Sunnyvale, CA)	11–44%	51–99%
CEDIA-Amphetamine	Microgenics (Fremont, CA)	57–68%	0%
EMIT d.a.u Monoclonal	Syva (Cupertino, CA)	7–46%	175–214%
SYNCHRON-CX-Amphetamine	Beckman (Fullerton, CA)	49–100%	58–73%
COBAS Integra Amphetamine	Roche Diagnostics (Indianapolis, IN)	38–48%	24–42%
Amphetamine/MDMA Immunoassay			
CEDIA-Amphetamine/MDMA	Microgenics (Fremont, CA)	100%	36–63%
On-Line Amphetamine/MDMA	Roche Diagnostics	111–142%	75–100%

*Higher cross reactivities were observed with lower drug concentration (MDMA and MDA concentration ranges tested 300–1000 ng/mL; reference 4).

laboratories may desire immunoassays directed toward the broad spectrum of sympathomimetic amines, including OTC drugs. The desire for a broad-spectrum assay is articulated as a recommendation in the National Academy of Clinical Biochemistry's Laboratory Medicine Practice Guidelines that the optimum immunoassays for amphetamines testing in emergency department patients are those directed toward phenethylamine as a drug class (7). These guidelines also recommend that the name of the test should be changed from "amphetamines" to sympathomimetic amines" or "stimulant amines." The cross-reactivity of various phenethylamine in different immunoassays has been extensively reported (8). However, it is possible that current assays produced by the manufacturers may have different cross-reactivities from those reported due to changes in antibodies and reagent composition. The analyst should consult immunoassay package inserts or contact the manufacturer for up-to-date cross-reactivity data. In addition, it is also possible that potential interfering substances used during studies are lower than those encountered in the clinical setting and that manufacturers often do not test for cross-reactivity of endogenous metabolites (9,10). Lot-to-lot variability concerning sensitivity to targeted analytes has been reported (11).

3.2.1. Interferences of Sympathomimetic Amines with Amphetamine Immunoassay

Interference of various sympathomimetic amines in amphetamine immunoassay is well recognized. Various OTC cold medications contain these sympathomimetic amines, and use of such medications may cause positive test results in amphetamine immunoassay screening. However, a GC/MS confirmation test must be able to identify the sympathomimetic amine causing the false-positive test result. Because if the screening assay is positive but confirmation is negative, the drugs of abuse testing must be considered as negative. Taking cold medication is not a problem in workplace drug testing but may cause confusion during medical drug testing especially if a confirmation test is not ordered or the small hospital laboratory does not have the capability of GC/MS confirmation. Unfortunately, misidentification of ephedrine/pseudoephedrine as methamphetamine has been reported during GC/MS confirmation (12). These are the causes of misidentification:

1. Ephedrine and pseudoephedrine can be thermally dehydrated to methamphetamine at an injector port temperature over 185°C if used in the gas chromatograph of the GC/MS.
2. Electron impact mass spectrum of trifluoroacetyl, pentafluoropropyl, heptafluorobutyl, 4-carbethoxyhexafluorobutyryl, and various carbamate derivatives of methamphetamine are similar to the corresponding derivatized ephedrine/pseudoephedrine. Because ephedrine and pseudoephedrine are eluted just after the methamphetamine peak, if the mass spectral analysis is not done carefully, ephedrine/pseudoephedrine can be misidentified as methamphetamine.

The chemical ionization mass spectrum of derivatized methamphetamine is very different from the derivatized ephedrine/pseudoephedrine, and misidentification problems can be completely eliminated by operating the mass spectrometer in the positive chemical ionization mode using methane as the reaction gas (13). In the electron ionization mode, usually the molecular ion of derivatized amphetamine is present as a very weak peak. In contrast, protonated molecular ion is the base peak (100% abundance) in the chemical ionization mass spectra of derivatized amphetamine and methamphetamine. This misidentification problem also exists with propyl carbamate derivatives of methamphetamine. In the electron ionization mode, both methamphetamine propyl carbamate and ephedrine propyl carbamate showed almost identical mass spectral fragmentation patterns. Again using chemical ionization where methamphetamine propyl carbamate showed a base peak at m/z 236 can circumvent this problem, and the ephedrine propyl carbamate showed a base peak at m/z 192. The major mass spectral fragmentation pattern differences are given in Table 3.2. Recently Ojanpera et al described that using positive chemical ionization mass spectrometry allowed the authors to observe protonated molecular ion peak of new amphetamine like designer drug 3,4-methylenedioxypyrovalerone in urine at m/z 276 (14).

Ephedrine or pseudoephedrine, which is present in many OTC cold medications, is responsible for most of the false-positive results in amphetamine immunoassay screening tests. These sympathomimetic amines may be present in large amounts in urine specimens that are initially tested positive by amphet-

TABLE 3.2. Electron Ionization and Chemical Ionization Mass Spectral Features of Derivatized Methamphetamine and Ephedrine

Compound	Electron Impact (m/z)	Chemical Ionization (m/z)
Methamphetamine	Base: 144	Base: 236 $(M + 1)^+$
Propyl carbamate	Other peaks: 102, 91, 58	Other peaks: 176, 144, 119, 58
Ephedrine propyl	Base: 144	Base: 192
Carbamate	Other peaks: 102, 77, 58	Other peaks: 220, 148, 58
Methamphetamine	Base: 154	Base: 246 $(M+1)^+$
Trifluoroacetyl	Other peaks: 118, 58	Other peaks: 154, 119
Ephedrine	Base: 154	Base: 244
Trifluoroacetyl	Other peaks: 118, 58	Other peaks: 276, 158
Methamphetamine	Base: 315	Base: 400 $(M + 1)^+$
4-carbethoxy	Other peaks: 118, 91	Other peaks: 308, 119
Ephedrine	Base: 308	Base: 398
4-Carboxy	Other peaks: 118, 91	Other peaks: 121

TABLE 3.3. Drugs That Interfere with Amphetamine Immunoassays

Drug Class	Individual Drugs
Sympathomimetic Amines	Brompheniramine, Benzphetamine, Ephedrine, Isometheptene, Mephentermine, Methylphenidate, Pseudoephedrine, Phentermine, Phenylpropanolamine*, Propylhexedrine, Phenylephrine, Phenmetrazine, Tyramine
Antidepressant/Antipsychotic	Chlorpromazine, Bupropion, Desipramine, Fluoxetine, Perazine, Thioridazine, Trimipramine, Trazodone
Antihistamine	Promethazine, Ranitidine
Antiprotozoal	Quinacrine, Chloroquine
Antiemetic	Trimethobenzamide
β-Blocker	Labetalol
Cardioactive	N-Acetyl procainamide (metabolite of procainamide) Mexiletine
Nonsteroidal anti-inflammatory	Tolmetin
Vasodilator	Isoxuprine
Tocolytic agent	Ritodrine

* Removed from the U.S. market in 2005.

amine immunoassays. Stout et al, based on the analysis of 27,400 randomly collected urine specimens, reported that 833 of 1104 urine specimens (92%) that failed to confirm amphetamine, methamphetamine, MDMA, or MDA by GC/MS contained pseudoephedrine (15). Commonly encountered medications that interfere with amphetamine screening assays are listed in Table 3.3.

3.2.2. Vicks Inhaler and False Positives in Amphetamine Immunoassays

Amphetamine and methamphetamine have optical isomers designated *d* (or +) for dextrorotatory and *l* (or –) for levorotatory, and the *d* isomers, the more physiologically active compounds, are the intended targets of immunoassays because *d* isomers are abused. Ingestion of medications containing the *l* isomer can cause false-positive results. For example, the Vicks inhaler contains the active ingredient *l*-methamphetamine (16), and extensive use of this product may cause false-positive results for immunoassay screening. Specific isomer resolution procedures must be performed to differentiate the *d* and *l* isomers because routine confirmation by GC/MS does not determine isomer composition. Poklis et al reported the relatively high concentrations of l-methamphetamine observed in two subjects were 1390 ng/mL and 740 ng/mL, respectively, after extensively inhaling Vicks inhaler every hour for several hours. However, urine specimens tested negative by the enzyme-multiplied immunoassay technique (EMIT II amphetamine/methamphetamine assay; Dade Behring) even after such extensive use of Vicks inhaler (17). In another study, the authors reported that nasal inhalation of Vicks inhaler following the recommended dosage should not cause false-positive amphetamine/methamphetamine urine screening results using fluorescence polarization immunoassay (FPIA) for amphetamine/methamphetamine for application on TDxADX/FLx analyzer (Abbott Laboratories, Abbott Park, IL) because the antibody used in this immunoassay is stereoselective and recognizes d-methamphetamine or *d*-amphetamine if present in the urine specimens. However, when two subjects inhaled twice the recommended dosage, a positive test result was obtained by the FPIA screening assay. The concentrations of *l*-methamphetamine in urine specimens of these two subjects were 1560 ng/mL and 1530 ng/mL, respectively (18). However, in doubtful cases, *l*-methamphetamine can be distinguished from *d*-methamphetamine by using GC/MS and the chiral derivatization technique. If greater than 80% *l*-methamphetamine is found in the specimen, the result is considered consistent with use of the Vicks inhaler. Abbott Laboratories discontinued FPIA assays for amphetamine/ methamphetamine assay for application on the TDX analyzer in 2010.

Case Study

An emergency medical team responded to the residence of a 77-year-old man who was experiencing difficulty in breathing. His oxygen saturation was 69%. A breathing treatment of albuterol and ipratropium bromide was administered, and his oxygen saturation was improved to 96%. The patient was transported to the emergency department and admitted to the intensive care unit. Approximately 12 hours after admission, he was unresponsive and died. His medical history included heart failure, atrial flutter, and bronchial asthma. His urine drug testing (GC/MS) was positive for acetaminophen, nicotine, cotinine, caffeine, diltiazem, doxylamine, and methamphetamine. However, chiral analysis of methamphetamine was not initially performed, raising the possibility that the man had abused *d*-amphetamine. To address this question further, the authors performed chiral analysis using trifluoroacetyl-*l*-propyl chloride as the chiral derivatizing agent and identified the methamphetamine as the *l*-isomer. Further investigation revealed that the decedent frequently used Vicks inhaler for his bronchial asthma (19).

3.2.3. Dietary Weight Loss Products and Positive Amphetamine

The Food and Drug Administration (FDA) banned ephedra alkaloids containing all products including weight loss products in April 2004. The products that were banned include ma huang, *Sida cordifolia*, and Pinellia, but Chinese herbal products, herbal teas, and drugs containing synthetic ephedrine were not banned. Currently, there is a push for companies marketing herbal weight loss products to produce ephedra-free diet pills. The common ingredients of these ephedra-free diet pills may also contain phenylpropanolamine, tyramine, or phentermine, which like ephedrine interferes with the immunoassay screening of amphetamine/methamphetamine in urine specimens. However, other diet loss products that contain hydroxy citric acid or bitter orange do not interfere with drugs of abuse testing.

The FDA warned consumers in January 2006 that the Brazilian dietary supplements Emagrece Sim and Herbathin contain several active drug ingredients including fenproporex that is metabolized to amphetamine (20). Use of other weight loss or nutritional supplements containing fenproporex has resulted in the detection of *d*-amphetamine in the urine specimens of users (21).

Case Study

A 25-year-old woman presented to the emergency department with persistent abdominal pain of 2 weeks duration and a 1-day history of nausea and vomiting. Her medications included high-dose ibuprofen that she had discontinued 5 days prior to her presentation and a Brazilian weight loss product, Emagrece, purchased from the Internet, which she had also discontinued 2 weeks earlier. The patient discontinued the weight loss product due to feeling mildly tremulous after taking the product, and her friends also experienced similar symptoms. On admission to the hospital, results of routine blood test for electrolytes, complete blood count, amylase, lipase, and liver function tests were within normal limits. A urinary toxicology screen was positive for amphetamine. Analysis of the weight loss product after extraction using GC/MS confirmed the presence of 5-cyano-amphetamine, an amphetamine derivative used outside the United States for weight loss but considered illegal by the FDA. The list of ingredients only mentioned herbal components such as kava kava and ginkgo biloba, but there was no mention of the amphetamine-like compound. The patient was treated with metoclopramide, pantoprazole, and acetaminophen and was discharged from the hospital 3 days later when her symptoms resolved (22).

3.2.4. Other Drugs That May Cause Positive Amphetamine Screening

Ranitidine is a H_2 receptor blocking agent (antihistamine) that reduces acid production by the stomach and is available OTC without any prescription. Dietzen et al reported that ranitidine if present in urine at a concentration over 43 μg/mL may produce false-positive test results with an amphetamine screen using Beckman Synchron immunoassay reagents (Beckman Diagnostics, Brea, CA). This concentration of ranitidine is routinely exceeded in patients taking ranitidine (23). Poklis also reported that ranitidine interferes with the EMIT-d.a.u. amphetamine/methamphetamine assay if ranitidine concentration in urine exceeds 91 μg/mL. The concentrations of ranitidine in urine specimens collected from 23 patients receiving 150–300 mg ranitidine per day varied from 7 μg/mL to 271 μg/mL (24). Trazodone not only interferes with amphetamine immunoassays, but it also causes false positives in an immunoassay specific for MDMA. A series of patients who tested positive for MDMA (using Ecstasy EMIT II assay) did not show any presence of MDMA in urine when confirmed by a specific liquid chromatography combined with the tandem mass spectrometric method. However, all specimens showed the presence of trazodone and its metabolite meta-chlorophenylpiperazine. In addition, another hallucinogen, trifluoromethyl phenylpiperazine, also cross-reacts with the MDMA assay (25). Labetalol, a β-blocker commonly used for control of hypertension in pregnancy, causes a false-positive amphetamine screen using an immunoassay (26). Casey et al reported that bupropion, a monocyclic antidepressant and an aid for smoking cessation. may cause a false- positive screen using an amphetamine immunoassay. Of 234 specimens screened positive by an amphetamine immunoassay (EMIT II), 128 specimens did not show any presence of amphetamine during the GC/MS confirmation step. Three such specimens where amphetamine cannot be confirmed contained bupropion (27).

3.3. INTERFERENCES WITH OPIATE IMMUNOASSAY

Certain fluoroquinolone antibiotics may cause false-positive test results with opiate immunoassay screening (28). Baden et al evaluated the potential interference of 13 quinolones (levofloxacin, ofloxacin, pefloxacin, enoxacin, moxifloxacin, gatifloxacin, trovafloxacin, sparfloxacin, lomefloxacin, ciprofloxacin, clinafloxacin, norfloxacin, and nalidixic acid) with various opiate immunoassays and concluded that levofloxacin, ofloxacin, and pefloxacin administration would most likely cause a false-positive result with opiate immunoassays (29). Rifampicin is used in therapy for tuberculosis. In one report, the authors observed false-positive opiate immunoassay screening result in patients receiving rifampicin. Opiate immunoassay screens were conducted using the kinetic interaction of microparticle in solution methods assay on the Cobas Integra analyzer (Roche Diagnostics, Indianapolis, IN). The authors observed a 12%

cross-reactivity of rifampicin with the opiate assay over a concentration range of 156–5000 ng/mL. A false-positive result may be observed even after 18 hours of administration of a single oral dose of 600 mg of rifampicin (30).

Case Study

A 48-year-old man spent 12 days in an inpatient psychiatry facility for treatment of depression that began subsequent to his abuse of cocaine after 90 days of sobriety. He was then transferred to a residential facility. Eleven days after his transfer to the residential facility, the patient was prescribed gatifloxacin for treating a urinary tract infection. His urine drug screen before initiation of gatifloxacin therapy was negative. Six days after the start of gatifloxacin therapy, his urine screen was positive for opiate. The patient denied opiate abuse or consuming food containing poppy seeds. Moreover, the positive opiate screen did not show the presence of any opiate during GC/MS confirmation. Another opiate drug screen 24 days later was also positive for opiate during the drug screen, but the GC/MS result was negative. However, when another urine sample was analyzed 13 days after cessation of gatifloxacin therapy, the urine immunoassay screen was negative for opiate, indicating that the false-positive opiate screen was due to gatifloxacin (31).

3.4. FALSE-POSITIVE TEST RESULTS IN OTHER DRUGS OF ABUSE IMMUNOASSAYS

Other drugs may cause false-positive test results in various immunoassays used for screening the presence of abused drugs in urine specimens. Dextromethorphan is an antitussive agent found in many OTC cough and cold medications. Dextromethorphan is also abused in high dosages mostly by young adults. Ingesting high amounts of dextromethorphan (over 30 mg) may result in false-positive test results with opiate and phencyclidine (PCP) immunoassays. In another report, the authors observed three false-positive phencyclidine tests in pediatric urine specimens using an on-site testing device (Instant-View multitest drugs of abuse panel; Alka Scientific, Designs, Poway, CA). The authors concluded that false-positive PCP tests were due to the cross-reactivities of ibuprofen, metamizol, dextromethorphan, and their metabolites with the PCP assay. In addition, pheniramine and methylphenidate also produce a false-positive result with the PCP screen (32).

Studies indicate that nonsteroidal anti-inflammatory drugs may produce false-positive results in immunoassays screening tests for the presence of drugs of abuse in urine specimens. Rollins et al studied the effect of such drugs on drugs of abuse screening using 60 volunteers and 42 patients taking ibuprofen, naproxen, or fenoprofen. Of 510 urine specimens collected from 102 individuals, only two specimens showed false positive for cannabinoids using enzyme-mediated immunoassay, one in an individual who ingested 1200 mg ibuprofen in three divided doses and one in a patient taking naproxen regularly. Two urine specimens were false positive for barbiturates using the fluorescence polarization immunoassay, one in a patient taking ibuprofen and one in a patient taking naproxen. The authors concluded that there is a small likelihood of false-positive tests results in drugs of abuse screening assays in individuals taking nonsteroidal anti-inflammatory drugs (33). Joseph et al studied 14 nonsteroidal anti-inflammatory drugs for potential interference with EMIT and FPIA for various drugs of abuse and observed that tolmetin interferes with EMIT immunoassays at high concentrations (1800 μg/mL and higher) because of high molar absorptivity at 340 nm, the wavelength used for detection in EMIT technology. Samples containing cannabinoid and benzoylecgonine tested negative in the presence of tolmetin, but there was no effect on the FPIA assay because the detection wavelength was 525 nm (34).

Although most reports in the medical literature describe false-positive results in drugs of abuse testing due to the presence of OTC medications in urine specimens, Brunk reported a false-negative GC/MS confirmation of marijuana metabolite as methyl derivative due to the interference of ibuprofen with the methylation step. The urine specimen tested positive for marijuana metabolite using the EMIT-d.a.u. assay but showed a negative GC/MS confirmation result. When the specimen was analyzed by thin layer chromatography, the presence of marijuana metabolite was also observed (35). See Table 3.4 for a list of drugs that interfere with various drugs of abuse immunoassays.

Case Study

A 9-year-old boy presented to the pediatric emergency department with complaints of altered level of consciousness. He was receiving diphenhydramine for nasal congestion and sleeping difficulty in the previous week before presenting to the emergency department. The parents denied that

TABLE 3.4. Interferences in Various Immunoassays Used for Drugs of Abuse Testing

Immunoassay	Interfering Drugs
Benzodiazepines	Oxaprozin, Sertraline
Opiates	Diphenhydramine, Rifampin, Dextromethorphan, Verapamil, Fluoroquinolones, Quinine
Phencyclidine	Dextromethorphan, Diphenhydramine, Ibuprofen Imipramine, Ketamine, Meperidine, Thioridazine, Tramadol, Venlafaxine
Tetrahydrocannabinol	Non-steroidal antiinflammatory Drugs, Pantoprazole, Efavirenz
Methadone	Diphenhydramine

he took the medicine on the evening of going to the hospital. His physical examination showed dilated pupils, dry mucous membranes, confusion, and occasional hallucination. His urine drug screen showed positive for methadone, and the serum drug screen was positive for a tricyclic antidepressant. The patient denied taking any medication other than diphenhydramine and was admitted to the hospital. His symptoms resolved within 24 hours. The authors attributed the positive tricyclic antidepressant serum screen due to the interference of diphenhydramine with the assay consistent with literature reports, but only interference of diphenhydramine with EMIT opiate assay had been previously reported, and the assay used for the urine drug screen of the boy was performed with a point-of-care testing device (One-Step, Multi-Drug Multi-Line screen, ACON Laboratories, San Diego, CA). Therefore, the authors investigated the possible interference of diphenhydramine with this assay and when they supplemented a drug-free urine specimen with 100 μg/mL of diphenhydramine, the opiate test was positive for opiate. Therefore, the false-positive test result was due to the use of diphenhydramine (36).

3.5. CONCLUSIONS

Although false-positive test results are routinely encountered in urine drugs of abuse testing, such methods are widely used in clinical laboratories for initial screening for the presence of any illicit drugs in the urine in patients admitted to emergency departments and other departments. Although a positive screening of a urine specimen for the presence of any illicit drugs must be confirmed by GC/MS during workplace drug testing, such a confirmatory test may not be ordered by the clinician during a medical drug test. Melanson et al commented that directors should be aware of the characteristics of their laboratories' assays and should be able to communicate with ordering physicians regarding the proper interpretation of qualitative results. In addition, manufacturers' claim must be interpreted with caution and if possible should also be validated using the patient population of the hospital (37). However, false-positive results can be avoided in drugs of abuse testing by using more specific analytical methods such as GC/MS. Eichhorst et al described an improved cost-effective high throughput method for drugs of abuse screening by tandem mass spectrometry replacing immunoassay screening (38).

REFERENCES

1. Chen WJ, Fang CC, Shyu RS, Lin KC. Underreporting of illicit drug use by patients at emergency departments as revealed by two tiered urinalysis. Addict Behav 2006; 31: 2304–2308.

2. Standridge JB, Adams SM, Zotos AP. Urine drug screening: a valuable office procedure. Am Fam Physician 2010; 81: 635–640.
3. Dietzen DJ, Ecos K, Friedman D, Beason S. Positive predictive values of abused drug immunoassays on the Beckman SYNCHRON in a veteran population. J Anal Toxicol 2001; 25: 174–178.
4. Hsu J, Liu C, Liu CP, Tsay WI et al. Performance characteristics of selected immunoassays for preliminary test of 3,4-methylenedioxymethamphetamine. Methamphetamine and related drugs in urine specimens. J Anal Toxicol 2003; 27: 471–478.
5. Kunsman GW, Manno JE, Cockerham KR, Manno BR. Application of the Syva EMIT and Abbott TDx amphetamine immunoassays to detection of 3,4-methylenedioxymethamphetamine (MDMA) and 3,4-methylenedioxyethamphetamine (MDEA). J Anal Toxicol 1990; 14: 149–153.
6. Butler D, Guilbault GG. Analytical techniques for ecstasy. Anal Lett 2004; 37: 2003–2030.
7. Wu AHB, McKay C, Broussard LA, Hoffman RS, et al. National Academy of Clinical Biochemistry Laboratory Medicine Practice Guidelines: recommendations for the use of laboratory tests to support poisoned patients who present to the emergency department. Clin Chem 2003; 49: 357–379.
8. Moore KA. Amphetamines/Sympathomimetic amines. In: B. Levine, ed., *Principles of Forensic Toxicology*. Washington, DC: AACC Press, 2003:341–348.
9. Poklis A. Unavailability of drug metabolite reference material to evaluate false-positive results for monoclonal EMIT-d.a.u. assay of amphetamine. Clin Chem 1992; 38: 2580.
10. Williams RH, Erickson T, Broussard L. Evaluating sympathomimetic intoxication in an emergency setting. Lab Med 2000; 31: 497–507.
11. Singh J. Reagent lot-to-lot variability in sensitivity for amphetamine with the Syva EMIT II monoclonal amphetamine/methamphetamine assay. J Anal Toxicol 1997; 21: 174–175.
12. Hornbeck CL, Carrig JE, Czarny RJ. Detection of GC/MS artifact peak as methamphetamine. J Anal Toxicol 1993; 17: 257–263.
13. Dasgupta A, Gardner C. Distinguishing amphetamine and methamphetamine from other interfering sympathomimetic amines after various fluoro derivatization and analysis by gas chromatography-chemical ionization mass spectrometry. J Forensic Sci 1995; 40: 1077–1081.
14. Ojanpera IA, Heikman PK, Rasanen IJ. Urine analysis of 3,4-methyledioxypyrovalerone in opioid dependent patients by gas chromatography-mass spectrometry. Ther Drug Monit 2011; 33: 257–263.
15. Stout PR, Klette KL, Horn CK. Evaluation of ephedrine, pseudoephedrine, and phenylpropanolamine concentrations in human urine samples and comparison of the specificity of DRI amphetamines and Abuscreen online (KIMS) amphetamine screenings immunoassays. J Forensic Sci 2004; 49: 160–164.
16. Solomon MD, Wright JA. False-positive for (+)-methamphetamine. Clin Chem 1977; 23: 1504.
17. Poklis A, Jortani WSA, Brown CS, Crooks CR. Response of the EMIT II amphetamine/methamphetamine assay to specimens collected following use of Vicks inhalers. J Anal Toxicol 1993; 17: 284–286.

18. Poklis A, Moore KA. Stereoselectivity of the TDX/ADX/FLx amphetamine /methamphetamine II amphetamine/methamphetamine immunoassay response of urine specimens following nasal inhaler use. J Toxicol Clin Toxicol 1995; 33: 35–41.
19. Wyman JF, Cody JT. Determination of l-methamphetamine: a case history. J Anal Toxicol 2005; 29: 759–761.
20. FDA News P06–07, January 13, 2006.
21. Jemionek J, Bosy TJ, Jacobs A, Holler J, Magluli J, Dunkley C. Five cases of D-amphetamine positive urines resulting from ingestion of "Brazilian nutritional supplements" containing fenproporex have been reported. (ToxTalk SOFT newsletter 2006; 30: 11.
22. Nguten MH, Ormiston T, Kurani S, Woo DK. Amphetamine lacing of an Internet-marketed nutraceutical. Mayo Clin Proc 2006; 81: 1627–1629.
23. Dietzen DJ, Ecos K, Friedman D, Beason S. Positive predictive values of abused drug immunoassays on the Beckman Synchron in a veteran population. J Anal Toxicol 2001; 25: 174–178.
24. Poklis A, Hall KV, Still J, Binder SR. Ranitidine interference with the monoclonal EMIT d.a.u. amphetamine/methamphetamine assay. J Anal Toxicol 1991; 15: 101–103.
25. Logan BK, Costantino AG, Rieders EF, Sanders D. Trazodone, meta-chlorophenylpiperazine (an hallucinogenic drug and trazodone metabolite) and the hallucinogen trifluoromethylphenylpiperazine cross-react with EMITII ecstasy immunoassay in urine. J Anal Toxicol 2010; 34: 587–589.
26. Yee LM, Wu D. False positive amphetamine toxicology screen results in three pregnant women using labetalol. Obstet Gynecol 2011; 117: 503–506.
27. Casey ER, Scott MG, Tang S, Mullins ME. Frequency of false positive amphetamine screens due to bupropion using the Syva EMIT II immunoassay. J Med Toxicol 2011; 7: 105–108.
28. Zacher JL, Givone DM. False positive urine opiate screening associated with fluoroquinolone use. Ann Pharamacother 2004; 38: 1528–1529.
29. Baden LR, Horowitz G, Jacoby H, Eliopoulos GM. Quinolones and false positive urine screening for opiates by immunoassay technology. JAMA 2001; 286: 3115–3119.
30. De Paula M, Saiz LC, Gonzalez-Revalderia J, Pascual T et al. Rifampicin causes false positive immunoassay results for opiates. Clin Chem Lab Med 1998; 36: 241–243.
31. Straley CFM, Cecil EJ, Herriman MP. Gatifloxacin interference with opiate urine drug screen. Pharmacotherapy 2006; 26: 435–439.
32. Marchei E, Pellegrini M, Pichini S, Martin I et al. Are false positive phencyclidine immunoassay instant-view multi test results caused by overdose concentrations of ibuprofen, metamizol and dextromethorphan? Ther Drug Monit 2007; 29: 671–673.
33. Rollins DE, Jennison TA, Jones G. Investigation of interference by non-steroidal antiinflammatory drugs in urine tests for abused drugs. Clin Chem 1990; 36: 602–606.
34. Joseph R, Dickerson S, Willis R, Frankenfield D et al. Interference by nonsteroidal antiinflammatory drugs in EMIT and TDx assays for drugs of abuse. J Anal Toxicol 1995; 19: 1–7.

35. Brunk SD. False positive GC/MS assay for carboxy THC due to ibuprofen interference. J Anal Toxicol 1988; 12: 290–291.
36. Rogers SC, Pruitt CW, Crouch DJ, Caravati EM. Rapid urine drug screens: diphenhydramine and methadone cross-reactivity. Pediatr Emerg Care 2010; 26: 665–666.
37. Melanson SE, Baskin L, Magnani B, Knwong TC et al. Interpretation and utility of drugs of abuse immunoassays: lessons from laboratory drug testing surveys. Arch Pathol Lab Med 2010; 134: 735–739.
38. Eichhorst JC, Etter ML, Rousseaux N, Lehotay DC. Drugs of abuse testing by tandem mass spectrometry: a rapid, simple method to replace immunoassays. Clin Biochem 2009; 42: 1531–1542.

Chapter 4

True-Positive Drugs of Abuse Test Results Due to Use of Prescriptions and Nonprescription Drugs

4.1. INTRODUCTION

Many prescription and some nonprescription drugs may cause true analytical positive test results in drugs of abuse testing. This true analytical positive result may be troublesome in preemployment or workplace drug testing. Therefore, it is important for the donor during workplace drug testing to list all prescription and nonprescription medications he or she may be taking to avoid confusion during the interpretation of positive drug testing results by the medical review officer. Laboratory professionals and pathologists also should be familiar with various alternative reasons for positive drug testing results that may arise from taking certain medications or ingesting food containing poppy seeds or herbal tea. This chapter focuses on various sources of true-positive drug testing results.

4.2. POSITIVE AMPHETAMINE/METHAMPHETAMINE TEST RESULT

Various nonprescription cold and cough medications containing pseudoephedrine, ephedrine, or related compounds result in positive amphetamine

Resolving Erroneous Reports in Toxicology and Therapeutic Drug Monitoring: A Comprehensive Guide, First Edition. Amitava Dasgupta.

screening results due to the cross-reactivity of these substances with the antibody used in the amphetamine/methamphetamine screening immunoassay. However, during gas chromatography/mass spectrometric (GC/MS) confirmation step, such interfering substances can be identified and the urine drug test should be reported as negative for the presence of amphetamine or methamphetamine (see Chapter 3 for more details). However, there are several prescription medications that contain amphetamine or methamphetamine or these drugs may be metabolized to amphetamine or methamphetamine. Therefore, if a person is taking any of these prescription medications, the urine drug test should also be confirmed positive by GC/MS.

Methylphenidate is used in treating attention deficit disorder/hyperactivity especially in children. Methylphenidate contains a phenylethylamine nucleus and a piperidine ring. Therefore there are some structural similarities between methylphenidate and amphetamine, although methylphenidate is a cyclic amine whereas amphetamine is an aliphatic amine. Children taking prescription methylphenidate may show a positive amphetamine drug screen (1). Mebeverine is an antispasmodic drug structurally related to ethyl amphetamine. This drug is metabolized to para-methoxyamphetamine and may cause positive amphetamine screening results using the fluorescence polarization immunoassay (FPIA) amphetamine/methamphetamine screening assay (Abbott Laboratories, Abbott Park, IL). However, the GC/MS confirmation step should be able to distinguish between para-methoxyamphetamine and amphetamine, and thus the false-positive screening result can be reported as negative for the presence of amphetamine or methamphetamine (2).

Both amphetamine and methamphetamine are used clinically for treating attention deficit disorder. These medications contain *d*-amphetamine *d*-methamphetamine or a racemic mixture of *d* and *l* isomers. Adderall is a widely prescribed drug for treating attention deficit/hyperactive disorder (ADHD) and narcolepsy that contains a 3:1 mixture of *d* and *l*-enantiomer of amphetamine salt. Cody et al studied amphetamine excretion profile in five healthy volunteers after the administration of a single dose of 20 mg of Adderall. The peak amphetamine concentration ranged from 2645 ng/mL to 5948 ng/mL, and positive amphetamine was observed using the immunoassay at a 500 ng/mL cut-off up to 47.5 hours post dose. However, a substantial presence of *l*-isomer distinguishes prescription use of Adderall from amphetamine abuse because illicit amphetamine preparations mostly contain *d*-isomer (3). In addition to Adderall, there are also other prescription medications that contain amphetamine or methamphetamine (Table 4.1).

In 2002 the U.S. Food and Drug Administration approved Paremyd eye drops that contain hydroxyamphetamine and tropicamide. Hydroxyamphetamine is also used outside the United States as a topical nasal decongestant as well as for oral use. Because hydroxyamphetamine is not metabolized to amphetamine, using such a product should not a cause positive amphetamine or methamphetamine test result during the GC/MS confirmation step (4).

TABLE 4.1. Drugs That Contain Amphetamine or Methamphetamine or Metabolized to Amphetamine or Methamphetamine Causing True Analytical Positive Amphetamine/Methamphetamine Test Result

Active Ingredient	Comment
Amphetamine	Adderall is one prescription medication that contains amphetamine. There are also other products that contain either *d*-amphetamine or racemic mixture.
Lisdexamfetamine	Inactive water soluble prodrug that is readily concerted into *d*-amphetamine after oral administration.
Methamphetamine	Desoxyn is the brand name of a product that contain methamphetamine.
Amphetaminil	Aponeuron is metabolized to amphetamine, but it is mostly withdrawn from the market.
Clobenzorex	Metabolized to amphetamine.
Ethyl amphetamine	Metabolized to amphetamine.
Mefenorex	Metabolized to amphetamine.
Prenylamine	Metabolized to amphetamine.
Benzphetamine	Metabolized to methamphetamine and then further metabolized to amphetamine.
Famprofazone	Metabolized to methamphetamine and then to amphetamine.
Selegiline	Metabolized to *l*-methamphetamine.

Several prescription medications are metabolized to amphetamine or methamphetamine, thus producing positive amphetamine/methamphetamine test result in drugs of abuse testing. These drugs are listed in Table 4.1. Benzphetamine, clobenzorex, fencamine, fenproporex, mefenorex, and furfenorex are used for the treatment of obesity. Prenylamine, which is metabolized to amphetamine, is a calcium channel blocker clinically used for treating angina pectoris. Selegiline is used in treating Parkinson's disease, and this drug is metabolized to methamphetamine (5).

In one study, volunteers ingested a 120 mg oral dose of prenylamine, and GC/MS analysis confirmed the presence of both the *d* and *l* isomers of amphetamine because prenylamine is marketed as a racemic mixture. However, some specimens tested positive for amphetamine using the amphetamine immunoassay while other specimens tested negative (6). Grenhill et al studied the urine amphetamine and methamphetamine concentrations in subjects who ingested two Gewodin tablets containing a total of 50 mg of famprofazone. Using GC/MS, the authors identified both *d* and *l* methamphetamine as expected because famprofazone is metabolized to methamphetamine and then methamphetamine is further metabolized to amphetamine. The peak amphetamine concentrations varied from 148 to 2271 ng/mL, and peak methamphetamine concentrations varied from 615 to 7361 ng/mL 3–7 seven hours postingestion (7). Chan et al commented that famprofazone users may be interpreted as illicit users of methamphetamine. When six subjects received 25 mg of famprofazone and urine specimens were collected for 48 hours after administration and analyzed by amphetamine immunoassay (cut-off concentration: 500 ng/mL) in 18.8% of the specimens, positive amphetamine test results were observed. Further analysis of these specimens using GC/MS indicated that methamphetamine concentrations ranged from 901 to 2670 ng/mL and amphetamine concentrations ranged from 201 to 711 ng/mL, respectively. The positive methamphetamine results appeared within 2–34 hours postingestion (8).

Selegiline used in treating Parkinson's disease is metabolized to *l*-methamphetamine and then to *l*-amphetamine. Because of the stereospecificity of the antibody used in the design of immunoassays for the screening of amphetamine/methamphetamine in urine, a low dose of selegiline should not cause positive screening result. However, positive immunoassay screening results using the FPIA amphetamine/methamphetamine assay (Abbott Laboratories, Abbott Park, IL) has been reported for up to 2 days following ingestion of a 10-mg selegiline tablet. Nevertheless, the prescription use of selegiline can be easily distinguished from the illicit use of methamphetamine by stereospecific GC/MS analysis of methamphetamine because selegiline is metabolized to *l*-methamphetamine, whereas illicit methamphetamine is mostly *d*-isomer (9).

Lisdexamfetamine (Vyvanse) is a prodrug that is pharmacologically inactive because *d*-amphetamine is bonded to *l*-lysine. Lisdexamfetamine is a water-soluble compound that is readily converted *in vivo* into *l*-lysine and pharmacologically active *d*-amphetamine. This drug is used for treating ADHD as an alternative to an immediate-release stimulant such as Adderall (10). Chemical structure of lisdexamfetamine is given in Figure 4.1. Lisdexamfetamine may be more effective in treating ADHD than traditional amphetamine preparations (11). Because lisdexamfetamine is metabolized to *d*-amphetamine, taking this drug may cause (depending on dosage) positive amphetamine test results.

4.3. TRUE-POSITIVE MARIJUANA RESULT

Synthetic tetrahydrocannabinol (dronabinol), which is identical in structure to natural product Δ–9-tetrahydrocannibinol (THC; marijuana) is sold under the trade name of Marinol for treating nausea and vomiting in cancer patients

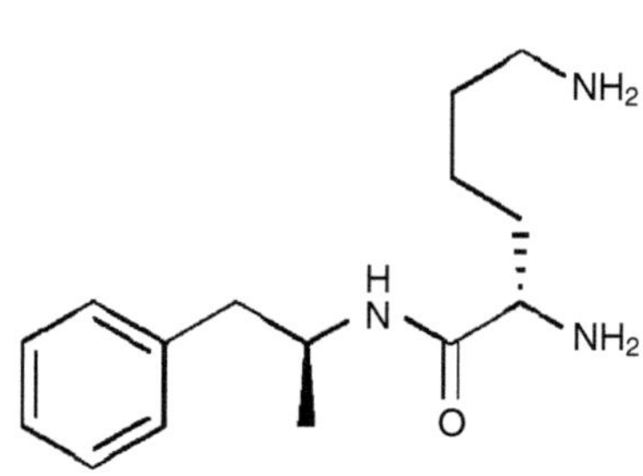

Figure 4.1. Chemical structure of lisdexamfetamine.

undergoing chemotherapy, as an appetite stimulant to patients with acquired immunodeficiency virus (AIDS) and also for treating various other conditions. Because Marinol is also converted into a marijuana metabolite (THC-COOH), use of Marinol produces a positive test for marijuana metabolites in drugs of abuse testing. However, delta-9-tetrahydrocannabivarin (THCV), which is a natural constituent of cannabis product, is absent in dronabinol. THVC is metabolized by human liver into THCV-COOH, and the presence of this metabolite in addition to THC-COOH in the urine indicates abuse of marijuana rather than the prescription use of dronabinol (12). However, Levin et al using 117 patients observed that a positive urine test for a marijuana metabolite for all urine specimens but 50% had an undetectable amount of THCV-COOH (<1 ng/mL), indicating that the absence of THCV-COOH in urine may not confirm the use of prescription dronabinol (13).

There are reports of positive marijuana test results following ingestion of hemp oil. Although many cannabinoids have been identified from the hemp plant including THC, industrial hemp is characterized by low THC and a high cannabidiol content. The THC level is so low in hemp products that it is unlikely for a person to experience euphoria following ingestion of hemp oil. In addition, cannabidiol present in the hemp product counteracts the effect of marijuana, and washed industrial quality hemp seeds contain no or very little THC coming from the resin sticking to the seed (14). Currently the use of hemp oil is legal, but use of cannabis flowers essential oil is illegal in the United States because the cannabis flower essential oil is produced from the flowering buds of the cannabis plant and is considered a marijuana product in the United States. However in some European countries such as Switzerland, England, France, and the Netherlands, the production of cannabis flower oil is legal.

Seeds and oils of hemp may contain a measurable amount of THC. In 1997 a survey found that the THC content of hemp oil was 11 to 117 μg/g (parts per million) (15). These seeds were imported from China. THC concentrations as high as 3568 μg/g were found in Swiss hemp oil products because growing higher THC-containing hemp plants (THC content over 0.3%) is legal in Switzerland. Proper cleaning of the seeds may not have been undertaken (16). Since 1998, a more thorough seed drying and cleaning process were undertaken that significantly reduced the level of THC in hemp products. Most THC is present on the outside of seed hulls. Studies conducted in 1998 or earlier reported significant amounts of THC metabolites after consumption of hemp oil because THC content in most hemp oil exceeded 50 μg/g, but now the THC content in most commercial hemp oil as low as 5 μg/g. Therefore studies published after 1998 concluded that ingestion of reasonable amount of hemp oil should not cause positive THC results in workplace drug testing. Leson et al reported that consumption of 125 mL of hemp oil (equivalent to the ingestion of 0.6 mg of THC, a very high consumption) produced the highest THC metabolite concentration of 5.2 ng/mL as determined by GC/MS. This value was below the 15 ng/mL cut-off of GC/MS confirmation. However several urinary specimens collected from volunteers after consumption of hemp oil tested positive by the immunoassay screen at the 20 ng/mL cut-off, and one specimen tested positive at the 50 ng/mL cut-off. Therefore programs that rely on immunoassay screening tests only and use a lower screening cut-off may still encounter occasional positives in subjects ingesting hemp food (17). Another study published in 2006 concluded that consumption of hemp oil currently produced by U.S. manufacturers according to the guidelines of the manufacturers should not lead to positive tests for THC in plasma or whole blood (18).

4.4. TRUE-POSITIVE OPIATE TESTS DUE TO USE OF MEDICATIONS AND INGESTION OF PRODUCTS CONTAINING POPPY SEEDS

A positive test result in a urine drugs test may be due to the use of prescription opioid or the ingestion of food containing poppy seeds. Despite the mortality associated with the use of a prescription opioid where mortality is strongly associated with the daily dose, these medications are widely used not only for management of pain in cancer patients but also patients with nonmalignant pain (19). Oxymorphone, which is structurally similar to hydromorphone, is the newest oral opioid used to treat moderate to severe persistent pain (20). Ingestion of fentanyl-containing drugs (which are also opioid) or use of the Duragesic fentanyl patch should not cause a positive opiate test result because fentanyl has very poor cross-reactivity with morphine-specific antibodies used in opiate screening assays. In addition, oxycodone has variable cross-reactivity with different opiate screening assays and there is also a specific immunoassay that recognizes the presence of oxycodone in urine (see Chapter 3). Although in federally mandated drug testing, oxycodone immunoassay is not usually used during the screening step, many private employers specifically test for oxycodone because this drug is widely abused. Therefore, if a person is taking prescription oxycodone for pain, it is important to disclose that use during workplace drug testing before submitting the urine specimen. In federally mandated drug testing, codeine- or morphine-containing analgesic medication produces a positive opiate test result. Therefore, it is important to disclose the use of such medication during workplace drug testing to avoid the negative consequence of a positive opiate test result.

Taking prescription medicine codeine or any drug that contains codeine should result in positive opiate test results for workplace drug testing because codeine is metabolized to morphine. However, the illicit use of heroin also results in confirmation of morphine in urine because heroin is first metabolized to 6-monoacetyl morphine and then into morphine, whereas the marker compound 6-monoacetyl morphine may not be confirmed in the urine of an illicit user of heroin because this compound can be detected only for a short time in heroin after heroin abuse. Stefanidou et al reported that a morphine-to-codeine ratio in blood exceeding 1 is strong evidence that the person had abused heroin as opposed to taking a prescription medication containing codeine (21).

Ethylmorphine, an antitussive agent, is metabolized to morphine. Ingestion of ethylmorphine produces positive test results with opiate screening in a specimen where both ethylmorphine and morphine are detected in urine specimens using GC/MS (22). Although morphine is primarily converted into morphine-glucuronide and then excreted in urine, hydromorphone may be detected in chronic pain patients who are taking morphine. Cone et al studied 13 patients who were chronically treated with morphine and other opioids and observed the presence of hydromorphone in the urine of 10 patients. These 10 patients showed a high amount of morphine, frequently exceeding 10,000 ng/mL; hydromorphone concentrations ranged from 120 to 1400 ng/mL. The ratio of hydromorphone to morphine in these 10 patients varied from 0.015 to 0.024. Morphine concentrations of three other patients where hydromorphone was not detected were significantly lower compared with 10 patients where hydromorphone was detected in urine. The authors concluded that their data suggested hydromorphone as a minor metabolite of morphine (23).

The presence of hydrocodone, oxycodone, hydromorphone, and oxymorphone is also tested as a part of the protocol for opiate testing, mostly in the workplace drug testing protocols used by private employers. Although oxycodone and oxymorphone may not be detected by opiate immunoassays, hydromorphone and hydrocodone may produce a positive result in the opiate screening assay depending on the type of antibody. However, all these drugs and their corresponding metabolites can be confirmed by using GC/MS. Interestingly, hydrocodone may be present in the urine of a person taking oxycodone. Oxycodone preparations are known to contain small amounts of hydrocodone as an impurity estimated to be less than 0.1%. In one study, the authors observed that of 187 specimens where oxycodone concentrations exceeded 100,000 ng/mL, 72% of the specimens also contained hydrocodone. The authors concluded that when oxycodone concentrations exceed 100,000 ng/mL, the hydrocodone concentration should be less than 1500 ng/mL, and the if oxycodone concentration is less than 100,000 ng/mL, hydrocodone concentration should be less than 500 ng/mL. Values of hydrocodone exceeding these limits indicate the nonprescription use of hydrocodone (24). Propoxyphene is an opioid that is also used in pain management. In addition, methadone is also used in pain management and for drug rehabilitation. These two drugs are not detected by routine opiate screening, but private employers may specifically include these two drugs that are also abused in a workplace drug testing panel. Prescription medications that may lead to a positive result in workplace drug testing are listed in Table 4.2.

Case Study

A 2-year-old girl child with a history of asthma played outside before entering her home and then started rubbing her mouth and also staggering. It appeared that the child had ingested some substance while playing, and later the child developed abdominal pain and was brought to the emergency

TABLE 4.2. Opioid That May Cause a Positive Test Result in Drugs of Abuse Testing

Opioid Drug	Comment
Codeine	Codeine-containing drugs should cause positive test result in SAMHSA-mandated drug program. Codeine with acetaminophen (Tylenol 3) is a prescription drug in the United States. Many cough medications may also contain codeine because it is a cough suppressant. Although most codeine-containing medications are prescription drugs. Kaodene with codeine may be available in some states without prescription.
Morphine	Morphine is poorly absorbed from the gut. but oral prescription drugs containing morphine are available. Some over-the counter antidiarrheal products such as Infantol Pink, Donnagel-PG, Kaodene with Paregoric, and Quiagel PG also contain morphine.
Oxycodone	OxyContin and Roxicodone are popular brand names. Because of poor cross-reactivity of oxycodone with opiate immunoassays, SAHMSA-mandated drug testing may be negative for a person taking either prescription oxycodone or abusing oxycodone. However, private employers often include oxycodone in s workplace drug testing panel.
Oxymorphone	Because of poor cross-reactivity of oxymorphone with opiate immunoassays, SAHMSA-mandated drug testing may be negative for a person taking either prescription oxymorphone or abusing oxymorphone. However, private employers often include oxymorphone in a workplace drug testing panel.
Hydrocodone	May be detected in a SAMHSA drug testing program if the antibody used in the opiate assay has significant cross-reactivity with hydrocodone.
Hydromorphone	May be detected in a SAMHSA drug testing program if the antibody used in the opiate assay has significant cross-reactivity with hydrocodone.
Propoxyphene	Not included in SAMHSA panel but a private employer may test for this drug in a workplace drug testing panel.

SAMHSA-mandated drug tests include amphetamine, cocaine, opiate, marijuana, and phencyclidine.

department for further evaluation. Physical examination showed a fatigued child with a minimal ataxic gait. Laboratory test results indicated mild anemia, but the urine toxicology screen was negative. Later her fatigue appeared to improve, and she was discharged with an instruction to her parents to return to the hospital if she developed fever, skin rash, lethargy, and any other condition. The next morning the child was found unresponsive, and the emergency medical team performed cardiopulmonary resuscitation. Despite all efforts the child was pronounced dead on arrival at the hospital. The body was transported to the office of the county coroner. During the postmortem investigation, oxycodone was found in her blood that was confirmed by GC/MS. The hospital toxicology screen for drugs of abuse and tricyclic antidepressants was negative because the Triage Drugs of Abuse Panel (Biosite, San Diego, CA) does not produce a positive opiate test result unless oxycodone concentration is 20,000 ng/mL or higher. The concentration of oxycodone in the heart blood was 1360 ng/mL, which is potentially lethal. This case indicated a potential limitation of the urine toxicology screen performed in hospital laboratories (25).

Poppy seeds contain both codeine and morphine. Therefore eating food containing poppy seeds may cause a positive opiate result in both medical and workplace drug testing. The amount of opiates in poppy seeds is not enough to produce a narcotic effect after consumption. However, the concentrations of morphine and codeine can vary widely among different brands of poppy seeds. The morphine content of the Australian poppy seeds ranges from 90 to 200 μg of morphine per gram of poppy seed, whereas Dutch and Turkish poppy seed contain only 4–5 μg of morphine per gram of poppy seed (26–29). Trafkowski et al reported that the morphine content varies from less than 0.1% to 450 μg/g of seed and codeine content varied from less than 0.1% to 57.1 μg/g of seed (30). Typically morphine concentrations are higher than codeine concentrations in urine after consumption of poppy seed products. In one study, the concentrations of morphine and codeine were 2797 ng/mL and 214 ng/mL, respectively, in one healthy volunteer who ingested three poppy seed bagels. Opiate was present in the urine 25 hours postingestion. No opiate was present 45 hours postingestion (31). Hayes et al also studied the effect of ingesting poppy seeds on morphine and codeine levels in urine. The poppy seeds used for this study contained 17–294 μg/g of morphine and 3–14 μg/g of codeine. The subjects ingested 40 g of poppy seeds that contained approximately 2.5 mg of morphine and 0.16 mg of codeine. The immunoassay screening of specimens by the enzyme-multiplied immunoassay technique (EMIT) showed typical positive screen (>300 ng/mL cut-off) up to 24 hours of ingestion of poppy seeds, and peak concentration of total morphine ranged from 700 to 2600 ng/mL as measured by GC/MS (32). In November 1998, the federal government (Department of Health and Human Services) increased the screening cut-off of opiate immunoassays from 300 ng/mL to 2000 ng/mL in an attempt to eliminate false-positive opiate results due to the ingestion of poppy seed products. Fraser and Worth reported that following the new guidelines, codeine and morphine positive results were reduced from 7.1% positive in 1994–1996 to 2.1% in 1998 (33). During the processing of poppy seed, morphine concentrations are reduced and normal consumption of food containing poppy seeds is considered safe (34). However, eating food containing poppy seed may affect urine drug testing, especially if a private employer uses the old 300 ng/mL cut-off for opiate screening.

Thebaine is a natural constituent of poppy seed, and there are suggestions that thebaine can be used as a marker for poppy seed consumption. In a case where morphine and codeine have been detected, the presence of thebaine would indicate poppy seed use by the subject. In one study thebaine was detected in concentrations ranging from 2 to 91 ng/mL in volunteers after consumption of 11 g of poppy seed (35). The elimination of thebaine varies widely between individuals, and the absence of thebaine in a specimen does not rule out the possibility of consumption of food containing poppy seed as the cause of positive opiate test result. Thebaine along with morphine, codeine, and related alkaloids in urine can be analyzed using GC/MS. The chemical structures of thebaine and morphine are given in Figure 4.2.

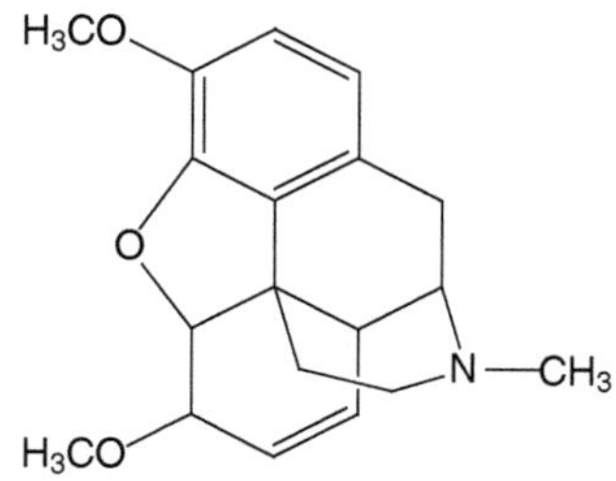

Thebaine

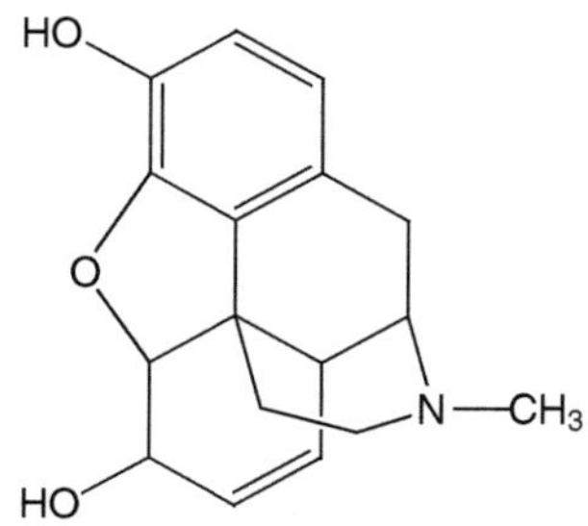

Morphine

Figure 4.2. Chemical structures of thebaine and morphine.

4.5. TRUE-POSITIVE BENZODIAZEPINE TEST RESULTS

Although benzodiazepines are not a part of five federally mandated drugs (drug test mandated by the Substance Abuse and Mental Health Services Association [SAMHSA]), private employers often include benzodiazepines in the workplace drug testing panel because they are one of the most frequently prescribed drugs in the United States and are used as tranquilizers, muscle relaxants, and anticonvulsants as well as to treat symptoms of anxiety, but these drugs are also widely abused. There are more than 50 different types of benzodiazepines, but not all drugs are available in the United States. The most commonly prescribed benzodiazepines in the United States are diazepam, temazepam, alprazolam, lorazepam, and clonazepam. Some of the drugs in this class, for example, estazolam, temazepam, halazepam, and quazepam, are derivatives of benzodiazepine. Several drugs in the benzodiazepine class, for example, flurazepam, is short acting, whereas diazepam, alprazolam, and other related drugs are long-acting benzodiazepines. Common benzodiazepines used in the United States are listed in Table 4.3.

Usually immunoassay screening for benzodiazepines recognizes the presence of common drugs and metabolites after medical use or abuse. Most benzodiazepine immunoassays are based on an antibody that recognizes oxaz-

TABLE 4.3. Common Benzodiazepines Available in the United States

Short-Acting Benzodiazepines
Estazolam
Flurazepam
Temazepam
Triazolam
Midazolam
Long-Acting Benzodiazepines
Alprazolam
Chlordiazepoxide
Clorazepate
Clonazepam
Diazepam
Halazepam
Lorazepam
Nitrazepam
Prazepam
Oxazepam
Quazepam

epam, and in addition various other benzodiazepines have significant cross-reactivity with the antibody. For example, the On-Line Benzodiazepine Plus assay by Roche Diagnostics (Indianapolis, IN) recognizes the presence of 22 common benzodiazepines and their metabolites including alprazolam, chlordiazepoxide, clonazepam, clorazepate, diazepam, estazolam, flunitrazepam, halazepam, lorazepam, midazolam, nitrazepam, oxazepam, temazepam, and triazolam. Garretty et al analyzed 100 urine specimens collected from subjects attending the Leeds Addiction Unit (Leeds, UK) for the presence of benzodiazepines using the EMIT assay (Dade Behring, Deerfield, IL) and later confirmed the presence of specific drugs using high-performance liquid chromatography. The authors detected nitrazepam in 61 specimens, but none of the subjects were prescribed that drug. Similarly, chlordiazepoxide was detected in 49 urine specimens but only five patients had a valid prescription for the drug. The authors concluded that multiple benzodiazepine abuse was evident among patients attending the addiction center (36). Medical use of benzodiazepines also leads to positive drugs of abuse testing result. Because of the wide diversity of this class of drug, immunoassay-positive specimens are usually confirmed by GC/MS for common drugs in this class including oxazepam, diazepam, temazepam, and alprazolam.

Case Study

A 20-year-old man with a history of drug abuse was accepted in a voluntary drug treatment facility after initial care in the intensive care unit due to initial symptoms of intoxication. Initially his urine specimens showed the presence of diazepam along with its metabolites oxazepam and 3-hydroxy diazepam. Because oxazepam was detected for at least 3 weeks following his acceptance in a drug treatment facility, continued diazepam abuse was suspected. However, creatinine corrected urine concentrations of diazepam metabolites over time verified that the person was no longer abusing diazepam. This conclusion was reached based on the long half-life of diazepam. Pharmacogenomics analysis revealed the person to be homozygotic CYP2C19*2/*2, suggesting a lack of activity of the CYP2C19 enzyme He was also homozygotic for CYP2C9*1/*3, indicating a reduced capability of this enzyme to metabolize certain drugs. Therefore, the person was identified as a poor metabolizer of diazepam, which may explain the symptoms of intoxication at hospital admission (37).

4.6. VARIOUS FACTORS THAT PRODUCE A POSITIVE COCAINE RESULT

Positive cocaine in workplace drug testing may result from drinking coca tea that may be contaminated with cocaine or after the medical use of cocaine. Coca tea is a common drink among people in South America. It is freely available in the market and may be also sold as "Health Inca Tea" or "Mate de Coca." The tea bags usually contain approximately 1 g of plant material, usually dried raw coca leaves that are also a natural source of cocaine. In one study authors found an average of 5.11 mg of cocaine per tea bag of coca tea originated from Peru and average 4.86 mg of cocaine per tea bag in Bolivian coca tea. When the tea was prepared, 1 cup of Peruvian coca tea had an average of 4.14 mg of cocaine, whereas 1 cup of Bolivian tea had an average of 4.29 mg of cocaine. Therefore as expected when one volunteer drank only 1 cup of Peruvian tea, a peak benzoylecgonine (major metabolite of cocaine) concentration of 3940 ng/mL was observed 10 hours after consumption. Similarly, consumption of only 1 cup of Bolivian tea by a volunteer resulted in a peak benzoylecgonine concentration of 4979 ng/mL 3.5 hours after consumption of the tea. These results clearly indicate that significant amounts of cocaine can be found in coca tea from Bolivian and Peru and that drinking such tea may cause positive cocaine results in workplace drug testing (38). Although U.S. custom regulations demand that no cocaine be present in any herbal tea, literature references indicate that some health Inca tea sold in the United States contains cocaine. Jackson et al reported urinary concentration of benzoylecgonine after ingestion of 1 cup of health Inca tea by volunteers. Benzoylecgonine was detected up to 26 hours postingestion. Maximum benzoylecgonine concentrations in urine specimens ranged from 1400 ng/mL to 2800 ng/mL after ingestion of health Inca tea. The total excretion of benzoylecgonine in 36 hours ranged from 1.05 to 1.45 mg, which correlated with 59–90% of the ingested cocaine dose (39). Drinking mate de coca also causes significant excretion of benzoylecgonine in urine. When five healthy volunteers consumed the tea, each participant showed a positive urine sample (over 300 ng/mL cutoff for cocaine metabolite) 2 hours after drinking the tea, and urine specimens from three of five volunteers showed positive test results up to 36 hours. Mean benzoylecgonine concentration was 1777 ng/mL (range: 1065–2495 ng/mL). The authors concluded that coca tea ingestion resulted in a positive urine test for the cocaine metabolite (40). Turner et al also reported positive tests for benzoylecgonine in subjects after drinking mate de coca tea. The tea was prepared by allowing one mate de coca tea bag to be immersed in 250 mL boiling water for 25 minutes. The bag was removed and squeezed in the tea to drain additional water. A 5-mL sample was taken for analysis, and volunteers drank the rest. Urinary samples were collected at 2, 5, 8, 15, 21, 24, 43, and 68 hours after drinking the tea. All urine samples tested positive for benzoylecgonine, the metabolite of cocaine by immunoassay. The amount of cocaine in the tea was estimated to be 2.5 mg (41).

Cocaine is used infrequently as a local anesthetic in ear, nose, and throat surgery and also is administered topically during ophthalmitic procedures. Positive drug testing results for cocaine (as benzoylecgonine) may be observed in a patient up to 72 hours after an otolaryngologic procedure where cocaine was used as an anesthetic. Benzoylecgonine may also be present in the urine of the physician who performed such a procedure (42). Jacobson et al studied the effect of the use of an ophthalmic solution containing cocaine on the urine excretion of benzoylecgonine in patients. Of 50 subjects studied, 47 subjects (94%) demonstrated a positive screening result for cocaine (as benzoylecgonine) 4–6 hours after receiving eye drops. In addition, 35 subjects (70%) showed positive results 24 hours after receiving eye drops (43). The combination of tetracaine, adrenalin, and cocaine is used as a topical anesthetic during the suture of simple skin lacerations. Application of such a topical anesthetic also produces a positive urine test for cocaine up to 2 days (44).

Case Study

A 54-year-old white man was involved in a physical altercation and suffered from a broken nose. Three days later a surgical procedure was performed to repair his broken nose and he was released from the hospital later that day with a prescription of Percocet (oxycodone 5 mg and acetaminophen 325 mg) for pain relief. He took one tablet at 4 P.M. and went to bed, but at 8 P.M. when his wife tried to wake him, up he was unconscious. She called the emergency medical team and despite performing cardiopulmonary resuscitation, he was pronounced dead at the scene at 8:40 P.M. Further investigation revealed that the man suffered from type 2 diabetes, hypertension, and high cholesterol, but there was no history of drug abuse. The subsequent postmortem examination revealed severe atherosclerotic right coronary artery disease with 70–80% limen stenosis, nasal bone and left maxillary sinus fractures but minimal subdural hematoma. However, toxicological analysis revealed the presence of cocaine (48 ng/mL) and benzoylecgonine (482 ng/mL) in the urine specimen, but no cocaine or its metabolite was detected in the subclavian blood specimen. In addition, oxycodone (73 ng/mL) was present in the blood as expected. Upon detection of cocaine and its major metabolite in the urine specimen, it was initially suspected that the subject abused illicit cocaine, but cocaine was also used during his nasal surgery as a local anesthetic. This case demonstrated that detection of cocaine and its metabolite in postmortem samples not necessarily indicates the illicit use of cocaine. An erroneous presumption of cocaine use by the subject may have had potential legal ramifications. In addition, family members may suffer from undue anguish if their loved one was falsely identified as an illicit drug abuser (45).

4.7. CONCLUSIONS

Positive workplace drug testing may have serious consequences because a person may be denied employment or may be fired from the position. Therefore, it is very important that not only medical review officers but also laboratory professionals including toxicologists and pathologists be aware of all sources of alternative explanations for positive GC/MS confirmation of an abused drug in urine in a subject undergoing workplace drug testing. Although the SAMHSA-mandated drug testing panel includes amphetamines, cocaine (as benzoylecgonine), opiates, marijuana, and phencyclidine, private employers may have a more stringent requirement of workplace drug testing while additional drugs such as benzodiazepines, propoxyphene, methadone, and oxycodone are also included. As mentioned earlier oxycodone may not be detected by an opiate immunoassay, and a person abusing oxycodone may pass a SAMHSA-mandated drug test. Urine toxicology screen ordered by physicians in emergency departments also may test positive for a drug of abuse where the patient in not an illicit drug abuser. However, a positive urine toxicology test ordered for medical purpose cannot be used against the patient because the information is protected by various laws protecting the privacy of the patient's medical record. Nevertheless, it may confuse the ordering physician whether his or her patient is an abuser of illicit drugs. Again laboratory professionals may play an important role in explaining to the physician that there might be alternative reasons for a positive urine toxicology report.

REFERENCES

1. Manzi S, Law T, Shannon MW. Methylphenidate produces a false positive urine amphetamine screen [letter to the editor]. Pediatr Emerg Care 2002; 18: 401.
2. Kraemer T, Wenning R, Maurer HH. The antispasmodic drug mebeverine leads to positive amphetamine results by fluorescence polarization immunoassay (FPIA)-

studies on toxicological analyses of urine by FPIA and GC-MS. J Anal Toxicol 2001; 25: 333–338.

3. Cody JT, Valtier S, Nelson SL. Amphetamine enantiomer excretion profile following administration of Adderall. J Anal Toxicol 2003; 27: 485–492.
4. Shults TF. Amphetamine review and update: technical and pharmaceutical developments. MRO Alert 2003; 14.
5. Musshoff F. Illegal or legitimate use? Precursor compounds to amphetamine and methamphetamine. Drug Metab Rev 2000; 32: 15–44.
6. Kraemer T, Roditis SK, Peters FT, Maurer HH. Amphetamine concentrations in human urine following single-dose administration of calcium antagonist prenylamine-studies using fluorescence polarization immunoassay (FPIA) and GC-MS. J Anal Toxicol 2003; 27: 68–73.
7. Grenhill B, Valtier S, Cody JT. Metabolic profile of amphetamine and methamphetamine following administration of the drug famprofazone. J Anal Toxicol 2003; 27: 479–484.
8. Chan KH, Hsu MC, Tseng CY, Chu WL. Famprofazone use can be misinterpreted as methamphetamine abuse. J Anal Toxicol 2010; 34: 347–353.
9. Maurer HH, Kraemer T. Toxicological detection of selegiline and its metabolites in urine using fluorescence polarization immunoassay (FPIA) and gas chromatography-mass spectrometry (GC-MS) and differentiation by enantioselective GC-MS of the intake of selegiline from abuse of methamphetamine or amphetamine. Arch Toxicol 1992; 66: 675–678.
10. Mattingly G. Lisdexamfetamine dimesylate: a prodrug stimulant for the treatment of ADHD in children and adults. CNS Spectr 2010; 15: 315–325.
11. Lopez FA, Scheckner B, Childress AC. Physician perception of clinical improvement in children with attention deficit/hyperactivity disorder: a post hoc comparison of lisdexamfetamine dimesylate and mixed amphetamine salts extended release crossover analog classroom study. Neuropsychiatr Dis Treat 2011; 7: 267–273.
12. ElSohly MA, deWit H, Wachtel SR, Feng S et al. Delta 9-tetrahydrocannabivarin as a marker for ingestion of marijuana versus Marinol: results of a clinical study. J Anal Toxicol 2001; 25: 565–571.
13. Levin FR, Mariani JJ, Brooks DJ, Xie S et al. Delta-9-tetrahydrocannabivarin testing may not have the sensitivity to detect marijuana use among individuals ingesting dronabinol. Drug Alcohol Depend 2010; 106: 65–68.
14. Schwab US, Callaway JC, Erkkilä AT, Gynther J, Uusitupa MI, Järvinen T. Effects of hempseed and flaxseed oils on the profile of serum lipids, serum total and lipoprotein lipid concentrations and haemostatic factors. Eur J Nutr 2006; 45:470–477.
15. Bosy TZ, Cole KA. Consumption and quantitation of delta9-tetrahydrocannabinol in commercially available hemp seed oil products. J Anal Toxicol 2000; 24: 562–566.
16. Mediavilla V, Derungs R, Kanzig A, Magert A. Qualitat von hanfsamenol aus der Schweiz. Agraforschung 1997; 4: 449–451.
17. Leson G, Pless P, Grotenhermen F, Kalant H, ElSohly MA. Evaluating the impact of hemp food consumption on workplace drug tests. J Anal Toxicol 2001; 25: 691–698.

18. Goodwin RS, Gustafson RA, Barnes A, Nebro W, Moolchan ET, Huestis MA. Delta(9)-tetrahydrocannabinol, 11-hydroxy-delta(9)-tetrahydrocannabinol and 11-nor-9-carboxy-delta(9)-tetrahydrocannabinol in human plasma after controlled oral administration of cannabinoids. Ther Drug Monit 2006; 28: 545–551.
19. Gomes T, Mamdani MM, Dhalla IA, Paterson JM et al. Opioid dose and drug related mortality in patients with nonmalignant pain. Arch Intern Med 2011; 171: 686–691.
20. Guay DR. Use of oral oxymorphone in the elderly. Consult Pharm 2007; 22: 417–430.
21. Stefanidou M, Athanaselis S, Spiliopoulou C, Dona A et al. Biomarkers of opiate use. Int J Clin Pract 2010; 64: 1712–1718.
22. Popa C, Beck O, Brodin K. Morphine formation from ethylmorphine: implications for drugs of abuse testing in urine. J Anal Toxicol 1998; 22: 142–147.
23. Cone EJ, Heit HA, Caplan YH, Gourlay D. Evidence of morphine metabolism to hydromorphone in pain patients chronically treated with morphine. J Anal Toxicol 2006; 30: 1–5.
24. West R, West C, Crews B, Almazan P et al. Anomalous observations of hydrocodone in patients on oxycodone. Clin Chim Acta 2011; 412: 29–32.
25. Armstrong EJ, Jenkins AJ, Sebrosky GF, Balraj EK. An unusual fatality in a child due to oxycodone. Am J Forensic Med Pathol 2004; 25: 338–341.
26. Schmidt J, Boettcher C, Kuhnt C, Zenk MH. Poppy alkaloid profiling by electrospray tandem mass spectrometry and electrospray FT-ICR mass spectrometry after [ring-13C]-tyramine feeding. Phytochemistry 2007; 68: 189–202.
27. Reid RG, Durham DG, Boyle SP, Low AS, Wangboonskul J. Differentiation of opium and poppy straw using capillary electrophoresis and pattern recognition techniques. Anal Chim Acta 2007; 605: 20–27.
28. Pelders M, Ross JJ. Poppy seeds: difference in morphine and codeine content and variation in inter-and intra individual excretion. J Forensic Sci 1996; 41: 209–212.
29. Hill V, Cairns T, Cheng CC, Schaffer M. Multiple aspects of hair analysis for opiates: methodology, clinical and workplace population, codeine and poppy seed ingestion. J Anal Toxicol 2005; 29: 696–703.
30. Trafkowski J, Madea B, Musshoff F. The significance of putative urinary markers of illicit heroin use after consumption of poppy seed products. Ther Drug Monit 2006; 28: 552–558.
31. Struempler RE. Excretion of codeine and morphine following ingestion of poppy seeds. J Anal Toxicol 1987; 11: 97–99.
32. Hayes LW, Krassolt WG, Mueggier PA. Concentrations of morphine and codeine in serum and urine after ingestion of poppy seeds. Clin Chem 1987; 33: 806–808.
33. Fraser AD, Worth D. Experience with a urine opiate screening and confirmation cutoff of 2000 ng/mL. J Anal Toxicol 1999; 23: 549–551.
34. Lachenmeier DW, Sproll C, Musshoff F. Poppy seed foods and opiate drug testing—where are we today? Ther Drug Monit 2010; 32: 11–18.
35. Cassella G, Wu AH, Shaw BR, Hill DW. The analysis of thebaine in urine for the detection of poppy seed consumption. J Anal Toxicol 1997; 21: 376–383.
36. Garretty DJ, Wolff K, Hay AW, Raistrick D. Benzodiazepines misuse by drug addicts. Ann Clin Biochem 1997; 34: 68–73.

37. Lennestal R, Lakso HA, Nilsson M, Mjorndal T. Urine monitoring of diazepam abuse—new intake or not? J Anal Toxicol 2008; 32: 402–407.
38. Jenkins AJ, Llosa T, Montoya I, Cone EJ. Identification and quantitation of alkaloids in coca tea. Forensic Sci Int 1996; 77: 179–189.
39. Jackson GF, Saady JJ, Poklis A. Urinary excretion of benzoylecgonine following ingestion of health Inca tea. Forensic Sci Int 1991; 49: 57–64.
40. Mazor SS, Mycyk MB, Wills BK, Brace LD, Gussow L, Erickson T. Coca tea consumption causes positive urine cocaine assay. Eur J Emerg Med 2006; 13: 340–341.
41. Turner M, McCrory P, Johnston A. Time for tea anyone? Br J Sports Med 2005; 39: e37.
42. Bruns AD, Zeiske LA, Jacobs AJ. Analysis of the cocaine metabolite in the urine of patients and physicians during clinical use. Otolaryngol Head Neck Surg 1994; 111: 722–726.
43. Jacobson DM, Berg R, Grinstead GF, Kruse JR. Duration of positive urine for cocaine metabolite after ophthalmic administration: implications for testing patients with suspected Horner syndrome using ophthalmic cocaine. Am J Ophthalmol 2001; 131: 742–747.
44. Altieri M, Bogema S, Schwartz RH. TAC topical anesthesia produces positive urine test for cocaine. Ann Emerg Med 1990; 19: 577–579.
45. Bailey KM, Clay DJ, Gebhardt MA, Schmidt MJ et al. Cocaine detection in postmortem samples following therapeutic administration. J Anal Toxicol 2009; 33: 550–552.

Chapter 5

When Toxicology Report Is Negative in a Suspected Overdosed Patient: The World of Designer Drugs

5.1. INTRODUCTION

Although not widely abused like marijuana, cocaine, heroin, and related drugs, designer drugs and rave party drugs are also a major public health and safety issue. Rave party drugs such as 3,4-methylenedioxymethamphetamine (MDMA; Ecstasy) is a designer drug, but fortunately it has significant cross-reactivity with the amphetamine immunoassay and can be easily identified during a routine toxicology screen when an amphetamine-positive sample is confirmed by gas chromatography/mass spectrometry (GC/MS). However, other rave party drugs such as ketamine and GHB (γ-hydroxybutyric acid) and most designer drugs cannot be identified by routine urine drug screens and are a challenge in the diagnosis of patients overdosed with these drugs. Fortunately, some of the National Reference Laboratories are beginning to offer testings for commonly abused designer drugs.

Designer drugs were initially synthesized by clandestine laboratories to bypass the legal consequences of manufacturing and selling illicit drugs. Then in 1986, the U.S. Controlled Substances Act was amended to make the manufacturing and selling of designer drugs illegal. Other countries also adopted

Resolving Erroneous Reports in Toxicology and Therapeutic Drug Monitoring: A Comprehensive Guide, First Edition. Amitava Dasgupta.

TABLE 5.1. Common Designer Drugs and Their Street Names

Designer Drug	Analog of Drug	Common Street Name
3,4-methylenedioxymethamphetamine (MDMA)	Methamphetamine	Ecstasy, XTC, Adam, X, Lover's Speed
3,4-methylenedioxyethylamphetamine MDEA	Amphetamine	Eve
4-Bromo-2,5-dimethoxy-Phenylethylamine (2-CB)	Phenylethylamine	Venus, Bromo, XTC Erox, Nexu
Para-methoxyamphetamine (PMMA)	Amphetamine	Killer, Dr. Killer, Mitsubishi Double
4-Methylthioamphetamine (4-MTA)	Amphetamine	Golden Eagle
Mephedrone	Beta-keto-amphetamine	Bath Salt
3,4-methylenedioxypyrovalerone	Beta-keto-amphetamine	Bath Salt
α-methylfentanyl	Fentanyl	China White
γ-Hydroxy butyrate (GHB)	Endogenous Substance	Liquid Ecstasy, G, Nitro
γ-Hydroxy butyrolactone (GBL)	GHB	Blue Nitro, GH Revitalizer
1,4-Butanediol	GHB	Weight Belt
Flunitrazepam	Benzodiazepine	Roofies, Roche, Red Forget-me-pill, Mexican Valium
JWH-018, JWH-073, JWH-250	Cannabinoid	Spice, Legal High
Ketamine	Not Applicable	Special K, Vitamin K New Ecstasy, Psychedelic Heroin

similar laws to ban designer drugs. The common types of designer drugs include amphetamine analogs, opiate analogs (including fentanyl derivatives), piperazine analogs, tryptamine-based hallucinogens, phencyclidine analogs, and GHB analogs. Common designer drugs are listed in Table 5.1. Most commonly encountered designer drugs are structurally related to phenethylamine, opioids, and tryptamines.

5.2. AMPHETAMINE-TYPE DESIGNER DRUGS

Amphetamine is a sympathomimetic amine that is structurally related to phenethylamine. Some designer drugs have a close structural similarity with amphetamine or methamphetamine. One of the most abused designer drugs that is also an analog of amphetamine is MDMA, commonly known as Ecstasy. This drug was synthesized in 1914. Another closely related designer drug, 3,4-methylenedioxy-amphetamine (MDA), was synthesized first in 1910. Chemical structures of MDMA and MDA are given in Figure 5.1. After 1986, a large number of amphetamine analogs were synthesized by clandestine laboratories to produce more potent effects after abuse. Amphetamine-like designer drugs can be classified under six broad categories:

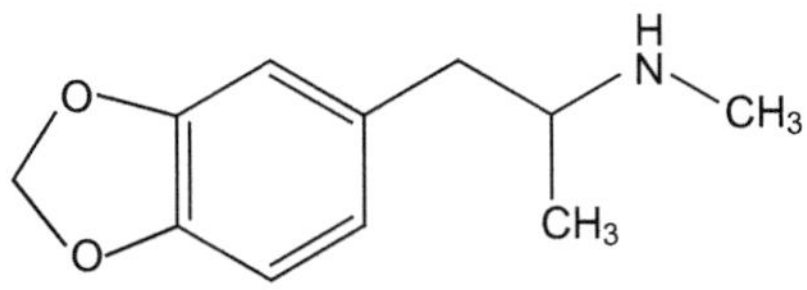

3,4-Methylenedioxymethamphetamine (MDMA)

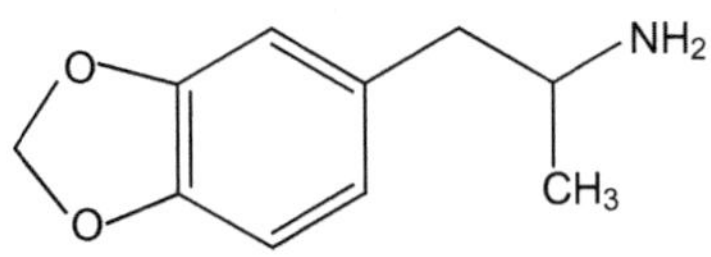

3,4-Methylenedioxyamphetamine (MDA)

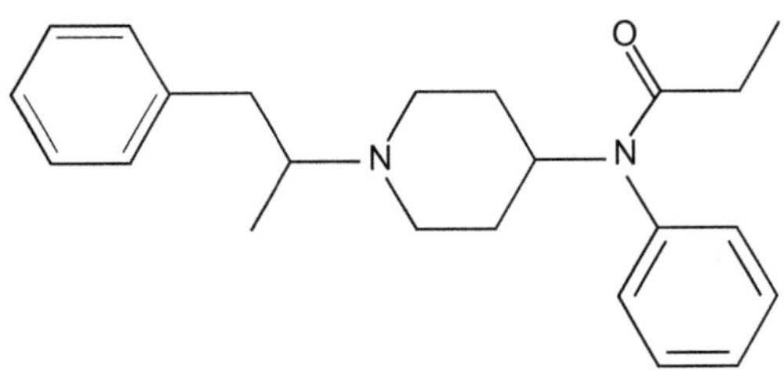

α-methylfentanyl

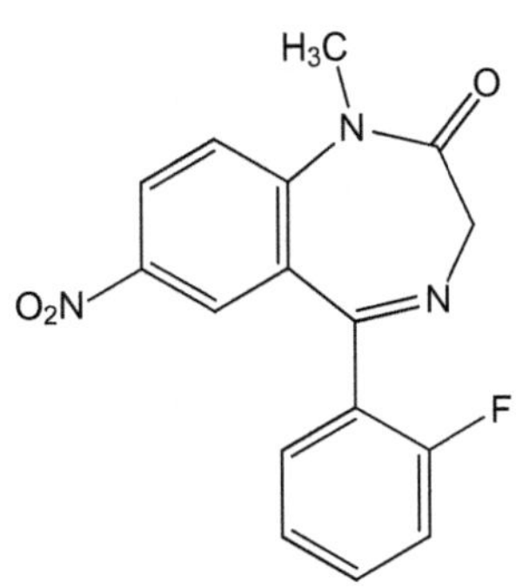

Flunitrazepam (Rohypnol)

Figure 5.1. Chemical structures of representative designer drugs and "club drugs."

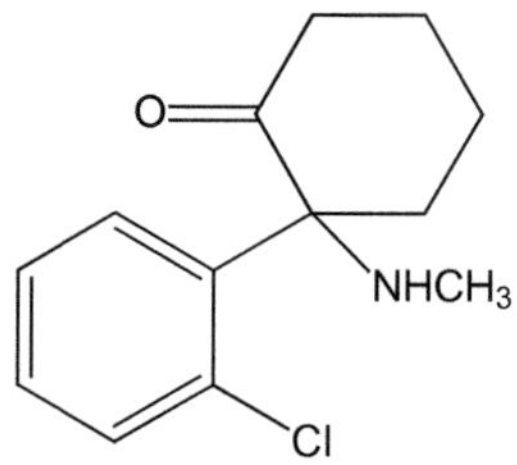

Ketamine

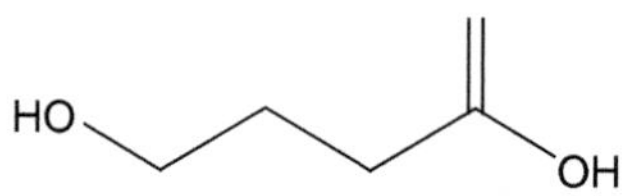

Gamma-hydroxybutyric acid

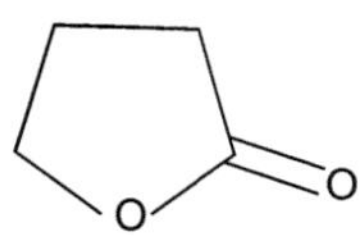

Gamma-butyrolactone

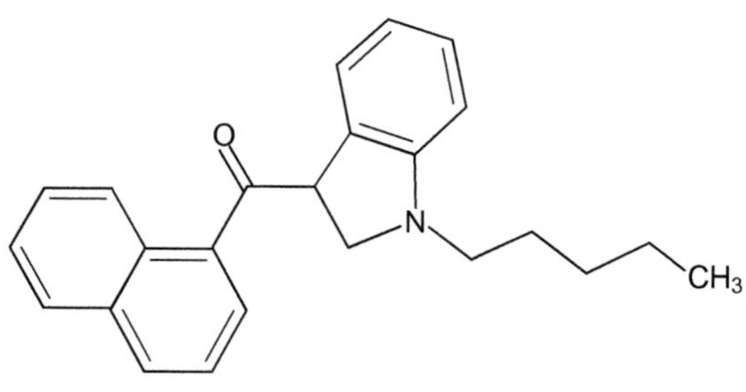

JWH-018

Figure 5.1. (*Continued*)

Phenylethylamine-type designer drugs
Substituted amphetamine-type designer drugs
Piperazine-type designer drugs
2,5-dimethoxy amphetamine–type designer drugs
β-Keto amphetamine–type designer drugs

The common examples of these designer drugs include para-methoxyamphetamine (PMA), para-methoxy-methamphetamine (PMMA), 3,4-methylenedioxy-N-ethylamphetamine (MDEA), 2,5-dimethoxy-4-methylamphetamine (DOM), and 2,5-dimethoxy-4-methylthioamphetamine (DOT) (1,2). Other designer drugs in this class include 4-iodo-2,5-dimethoxyamphetamine (DOI), 2,5-dimethoxy-4-bromo-amphetamine (DOB), 2,5-dimethoxy-4-bromo-methamphetamine (MDOB), N-methyl-1-(3,4)-methylenedioxy-phenyl)-2-butanamine (MBDB), and 3,4-(methylenedioxyphenyl)-2-butanamine (BDB) (1,2). In addition a chlorinated analog of MDMA was detected in the urine of a drug abuser (3). Bossong et al also described two new Ecstasy-like substances: methylone (3,4-methylenedioxymethcathinone) and meta-chlorophenylpiperazine (mCPPP). Methylone is the main ingredient of liquid designer drugs that appeared in the underground Dutch market (4). Designer drugs such as PMA, PMMA, and 4-methylthioamphetamine (4-MTA) have also been encountered at rave parties. 4-MTA is usually sold as "Ecstasy" or "flatliners" on the illegal drug market. MDMA, MDA, PMA, and related designer drugs are also widely used in rave parties along with ketamine and other drugs. Other designer drugs derived from phenylethylamine include 4-bromo-2,5-diemthoxy-β-phenylethylamine (2C-B), 2,5-dimethoxy-4ethylthio-β-phenylethylamine (2C-T-2), 2,5-dimethoxy-4 propylthio-β-phenylethylamine (2C-T-7), and related drugs that are also abused. These drugs are also toxic, and fatality from using 2C-T-7 has been reported. These designer drugs are mainly metabolized in the liver (5).

In recent years, a new class of designer drugs has emerged in many countries that are beta keto (bk) derivatives of amphetamine and related compounds. These drugs include mephedrone (bk-4-methylmethamohetamine), butylone, and methylone (bk-3,4-methylenedioxyamphetamine) (6).

Another group of designer drugs derived from piperazine has also entered the illicit drug market and can be divided into two groups: the benzylpiperazines 1-benzylpiperazine (BZP) itself and its methylenedioxy analog 1-(3,4-methylenedioxybenzyl)-piperazine (MDBP) and also the phenylpiperazines 1-(3-trifluoromethylphenyl)piperazine (TFMPP), 1-(3-chlorophenyl) piperazine (mCPP), and 1-(4-methoxyphenyl)piperazine (MeOPP) (7,8). The effects of BZP are claimed to be comparable with those of amphetamine.

5.2.1. Metabolism and Toxicity of Designer Drugs Related to Amphetamines

MDMA is metabolized to MDA and variety of other compounds including 4-hydroxy-3-methoxymethamphetamine (major metabolite), 3,4 dihydroxymethamphetamine, and 3-hydroxy-4-methoxymethamphetamine. Most 4-hydroxy-3-methoxymethamphetamine is excreted in urine as conjugated with glucuronide or sulfate. The designer drug DOB is metabolized by O-demethylation followed by an oxidative deamination to the respective ketones as well as deamination followed by a reduction to the corresponding alcohols. Other metabolic pathways involve O,O-bisdemethylation or hydroxylation of the side chain followed by O-demethylation and deamination to the corresponding alcohol. All metabolites carrying the hydroxyl group were found to be in conjugated form in rat urine (8). The designer drug DOI is also extensively metabolized by O-demethylation based on the study of metabolism of this drug in rats (9). The newer class of designer drugs such as mephedrone, butylone, and methylone (beta-keto derivative of amphetamine and related drugs) are metabolized by N-demethylation to the primary amine, reduction of keto moiety to the respective alcohol, and oxidation of tolyl moiety to the corresponding alcohols and carboxylic acid (6). Mephedrone and a related compound, 3,4-methylenedioxypyrovalerone, are sold as bath salts, although unlike traditional cosmetic bath salts that are packaged and sold for addition to bath water for soaking and cleaning, the designer drugs sold as bath salts have no legitimate use for bathing and are intended for abuse. According to one report, 35 persons who had ingested, inhaled, or injected bath salts visited emergency departments in Michigan from November 13, 2010, to March 31, 2011. Seventeen patients were hospitalized and one died. The most common symptoms among these patients were hallucination, agitation, and tachycardia (10).

MDMA, MDA, and related designer drugs are also toxic and may even cause fatality (11). Klys et al described a fatal case of MDMA abuse in a 22-year-old man with a history of drug abuse. Using liquid chromatography combined with atmospheric pressure chemical ionization tandem mass spectrometry, the authors confirmed the presence of MDMA (1420 ng/mL) and its metabolite 3, 4-methylenedioxamphetamine (MDA; 170 ng/mL) in his blood (12). Abuse of PMA and PMMA may also cause death. Johansen et al reported three fatal cases of PMA and PMMA poisoning in Denmark (13). A closely related designer drug, para-methoxyethylamphetamine (PMEA), is also associated with fatality. This drug is extensively metabolized to para-methoxyamphetamine and para-hydroxyethyl-amphetamine. In one fatal case involving PMEA, the postmortem blood concentration of PMEA was 12.2 μg/mL (14).

Case Report

A 39-year-old African American woman presented to the emergency department after a night of partying. The patient showed rapidly diminishing mental status, agitation, hypothermia, severe hypertension, and vasoconstriction. Her friend stated that she was taking cocaine, alcohol, and MDMA, along with another designer drug, 2CI (2,5-dimethoxy-4-iodophenethyamine) the night before and the day of admission. The urinary drug screen was positive for amphetamines. The liquid chromatography combined with tandem mass spectrometric analysis confirmed the presence of MDA (3,4-methylenedioxy-amphetamine; 5.56 mg/L) and 2CI (0.311 mg/L). The positive amphetamine immunoassay result was due to

MDA because further investigation showed that 2CI had no cross-reactivity (up to 100,000 ng/mL concentration tested) with the amphetamine screening assay (cloned enzyme donor immunoassay [CEDIA], Microgenic, Freemont, CA). The patient suffered severe disability, and after 4 months in and out of the hospital she was discharged to a nursing home facility (15).

5.2.2. Analysis of Designer Drugs Related to Phenethylamine

The recent abuse of MDMA or MDA can be detected by a urine drug screen using amphetamine immunoassays (see Chapter 3). However, amphetamine immunoassays are not suitable for the detection of most designer drugs structurally related to phenethylamine. Thus a person abusing such drug may produce a false-negative drug screen. Currently, only chromatographic techniques are available for the analysis of these designer drugs in biological fluids. Pichini et al described a liquid chromatography combined with tandem mass spectrometric determination of MDMA along with related designer drugs 2,5-diemethoxy-4-methyl-phenethylamine (2C-D), 4-bromo-2,5-dimethoxy-beta- phenylethylamine (2C-B), 1-(8-bromo-2,3,6,7-tetrahydrobenzo [1,2-b:4,5-b′] difuran-4-tl)-2-aminoethane (2CB-Fly), 4-ethylthio 2,5-dimethoxy-beta-phenylethylamine (2CT-2), 4-iodo-2,5-diemethoxy-beta-phenylamine (2C-I), 4-ethyl-2,5-dimethoxy-beta-phenyethalmine (2C-E), 1-(m-chlorophenyl)piperazine (m-CPP), 4-hydroxy-N,N′-disopropyltryptamine (4-OH-DIPT), and 4-acetoxy-N-N′-diisopropyltryptamine (4-acetoxy-DIPT) in human urine. The authors used 3,4-methylenedioxypropylamphetamine as the internal standard. Extraction was performed using solid phase extraction, and chromatographic separation was achieved by using a reverse phase C-18 column using a linear gradient of 10 mM ammonium bicarbonate, pH 7.3, and acetonitrile as a mobile phase. Separated analytes were identified by single ion monitoring using atmospheric pressure ionization-electrospray ionization interface (16). Ewald et al analyzed designer drugs DOB and MDOB using gas chromatography/mass spectrometry (17). Takahashi et al created a psychoactive drug data library by performing analysis using liquid chromatography with photodiode array spectrophotometry and GC/MS. This library has data on 104 drugs with the potential for abuse (18).

5.3. OPIOID DESIGNER DRUG

Heroin (diacetylmorphine), which was synthesized from morphine in 1874, can be considered the first designer drug that belongs to the opioid class. Heroin is a class I scheduled drug in the United States with no known medical use, but this drug is highly abused throughout the world. Heroin is metabolized to 6-monoacetylmorphine and then finally to morphine. Fortunately, a heroin abuser can be easily identified by an opiate assay during a urine drug screen.

Fentanyl is a widely used synthetic narcotic analgesic that is approximately 75–100 times more potent than morphine. Several analogs of fentanyl such as sufentanil, alfentanil, lofentanil, and remifentanil have been synthesized by pharmaceutical industries and are used clinically. Currently in the United States, fentanyl is a schedule II drug used as an anesthetic. Injection of 50–100 µg fentanyl produces rapid analgesic effect and unconsciousness. Fentanyl is also available as lozenges and transdermal patches (Duragesic) for pain management. Oral transmucosal fentanyl citrate (Actiq) is a relatively new formulation where fentanyl is incorporated into a sweetened matrix to produce fentanyl lozenge. This lozenge can be allowed to dissolve in the mouth (takes approximately 15 minutes) for the rapid absorption of fentanyl through the mucosa. Fentanyl lozenges are used for the control of breakthrough pain in cancer patients and also for control of severe pain. The therapeutic range of fentanyl is considered as 1–3 ng/mL in serum, and the toxicity of fentanyl is similar to opiate toxicity (19).

The Duragesic patches deliver 25, 50, 75, or 100 µg of fentanyl per hour through the transdermal route. The patch provides a continuous systemic delivery of fentanyl for the intended duration of application of 72 hours. Fentanyl patches should only be given to patients who are tolerant to opioid therapy because toxicity and even death can be associated with the abuse of the transdermal patch. Anderson and Mutto reported in 2000 their experience of investigating 25 deaths over the past 3 years in Los Angeles County that involved Duragesic patches. The mode of death included 15 accidental, 5 natural, 3 suicidal, and 2 undetermined causes of death. The concentration of fentanyl in the heart blood ranged from 1.8 to 139 ng/mL, femoral blood ranged from 3.1 to 43 ng/mL, and the urinary concentration of fentanyl ranged from 2.9 to 895 ng/ml (20).

The major route of fentanyl abuse is by intravenous injection. An overdose from fentanyl can cause coma, and the patient may stop breathing. Death from fentanyl overdose is common. Martin et al reported that 54 deaths in Ontario, Canada, from 2002 to 2004 were related to fentanyl. The mean fentanyl concentration in blood was 25 ng/mL (3.0–383 ng/mL) (21). Schumann et al reported a fentanyl epidemic in Chicago. The city and surrounding Cook County experienced an outbreak of 342 fentanyl-related deaths between April 2005 and December 2006 (22). Fentanyl analogs are available as designer drugs and also abused.

5.3.1. Fentanyl Analogs as Designer Drugs

Fentanyl analog designer drug China White appeared in the underground market of California in 1979. The active ingredient of China White is α-methylfentanyl, a very potent analog of fentanyl. Chemical structure of α-methylfentanyl is given in Figure 5.1. Abuse of China White caused over 100 deaths in California. Gillespie et al determined postmortem blood, bile, and liver concentrations of α-methylfentanyl in a drug overdose victim. The blood concentration of α-methylfentanyl was 3.1 ng/mL, bile concentration was 6.4 ng/mL, and the level in the liver was 78 ng per gram of liver tissue (23). In 1984, another illicit designer drug, 3-methylfentanyl, appeared as a street drug in California that was also related to a fatal drug overdose. During 1988, 3-methylfentanyl was identified in 16 fetal overdose cases in Allegheny County in Pennsylvania. In addition to 3-methylfentanyl, morphine was detected in the blood of five individuals, and cocaine was detected in the blood of three persons (24). Ojanpera et al described three fetal poisoning cases involving 3-methylfentanyl. The mean blood level of 3-methylfentanyl was 0.5 ng/mL (range: 0.3–0.9 ng/ml). The blood fentanyl levels were measured using a specific tandem mass spectrometric technique (25). Ojanpera et al later reported an epidemic of 3-methylfentanyl poisoning in Estonia where 46 deaths were reported in 2005 and 71 deaths reported in 2006. The mean total concentration (both stereoisomers) of 3-methylfentanyl was 1.9 ng/mL (26).

Neither fentanyl nor its analogs can be detected in routine urine drug testing by regular opiate screening assays because these compounds do cross-react with the antibody used in opiate assays that are specific for morphine, The concentrations of fentanyl and its analogs can be measured in serum, urine, and other biological matrix using GC/MS or liquid chromatography combined with tandem mass spectrometry. Sato et al described a method for the quantification of alpha-methyl fentanyl in rat urine (27).

5.3.2. Meperidine Analogs as Designer Drugs

Meperidine (Demerol) is a synthetic opioid used as a narcotic analgesic It is available both as an injectable form and oral use supplied as hydrochloride salt. Both meperidine and several of its analogs are abused. A common designer drug that is a meperidine analog is MPPP (1-methyl-4-phenyl-4-propionoxypiperdine). An impurity found in some illicitly synthesized MPPP preparation identified as 1-methyl-4-phenyl-1,2,3,6-tetrahydropyridine (MPTP) caused permanent parkinsonism in a number of intravenous drug abusers (28).

Meperidine and its analogs have negligible cross-reactivity with the opiate screening assay for drugs of abuse, and if present in urine they cannot be detected by routine urine drugs of abuse screening. A more sophisticated technique such as GC/MS needs to be applied to determine the presence of such drugs and their metabolites in blood, urine, or other biological specimens.

5.4. PHENCYCLIDINE ANALOGS AS DESIGNER DRUGS

Phencyclidine (PCP) was introduced in the 1950s as a human anesthetic but was discontinued due to significant side effects. PCP undergoes extensive metabolism by liver cytochrome P450 enzymes (especially CYP3A4) into several hydroxy metabolites.

A number of designer drugs are analogs of phencyclidine. These drugs include thienylcyclohexylpiperidien (TCP), (1-phencyclohexyl0ethylamine (PCE), 1-(1-phencyclohexyl) pyrrolidine, N-(1-phenylcyclohexyl)-2-ethoxyethanamine (PCEEA), and N-(1-phencyclohexyl)-2-methoxyethanamine (PCMEA). Metabolism of these designer drugs has been mostly studied in rat models (29).

5.5. GHB AND ITS ANALOGS

GHB is an endogenous constituent of the mammalian brain that is present in nanomolar concentration and acts as a neurotransmitter. Chemical structure of GHB is given in Figure 5.1. Although GHB was found in health food stores before as a safe body building and fat burning compound, 16 cases of adverse effects due to GHB-containing health products were reported to the San Francisco Bay Area Regional Poison Control Center from June 1990 to October 1990. Use of GHB caused coma in four patients and tonic-clonic seizure in two patients for dosages ranging from a quarter of a teaspoon to 4 tablespoons (30). Following that report, the Food and Drug Administration banned the over-the-counter sale of GHB in November 1990, and now GHB is a schedule I controlled substance in the United States. The management of patients addicted to GHB is complicated due to the high dosage of medications they require to control withdrawal symptoms (31). Even the use of GHB for 7 days may produce severe withdrawal (32).

Case Study

A 23-year-old woman presented to the emergency department with paranoid delusions as well as auditory hallucination. Previously she had two episodes of psychosis induced by amphetamine and marijuana. During the past 6 months prior to presentation, she exhibited bizarre behavior and was unable to perform daily chores. The patient admitted the occasional use of GHB. After admission, her toxicology screen was negative. Therefore, the attending physician suspected that the patient was suffering from a schizophrenic disorder and prescribed antipsychotic medication. A few hours after admission, she developed tachycardia and hypertension and exhibited severe agitation and confusion. Physical restrain was instituted and she was treated with 7 mg of lorazepam. Her confusion resolved in less than 72 hours, and at that point the patient admitted she had been taking GHB for the past 6 months as a self-medication for insomnia and anxiety and abruptly stopped GHB 24 hours prior to this episode. The authors concluded that withdrawal of GHB mimics schizophrenia (33).

Case Study

A 35-year-old man was partying with friends and consumed an unknown quantity of GHB and wine. One woman present in the party reported that the man was alive before she passed out. When she was awakened, she found the decedent cold and stiff. Analysis of postmortem specimens showed no presence of alcohol or any common drugs. On more specific examination of postmortem specimens, GHB in urine, brain, vitreous fluid, femoral blood, heart blood, and liver were detected in the amounts of 1665 mg/L, 102 mg/kg, 48 mg/L, 461 mg/L, 276 mg/L, and 52 mg/kg, respectively. The cause of death was determined to be an accidental GHB overdose (34).

Although the sale of GHB and its analogs for human consumption is against the law in the United States, several GHB analogs are commercially available as industrial solvents and for the manufacture of plastic and other products. In addition, analogs of GHB are sold in the underground markets and worldwide. GHB analogs that are found as street drugs include GBL (γ-butyrolactone), 1, 4-BD (1, 4-butanediol), GHV (γ-hydroxy valeric acid), and GVL (γ-valerolactone). GBL and 1,4-BD are converted endogenously after abuse into GHB and exert similar effects of GHB abuse. Therefore, clinical manifestation and management of BDL and 1,4-BD intoxication is similar to GHB intoxication (35). However, self-reported GHB ingestion was much more common than GBL ingestion, but GBL was more commonly found in seized samples. In one report, the authors stated that of the 418 seized samples, 225 were in liquid form, of which 140 contained GBL and only 85 contained GHB (36).

Neither GBL nor 1,4-BD directly binds to GHB receptors (GABA receptors), but GHV binds directly to GHB receptors and has a similar effect as GHB. However, GVL does not metabolize to GHB. GVL is a precursor of GHV and also can be found as a street drug. GVL, like GHB, produces sedation, catalepsy, and ataxia but requires a larger dosage than GHB to produce similar effects. A larger dose of GHV is also lethal (37). There is no antidote for treating a GHB overdose, and management is mostly supportive with the emphasis on airway management. Usually a dosage of 20–30 mg/kg of GHB causes sleep, and doses of 50 mg/kg or higher produces anesthesia. The usual dose of GHB in the cases of abuse is over 35 mg/kg. Both GHB and its analog 1,4-BD have a short half-life in humans. Thai et al studied the pharmacology of 1,4-BD in humans and observed that 1,4-BD is quickly absorbed and cleared with a maximum plasma concentration achieved at an average of 24 minutes after ingestion; the elimination half-life was 39.3 minutes on average. 1,4-DB is also extensively converted into GHB endogenously, and the average half-life of GHB was 32.3 minutes (38). Because GHB has no distinct taste, it can be slipped into a drink and has been used in date rape situations.

It is usually considered that the concentration of GHB in blood should not exceed 50 mg/L and in urine should not exceed 10 mg/L due to the endogenous production of GHB. Yeatman and Reid studied the urinary concentrations of endogenous GHB in 55 volunteers and observed that the endogenous urinary concentrations of GHB ranged from 0.9 to 3.5 mg/L with a mean urinary concentration of 1.56 mg/L. The authors concluded that their results confirm that the suggested 10 mg/L cut-off value for endogenous GHB concentration in forensic analysis is justified (39).

Currently, there is no immunoassay for the routine screening of GHB in urine or any other biological matrix. Unfortunately, GHB cannot be detected by the routine drugs of the abuse testing protocol. In the case of suspected overdose of GHB, a more sophisticated analytical technique such as GC/MS should be used to confirm the presence of GHB in blood or urine. GHB in blood can be determined using GC/MS after liquid–liquid extraction and ditrimethylsilyl derivatization (40). Couper and Logan also used a simple liquid–liquid extraction for the extraction of GHB from biological fluid and GC/MS analysis after forming ditrimethylsilyl derivatization. Using the technique the authors determined GHB concentration of 3.2 mg/L in the blood of a sexual assault victim. In addition, GHB levels of two drivers who were accused of driving while intoxicated were 33 and 34 mg/L (41). GHB in human serum or urine can also be determined by liquid chromatography combined with mass spectrometry. Kaufmann and Alt reported the analysis of GHB in human serum and urine using this technique after the conversion of GHB to its respective n-butyl ester derivative. The authors used hexadeutero GHB as the internal standard (42).

Major hospital-based laboratories and National Reference Laboratories offer the testing of GHB in blood and urine. Moreover, most forensic toxicology laboratories in the medical examiner or coroner's office offer the testing of GHB. Therefore, in the case of a suspected GHB overdose, specimens can be sent to these laboratories for analysis.

5.6. MARIJUANA ANALOGS AS DESIGNER DRUGS

More recently, a series of new designer drugs emerged in the market known as "legal highs" or "herbal highs." These drugs include both herbal substances but more commonly synthetic designer drugs that can be purchased through Internet sites and mimic the psychoactive effects of illicit drugs. These "legal high" drugs can be broadly classified as amphetamine and Ecstasy-like psychostimulants as well as synthetic cannabinoids also known as "spice" (43). Although "spice" was classified as an incense blend that releases a rich aroma and is not intended for human consumption, when smoked by humans, it has a marijuana-like hallucinogenic effect. Synthetic cannabinoids that have been reported as spice include JWH-018, JWH-073, JWH-250, HU-210, CP-47,497, and its homologs. These compounds are lipid soluble, typically containing 20–26 carbon atoms and are more potent than marijuana. JWH-018 was the first compound of this class reported in 2008 and is an effective cannabinoid receptor (CB1) agonist. In the 1980s, scientists at Pfizer developed CP-47,497 and CP-55,940, which are more potent than marijuana. HU-20 is a synthetic agonist analog of marijuana (44). The chemical structure of JWH-018 is given in Figure 5.1.

Case Study

A 25-year-old man presented to the hospital with increased anxiety over the past month. The patient thought that a chip was implanted into his abdomen that was controlling him. The past medical history obtained from his mother revealed that the patient had experienced recurrent psychotic episodes from age 18 probably due to cannabis abuse. However, for the past 5 years he was stable with 800 mg per day amisulpride and did not abuse any drug or alcohol. However, a month prior to the episode, he abused "spice" on three different occasions, consuming approximately 3 g each time. After abusing "spice" he developed paranoid hallucinations, a symptom he never experienced before. Urine and hair drug analysis for cannabinoids revealed negative results. A magnetic resonance examination of the brain was normal (45).

Synthetic cannabinoids usually do not cross-react with the marijuana immunoassay. However, such compounds can be analyzed by chromatographic methods. Dresen et al described a protocol for the analysis of JWH-015, JWH-018, JWH-073, JWH-081, JWH 200, JWH-250, and related compounds after liquid–liquid extraction from serum and quantification by liquid chromatography combined with tandem mass spectrometry. The authors analyzed 101 serum specimens from 80 subjects provided by hospitals, detoxification and therapy centers, forensic psychiatry centers, and police authorities and observed that 57 specimens were positive for at least one of these compounds (46).

5.7. CONCLUSIONS

Although so-called club drugs such as MDMA can be detected in most routine toxicology screens as positive for amphetamines and Rohypnol may trigger positive benzodiazepine screens if present in sufficient amount, other club drugs such as GHB and ketamine cannot be detected by a routine toxicology drug screen. These club drugs and other designer drugs pose a challenge to health care providers because they escape detection in a routine toxicological analysis. Fortunately, some reference laboratories have initiated efforts to provide an analysis of GHB and other designer drugs. It is important for health care providers to be aware of such situations so that in the case of a suspected drug overdose or sexual assault where a urine toxicology screen is negative, proper steps can be taken to send the specimen to an appropriate reference laboratory for further testing.

REFERENCES

1. Christophersen AS. Amphetamine designer drugs—an overview and epidemiology. Toxicol Lett 2000; 112–113: 127–131.

2. Johansen SS, Hansen AC, Muller IB, Lundemose JB et al. Three fetal cases of PMA and PMMA poisoning in Denmark. J Anal Toxicol 2003; 27: 253–256.
3. Maresove V, Hampl J, Chundela Z, Zrcek F et al. The identification of a chlorinated MDMA. J Anal Toxicol 2005; 29: 353–358.
4. Bossong MG, Van Dijk JP, Niesink RJ. Methylone and mCPP, two new drugs of abuse. Addict Biol 2005; 10: 321–323.
5. Theobald DS, Fehn S, Maurer HH. New designer drug 2,5-dimethoxy-4propylthio-β-phenylethylamine (2C-T-7): studies on its metabolism and toxicological determination in rat urine using gas chromatography/mass spectrometry. J Mass Spectrom 2005; 40: 105–116.
6. Meyer MR, Wilhelm J, Peters FT, Maurer HH. Beta-keto amphetamines: studies on the metabolism of the designer drug mephedrone and toxicological detection of mephedrone, butylone and methylone in urine using gas chromatography-mass spectrometry. Anal Bioanal Chem 2010; 397: 1225–1233.
7. de Boer D, Bosman IJ, Hidvegi E, Manzoni C, et al. Piperazine-like compounds: a new group of designer drugs-of-abuse on the European market. Forensic Sci. Int 2001; 121: 47–56.
8. Balmelli C, Kupferschmidt H, Rentsch K, Schneemann M. Fatal brain edema after ingestion of ecstasy and benzylpiperazine [in German]. Dtsch Med Wochenschr 2001; 126: 809–811.
9. Szucks Z, Szentendrei T, Fekete MI. The effect of EGYT-475 (trelibet) and its metabolites on the potassium-stimulated 3H-noradrenaline release from cortical slices of rat brain. Pol J Pharmacol Pharm 1987; 39: 185–193.
10. Centers for Disease Control and Prevention. Emergency department visits after use of a drug sold as bath salts-Michigan, November 13, 2010-march 31, 2011. MMWR Morb Mortal Wkly Rep 2011; 60: 624–626.
11. Murphy DL, Lesch KP, Aulakh CS, Pigott TA. Serotonin-selective arylpiperazines with neuroendocrine, behavioral, temperature, and cardiovascular effects in humans. Pharmacol Rev 1991; 43: 527–552.
12. Klys M, Rojek S, Wozniak K, Rzepecka-Wozniak E. Fatality due to use of a designer drug MDMA (Ecstasy). Leg Med (Tokyo) 2007; 9: 185–191.
13. Johansen SS, Hansen AC, Muller IB, Lundemose JB et al. Three fatal cases of PMA and PMMA poisoning in Denmark. J Anal Toxicol 2003; 27: 253–256.
14. Zaotsu K, Katagi M, Kamata H, Shima N et al. Determination of a newly encountered designer drug "p-methoxyethylamphetamine" and its metabolites in human urine and blood. Forensic Sci Int 2008; 177: 77–84.
15. Dress JC, Stone JA, Wu AH. Morbidity involving the hallucinogenic designer amines MDA and 2CI. J Forensic Sci 2009; 54: 1485–1487.
16. Pichini S, Pujadas M, Marchei E, Pellegrini M et al. Liquid chromatography-atmospheric pressure ionization electrospray mass spectrometric of hallucinogenic designer drugs in urine of consumers. J Pharm Biomed Appl 2008; 47: 335–342.
17. Ewald AH, Fritschi G, Bork WR, Maurer HH. Designer drugs 2,5-dimethoxy-4-bromo-amphetamine (DOB) and 2,5-diemthoxy-4-bromo-methamaphetamine (MDOB): studies of their metabolite and toxicological detection in rat urine using gas chromatography/mass spectrometric techniques. J Mass Spectrom 2006; 41: 487–498.

18. Takahashi M, Nagashima M, Suzuki J, Seto T et al. Creation and application of psychoactive designer drugs data library using liquid chromatography with photodiode array spectrometry detector and gas chromatography-mass spectrometry. Talanta 2009; 77: 1245–1272.

19. Mystakidou K, Katsouda E, Parpa E, Vlahos L et al. Oral transmucosal fentanyl citrate: overview of pharmacological and clinical characteristics. Drug Deliv 2006; 13: 269–276.

20. Anderson DT, Mutto JJ. Duragesic transdermal patch: postmortem tissue distribution of fentanyl in 25 cases. J Anal Toxicol 2000; 24: 627–634.

21. Martin TL, Woodall KL, McLellan BA. Fentanyl related deaths in Ontario, Canada: toxicological findings and circumstances of death in 112 cases (2002–2004). J Anal Toxicol 2006; 30: 603–610.

22. Schumann H, Erickson T, Thompson TM, Zautcke JL et al. Fentanyl epidemic in Chicago, Illinois and surrounding Cook County. Clin Toxicol (Phila) 2008; 46: 501–506.

23. Gillespie TJ, Gandolfi AUJ, Davis TP, Morano RA. Identification and quantification of alpha-methylfentanyl in post mortem specimens. J Anal Toxicol 1982; 6: 139–142.

24. Hibbs J, Perper J, Winek CL. An outbreak of designer drug-related deaths in Pennsylvania. JAMA 1991; 265: 1011–1013.

25. Ojanpera I, Gergov M, Rasanen I, Lunetta P et al. Blood levels of 3-methylfentanyl in 3 fetal poisoning cases. Am J Forensic Sci 2006; 27: 328–331.

26. Ojanpera I, Gergov M, Loiv M, Riikoja A et al. An epidemic of fetal 3-methylfentanyl poisoning in Estonia. Int J Legal Med 2008; 122: 395–400.

27. Sato S, Suzuki S, Lee XP, Sato K. Studies on 1-(2-phenethyl)-4-(N-propionyanilino) piperidine (fentanyl) and related compounds VII. Quantification of alpha-methylfentanyl metabolites excreted in rat urine. Forensic Sci Int 2010; 25: 68–72.

28. Johannessen JN, Markey SP. Assessment of the opiate properties of two constituents of a toxic illicit drug mixture. Drug Alcohol Depend 1984; 13: 367–374.

29. Sauer C, Peters FT, Staack RF, Fritschi G et al. New designer drugs N-(1-phenylcyclohexyl)-2-ethoxyethanamine (PCEEA) and n-(1-phenylcyclohexyl)-2-methoxyethanamine (PCMEA): studies on their metabolism and toxicological detection in rat urine using gas chromatography/mass spectrometric techniques. J Mass Spectrom 2008; 43: 305–316.

30. Dyer JE. Gamma-hydroxybutyrate: a health food product producing coma and seizure like activity. Am J Emerg Med 1991; 9: 321–324.

31. Gonzalez A, Nutt DJ. Gamma hydroxy butyrate abuse and dependency. J Psychopharmacol 2005; 19: 195–204.

32. Perez E, Chu J, Bania T. Seven days of gamma-hydroxybutyrate (GHB) use produces severe withdrawal. Ann Emerg Med 2006; 48: 219–220.

33. Constantinides P, Vincent P. Chronic gamma-hydroxybutyric acid use followed by gamma-hydroxybutyric acid withdrawal mimic schizophrenia: a case report. Cases J 2009; 2: 7520.

34. Mazaar-Pro S, Kerrigan S. Distribution of GHB in tissues and fluids following a fatal overdose. J Anal Toxicol 2005; 29: 398–400.

35. Palmer RB. Gamma-hydroxybutyrolactone and 1,4-butanediol: abused analogs of gamma-hydroxybutyrate. Toxicol Rev 2004; 23: 21–31.
36. Wood DM, Warren-Gash C, Ashraf T, Greene SL et al. Medical and legal confusion surrounding gamma-hydroxybutyrate (GHB) and its precursors gamma-butyrolactone (GBL) and 1,4-butanediol (1,4-BD). QKM 2008; 101: 23–29.
37. Carter LP, Chen W, Wu H, Mehta AK et al. Comparison of the behavioral effects of gamma-hydroxybutyric acid (GHB) and its 4-methyl substituted analog, gamma-hydroxyvaleric acid (GHV). Drug Alcohol Depend 2005; 78: 91–99.
38. Thai D, Dyer JE, Jacob P, Haller CA. Clinical pharmacology of 1,4-butanediol and gamma-hydroxybutyrate after oral 1,4-butanediol administration to healthy volunteers. Clin Pharmacol Ther 2007; 81: 178–184.
39. Yeatmen DT, Reid K. A study of urinary endogenous gamma-hydroxybutyrate (GHB) level. J Anal Toxicol 2003; 27: 40–42.
40. Elian AA. GC-MS determination of gamma-hydroxybutyric acid (GHB) in blood. Forensic Sci Int 2001; 122: 43–47.
41. Couper FJ, Logan BK. Determination of gamma-hydroxybutyrate (GHB) in biological specimens by gas chromatography-mass spectrometry. J Anal Toxicol 2000; 24: 1–7.
42. Kaufmann E, Alt A. Determination of GHB in urine and serum by LC/MS using a simple one-step derivative. Forensic Sci Int 2007: 168: 133–177.
43. Zawilska JB. Legal highs—new players in the old drama. Curr Drug Abuse Rev 2011; 4: 122–130.
44. Vardakou I, Pistos C, Spiliopoulou CH. Spice drugs as a new trend: mode of action, identification and legislation. Toxicol Lett 2010; 197: 157–162.
45. Muller H, Sperling W, Kohrmann M, Huttner HH et al. The synthetic cannabinoid spice as a trigger for acute exacerbation of cannabis induced recurrent psychotic episodes. Schizophr Res 2010; 118: 309–310.
46. Dresen S, Kneisel S, Weinmann W, Zimmermann R et al. Development and validation of a liquid chromatography-tandem mass spectrometry method for the quantitation of synthetic cannabinoids of the aminoalkylindole type and methanandamide in serum and its application to forensic samples. J Mass Spectrom 2011; 46: 163–171.

Chapter 6

Abuse of Magic Mushrooms, Peyote Cactus, Khat, and Solvents: No Readily Available Laboratory Tests

6.1. INTRODUCTION

Although life-threatening poisoning due to the abuse of magic mushrooms and peyote cactus is infrequently encountered in emergency departments, solvent and glue abuse is becoming a serious problem among the adolescence population due to its potentially life-threatening toxicity. Even death may result from the abuse of solvents and glue. In general, mushrooms and peyote cactus are hallucinogens. Under the influence of hallucinogens, people see images and feel sensations that are unreal. In addition, subjects may also experience mood swings and feelings of spiritual energy. Solvent and glue abuse may also cause euphoria. Unfortunately, there is no readily available test for the diagnosis of poisoning resulting from the abuse of magic mushrooms, peyote cactus, and solvents. The active components of magic mushrooms and peyote cactus can only be identified by various chromatographic techniques,

Resolving Erroneous Reports in Toxicology and Therapeutic Drug Monitoring: A Comprehensive Guide, First Edition. Amitava Dasgupta.

especially gas chromatography (GC) combined with mass spectrometry (MS), but such sophisticated methods may not be available in community-based hospital laboratories. Certain solvents can be detected using headspace GC, but the indirect determination of serum osmolarity may be helpful in identifying a potential poisoning due to solvent abuse.

6.2. MAGIC MUSHROOM ABUSE

Magic mushrooms (psychoactive fungi) that grow in the United States, Mexico, South America, and many other parts of the world contain the hallucinogenic compounds psilocybin and psilocin. Chemical structures of these two compounds are given in Figure 6.1. Both psilocybin and psilocin along with related compounds in the "tryptamine" class of drugs are classified under class I controlled substances by the U.S. Food and Drug Administration because these compounds have no known medical use but have a very high abuse potential. Unlawful possession of a class I controlled substance is a felony by law in the United States. In 2005 the UK government changed the law and included dried or cooked psychoactive mushrooms under class A drugs. Therefore, it is a criminal offense to possess or sell these products.

There are many species of magic mushrooms containing hallucinogenic compounds, and almost all of them are brown or dark tan. Magic mushrooms can be eaten raw, cooked with food, or dried and then consumed. These mushrooms can be mistaken for other nonhallucinogenic mushrooms or even poisonous mushrooms such as those in the *Amanita* class. Therefore it is a dangerous practice to try to identify such mushrooms for human consumption. There are reported deaths from the ingestion of *Amanita* mushrooms due to mistaken identity (1).

Psilocybin and psilocin are found in over 150 species of mushrooms. The group of psychoactive mushrooms includes species of the genera of *Conocybe, Gymnopilus, Panaeolus, Pluteus, Psilocybe*, and *Stropharia* (2). *Panaeolus cyanescecens* usually contains the highest amounts of psilocybin and psilocin; the

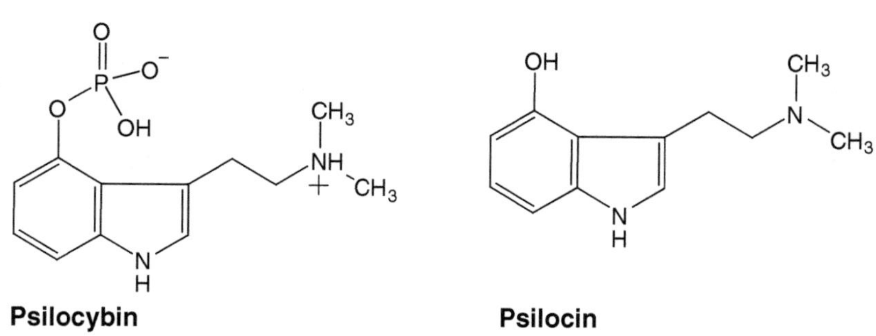

Figure 6.1. Chemical structures of psilocybin and psilocin, the active ingredients in magic mushrooms.

most common type of magic mushroom in Germany is *Psilocybe semilanceata,* which contains less than 0.003–1.15% psilocybin and 0.01–0.90% psilocin (3). The psilocybin contents in *Psilocybe cubeneis* is 0.44–1.35% in the cap of the mushroom and 0.05–1.27% in the stem; the corresponding psilocin concentrations are 0.17–0.78% and 0.09–0.30%, respectively. In general, hallucinogenic alkaloids are found more in the cap of the mushroom than the stem (4). Riley and Blackman surveyed 174 participants in 2004 in the United Kingdom when the sale of magic mushrooms was not prohibited there. Participants reported infrequent but heavy consumption of magic mushrooms (47% used magic mushrooms 4–12 times in a year), and the average consumption in one setting was about 12 g. Although most participants experienced hallucinations, a feeling of altered perception (41–74%) and feeling closer to nature (49%) were common. However, a significant number of participants also had negative experiences including paranoia (35%) and anxiety (32%) (5).

After ingestion, psilocybin, often the major component of the mushrooms, is rapidly converted by dephosphorylation into psilocin, which has psychoactive effects similar to lysergic acid diethylamide (LSD), but the duration of the effect is short (0.5–2 hours after ingestion) (6). In general, the duration of the trip after abusing magic mushrooms may last between 2 and 6 hours, and effects range from the intended feelings of relaxation, uncontrollable laughter, joy, euphoria, and the visual enhancement of colors to hallucination and altered perception of real events. However, some abusers of magic mushrooms may also experience negative effects ("bad trip"), which is probably the main reason a person seeks medical attention at the emergency department. A bad trip is usually followed by faintness, sadness, depression, and paranoid interpretations that may even last for weeks. It has also been speculated that the combined use of magic mushrooms with other psychoactive drugs or alcohol increases the risk of a bad trip (7). Table 6.1 lists the active components of magic mushrooms and their street names.

6.2.1. Toxicity and Fatality from Using Magic Mushrooms

Intoxication from the use of magic mushrooms is common. Some of the species of magic mushrooms contain phenylethylamine that may cause cardiac toxicity. Beck et al reported that the amount of phenylethylamine may vary much more than psilocybin in magic mushrooms. Three young men were hospitalized

TABLE 6.1. Active Components and Street Names of Magic Mushrooms, Peyote Cactus, and Khat

Abused Substance	Active Ingredient	Street Names
Magic mushroom	Psilocybin, psilocin	Mushies, Shroom, Liberty cap, Fly Agaric
Peyote cactus	Mescaline	Cactus, Buttons, Peyote, Mesc
Khat	Cathinone, cathine	Khat, Qat, Kat, Chat, Miraa

due to an adverse reaction from using magic mushrooms. The highest amount of phenylethylamine found in the mushrooms (*Psilocybe semilanceata*) was 146 μg/g wet weight (8).

Case Study

A 20-year-old woman was admitted to the hospital due to nausea, vomiting, and abdominal pain that started 5 days earlier after ingesting magic mushrooms. However, she did not experience any of the desired hallucinogenic effects. She visited her family physician 2 days after eating magic mushrooms due to abdominal pain and was diagnosed with a urinary tract infection. The physician prescribed trimethoprim in combination with sulfamethoxazole. Because her symptoms did not improve after 2 days of treatment, she decided to visit the emergency department. She was given ampicillin intravenously and sent home. The patient returned to the hospital the next day with worsening symptoms and was admitted. Her electrolytes were normal except for potassium (5.3 mmol/L), and her liver function tests were normal. However, the patient was uremic during admission with a serum creatinine level of 356 μmol/L (4.7 mg/dL) and blood urea nitrogen of 10.1 mmol/L (60.2 mg/dL). Over the next week the patient's diuresis increased, and serum creatinine as well as urea levels decreased to normal levels (creatinine: 99 μmol/L, 1.3 mg/dL; urea: 6.0 mmol/L, 36.1 mg/dL); her body weight was reduced by 4 kg. The patient was discharged with normal renal function and mild proteinuria (9).

Case Study

A 25-year-old man infected with hepatitis C with a history of heroin, opiate, and cannabis abuse experienced severe rhabdomyolysis, acute renal failure, and posterior encephalopathy after the abuse of the magic mushroom *Psilocybe cubensis*. The patient was first accompanied to the medical center by his mother, and he experienced abdominal discomfort, hallucinations, and was unusually aggressive. The initial diagnosis was intoxication with Angel's trumpet (*Datura swaveolens*), and he was treated with physostigmine (2 mg) and midazolam (5 mg), both administered intravenously. But the patient also had high creatinine kinase (500,000 U/L) and creatinine (4.6 mg/dL). His parents were advised to search the house for illicit drugs, and they found the magic mushroom identified as *P. cubensis*. This species was one of the most commonly available magic mushrooms in the German market. Methadone was administered to avoid withdrawal symptoms, and the patient was also treated with isovolemic alkaline diuresis for rhabdomyolysis. In addition, furosemide was administered to achieve volume homeostasis. On the seventh day of admission, the patient complained of a complete loss of vision in both eyes. The cranial computed tomography showed bilateral hypodense lesions involving white and gray matter in the parieto-occipital and frontal region with general cerebral edema and flat gyral structure. The patient was put on dialysis, and on day 15 of admission the frequency of the dialysis treatments was reduced. Six months later, the patient was fully oriented, regained his vision, and was participating in an ambulatory methadone rehabilitation program. The authors commented that severe rhabdomyolysis and myocarditis is rarely encountered after *P. cubensis* abuse, but such serious toxicity is the risk of abusing magic mushrooms (10).

Fatality from abusing magic mushrooms alone is rare. Nevertheless there is also a report of a fatal case due to magic mushroom poisoning (11). McClintock et al commented that recognizing the abuse of magic mushrooms is difficult because they are perceived as a natural substance, and toxicity from using such products is underreported. However, toxicity from psilocybin may resemble the abuse of LSD because psilocybin also acts by inhibiting the firing rate of serotonergic neurons. Although subjects may feel alteration in normal perception, mood, and behavior within 15–30 minutes of abusing magic mushrooms, toxic side effects such as nausea, vomiting, diaphoresis, and mydriasis may also be observed after abuse. In addition, more serious toxic effects such as tachycardia, hyperthermia, and convulsion are also possible, and such effects can last as long as 24 hours after ingestion, although most effects are resolved 12 hours after ingestion. Unfortunately, there is no antidote for magic mushroom poisoning, and only supportive care is the standard medical practice for treating magic mushroom poisoning (12). Amsterdam et al recently reviewed the literature on the physical and psychological dependence of magic mushrooms along with their acute and chronic toxicity (13).

6.2.2. Diagnosis of Magic Mushroom Poisoning: Role of the Toxicology Laboratory

Unfortunately, there is no commercially available immunoassay for detecting the active ingredients of magic mushrooms in serum or urine. However, Tiscione and Miller identified psilocin in a urine specimen during a routine investigation of driving under the influence using routine immunoassays for drugs of abuse testing. The urine specimen initially showed a positive result with the fluorescence polarization immunoassay (FPIA) for screening for amphetamine/methamphetamine in urine. The authors determined that at a concentration of 50 mg/L, the cross-reactivity of psilocin with the amphetamine immunoassay is 1.3%. The authors also confirmed the presence of psilocin in the urine specimen using GC/MS. Interestingly, the driver whose urine was tested did not exhibit any response to the automobile crash, indicating he may have been unaware of the severity of the situation or his immediate surroundings. The urine concentration of psilocin decreased rapidly despite the fact that the specimen was stored at 4°C (14).

Other than this report, there is no other report in the literature regarding using a commercially available immunoassay for detecting an active ingredient of magic mushrooms. Therefore, only chromatographic methods are available for the diagnosis of magic mushroom poisoning using laboratory tests. Both GC and high performance liquid chromatographic (HPLC) methods are available for the analysis of the active ingredient of magic mushrooms. Unfortunately, such specialized methods are available only in a few reference laboratories and major academic medical centers. Most hospital-based laboratories do not have the capability of offering such a speciality technique. The active component of magic mushroom psilocybin is rapidly converted into psilocin, and eventually psilocin is conjugated in the liver and excreted as glucuronide in the urine. In one report, the concentration of total psilocin in serum was 52 ng/mL and the total psilocin concentration in urine was 1760 ng/mL in one person who consumed magic mushrooms (15). In another report the authors used enzymatic hydrolysis using the enzyme β-glucuronidase to liberate free psilocin from its conjugated form in urine specimens and then used N-methyl-N-trimethylsilyltrifluoroacetamide (MSTFA) for the derivatization of psilocin to its trimethylsilyl derivative to determine the concentration of psilocin by GC/MS (16). Psilocin can also be analyzed by GC/MS after derivatization with a BSTFA reagent (N,O-bis-trimethylsilyl-acetamide containing 1% trimethylchlorosilane) to covert psilocin to its trimethylsilyl ester (psilocin-1,4-di-trimethylsilyl). Figure 6.2 shows the full scan mass spectrum of trimethylsilyl derivative of psilocin. In the mass spectrum, a distinct molecular ion peak at m/z 348 was observed. Another strong peak at m/z 290 (M^+-58) was due to a loss of the $N(CH_3)_2CH_2$ group from the side chain of the derivatized psilocin.

However, there are only a few reports of an analysis of psilocin using GC, whereas many investigators have used HPLC techniques to analyze psilocin in serum, urine, and other body fluids. Kamata et al used liquid chromatography combined with MS or tandem mass spectrometry for the analysis of psilocin and psilocybin in magic mushroom samples, and the authors used a C-18 reverse phase column for liquid chromatographic analysis. The detection limits ranged from 1 to 25 pg by operating the mass spectrometer in selected ion monitoring mode (17). For the identification of psilocin in urine, a hydrolysis step is recommended to obtain free psilocin by breaking the glucuronide complex. The hydrolysis of psilocin glucuronide can be achieved by the acid, alkaline, or enzymatic method. However, in one report, the authors identified psilocin glucuronide in the urine of a magic mushroom user using both liquid chromatography combined with mass spectrometry (LC-MS) and liquid chromatography combined with tandem mass spectrometry. However, after enzymatic hydrolysis of the psilocin glucuronide using β-glucuronidase enzyme, the authors found 3.55 μg/mL of psilocin in the urine of a subject who abused magic mushrooms (18).

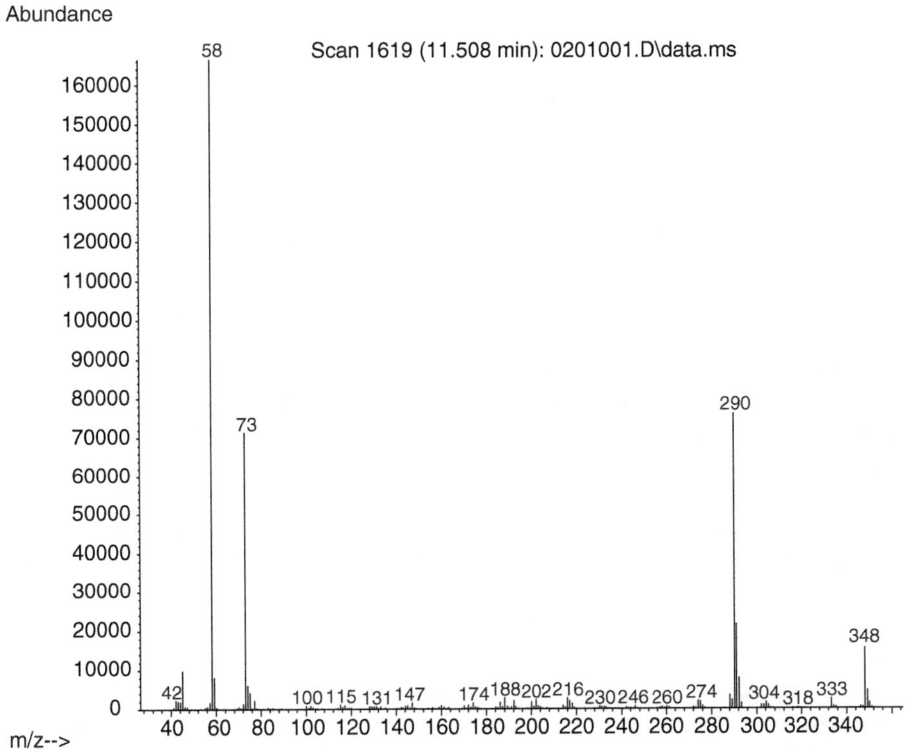

Figure 6.2. Mass spectrum of trimethylsilyl derivative of psilocin. Courtesy of Dr. Buddha D. Paul, Office of the Armed Forces Medical Examiner (OAFME), Forensic Toxicology Division of the Armed Forces Institute of Pathology, Rockville, MD.

6.3. PEYOTE CACTUS: USE AND ABUSE

Peyote cactus (*Lophophora williamsii*) is a small spineless cactus that grows in the U.S. Southwest and Mexico. Peyote cactus is round without sharp spines, and the top of the cactus above ground is called the crown, which consists of disk-shaped buttons where the psychoactive compound mescaline (3, 4, 5-trimethoxyphenethylamine) can be found. The chemical structure of mescaline is given in Figure 6.3. Because of the known psychoactive properties of peyote cactus, it is used worldwide as a supplement to various spiritual rituals and transcendence practices including meditation and psychedelic psychotherapy. Native North American Indians recognized the psychotropic properties of peyote as long as 5700 years ago. El-Seedi et al analyzed two archaeological specimens of peyote buttons from the collection of the White Museum in San Antonio, Texas, using carbon dating and observed that the buttons dated back to 3780–3660 BC. The authors also extracted alkaloids from

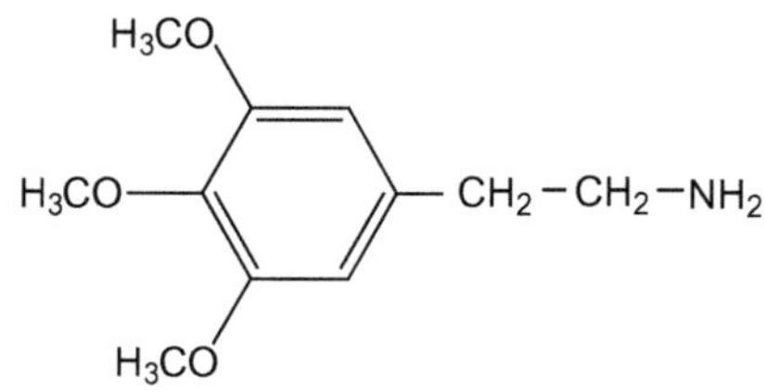

Figure 6.3. Chemical structure of mescaline, the active component of peyote cactus.

the buttons and demonstrated the presence of mescaline, the active ingredient of peyote (19). Peyote cactus flowers from March through May and sometimes as late as September. After flowering, the peyote cactus produces small edible fruits. Mescaline can be extracted from the peyote cactus or buttons. Alternatively, it can be chewed or soaked in water to produce a psychoactive liquid. Mescaline is classified as a class I controlled substance, but approximately 300,000 members of the Native Americans Church can ingest peyote cactus legally as a religious sacrament during all-night prayer. The members attend the prayer ceremonies as regularly as two to three times a week or as infrequently as once a year (20). The mescaline content of peyote cactus is usually 0.4% in fresh cactus and 3–6% in dried cactus. Ogunbodede et al analyzed the mescaline concentrations in 14 species of cactus and observed that mescaline concentrations varied between 0.053% and 4.7% by dry weight. The authors determined mescaline concentrations by Soxhlet extraction with methanol followed by acid-base extraction with water and dichloromethane and then analysis by HPLC (21). Although mescaline has the lowest potency among naturally occurring hallucinogens, a full dose (200–400 mg) has a long duration of action. The highest psychedelic effect may be achieved within 2 hours of ingestion, but the effect may last as long as 8 hours (22). The psychoactive effects of mescaline are similar to LSD including deeply mystical feelings. Abuse of peyote cactus may cause serious toxicity requiring medical attention and may even cause fatality. Botulism can be a consequence if peyote buttons are stored in water (23). Carstairs and Cantrell performed a retrospective review of the California Poison Control system database for the years 1997–2008 for all cases of single substance human exposure of peyote cactus and mescaline. There were 31 cases of exposure to peyote or mescaline; 97% of such exposures (30 subjects) were intentional, and subjects took such substances orally. One subject insufflated mescaline powder. Of 31 subjects, 16 were managed at home; the rest were treated at medical facilities. Commonly reported symptoms were hallucination, tachycardia, agitation, and mydriasis; vomiting was reported in one subject. The authors concluded that although abuse of peyote or mescaline is less common compared with other illicit drugs, abuse of mescaline is associated with significant toxicity in a number of patients requiring treatment in a medical facility (24). Table 6.1 lists the active components and street names of peyote cactus.

6.3.1. Diagnosis of Peyote Cactus Poisoning: Role of the Toxicology Laboratory

Currently, there is no commercially available immunoassay for the analysis of mescaline in body fluids, but GC/MS or LC-MS can be used to determine the concentrations of mescaline in the biological matrix. Alternatively, if anyone accompanying the patient brings in the suspected species, peyote cactus can be identified based on differences in morphology, although that might need the expertise of a botanist. Various peyote cactus species can also be identified based on mescaline content (25), but again such a process requires sophisticated chromatographic methods not readily available in many hospital laboratories.

Although mescaline is metabolized to several different metabolites, a large amount of mescaline can be recovered unchanged in urine. Therefore, the detection of mescaline in serum or urine can be used for establishing the abuse of magic mushrooms. For a death investigation, confirmation of the presence of mescaline in body fluid is essential to establish the cause of death. Severe toxicity and even death may occur from mescaline overdose.

Case Study

A 53-year-old American Indian man who was purportedly using peyote cactus during a religious ceremony with other participants was shot multiple times by another participant. The cause of death was a gunshot wound. His mescaline level in blood was lower than the concentration of mescaline found in a fatal overdose. His femoral blood showed a mescaline concentration of 2.95 mg/L, vitreous had a concentration of 2.36 mg/L, and the concentrations of mescaline in liver and brain tissues were 8.2 mg/kg and 2.2 mg/kg, respectively. The authors confirmed the presence of mescaline in his blood and tissues using GC/MS, and quantitative analyses were performed using GC combined with a nitrogen-phosphorus detector. The authors extracted mescaline from blood or tissue specimens using butyl chloride. For identification and qualitative analysis of mescaline, the authors operated a mass spectrometer in the selected ion monitoring mode. The primary ion used for positive identification was m/z 182 with secondary ions at m/z 167 and 181. For quantitative analysis, the authors used GC combined with a nitrogen phosphorus detector (26).

Another person who died under the influence of mescaline showed 9.7 μg/mL of drug in serum and 1163 μg/mL of drug in urine. The corresponding drug concentration in liver was 70.8 μg/g (27). Nolte and Zumwalt reported a case of a 32-year-old Native American man with a history of alcohol abuse who ingested peyote tea. Later he developed respiratory distress and suddenly collapsed. His antemortem blood specimen showed the presence of mescaline (0.48 mg/L), and the concentration of mescaline in urine was 61 mg/L. Other than a trace amount of chlordiazepoxide, no other drug or ethanol was in his blood or urine (28).

Mescaline can be easily analyzed by GC/MS without derivatization. The electron impact mass spectrum of mescaline is shown in Fig 6.4. A distinct molecular ion peak was observed at m/z 211. However, liquid chromatography combined with mass spectrometry can also be used by detection and quantification of mescaline in biological fluids. In one report, the authors achieved liquid chromatographic analysis of mescaline using a reverse phase C-18 column. The mobile phase was a methanol gradient in an ammonium acetate buffer. The tandem mass spectrometric identification of mescaline and the deuterated internal standard (mescaline-d9) was achieved by tandem mass spectrometry operated in the selected reaction monitoring mode. The transition monitored for mescaline was m/z 212.3 → m/z 180.3, and for the deuterated standard the transition ions were m/z 221.3 → m/z 186.3. The detection limit of mescaline in urine was 3.5 ng/mL (29). Sergi et al described a HPLC combined with the tandem mass spectrometric method for a simultaneous determination of mescaline and psilocybin along with amphetamine, methamphetamine, methylenedioxymethamphetamine, methylenedioxyethylamphetamine, cocaine, benzoylecgonine, and ketamine in oral fluid. The quantitation was performed by detecting analytes in positive ionization mode by means of an electrospray source (30).

6.4. ABUSE OF KHAT AND RELATED COMPOUNDS

Khat (*Catha edulis Forsk*) is a flowering evergreen shrub native to East Africa that is also found in Arabian countries. It has been used widely for recreational purposes since the thirteenth century by native people in East Africa and the Arabian Peninsula mostly by chewing fresh or dried leaves or shoots like tobacco. These plant materials are bitter, and abusers often ingest water or a sweet drink to counteract the taste. Rarely people smoke dried leaves or shoots. Although sold in African countries in the open market, khat is illegal in the United States because it has no known medical use. Abuse of khat in the United States is mostly by immigrants from Somalia, Ethiopia, and Yemen.

The major active constituents of khat are cathine and cathinone [S(-)-alpha-aminopropiophenone), a sympathomimetic amine structurally similar to amphetamine. Cathine, also known as d-norpseudoephedrine, is approximately 10 times less potent than cathinone as a central nervous system (CNS) stimulant. Chemical structure of cathinone is given in Fig 6.5. Cathinone is a schedule I drug and cathine is a schedule IV drug per the U.S. Controlled Substance Act.

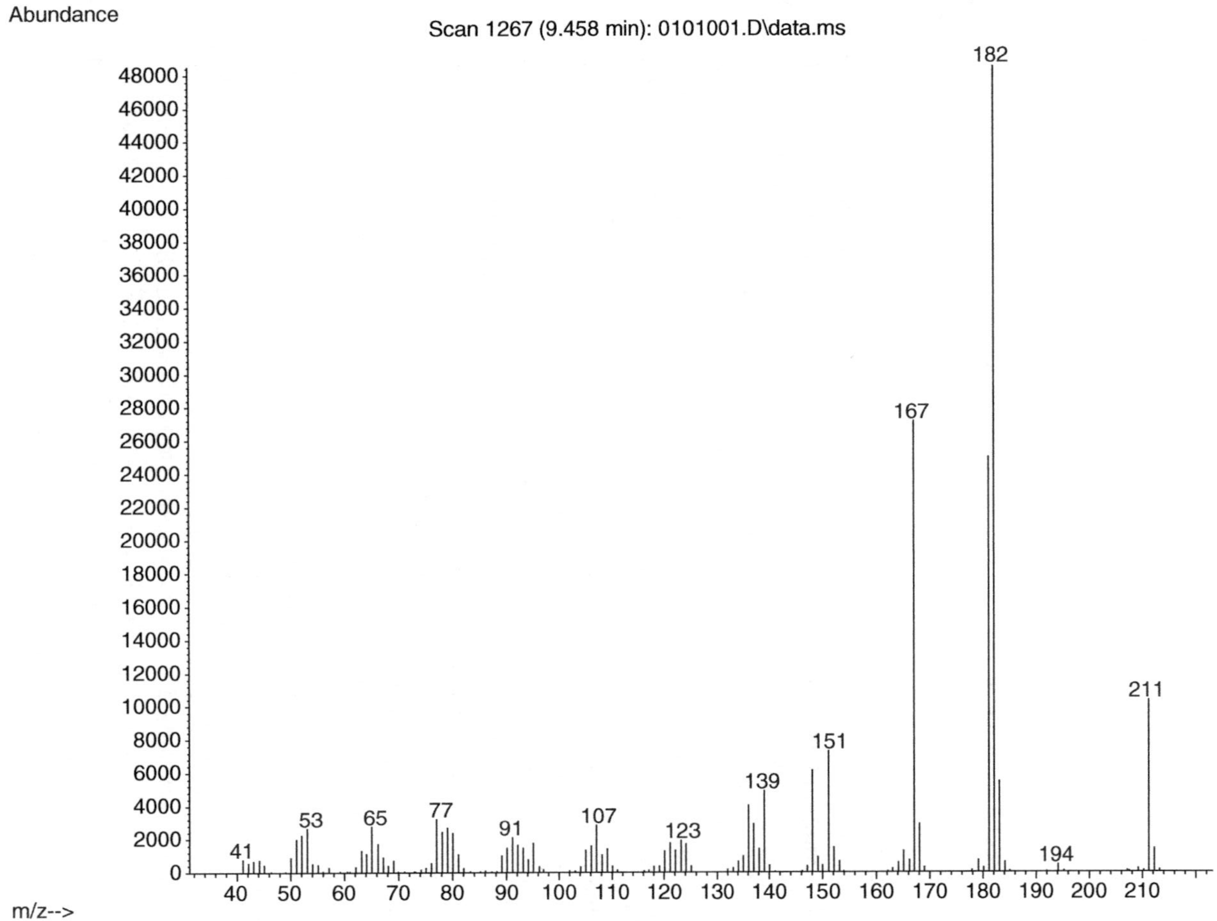

Figure 6.4. Mass spectrum of mescaline. Courtesy of Dr. Buddha D. Paul, Office of the Armed Forces Medical Examiner (OAFME), Forensic Toxicology Division of the Armed Forces Institute of Pathology, Rockville, MD.

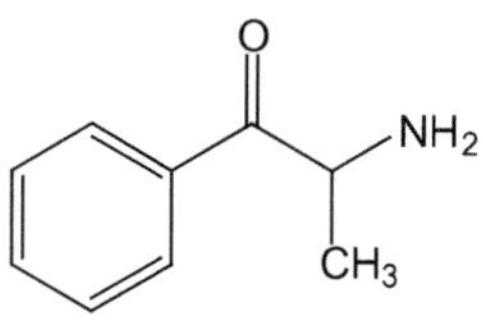

Cathinone

Figure 6.5. Chemical structure of cathinone, the active component of khat and amphetamine.

The concentration of cathinone is highest in the fresh-cut khat plant, and then the concentration starts declining with time. In addition to these most active components, khat also contains other weaker psychoactive compounds such as phenylpropanolamine, norephedrine, and norpseudoephedrine. Another name for khat is "herbal ecstasy" because of its CNS stimulant effects similar to amphetamine. Khat abuse produces amphetamine-like effects such as euphoria, a feeling of increased alertness and energy, hyperactivity, anorexia, and lack of fatigue. The effects usually last from 90 minutes to 3 hours, but toxicity from khat abuse also has similar features of amphetamine abuse. Cardiovascular complications from khat use are also similar to those observed with amphetamine abuse (31). Increased blood pressure and increased heart rates are also observed in individuals abusing khat (32). Hagigat capsules (containing 200 mg cathinone produced illicitly) are available in the streets in Israel as a natural stimulant and aphrodisiac. Bentur et al studied the data of 34 consecutive patients (age 16–54 years) who consumed half to six capsules (ingestion-32, snifinf-2). The major complications of use of cathinone were myocardial ischemia, pulmonary edema, and intracerebral hemorrhage. The authors concluded that exposure to illicitly synthesized cathinone may cause serious cardiovascular and neurological toxicity even in young subjects (33). Severe acute liver injury may be associated with khat abuse (34). Khat chewing can also cause stroke (35). In another report, based on the study of 8176 patients including 934 patients who were khat chewers, the authors concluded that khat chewing was associated with a significantly higher risk of cardiogenic shock, stroke, and mortality. Khat chewing as also an independent risk factor for in-house mortality (36). Street names of Khat are given in Table 6.1.

Methcathinone is an illicit drug also known on the street as "cat" or "ephedrine." This illicit drug is a methyl derivative of cathinone, the natural product found in the khat plant. Methcathinone can be synthesized by the oxidation of pseudoephedrine in clandestine laboratories. Abuse of methcathinone can produce neuropsychiatric symptoms including agitation, insomnia, and tremors. Belhadj-Tahar and Sadeg reported the case of a 29-year-old woman who was admitted to the hospital with coma due to drug toxicity. The family indicated that she was taking an amphetamine-like drug. Her blood alcohol level was 167 mg/dL, and analysis of her urine showed the presence of benzodiazepines and high levels of amphetamines (methcathinone, 17.24 mg/L; ephedrine, 11.60 mg/L; and methylephedrine, 11.10 mg/L). The analysis of serum also showed the presence of methcathinone (0.50 mg/L), methylephedrine (0.19 mg/L), and bromazepam (8.89 mg/L). The authors concluded that ingestion of alcohol and bromazepam altered the typical symptoms of methcathinone-induced intoxication, namely hypertension and convulsion. The authors further stated that the identification of methylephedrine, which is an impurity in the synthesis of methcathinone, can serve as a tag indicating fraudulent synthetic origin of methcathinone (37). Mephedrone (4-methylmethcathinone), another synthetic product structurally similar to cathinone, is also abused.

Case Study

A 22-year-old man purchased mephedrone powder over the Internet from a chemical supplier based in China. He initially ingested 200 mg powder but experienced no euphoria. Therefore, he decided to inject the remaining 3.8 g powder into his thigh intramuscularly. Shortly after injecting the powder he experienced severe toxic reactions including palpitation, blurred tunnel vision, chest pressure, and sweating. After admission to the hospital, he received a single dose of oral lorazepam and felt better over the next 4 hours. Qualitative analysis of his urine and serum samples by GC/MS identified the presence of 4-methylmethcathinone. Quantitative analysis of his serum sample using liquid chromatography combined with MS showed a 4-methylmethcathinone concentration of 0.15 mg/L (38).

6.4.1. Detecting Khat and Related Substances in Body Fluids: Role of the Toxicology Laboratory

Cathinone and methcathinone cross-react with some amphetamine immunoassays because of the structural similarities of these compounds with amphetamine. The presumptive presence of cathinone and methcathinone in urine should be confirmed by using GC/MS. Paul and Cole reported the presence of cathinone in two urine specimens (range: 118–3266 ng/mL) screened positive for amphetamines and methcathinone in six urine specimens (range: 13–91 ng/mL), which also screened positive for amphetamines using immunoassay (39). However, cathinone and methcathinone if present in urine do not produce a positive screen with all amphetamine assays. In one report, authors observed a negative response in all urine specimens except one using the FPIA screening test for amphetamines (Abbott Laboratories) where the presence of cathinone in urine was suspected. In contrast to the FPIA assay, the Mahsan-AMP(300) on-site immunoassay (Mahsan Diagnostika, Reinbeck, Germany) provided positive tests for amphetamine in seven of eight cases. The concentration of cathinone varied from 0.1 mg/L to 28.8 mg/L in urine specimens. The authors commented that the on-site test for amphetamine is more sensitive than FPIA for detecting the presence of cathinone in urine specimens (40).

6.5. SOLVENT AND GLUE ABUSE

Solvent and glue sniffing are common among adolescents throughout the world, but this problem is often overlooked. In the United States, approximately 20% of adolescents have tried inhalants at least once by the time they reach eighth grade (41). Inhalants abused include solvents, glues, adhesives, and paint thinners, fuels, and propellants (petroleum products), and such compounds can be abused either by breathing directly from a container or by soaking a rag with the solvent and then placing it over the nose and mouth. Alternatively, fumes from such solvents can be inhaled indirectly. However, all such methods result in euphoria of different degrees, and the euphoric effect can even be intensified if the abuser can become hypoxic by rebreathing from a bag (42).

Various readily available household and office products abused include glue, adhesives, nail polish, nail polish remover, cigarette lighter fluid, butane gas, gas (petrol), air fresheners, deodorant, hair spray, pain-relieving spray, typewriter correction fluid, paint thinners, paint removers, and a variety of other agents. These household and office products contain toxic solvents such as toluene (paint, spray paint, adhesives, paint thinner, shoe polish), acetone (nail polish remover, typewriter correction fluid and markers), hexane (glue, rubber cement), chlorinated hydrocarbon (spot and grease removers), xylene (permanent markers), propane gas (gas to light the grill, spray paints), butane gas (lighter fluid, spray paint), and fluorocarbons (hair spray, analgesic spray, refrigerator coolant such as Freon). Chemical compositions of commonly abused volatile substances and inhalants are summarized in Table 6.2.

In one study the authors investigated 318,393 exposures and concluded that exposure to hydrocarbons that are systematically absorbed and have low viscosity such as benzene, toluene, xylene, halogenated hydrocarbon, kerosene, and lamp oil caused the highest hazard values (43). Chronic abuse of toluene-containing products and chlorinated solvents can produce severe liver, kidney, and brain damage (44). Recently, Gupta et al reported that occupational or purposeful exposure to toluene can lead to neurological abnormalities. Timely diagnosis of toluene exposure as the cause is important because patients may experience improvements in neurological and ocular manifestation of toluene toxicity after discontinuation of exposure. In contrast, continued inhalation of toluene or abuse may result in the permanent loss of neurological functions (45).

Case Study

After running away from home at age 10, a 22-year-old male patient mostly lived on the streets of Istanbul. During the last 10 years he abused volatile

TABLE 6.2. Commonly Abused Volatiles and Inhalants

Volatile/Inhalant	Chemical Composition
Spray paint	Propane, butane, toluene, hydrocarbons fluorocarbons
Hair spray	Propane, butane, fluorocarbons
Room freshner	Proane, butane, fluorocarbons
Analgesic spray	Fluorocarbons, isopropyl alcohol
Nail polish remover	Acetone, toluene
Paint thinner	Toluene, methyl chlorodie, methanol
Lighter fluid	Butane, isopropane
Fuel	Butane
Gasoline	Combination of aliphatic and aromatic hydrocarbons and other organic compounds with low molecular weight
Glue	Toluene, hexane
Correction fluid	Acetone
Rubber cement	Toluene, hexane, methyl chloride, acetone
Spot remover	Trichloroethane, trifluroethane, trichloroethylene
Refrigerator fluid	Trichlorofluoromethane (Freon)

From Dasgupta A. Abuse of magic mushrooms, peyote cactus, khat, and volatiles. In: Dasgupta A, ed., *Critical Issues in Alcohol and Drugs of Abuse Testing.* Washington, DC: AACC Press, 2009:285. Reprinted with permission.

substances (paint thinner and glue) almost every day. At the time of his first examination he was living under the auspices of a nonprofit organization and presented at the infirmary there with a complaint of shortness of breath during activities and swelling as well as pain in his abdomen. The infirmary medical staff, noticing his irregular pulse, referred him to a hospital. His blood pressure on admission was 110/60 mm Hg; his pulse rate was irregular at 55 beats/minute. His blood tests did not reveal any abnormality, but based on electrocardiogram and echocardiogram analyses, he was diagnosed with cardiomyopathy. His cardiomyopathy was probably related to solvent abuse because both paint thinners and glue in Turkey have 50–70% and 35–40% toluene, respectively, although the institution where he was living during this episode claimed he had not abused any solvent in the last 6 months. His condition improved in the next 6 months as he stopped abusing any solvent. Toluene has direct cardiac toxicity and may also render the heart susceptible to endogenous catecholamines or lipid peroxidation. In addition, coexisting electrolyte abnormalities caused by renal toxicity associated with toluene abuse may contribute to the development of arrhythmias that may even be fatal (46).

Case Study

A 14-year-old healthy girl was found by her parents in a state of confusion. She was laughing and crying without any reason and lost perception of time and place. She was taken to the emergency department, and 3 hours later she regained consciousness and was cooperative but also complained about weakness, dizziness, and headache. Her blood pressure was 120/70 mm Hg, pulse 80 beats/minute and regular, and the rest of her physical examination and electrocardiogram were unremarkable. She admitted sniffing glue for the last 5 days prior to admission but denied abuse of any illicit drug. Her urinary hippuric acid level 4 hours after her parents found her disoriented was 93.9 g/g of creatinine. However, her serum electrolytes, creatinine, urea, anion gap, osmolality, and blood gas results were all within normal ranges. She was discharged 24 hours after admission with a referral for psychiatric follow-up. Glue contains toluene that is absorbed with a peak concentration achieved within 15–30 minutes after exposure. About 80% of toluene is metabolized to benzyl alcohol by the liver, which is then oxidized to benzoic acid and conjugated with glycine and finally excreted in the urine as hippuric acid. Because blood or urinary toluene levels are transient and difficult to measure, the authors commented that urinary excretion of hippuric acid may aid in the diagnosis of toluene exposure (47).

Solvent abusers often present with nonspecific symptoms, but long-term abusers may demonstrate a wide range of neuropsychiatric symptoms. The common toxicities encountered in solvent abusers are listed in Table 6.3. The exact mechanism of action for the volatiles and inhalants is not completely understood, but volatiles in general produce a slowing of the axonal ion channel transport by increasing fluidity of membranes. Another possibility is that volatiles potentiate hyperpolarization of γ-aminobutyric acid receptors (GABA receptors). Due to their lipophilic nature, inhalants can quickly produce CNS depression that results in slurred speech, diplopia, ataxia, visual hallucination, and disorientation. Further CNS depression can cause respiratory arrest, seizures, and coma (48). Shu and Tsai reported six cases where subjects sniffed glue for a long time. The authors observed that long-term abuse of glue is often associated with violent behavior and/or self-mutilation. Functional deterioration characterized by impairment of intelligence was also observed among these subjects, and some abusers also developed psychosis

TABLE 6.3. Adverse Effects of Solvent and Inhalant Abuse

Toxicity	Agents
Cardiotoxicity	Inhalants such as fluorocarbons, gasoline, benzene, toluene, butane, propane (acute effect of inhalants on the heart may be fatal)
Pulmonary	Some hydrocarbons may cause chemical pneumonitis
Renal	Toluene, hydrocarbons
Hepatic	Halogenated hydrocarbons such as carbon tetrachloride, trichloroethane, toluene
Teratogenic	Inhalants due to lipophilic nature cross placenta
Neurological	Various inhalants such as toluene, hexane, methyl isobutyl ketone

From Dasgupta A. Abuse of magic mushrooms, peyote cactus, khat, and volatiles. In: Dasgupta A, ed., *Critical Issues in Alcohol and Drugs of Abuse Testing*. Washington, DC: AACC Press, 2009:285. Reprinted with permission.

(49). The most serious consequence of solvent abuse is death, which may occur secondary to aspiration or asphyxia. Steffee et al reported two cases of fatal volatile solvent inhalation abuse: gasoline sniffing in a 20-year-old man and aerosol air freshener inhalation in a 16-year-old girl (50). Schloneger et al reported a case of an 18-year-old man who was presented to a community hospital emergency department with shortness of breath and fever. Despite treatment his condition deteriorated, and he admitted deliberate inhalation of keyboard cleaner. He died on day 11 from alveolar hemorrhage, a known complication of inhalant abuse (51).

Case Study

A 42-year-old man was found dead after repeated exposure to 1,1-difluoroethane, a propellant found in a duster intended for removal of dust and lint. The postmortem toxicology analysis showed a femoral blood level of 136.3 mg/L; concentrations of 1,1-difluroethane in brain, liver, lung, and adipose tissue were 117.5 mg/kg, 87.3 mg/kg, 60.3 mg/kg, and 235.7 mg/kg, respectively. The authors also detected 1,1-difluoroethane in the vitreous fluid (25.1 mg/L). The cause of death was established as cardiac arrhythmia due to intoxication with 1,1-difluoroethane. The authors also commented that due to the accumulation of 1,1-difluoroethane in adipose tissue, this may be a valuable alternative specimen for forensic investigation where loss of elimination of blood may occur (52).

6.6. CONCLUSIONS

Although less common than abuse of common illicit drugs such as marijuana, cocaine, and heroin, recreational abuse of magic mushrooms and peyote cactus can cause serious toxicity including fatality. Unfortunately, there is no readily available test in clinical toxicology laboratories for the rapid diagnosis of such poisoning. Similarly, solvent abuse is a less publicized problem, although it can cause significant organ damage including death. Again blood tests are often unremarkable in patients abusing solvents. Although hippuric acid in urine can be used for the diagnosis of toluene abuse, this specialized test using chromatography is available in only a few major academic medical centers and reference laboratories.

REFERENCES

1. Madhok M. Amanita bisporgera. Ingestion and death from mistaken identity. Minn Med 2007; 90: 48–50.
2. Reingardiene D, Vilcinskaite J, Lazauskas R. Hallucinogen mushrooms. Medicina (Kaunas) 2005; 41: 1067–1070.

3. Musshoff F, Madea B, Beike J. Hallucinogenic mushrooms on the German market-simple instructions for examination and identification. Forensic Sci Int 2000; 113: 389–395.
4. Tsujikawa K, Kannamori Y, Iwata Y, Ohmae Y et al. Morphological and chemical analysis of magic mushrooms in Japan. Forensic Sci Int 2003; 138: 85–90.
5. Riley SC, Blackman G. Between prohibitions: patterns and meaning of magic mushroom use in the UK. Subst Use Misuse 2008; 43: 55–71.
6. Anastos N, Barnett NW, Lewis SW, Gathergood N et al. Determination of psilocin and psilocybin using flow injection and tris (2,2′-bipyridyl) ruthenium (II) chemiluminescence detection respectively. Talanta 2005; 67: 354–359.
7. Sarato L, Goszcz H, Ciszowski K. Poisoning resulting from the ingestion of magic mushrooms in Krakow. Przegl Lek 2005; 62: 394–396.
8. Beck O, Helander A, Karlson-Stiber C, Stephansson N. Presence of phenylethylamine in hallucinogenic Psilocybe mushroom: possible role in adverse reactions. J Anal Toxicol 1998; 22: 45–49.
9. Raff E, Halloran PF, Kjellstrand CM. Renal failure after eating "magic" mushrooms. Can Med Asso J 1992; 147: 1339–1441.
10. Bickel M, Ditting T, Watz H, Roesler A et al. Severe rhabdomyolysis, acute renal failure and posterior encephalopathy after magic mushroom abuse. Eur J Emerg Med 2005; 12: 306–308.
11. Gonomori K, Yoshioka N. The examination of mushroom poisonings at Akita University. Leg Med (Tokyo) 2003; 5 (Suppl 1): S83-S86.
12. McClintock RL, Watts DJ, Melanson S. Unrecognized magic mushroom abuse in a 28 year old man. Am J Emerg Med 2008; 26: 972.e3–4.
13. Amsterdam J, Oppenhuizen A, Brink W. Harm potential of magic mushroom: a review. Regul Toxicol Pharamcol 2011; 59: 423–429.
14. Tiscione NB, Miller MI. Psilocin identified in a DUID investigation. J Anal Toxicol 2006; 30: 342–345.
15. Sticht G, Kaferstein H. Detection of psilocin in body fluids. Forensic Sci Int 2000; 113: 403–407.
16. Grieshaber AF, Moore KA, Levine B. The detection of psilocin in human urine. J Forensic Sci 2001; 46: 627–630.
17. Kamata T, Nishikawa M, Katagi M, Tsuchihashi H. Liquid chromatography-mass spectrometric and liquid chromatographic-tandem mass spectrometric determination of hallucinogenic indoles psilocin and psilocybin in magic mushroom samples. J Forensic Sci 2005; 50: 336–340.
18. Kamata T, Nishikawa M, Katagi M, Tsuchihashi H. Optimized glucuronide hydrolysis for the detection of psilocin in human urine samples. J Chromtaogr B Analyt Technol Biomed Life Sci 2003; 796: 421–427.
19. El-Seedi HR, DeSmet PA, Beck O, Possnert G et al. Prehistoric peyote use: alkaloids and radiocarbon dating of archaeological specimens of Lophophora from Texas. J Ethnopharmacol 2005; 101: 238–242.
20. Halpern JH, Sherwood AR, Hudson JI, Yugerlum-Todd D et al. Psychological and cognitive effects of long term peyote use among Native Americans. Biol Psychiatry 2005; 58: 624–631.

21. Ogunbodede O, McCombs D, Trout K, Daley P et al. New mescaline concentrations from 14 taxa/cultivars of Echinopsis spp (cactaceae) ("San Pedro") and their relevance to Shamanic practice. J Enthnopharmacol 2010; 131: 356–362.
22. Nichols DE. Hallucinogens. Pharmacol Ther 2004; 101: 131–181.
23. Hashimoto H, Clyde VJ, Parko KL. Botulism from peyote [letter]. N Eng J Med 1998; 339: 203–204.
24. Carstairs SD, Cantrell FL. Peyote and mescaline exposure: a 12 year review of a statewide center database. Clin Toxicol (Phila) 2010; 48: 350–353.
25. Aragane M, Sasaki Y, Nakajima J, Fukumori N et al. Peyote identification on the basis of differences in morphology, mescaline content and trn/trnF sequence between *Lophophora williamsi* and *L. diffusa*. J Nat Med 2011; 65: 103–110.
26. Henry JL, Epley J, Rohrig TP. The analysis and distribution of mescaline in postmortem tissues. J Anal Toxicol 2003; 27: 381–382.
27. Reynolds PC, Jindrich EJ. A mescaline associated fatality. J Anal Toxicol 1985; 9: 183–184.
28. Nolte KB, Zumwalt RE. Fetal peyote ingestion associated with Mallory-Weiss lacerations. West J Med 1999; 170: 328.
29. Bjornstad K, Helander A, Beck O. Development and clinical application of an LC-MS method for mescaline in urine. J Anal Toxicol 2008; 32: 227–231.
30. Sergi M, Compagnone D, Curini R, D'Ascenzo G et al. Micro-solid extraction coupled with high performance liquid chromatography-tandem mass spectrometry for the determination of stimulants, hallucinogens, ketamine and phencyclidine in oral fluids. Anal Chim Acta 2010; 675: 132–137.
31. Kuczkowski KM. Herbal ecstasy: cardiovascular complications of khat chewing in pregnancy. Acta Anaesthesiol Belg 2005; 56: 19–21.
32. Al-Motarreb A, Baker K, Broadley KJ. Khat: pharmacological and medical aspects and its social use in Yemen. Phytother Res 2002; 16: 403–413.
33. Bentur Y, Bloom-Krasik A, Raikhlin-Eisenkraft B. Illicit cathinone (Hagigat) poisoning. Clin Toxicol (Phila) 2008; 46: 206–210.
34. Chapman MH, Kajihara M, Borges G, O'Berine J et al. Severe acute liver injury and khat leaves. N Eng J Med 2010; 362: 1642–1644.
35. Vanwallehhem IE, Vanwalleghem PW, De Bleecker JL. Khat chewing can cause stroke. Cerebrovasc Dis 2006; 22: 198–200.
36. Ali WM, Zubaid M, Al-Motarred A, Singh R et al. Association of khat chewing with increased risk of stroke and death in patients presenting with acute coronary syndrome. Mayo Clin Proc 2010; 85: 974–980.
37. Belhadj-Tahar H, Sadeg N. Methcathinone; a new postindustrial drug. Forensic Sci Int 2005; 153: 99–101.
38. Wood DM, Davies S, Puchnarewicz M, Button J et al. Recreational use of mephedrone (4-methylmethcathinone) with associated sympathomimetic toxicity. J Med Toxicol 2010; 6: 327–330.
39. Paul BD, Cole KA. Cathinone (Khat) and methcathinone (CAT) in urine specimens: a gas chromatography-mass spectrometric procedure. J Anal Toxicol 2001; 25: 525–530.
40. Toennes SW, Kauert GF. Driving under the influence of khat-alkaloid concentrations and observations in forensic cases. Forensic Sci Int 2004; 140: 85–90.

41. 2001: Monitoring the future survey results: smoking among teenagers decreases sharply and increase on ecstasy use slows. US Department of Health and Human Services News. December 19, 2001. http://www.nida.nig.gov/MedAdv/01/NR-12-19.html.
42. Anderson CE, Loomis GA. Recognition and prevention of inhalant abuse. Am Fam Physicians 2003; 68: 869–874.
43. Cobaugh DJ, Seger DL, Krenzelok EP. Hydrocarbon toxicity: an analysis of AAPCC TESS data. Przegl Lek 2007; 64: 194–196.
44. Flanagan RJ, Ives RJ. Volatile substance abuse. Bull Narc 1994; 46: 49–78.
45. Gupta SR, Palmer CA, Cure JK, Balos LL et al. Toluene optic neurotoxicity: magnetic resonance imaging and pathological features. Hum Pathol 2011; 42: 295–298.
46. Vurul M, Ogel K. Dilated cardiomyopathy associated with toluene abuse. Cardiology 2006; 105: 158–161.
47. Raikhlin-Eisenkraft B, Hoffer E, Baum Y, Bentur Y. Determination of urinary hippuric acid in toluene abuse. J Toxicol Clin Toxicol 2001; 39: 73–76.
48. Lorens JD. Inhalant abuse in the pediatric population: a persistent challenge. Curr Opin Pediatr 2003; 15: 204–209.
49. Shu LR, Tsai SJ. Long term glue sniffing: report of six cases. Int J Psychiatry Med 2003; 33: 163–168.
50. Steffee CH, Davis GJ, Nicol KK. A whiff of death: fatal volatile solvent inhalation abuse. South Med J 1996; 89: 879–884.
51. Schloneger M, Stull A, Singer JI. Inhalant abuse: a case of hemoptysis associated with halogenated hydrocarbon. Pediatr Emerg Care 2009; 25: 754–757.
52. Avella J, Wilson JC, Lehrer M. Fatal cardiac arrhythmia after repeated exposure to 1,1-difluoroethane (DFE). Am J Forensic Med Pathol 2006; 27: 58–60.

Chapter 7

Limitations of Blood Alcohol Measurements Using Automated Analyzers and Breath Analyzers

7.1. INTRODUCTION

Ethyl alcohol, more commonly referred to as alcohol, has some health benefits if consumed in moderation, but all such health benefits eventually disappear with heavy drinking. Humans have used alcohol from prehistoric times (10,000 B.C.). Alcoholic beverages can be classified under three broad categories: beer, wine, and spirits. Beer and wine are fermented beverages produced from sugar- or starch-containing plant materials. The normal fermentation process that uses yeast cannot produce alcohol beverages with an alcohol content over 14%. Therefore, hard liquors or spirits are produced using fermentation followed by distillation. The alcohol content of various alcoholic beverages varies widely. The average alcohol content of beer is 5%, the average alcohol content of wine is 10%, and the average alcohol content of whiskey is 40%. However, beer usually comes in a 10- or 12-ounce bottle, whereas a shot of tequila in a mixed drink is only 1.5 ounces. Thus regardless of the alcoholic beverage, a standard drink contains roughly the same amount of alcohol. In the United

Resolving Erroneous Reports in Toxicology and Therapeutic Drug Monitoring: A Comprehensive Guide, First Edition. Amitava Dasgupta.

States, a standard drink is defined as a bottle of beer (12 ounces) containing 5% alcohol, 8.5 ounces of malt liquor containing 7% alcohol, a 5-ounce glass of wine containing 12% alcohol, 3.5 ounces of fortified wine like sherry or port containing about 17% alcohol, 2.5 ounces of cordial, liqueur containing 24% alcohol, or one shot of a distilled spirit such as gin, rum, vodka, or whiskey (1.5 ounces). Each standard drink contains approximately the same amount of alcohol: 0.6 ounce (1). Therefore, blood alcohol content can be calculated based on the number of drinks consumed regardless of the type of alcoholic beverage consumed.

7.2. ALCOHOL METABOLISM

Alcohol after consumption is absorbed from the stomach and metabolized by the liver. A small amount of alcohol not absorbed is found in the breath and is the basis of the breath analysis of alcohol in suspected drivers driving with impairment. Factors affecting alcohol metabolism include age, gender, ethnicity, body weight, and if alcohol has been consumed along with food. When alcohol is consumed, about 20% is absorbed by the stomach, and the rest is absorbed from the small intestine. When alcohol is consumed on an empty stomach, the peak level of blood alcohol is observed between 15 and 90 minutes after consumption. Food substantially slows down the absorption of alcohol, and food can even reduce the rate of absorption of alcohol for 4–6 hours. Sipping alcohol instead of drinking also slows the absorption. The effect of food on the absorption and metabolism of alcohol has been well studied and reported in the medical literature. In one study, 10 healthy men drank a moderate amount of alcohol (0.80 g of alcohol per kilogram of body weight) in the morning after an overnight fast or immediately after breakfast (two cheese sandwiches, one boiled egg, orange juice, and fruit yogurt). Subjects who drank alcohol on an empty stomach felt more intoxicated than the subjects who drank the same amount of alcohol after breakfast. The average peak blood alcohol in subjects who drank on an empty stomach was 104 mg/dL (0.104%), which is above the legal level of intoxication (0.08%). In contrast, the average peak blood alcohol in subjects who drank alcohol after eating breakfast was 67 mg/dL (0.067%). The time required to metabolize all alcohol was on average 2 hours shorter in subjects who drank alcohol after eating breakfast compared with subjects who drank on an empty stomach. The authors concluded that food in the stomach before drinking not only reduces the peak blood alcohol concentration but also boosts the rate at which the body eliminates alcohol (2). The effect of the nature of food such as high fat versus high protein or high carbohydrate on the magnitude of the reduction of absorption of alcohol has also been studied. However, drinking ethanol after eating a meal regardless of the composition decreases the absorption of alcohol (3).

A small amount of alcohol is metabolized by the enzyme present in the gastric mucosa, and also a small amount of alcohol is metabolized by the liver before it can enter the main bloodstream through first-pass metabolism. Then the rest of the alcohol enters the bloodstream known as systematic circulation. After drinking the same amount of alcohol, a man would have a lower peak blood alcohol level compared with a woman with the same body weight. This gender difference in the blood alcohol level is related to the different body water content between a male and a female. Alcohol is freely soluble in water and distributes into the serum. Because a woman has a lesser amount of body water (52% average) than a man (61% average), a lesser amount of total body water is available in a woman to dissolve the same amount of alcohol compared with a man. Some studies also report that women are more susceptible than men to alcohol-related impairment of cognitive functions (4).

Women also metabolize alcohol slower than men because the concentration of alcohol dehydrogenase is usually lower in women compared with men. Hormonal changes also play a role in the metabolism of alcohol in women, although this finding has been disputed in the medical literature. Some publications indicated that women metabolize alcohol at a higher rate during the luteal phase of the menstrual cycle (19–22 days of the cycle), but a few days before getting a period, a woman's alcohol metabolism may slow down (5). However, it has been well documented that alcohol addiction causes disturbances in the menstrual cycle, and such disturbances are more prominent during the middle part of the cycle (6). The human liver metabolizes alcohol using zero order kinetics. Several enzyme systems are involved in the metabolism of ethanol, namely alcohol dehydrogenase (ADH), microsomal ethanol oxidizing system, and catalase (7). These enzymes also metabolize other similar compounds such as methanol, isopropyl alcohol, and ethylene glycol.

The first and most important of these, alcohol dehydrogenase, is found primarily in hepatocytes.

$$CH_3CH_2OH \xrightarrow[\text{NAD} \quad \rightarrow \quad \text{NADH}]{\text{Alcohol Dehydrogenase}} CH_3CHO$$

At least five classes of ADH are found in humans. ADH activity is greatly influenced by the frequency of ethanol consumption. Adults who consume two to three alcoholic beverages per week metabolize ethanol at a rate much lower than alcoholics. For medium-size adults, the blood ethanol level declines at an average rate of 15–20 mg/dL per hour. The major drug metabolizing family of enzymes found in the liver is the cytochrome P450 mixed function oxidase. Many members of this family of enzymes, most notably the CYP3A4, CYP1A2, CYP2C19, and CYP2E1 isoenzymes, play vital roles in the metabolism of many drugs. For nonalcoholics, this metabolic pathway is considered a minor secondary route, but it becomes much more important in alcoholics, and the CYP2EI isoenzyme plays a major role in metabolizing alcohol in addition to ADH. Because of the additional participation of CYP2E1, alcoholics can get rid of alcohol faster from their bodies compared with nonalcoholics.

$$CH_3CH_2OH \xrightarrow{\text{CYP2E1}} CH_3CHO$$

The acetaldehyde produced due to the metabolism of alcohol, regardless of pathway, is subsequently converted to acetate as the result of the action of mitochondrial aldehyde dehydrogenase (ALDH2). Acetaldehyde is fairly toxic compared with ethanol and must be metabolized quickly.

$$CH_3CHO \xrightarrow[\text{NAD} \;\rightarrow\; \text{NADH}]{\text{Acetyl Dehydrogenase 2}} CH_3COOH$$

Acetate or acetic acid then enters the citric acid cycle, which is a normal metabolic cycle of living cells and converted into carbon dioxide and water. From the chemical point of view, the body oxidizes alcohol into carbon dioxide and water, and this process generates calories. Therefore, alcoholic drinks are high in calories.

The metabolism of alcohol changes with advancing age because the activity of the enzymes involved in alcohol metabolism is reduced with age. Therefore, an elderly person has a higher blood alcohol level from consumption of the same amount of alcohol compared with a younger person of the same gender. Elderly persons consume more medications than younger people, and a medication may interact with alcohol. It is safe for elderly persons to consume one drink a day provided it does not interact with any medication the person is taking (8). Moderate alcohol consumption as defined by the U.S. government to include no more than two standard drinks per day for men, no more than one standard drink per day for women, but for individuals over 65 years of age, only one standard drink per day regardless of gender (9).

Drinking more than recommended can invite problems because the health benefits of drinking in moderation disappear fast when consuming more than three to four drinks a day. Theoretically, men drinking more than three drinks a day and women drinking more than two drinks a day can be considered heavy drinkers. For all practical purposes, the National Institute of Alcohol Abuse and Alcoholism sets this threshold at more than 14 drinks per week for men (or more than 4 drinks per occasion) and more than 7 drinks per week for women (or more than 3 drinks per occasion). Individuals whose drinking exceeds these guidelines are at increased risk for adverse health effects. Hazardous drinking is defined as the quantity or pattern of alcohol consumption that places individuals at high risk for alcohol-related disorders. Usually hazardous drinking is defined as 21 or more drinks per week by men or more than 7 drinks per occasion at least three times a week. For women, more than 14 drinks per week or drinking more than 5 drinks in one occasion at least three times a week is considered hazardous drinking (10). Alcohol abuse is a leading cause of mortality and morbidity internationally and ranked by the World Health Organization as one of the top five risk factors for disease burden.

Without treatment, approximately 16% of all hazardous or heavy alcohol consumers progress to alcoholism (11). Heavy consumption of alcohol not only leads to increased domestic violence, decreased productivity, increased risk of motor vehicle as well as job-related accidents, but also to increased mortality from liver cirrhosis, stroke, and cancer.

Binge drinking means heavy consumption of alcohol within a short period of time with an intention to become intoxicated. Although there is no universally accepted definition for binge drinking, usually consumption of five or more drinks by males and four or more drinks by females is considered "binge drinking," and such drinking patterns always result in blood alcohol levels above 0.08%, the legal limit of driving. Despite the legal drinking age of 21 in the United States, binge drinking is very popular among U.S. college students. In one study the authors found that 74.4% of binge drinkers consumed beer exclusively or predominantly, and 80.5% of binge drinkers consumed at least some beer. Wine accounted for only 10.9% of binge drinks consumed (12).

7.3. MODERATE DRINKING VERSUS ALCOHOL ABUSE

There are many benefits of drinking in moderation, most notably a reduced risk of cardiovascular disease. The benefits of drinking in moderation are listed in Table 7.1. The term *French paradox* was coined in 1992 to describe the unexpected relatively low incidence of cardiovascular disease in the French

TABLE 7.1. Benefits of Consuming Moderate Alcohol

Benefits	Comments
Reduced risk of cardiovascular disease	Alcohol increases high-density lipoprotein cholesterol. Drinking red wine provides further defense due to polyphenolic antioxidants.
Reduced risk of stroke	Moderate alcohol consumption prevents Blood clot formation and fibrinolysis.
Reduced risk of developing diabetes	Alcohol at low concentration reduces insulin resistance.
Reduced risk of developing arthritis	Mechanism not clearly defined.
Reduced risk of developing age-related dementia and Alzheimer's disease.	Resveratrol, an antioxidant found in red wine protests neurons from amyloid-induced toxicity during development of Alzheimer's disease.
Reduced risk of certain types of cancer	Moderate drinking especially red wine consumption may reduce risks of certain cancer probably due to the antioxidant effect.
Increased longevity	Moderate drinking increases longevity by reducing risks of various diseases and also improves quality of life.
Less chance of getting common cold	Moderate consumption of red wine may prevent the incidence of common cold in winter months.

population despite high dietary intake of fatty food containing saturated fats. The lower incidence of cardiovascular disease was attributed to regular consumption of red wine by the French people. Extensive research has indicated that light to moderate consumption of alcohol, especially red wine (10–30 g of alcohol per day or up to two standard drinks), has beneficial effects by reducing all phases of atherogenesis as well as vessel occlusion (flow-mediated dilatation). Various components of red wine including ethyl alcohol, resveratrol, and other polyphenolic compounds may be effective in reducing oxidative stress and also in enhancing the cholesterol efflux from vessel walls through increasing high-density lipoprotein. These compounds can also inhibit lipoprotein oxidation, macrophage cholesterol accumulation, and foam cell formation. In contrast, heavy alcohol consumption and binge drinking are associated with an increased risk of cardiovascular disease (13). Studies have indicated that red wine intake is more cardioprotective than drinking other alcoholic beverages (14). Drinking red wine may also induce longevity genes (15).

7.4. BLOOD ALCOHOL DETERMINATION

Emergency department physicians routinely order blood alcohol tests for patients suspected of drug and/or alcohol overdose as well as patients who arrive in the emergency department following a motor vehicle accident where the driver involved in the accident may be under the influence of a drug or alcohol. DWI stands for "driving with impairment." Although impairment may also be drug related, alcohol is the major cause of DWI not only in the United States but also worldwide. Alcohol-related motor vehicle accidents kill approximately 17,000 Americans annually and are associated with more than $51 billion in total costs annually. There is a strong correlation between binge drinkers and alcohol-impaired drivers in the United States. In one study, the authors found that overall, 84% of all alcohol-impaired drivers are binge drinkers. Nonheavy drinkers are also involved in alcohol-related motor vehicle accidents (16). Currently, in all U.S. states, the legal limit of driving is 0.08% alcohol in the blood. Serum concentration of alcohol is more than whole blood concentration of alcohol. To calculate whole blood concentration of alcohol, the measured serum concentration must be multiplied by a factor generally considered to be 0.85. Therefore, if serum alcohol concentration is 100 mg/dL (0.1%), the whole blood concentration is 85 mg/dL (0.085%). In most states, the legal limit of intoxication is defined as 0.08% alcohol in whole blood rather than serum alcohol concentration. In the United Kingdom and Canada, the legal limit of driving is also 0.08%, but in other countries, lower levels of alcohol are mandated as the acceptable upper limit of driving under the influence of alcohol. In Switzerland, Denmark, Italy, Netherlands, Austria, Australia, China, Thailand, and Turkey, the upper limit is 0.05% alcohol. In Japan, the upper acceptable limit is only 0.03%, and in certain countries such as various

Middle Eastern countries, Hungary, Romania, and Georgia, there is a zero tolerance for blood alcohol in drivers.

7.4.1. Calculation of Blood Alcohol Concentration: Widmark Formula

In 1932, the Swedish scientist Eric P. Widmark developed a formula that is still used today to the calculate the amount of alcohol ingested and to assess the concentration of alcohol at some point before a blood alcohol analysis (17). The Widmark formula suggests estimating the blood alcohol level on a given amount of alcohol administration knowing the subject's body weight and gender:

$$A = C \times W \times r$$

Where A represents total amount of alcohol consumed by the person in grams, C is the blood alcohol concentration in grams per liter, W is the body weight of the person expressed in kilograms, and r is a constant that is assumed to be roughly 0.7 for men and 0.6 for women. The modern form of the formula to calculate a blood alcohol level from the amount of alcohol consumed by the individual, the body weight, and gender is as follows:

$$C = (A/W \times r) - 0.015\,t, \text{ where t represents time passed since the beginning of drinking}$$

In the United States, one standard drink of alcohol has 0.6 ounce of alcohol, and the weight of a person is expressed in pounds. However, blood alcohol concentration is expressed as milligrams per deciliter. Taking into accounts all these factors, this formula can be modified for calculating blood alcohol concentration as follows:

$$C = (\text{Total amount of alcohol consumed in ounces} \times 5.14/\text{weight in pounds} \times r) - 0.015\,t$$

C is the blood alcohol expressed as a percentage. Assuming each drink contains 0.6 ounce of alcohol, this equation can be further modified to:

$$C = (\text{Number of drinks} \times 3.1/\text{Weight in pounds} \times r) - 0.015\,t$$

Because most standard drinks contain approximately the same amount of alcohol, it is only important to know how many drinks one person consumes. The type of drink does not matter, which makes the calculation easy. For example, if a 170-pound man drinks five beers in a 2-hour period, his blood alcohol at the end of drinking would be:

$$\begin{aligned} C &= (5 \times 3.1/170 \times 0.7) - 0.015 \times 2 \\ &= 0.130 - 0.030 \\ &= 0.1\%\ (100\ \text{mg/dL}) \end{aligned}$$

Many Internet sites calculate the blood alcohol level based on the number of drinks consumed, gender, and body weight. In addition, many charts are also available on Internet sites that provide approximate blood alcohol levels based on body weight, gender, and number of drinks consumed. It has been demonstrated that Widmark's formula is a practical and reliable method of estimating blood alcohol from number of drinks, body weight, and gender if alcohol is consumed in the fasting state, but if alcohol is consumed with food, this formula may not be accurate to estimate blood alcohol (18).

It is a very popular DWI defense that the person never drank alcohol but felt tipsy after eating a big meal and that caused the accident. Although substantial alcohol may be produced endogenously in a decomposed body by the action of various microorganisms, the human body does not produce enough endogenous alcohol. In healthy individuals who do not drink, usually endogenous alcohol levels are way below the detection level. There are reports of measurable endogenous ethanol production in patients with liver cirrhosis. In one report, after a meal in such patients, negligible alcohol levels of 11.3 mg/dL (0.01%) and 8.2 mg/dL (0.008%) were detected in two of eight patients. Small intestinal bacterial overgrowth generates such small amounts of endogenous alcohol. Patients with liver cirrhosis often have small intestinal bacterial overgrowth (19).

7.5. BREATH ALCOHOL VERSUS BLOOD ALCOHOL DETERMINATION

For prosecution of DWI depending on the state, blood alcohol determined by an evidentiary breath analyzer may be admitted in a court of law for prosecution. In general, blood alcohol determined by gas chromatography (GC) is subjected to virtually no interference, but breath alcohol may be affected by certain interfering substances. A very small amount of alcohol is found in human breath. Only air in the deepest portion of the lung known as the alveolar sacs comes in contact with alcohol if present in the blood, and there is equilibrium between the alcohol level in the exhaled air and blood alcohol. The estimated ratio between breath alcohol and blood alcohol is 1:2100, and this ratio is used in various breath alcohol analyzers to calculate blood alcohol level based on the concentration of ethanol in exhaled air. This process is based on Henry's law, which states that the ratio between alcohol in blood and alcohol in deep lung air is constant. After measuring the concentration of alcohol in the exhaled air, the microprocessor of the instrument automatically converts the value to the calculated blood alcohol level. In contrast, the blood alcohol measurement is a direct measurement, and there are established guidelines for assessing the degree of impairment of an individual based on blood alcohol level and drinking history.

7.5.1. Chemical Principle of Breath Alcohol Analyzers

Breath analyzer uses one of the following technologies:

- Analyzers that utilize color change due to a chemical reaction to determine alcohol level
- Analyzers based on detecting ethanol in breath based on infrared spectroscopy
- Analyzers based on fuel cell technology
- Analyzers based on mixed technology (infrared and fuel cell) and other techniques

The earliest developed breath analyzer was based on a chemical principle when exhaled air was allowed to pass through a cocktail of chemicals containing sulfuric acid, potassium dichromate, silver nitrate, and water. Silver nitrate acts as a chemical catalyst in the reaction where alcohol in the presence of sulfuric acid (sulfuric acid absorbs alcohol from air and also provides an acidic medium for facilitating the reaction) turns orange potassium dichromate solution green due to the conversion of potassium dichromate into chromium sulfate, which is green in color. The intensity of the green color can be used to estimate the amount of alcohol in the exhaled air. Captain Robert Borkenstein of the Indiana State Police used this chemical principle to develop breath analyzers in 1954, and some breath alcohol analyzers still use this principle today. Breathalyzer is the oldest technology of breath alcohol analyzer based on the principle of color change of potassium dichromate solution in the presence of alcohol and then analysis by using spectroscopy after a specified time to ensure a complete reaction. The analyzer contains two vials of a chemical cocktail. After a subject exhales into the device, the air is passed through one vial, and if alcohol is present in the exhale, a color change occurs. A system of photocells connected to a meter to measure color change associated with the chemical reaction by comparing the response from the second vial (where no air is passed through) thus produces an electrical signal proportional to the color change in the reaction vial. This electrical signal can move the meter (more alcohol, more signal and higher reading), and the alcohol level in the subject can be determined. Breathalyzer was the brand name originally developed and marketed by Smith and Wesson, and the company then sold that brand to Dräger, a German company. The old Breathalyzer 900 model was replaced by newer versions such as model 1100, but this technology is subject to interferences from a variety of substances, so other companies have focused on developing a more robust technology for breath alcohol analysis.

Several evidentiary breath alcohol analyzers are based on the principle of infrared (IR) spectroscopy for the quantitative determination of alcohol in exhaled air. The Intoxilyzer was originally developed by Omicron in Palo Alto, California, and later sold to CMI, Inc., in Owensboro, Kentucky. The earlier models were 4011 A, 4011 S, and then Intoxilyzer 5000, which is used as an evidentiary breath alcohol analyzer (evidence admissible in a court of law), and more recently a Intoxilyzer 8000 model became commercially available. Many states use this analyzer as the evidentiary breath analyzer. In addition to the Intoxilyzer, Data Master cdm (National Patent Analytical System, Mansfield, OH), which is also used in many states as the evidentiary breath alcohol analyzer, is based on IR spectrum technology. The newer version of Intoxilyzer (Intoxilyzer 5000) uses a five wavelength filter at 3.36, 3.4, 3.47, 3.52, and 3.8 microns, and thus it can differentiate between ethanol and common interferences in exhaled air such as acetone, acetaldehyde, and toluene. The 3.4-micron wavelength is used to detect alcohol; 3.47 identifies interfering substances, and the 3.9 micron is used as the reference wavelength. The latest model Intoxilyzer 8000 uses a pulsed IR source instead of a moving wavelength filter and uses dual wavelength for measuring alcohol (3.4 and 9.36 microns) in the breath. It also has more advanced computer technology to provide accurate alcohol level results. Data Master cdm, an evidentiary breath analyzer widely used by police officers in many states, is also based on the principle of IR spectra where alcohol is detected using two different wavelengths (3.37 and 3.44 microns). Kechagias et al compared blood alcohol values with values obtained by a breath alcohol analyzer (Data Master) in patients suffering from gastroesophageal reflux disease and concluded that breath alcohol analyzers can overestimate true blood alcohol value due to the eruption of alcohol from the stomach to the mouth from gastric reflux (20). There are several different brands of evidentiary breath alcohol analyzers based on the principle of fuel cell technology, such as Alcotest Models 6510, 6810, and 7410 (National Drager, Durango, CO) and Alco-Sensor III and IV (Intoximeters Inc., St. Louis, MO), which are evidentiary breath analyzers. The fuel cell is a porous disk coated with platinum oxide (also called platinum black) on both sides. The porous layer is impregnated with an acidic solution containing various electrolytes so that charged particles such as hydrogen ions can travel through that medium. In addition, both sides of the disk containing platinum oxide are connected through a platinum wire. The manufacturer mounts this fuel cell in a case along with the entire assembly so that when a person blows into the disposable mouthpiece, the air can travel through the fuel cell. If any alcohol is present in the exhaled air, the alcohol is converted into acetic acid, hydrogen ion, and electrons on the top surface by the platinum oxide. Then hydrogen ions travel to the bottom surface (also containing platinum oxide) and are converted into water by combining with the oxygen present in the air. In this process, electrons are removed from the platinum oxide. Because there is an electron excess on the top surface and electron deficit on the bottom surface, electrons flow from one surface to another, generating an electric current that flows through the platinum wire, and the intensity of the current is proportional to the amount of alcohol present in the exhaled air. The microprocessor of the instrument then converts that current to equivalent blood alcohol.

Some evidentiary breath analyzers are based on both fuel cell and IR spectroscopy technology giving them good sensitivity and specificity to analyze alcohol on the breath. Various models of Intox EC/IR 1 desktop evidentiary alcohol breath analyzers, also manufactured by Intoximeters, Inc., combine reliable fuel cell analysis with the real-time analytical advantages of IR technology.

7.5.2. Interferences in Measuring Alcohol Using Various Breath Analyzers

Sometimes a driver stopped by the police may use mouthwash to hide any alcoholic breath. Because some mouthwashes contain alcohol, use of a mouthwash prior to taking a breath alcohol analysis may cause a falsely elevated alcohol result. However, residual alcohol evaporates from the mouth rapidly, which is the reason for waiting for 15 minutes in a police station under supervision so the suspect cannot take anything by mouth during the waiting period. Fessler et al studied the effect of alcohol-based substances such as mouthwash, cough mixture, and breath spray just prior to breath alcohol measurement using a Drager Evidentiary portable breath alcohol analyzer and 25 volunteers and concluded that a 15-minute waiting period is necessary to ensure there is no residual alcohol in the mouth following the use of mouthwash and other alcohol-containing products. Otherwise alcohol from mouthwash may interfere in breath alcohol analysis (21). Drinking an energy drink while driving a car is not against the law, but some energy drinks contain very low levels of alcohol. When volunteers drank various energy drinks, 11 of 27 persons showed positive results using evidentiary breath analyzers when testing was done just after drinking. However, after a 15-minute waiting period, all breath alcohol analysis reports were negative. The authors concluded that a 15-minute waiting period eliminates the possibility of testing false positive after consuming an energy drink with low alcohol content (22). Laakso et al studied the effect of various volatile solvents for potential interference with breath alcohol analysis using the Drager 7110 evidentiary breath analyzer and concluded that acetone, methyl ethyl ketone, methyl isobutyl ketone, ethyl acetate, and diethyl ether do not influence breath alcohol measurement significantly, but propyl alcohol and isopropyl alcohol have a significant effect on breath alcohol measurement (23). Jones and Rossner described a case in which a 59-year-old man undergoing a weight loss program using a ketogenic diet attempted to drive a car fitted with an alcohol ignition interlock device, but the vehicle did not start. Because he completely stopped drinking, he was surprised and upset. A ketogenic diet used for treating obesity and controlling seizure in some epileptic children is high in fat, very low in carbohydrate, and also has adequate protein. The goal is to burn fat to get energy rather than getting it from the glucose formed by carbohydrate metabolism. However, consuming the ketogenic diet led to a stage called ketonemia where concentrations of acetone, acetoacetic acid, and beta hydroxybutyric acid are high. This high amount of acetone may be found in the exhaled air. The interlock device in the car determines alcohol by the electrochemical oxidation method, and acetone does not interfere with the process. However, acetone is known to be converted into isopropyl alcohol by the action of liver alcohol dehydrogenase, and isopropyl alcohol can be falsely identified as ethanol by the ignition interlock device. In addition, methanol and propanol can also be falsely identified as alcohol. The authors concluded that the side effects of ketogenic diets need further evaluation by authorities especially for people involved in safety-sensitive positions such as airline pilots and bus drivers who are subjected to much tougher alcohol tolerance policies (24).

Methanol poisoning is dangerous because it may cause death or blindness, but drinking methanol is not against the law. A recent article reported the case of a 47-year-old man who was found at a public park acting intoxicated (methanol poisoning may cause intoxication). A breath analyzer (Intoxilyzer 5000 EN) measured 0.28% blood alcohol (legal limit: 0.08% blood alcohol). In the emergency department, the patient admitted drinking gas line antifreeze that contains 99% methanol. The patient was subsequently treated and survived, but this case indicates that methanol poisoning can be mistaken by a breath analyzer as alcohol poisoning (25). In another report the authors found that toluene, xylene, methanol, and isopropyl alcohol in exhaled air can be mistakenly identified as breath alcohol by the Intoxilyzer 5000 evidentiary breath alcohol analyzer (26).

7.6. BLOOD ALCOHOL DETERMINATION

In general, whole blood alcohol level is determined using GC in a crime or forensic laboratory, whereas serum or plasma blood alcohol is determined in a hospital laboratory either by an enzymatic method or using GC. In one study, using 212 consecutive patients admitted to a hospital trauma center, serum was analyzed for ethanol using an enzymatic method and whole blood was analyzed using GC in a forensic toxicology laboratory. The authors observed that the serum to whole blood alcohol ratio depended on the alcohol concentration, but the average value was 1.12 up to an alcohol concentration of 300 mg/dL (0.3%) but may be as high as 1.18. Therefore, a serum blood alcohol of 103 mg/dL (0.13%) corresponds to a whole blood alcohol level of 80 mg/dL (0.08%) (27). Rainey reported that the ratio between serum and whole blood alcohol ranged from 0.88 to 1.59, but the median (most commonly occurred value) was 1.15. Thus dividing the serum alcohol value by 1.15 would calculate the whole blood alcohol concentration. The serum to whole blood alcohol ratio was independent of the serum alcohol concentration and hematocrit (28).

In hospital laboratories, ethyl alcohol is also analyzed using enzymatic methods and automated analyzers. Several different automated analyzers are available from various diagnostic companies that are capable of analyzing alcohol in serum or plasma. Enzyme-based automated methods are generally not applicable for the analysis of whole blood, although modified methods are available for the analysis of alcohol in urine specimens. Enzymatic automated analysis of alcohol is based on the principle of conversion of alcohol to acetaldehyde by alcohol dehydrogenase, and in this process nicotinamide adenine dinucleotide (NAD) is converted into NADH. NAD has no absorption at ultraviolet light at 340 nm wavelength, whereas NADH absorbs at 340 nm. Therefore, an absorption peak is absorbed when alcohol is converted into acetaldehyde because NAD is also converted into NADH. The intensity of the peak is proportional to the amount of alcohol present in the specimen. If no alcohol is present, no peak is absorbed. Usually methanol, isopropyl alcohol, ethylene glycol, and acetone have a negligible effect on alcohol determination using enzymatic methods, but propanol if present may cause a 15–20% cross-reactivity with alcohol assay. Although isopropyl alcohol used as rubbing alcohol is common in household products, propanol is used with much less frequency.

However, interference of lactate dehydrogenase (LDH) and the lactate in the enzymatic method of alcohol determination is significant and may cause misinterpretation of the alcohol value over the legal limit in a patient with lactic acidosis. In addition, enzymatic alcohol assay is unsuitable for the determination of alcohol in postmortem blood because high concentrations of LDH and lactate-only GC can be used for measuring alcohol in postmortem blood. In one report, the authors observed 690 mg/dL (0.69%) of alcohol in serum using an enzymatic method for alcohol in a patient, but the GC did not show any alcohol in the serum. This patient had end-stage renal disease and received a kidney transplant. At the time when blood was drawn, she had severe metabolic acidosis and was admitted to the hospital. Her LDH concentration was 27,000 U/L and her lactate concentration was 15.0 mmol/L. However, the authors observed no apparent alcohol level in any specimen containing normal levels of LDH and lactate (29). Lactate concentrations also tend to increase in trauma patients. Dunne et al reported that 27% of 13,102 patients they studied had a positive alcohol screen (Mean alcohol: 141 mg/dL; range: 10–508 mg/dL) (30). In contrast, Winek et al compared alcohol concentration obtained by an enzyme assay (Dimension Analyzer, Siemens Diagnostics, Deerfield, IL) and GC in trauma patients and observed no false positives by immunoassay. Alcohol concentrations obtained by immunoassays correlated well with GC values, and only in 6 specimens (of 27) did the differences between GC and immunoassay values exceed 10%. The highest difference was 22%. The authors concluded that the immunoassay method can be used in the hospital laboratory to determine alcohol concentrations in trauma patients (31). Contrasting results may be related to differences in the lactate concentrations in the patients in two different studies.

The gold standard for the determination of blood alcohol is GC or, as recently described, using GC/mass spectrometry. Jain described a GC method in which blood can be injected directly to a column after the addition of isobutanol as the internal standard. Ethanol (retention time: 3.5 minutes) is eluted from the column before isobutanol (retention time: 7.25 minutes). Baseline separation is also observed for acetone, isopropyl alcohol toluene, and xylene (32). Smith et al described the determination of ethanol along with methanol, isopropanol, and acetone using capillary GC after direct sample injection. The author used 1-propanol as the internal standard. Serum (200 μL) was mixed with the 200-μL internal standard solution (prepared in deionized water) along with 200 μL of 200 mmol/L of sodium tungstate and 200 μL of 200 mmol/L cupric sulfate. After vortex mixing and centrifugation to sediment the precipitate, 1 μL of aqueous supernatant was directly injected into the gas chromatograph (33). Baseline separation was obtained between ethanol, other analytes, and the internal standard (Fig. 7.1). Maleki et al described the analysis of methanol and ethanol in human whole blood, urine, and saliva samples based on headspace solid-phase microextraction using silver sulfide and polyvinyl chloride–coated silver wire. Unlike commercially fibers that are coated with fused silica, due to metallic base of this fiber as described by the authors, this fiber is very durable and thermally stable up to 250°C (34). Tiscione et al

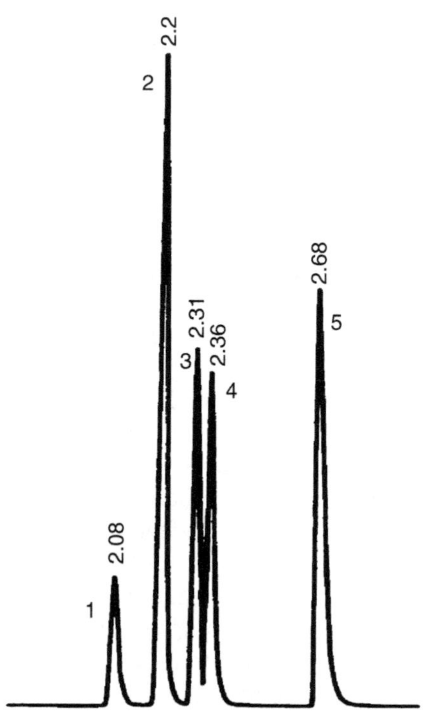

Figure 7.1. Gas chromatographic analysis of various alcohols. Peak 1: methanol; peak 2, ethanol (alcohol); peak 3, acetone; peak 4, isopropyl alcohol; peak 5; propyl alcohol (internal standard). Source: Smith NB et al. Clin Chem 1984; 30: 1672–1674. Reprinted with permission from AACC.

recently described ethanol analysis by headspace GC with simultaneous flame ionization and mass spectrometry detection (35).

7.7. CONCLUSIONS

Alcohol if consumed in moderation has some health benefits, but alcohol abuse is detrimental to health. Drivers under the influence of alcohol are responsible for most of the fatalities from accidents. In addition, alcohol poisoning may require immediate medical care. Although breath analyzers are widely used by law enforcement officers to identify drivers driving under the influence of alcohol, breath analyzers results may be affected by interference. Blood alcohol determination in a toxicology laboratory especially using a gas chromatograph produced a more accurate result than breath analyzers as well as alcohol determination using alcohol dehydrogenase and an automated analyzer.

REFERENCES

1. Kerr WC, Greenfield TK, Tujague J, Brown SE. A drink is a drink? Variation in the amount of alcohol contained in beer, wine and spirits drinks in a US methodological sample. Alcohol Clin Exp Res 2005; 29: 2015–2021.
2. Jones AW, Jonsson KA. Food-induced lowering of blood ethanol profiles and increased rate of elimination immediately after a meal. J Forensic Sci 1994; 39:1084–1093.
3. Jones AW, Jonsson KA, Kechagias S. Effect of high fat, high protein and high carbohydrate meals on the pharmacokinetics of a small dose of alcohol. Br J Clin Pharamacol 1997; 44: 521–526.
4. Mumenthaler MS, Taylor JL, O'Hara R, Yesavage JA. Gender difference in moderate drinking effects. Alcohol Res Health 1999; 23: 55–64.
5. Gill J. Women alcohol and the menstrual cycle. Alcohol Alcohol 1997; 3: 435–441.
6. Augustynska B, Ziolkowski M, Odrowaz-Sypmiewska G, Kielpinski A et al. Menstrual cycle in women addicted to alcohol during the first week following drinking cessation—changes in sex hormones levels in relation to detected clinical features. Alcohol Alcohol 2007; 42: 80–83.
7. Zakhari S. Overview: How is alcohol metabolized by the body? Alcohol Res Health 2006; 9: 245–254.
8. Meier P, Seitz HK. Age, alcohol metabolism and liver disease. Curr Opin Clin Nutri Metabol Care 2008; 11: 21–26.
9. U.S. Department of Agriculture and U.S. Department of Health and Human Services. In: *Dietary Guidelines for Americans*. Chapter 9: Alcoholic Beverages. Washington, DC: US Government Printing Office; 2005:43–46. Available at: http://www.health.gov/DIETARYGUIDELINES/dga2005/document/html/chapter9.htm. Accessed July 1, 2011.

10. Reid MC, Fiellin DA, O'Connor PG. Hazardous and harmful alcohol consumption in primary care. Arch Intern Med 2008; 159: 1681–1689.
11. Coulton S. Alcohol misuse. Clinical Evidence (online journal), August 27, 2008; pii: 1017.
12. Naimi TS, Brewer RD, Miller JW, Okoro C et al. What do binge drinkers drink? Implications for alcohol control policy. Am J Prev Med 2004; 33: 188–193.
13. Lippi G, Franchini M, Favaloro EJ, Targher G. Moderate red wine consumption and cardiovascular disease risk: beyond the "French paradox." Semin Thromb Hemost 2010; 36: 59–70.
14. Wu JM, Hsieh TC. Resveratrol: a cardioprotective substance. Ann NY Acad Sci 2011; 1215: 16–21.
15. Bertelli AA, Das DK. Grapes, wines, resveratrol and heart disease. J Cardiovasc Pharmacol 2009; 54: 468–476.
16. Flowers NT, Naimi TS, Brewer RD, Elder RW et al. Patterns of alcohol consumption and alcohol-impaired driving in the United States. Alcohol Clin Exp Res 2008: 32: 639–644.
17. Brouwer IG. The Windmark formula for alcohol quantification. SADJ 2004; 59: 427–428.
18. Jones AW, Neri A. Reinvestigation of Widmark's method for quantitative evaluation of blood ethanol profiles: influence of alcohol dose and mode of drinking. Clin Chem 1987; 33: 1469.
19. Torrens J, Riu-Aumatell M, Lopez-Tamames E, Buxaderas S. Volatile compounds of red and white wines by head-space-solid phase microextraction using different fibers. J Chromatogr Sci 2004; 42: 310–316.
20. Kechagias S, Jonsson KA, Franzen T, Anderson L et al. Reliability of breath alcohol analysis in individuals with gastroesophageal reflux disease. J Forensic Sci 1999; 44: 814–818.
21. Fessler CC, Tulleners FA, Howitt DG, Richrads JR. Determination of mouth alcohol using the Dräger Evidential Portable Alcohol System. Sci Justice 2008; 48: 16–23.
22. Lutmer B, Zurfluh C, Long C Potential effect of alcohol content in energy drinks on breath alcohol testing. J Anal Toxicol 2009; 33: 167–169.
23. Laakso O, Pennanen T, Himbwerg K, Kuitunen T et al. Effect of eight solvents on ethanol analysis by Drager 7110 evidentiary breath analyzer. J Forensic Sci 2004; 49: 1113–1116.
24. Jones AW, Rossner S. False-positive breath alcohol test after a ketogenic diet. Int J Obes (Lond) 2007; 31: 559–561.
25. Caravati EM, Andrson KT. Breath alcohol analyzer mistakes methanol poisoning for alcohol intoxication. Annal Int Med 2010; 55: 198–200.
26. Caldwell JP, Kim ND. The response of the Intoxilyzer 5000 to five potential interfering substances. J Forensic Sci 1997; 42: 1080–1087.
27. Barnholl MT Jr, Herbert D, Wells DJ Jr. Comparison of hospital laboratory serum alcohol levels obtained by an enzymatic method with whole blood levels forensically determined by gas chromatography. J Anal Toxicol 2007; 31: 23–30.
28. Rainey P. Relation between serum and whole blood ethanol concentrations. Clin Chem 1999; 39: 2288–2292.

29. Thompson WT, Malhotra D, Schammel DP, Blackwell W et al. False positive ethanol in clinical and postmortem sera by enzymatic assay: elimination of interference by measuring alcohol in protein free ultrafiltrate. Clin Chem 1994; 40: 1594–1595.
30. Dunne JR, Tracy JK, Scalea TM, Napolitano L. Lactate and base deficit in trauma: does alcohol or drug use impair predictive accuracy? J Trauma 2005; 58: 959–966.
31. Winek CL, Wahba WW, Windisch R, Winek CL. Serum alcohol concentrations in trauma patients determined by immunoassays versus gas chromatography. Forensic Sci Int 2004; 139: 1–3.
32. Jain NC. Direct blood injection method for gas chromatographic determination of alcohols and other volatile compounds. Clin Chem 1971; 17: 82–85.
33. Smith NB. Determination of volatile alcohols and acetone in serum by non-polar capillary gas chromatograph after direct sample injection. Clin Chem 1984; 30: 1672–1674.
34. Maleki R, Farhadi K, Matin AA. Analysis of ethanol and methanol in human fluids by headspaces solid phase microextraction coupled with capillary gas chromatography. Anal Sci 2006; 22: 1253–1255.
35. Tiscione NB, Alford I, Yeatman DT, Shan X. Ethanol analysis by headspace gas chromatography with simultaneous flame-ionization and mass spectrometry detection. J Anal Toxicol 2011; 35: 501–511.

Chapter 8

Role of the Laboratory in Detecting Other Poisoning, Including Pesticides, Ethylene Glycol, and Methanol

8.1. INTRODUCTION

Although pesticide poisoning, especially organophosphate poisoning, can be detected using a serum cholinesterase test, ethylene glycol and methanol poisoning are detected by directly analyzing these volatile compounds using gas chromatography. However, ethylene glycol concentration in the serum or plasma can also be determined by using an enzymatic method. In addition, high anion gap and osmolality determination may indirectly confirm a suspected methanol or ethylene glycol poisoning.

These individual poisons are grouped in one chapter because there is no readily available clinical laboratory test for the diagnosis or the follow-up of therapy for them. However, for certain pesticide poisonings, the cholinesterase test may be useful for diagnosis. For methanol and ethylene glycol poisoning, determining the serum levels of such toxins is useful in making clinical decisions such as whether administration of ethyl alcohol intravenously would be sufficient for therapy or hemodialysis treatment should be initiated. The

Resolving Erroneous Reports in Toxicology and Therapeutic Drug Monitoring: A Comprehensive Guide, First Edition. Amitava Dasgupta.

progress of therapy can also be followed by demonstrating a reduction in the serum methanol or ethylene glycol level. This chapter discusses the role of the laboratory in the diagnosis of certain pesticide poisonings with an emphasis on organophosphate and carbamate-based pesticides as well as ethylene glycol and methanol poisoning.

8.2. PESTICIDE POISONING

Pesticides are widely used in agriculture to control insects, fungi, weeds, and microorganisms in order to increase crop yield as well as to decrease the manual labor involved in agriculture (1). Chemicals in pesticides may cause toxicity in humans if the chemical comes in contact with the skin or eyes or is inhaled or swallowed (more commonly in a suicide attempt). Pesticides can be absorbed from the skin after exposure, but they can be absorbed more readily from the eyes or through the lungs after breathing contaminated air because many pesticides form droplets or fumes. Exposure to pesticides may be acute or chronic. Acute pesticide toxicity may occur within hours of exposure to a relatively large dose of pesticide.

Farming is an essential component of the U.S. economy. Agricultural workers suffer from more injuries compared with other workers, and pesticides are responsible for many injuries and illnesses experienced by such workers. Farmers and farm workers may be exposed to pesticides during their preparation for application as well as during application such as weeding, harvesting, irrigating, and planting (2). Recognizing such risks, the U.S. Environmental Protection Agency developed rules in 1974 to protect farm workers from unnecessary exposure to pesticides known as the Workers Protection Standard for Agricultural Pesticides.

Pesticides can be grouped according to chemical classes. The most commonly used classes include organophosphates (also called organophosphorus), organochlorines, and carbamates. A commonly known organochlorine compound is dichlorodiphenyl trichloroethane (DDT), but due to the high toxicity of these classes of compounds, they have been phased out. Both organophosphate and carbamates replace such organochlorine compounds. However, organophosphates in general are more toxic than carbamates. Thiocarbamate and dithiocarbamate are a subclass of carbamate pesticides. Table 8.1 lists the active ingredients of various pesticides that may be harmful.

Approximately 75% of the pesticide use in the United States occurs in agriculture. Calvert et al researched acute poisoning among agricultural workers in the United States by studying 3271 cases of exposure between 1998 and 2005. Of the subjects they studied, 2334 (71%) were employed as farm workers. The remaining cases involved personnel engaged in processing/packing (12%), farmers (3%), and other miscellaneous agriculture workers (19%). Most of the workers experienced low-severity illness (2848 cases; 87%), another 402 cases (12%) involved medium severity, and only 20 workers (0.6%) had a major illness due to pesticide exposure. One case was fatal. Low-

TABLE 8.1. Active Ingredients of Various Toxic Pesticides

Chemical Class	Active Ingredient	Function
Organophosphate	Chlorpyrifos	Insecticide
	Methamidophos	Insecticide
	Dimethoate	Insecticide
	Malathion	Insecticide
	Parathion	Insecticide
	Methamidophos	Insecticide
	Omethoate	Insecticide
Dithiocarbamate	Metam-sodium	Fumigant
	Mancozeb	Fungicide
N-methyl carbamate	Methomyl	Insecticide
Triazole	Myclobutanil	Fungicide
Sulphur ester	Propargite	Insecticide
Halocarbon	Methyl bromide	Fumigant
Halogen	Sodium Hypochlorite	Disinfect

severity illness was defined by the authors as a resolution of symptoms without treatment and less than 3 working days were lost; moderate severity was defined as a non-life-threatening illness requiring medical treatment but no disability resulted from such exposure with less than 6 days of work loss. A high-severity illness consisted of a life-threatening illness requiring hospitalization and might result in permanent injury or disability. Time lost from work was more than 5 days (3).

Liu et al studied pesticide poisoning in Huzhou, China, from 2006 to 2009 involving 2298 patients and reported that occupational poisoning accounted for 588 patients (25.6%); nonoccupational poisoning was observed in 1710 patients (74.4%). Most occupational pesticide poisoning occurred in the summer and autumn months, and there were four fatalities among patients who were exposed to pesticides due to their occupation. Most of the fatalities (112 patients) occurred in the nonoccupational patient group, however. Organophosphate compounds including methamidophos, parathion, and omethoate comprised a higher proportion (80%) of all insecticide poisonings. The authors concluded that nonoccupational exposure to insecticide poisoning is a major public health issue (4).

8.3. ORGANOPHOSPHATE AND CARBAMATE POISONING: MECHANISM OF POISONING AND SYMPTOMS

Other than exposure by agricultural workers, self-poisoning with organophosphate compounds is a serious problem in rural developing countries, accounting for an estimated 200,000 fatalities every year. Unintentional poisoning kills far fewer people but is still a problem where highly toxic organophosphate insecticides are available. In general, a 15% fatality rate is associated with such poisoning (5). Davies et al reported that organophosphate pesticide dimethoate has a human case fatality threefold higher than poisoning with chlorpyrifos despite similar animal toxicity. Interestingly, the clinical symptoms of dimethoate poisoning are distinct from chlorpyrifos poisoning because many patients with dimethoate poisoning present with hypotension that progresses to shock and death within 12–48 hours postingestion. In addition, the authors described three cases of dimethoate poisoning where despite treatment with atropine, intravenous fluids, pralidoxime chloride, and inotrope, all patients died 2.5–32 hours postadmission. Continuous cardiac monitoring and troponin T measurement provided little evidence of the cardiotoxic effect of dimethoate (6). Carbamate insecticide such as aldicarb can also cause serious poisoning and even fatality despite treatment with atropine (7).

Case Report

A 22-year-old woman in her 29th week of pregnancy was admitted to the hospital after experiencing multiple tonic-clonic seizures at home. The patient was drowsy at admission with a Glasgow coma scale score of 14/15. An initial diagnosis of seizures in pregnancy was made, and she was managed as an atypical eclamptic patient and given intravenous magnesium sulfate. Ten hours following admission, her condition deteriorated due to airways obstruction related to copious secretion and hypersalivation. Atropine was administered, and the patient responded to the therapy. The patient later confirmed that she had ingested insecticide for the purpose of suicide after an altercation about the pregnancy with her partner. She had an unplanned pregnancy and had been clinically depressed since the death of her only child approximately 12 months earlier. She was positive for human immunodeficiency virus but was not receiving any antiretroviral therapy. Two days after admission she delivered a live infant who died 2 days later. The insecticide was identified as chlorpyrifos (see Fig. 8.1 for chemical structure), an organophosphate insecticide, but this case presented a diagnostic dilemma because organophosphate poisoning is uncommon in pregnancy. However, the mother displayed typical symptoms of organophosphate poisoning including seizure. A seizure in pregnancy is typically considered to be eclampsia unless proven otherwise (8).

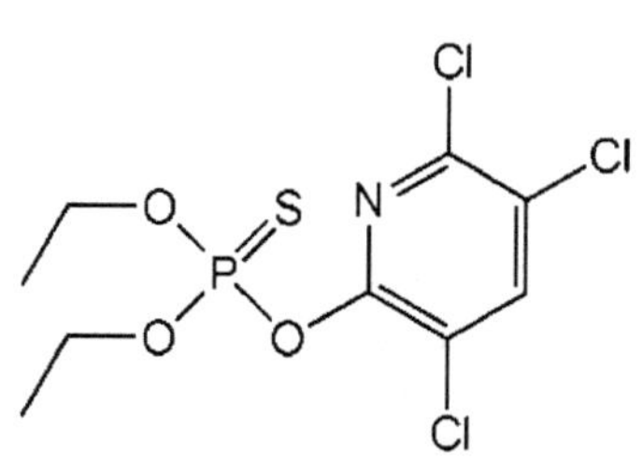

Figure 8.1. Chemical structure of chlorpyrifos.

Case Report

A 29-year-old man presented to a rural hospital after ingesting 50–100 mL of pesticide from a bottle labeled methyl parathion. He presented to the hospital an hour after ingesting the organophosphate pesticide. He had a history of depression and was being treated with venlafaxine extended-release tablet 150 mg a day and olanzapine 7.5 mg a day. On admission he had blurred vision but a normal neurological examination, normal biochemistry parameters, and his initial red cell acetylcholinesterase level was also normal. The patient was treated with activated charcoal and pralidoxime and later transferred to a regional hospital 12 hours after ingestion. He was asymptomatic and had a normal physical examination. Pralidoxime was also administered in the regional hospital but was discontinued 30 hours postingestion. Interestingly, his acetylcholinesterase and total serum plasma cholinesterase levels were measured spectrophotometrically using butyrylthiocholine iodide/dithiobis (2-nitrobenzoic acid) reagent, and they started to decline and decreased to just below 25% of the normal value in 6 days. Analysis of serum showed the presence of methyl parathion and its metabolite 4-nitrophenol just above the limit of quantitation. Urinary analysis showed 2.8 mg/L of 4-nitrophenol in a 24-hour urine sample collected between 37 and 61 hours postingestion. The container patient drank from showed 24% methyl parathion. The patient survived and was eventually discharged from the hospital. This case is interesting because the patient had almost no clinical symptoms of pesticide poisoning but showed a prolonged suppression of acetylcholinesterase activity (9).

Organophosphate and carbamate pesticides, to some extent, inhibit acetylcholinesterase. Acetylcholine is an important neurotransmitter at the synaptic junction; acetylcholinesterase breaks down acetylcholine into its primary components: acetic acid and choline. Two major human cholinesterases are acetylcholinesterase found primarily in nerve tissues and erythrocytes; pseudocholinesterase is found in both serum and liver. Organophosphate pesticides inhibit cholinesterase by binding irreversibly to the active site of the enzyme forming phosphorylated enzyme, which is inactive. Because organophosphate and to a lesser extent carbamate insecticides act by inhibiting the activity of cholinesterases, such poisoning can be assessed by measuring the activity of red blood cell cholinesterase activity or serum pseudocholinesterase activity. Excess concentration of acetylcholine due to inhibition of cholinesterase initially excites and then paralyzes cholinergic synapses causing muscarine, nicotinic, and central nervous system effects as reflected in the cholinergic symptoms expressed by the patient (Table 8.2). Cholinesterase enzymes are considered a good marker for the evaluation of organophosphate poisoning (10).

Intermediate syndrome is a major cause of death from respiratory failure following acute organophosphate poisoning (11). The intermediate syndrome

TABLE 8.2. Clinical Symptoms of Organophosphate Poisoning

Type of Symptom	Specific Symptom
Muscarine effect	Sweating, lacrimation, excessive sweating, blurred vision, constricted pupils, cramps, vomiting, diarrhea, urinary incontinence, bradycardia, hypotension
Nicotinic effect	Fasciculations, paralysis, tachycardia
Central nervous system effect	Anxiety, restlessness, ataxia, convulsion, respiratory depression, lack of reflexes
	Cheyne-Stokes respiration
Dermatological	Pruritus, rash, erythema, skin irritation

was first described in the mid-1980s and has characteristic symptoms and signs that occur after an apparent recovery from initial cholinergic shock. However, intermediate syndrome appears before organophosphate-induced delayed polyneuropathy. Because this syndrome develops after initial cholinergic shock but before polyneuropathy, it is called the intermediate syndrome. Intermediate syndrome occurs in approximately 20% of patients following oral exposure to organophosphate pesticides in 2–4 days postexposure. The characteristic features are weakness of muscle involved in respiration (diaphragm, intercostal muscles, and accessory muscles including neck muscles) as well as the proximal limb muscle. Management is essentially that of rapidly developing respiratory distress and respiratory failure; delays in instituting ventilator care may cause death (12). Both organophosphate and carbamates are cholinesterase inhibitors, and pneumonia is a frequent complication of such poisoning and a risk factor for death (13).

Carbamates are a class of pesticides derived from carbamic acid. Although like organophosphates, carbamates are inhibitors of acetylcholinesterase, such inhibition is reversible because the carbamyl-enzyme complex dissociates far more rapidly than the organophosphate-enzyme complex. Although carbamate poisoning shows many of the clinical features of organophosphate poisoning, the prominent central nervous system effect such as convulsion is usually absent in carbamate poisoning. Because red cell cholinesterase values may return to normal more rapidly than organophosphate poisoning, such laboratory tests may be less reliable for the diagnosis of carbamate poisoning. However, like treating organophosphate poisoning, atropine can also be used as an antidote. Currently, a combination of an antimuscarine agent such as atropine and an acetylcholinesterase activator such as pyridinium oximes (pralidoxime, trimedoxime, obidoxime, etc.) and diazepam are used for treating organophosphate and carbamate poisoning. Fatality from such poisoning usually occurs from respiratory failure (14).

Although considered less toxic than organophosphate, severe toxicity may also occur from carbamate poisoning. Thiodicarb is a carbamate insecticide whose acetylcholinesterase inhibition activity is related to its main methomyl degradation product. A 40-year-old woman was found dead in her car with an empty package of medicine and an open bottle of Larvin containing thiodicarb. There was no sign of violence. The postmortem investigation revealed therapeutic concentrations of zolpidem (2.87 mg/L), bromazepam (2.39 mg/L), nordiazepam (4.21 mg/L), and levopromazine (0.64 mg/L) in her blood, but a high concentration of thiodicarb (24.3 mg/L) was detected in the gastric contents. In addition, methomyl metabolite was found in the gastric contents (19.9 mg/L), peripheral blood (0.7 mg/L), urine (8.5 mg/L), and bile (2.7 mg/L) as well as in liver, kidney, lung, heart, and brain tissue (15). Hu et al described the case of a 41-year-old woman who attempted suicide by drinking 200 mL of methomyl-alphamethrin insecticide. This insecticide is a mixture of carbamate and pyrethroid insecticide. She presented to the hospital with hypothermia, unconsciousness, and shock. The diagnosis of carbamate poisoning was based on clinical symptoms and decreased levels of cholinesterase as well as the presence of methomyl in her urine. She complained of blurred vision and blindness 4 days postexposure. On day 21 she had numbness in her lower limbs, progressive weakness, and right foot drop. Electrophysiological tests performed on day 27 demonstrated neuropathy of the bilateral peroneal nerves. A brain magnetic resonance study and visual evoked potential confirmed her diagnosis of cortical blindness. The authors concluded that the patient manifested severe carbamate insecticide poisoning that was ultimately responsible for her blindness (16).

Case Report

A 45-year-old military man was brought to the hospital after consuming 250 mL of Furadan (carbofuran, a carbamate insecticide). He was gasping for breath with secretions in his mouth, throat, and nostrils. On admission he became unconscious and cyanotic and was salivating and lacrimating excessively. His serum potassium was 2.1 mmol/L, and he was given a potassium supplement. His plasma cholinesterase level was 2005 U/L on admission, and he was treated for carbamate poisoning using pralidoxime. Although his plasma cholinesterase level returned to normal range in 2 days and was normal (10,768 U/L) on day 6, his clinical condition worsened, and despite repeated treatment with the antidote pralidoxime and treating the convulsions with phenytoin as well as putting him on a ventilator, he died on day 7 of admission due to pesticide poisoning, right lobe pneumonia, and posthypoxic injury to the brain (17).

8.4. CHALLENGES IN USING CHOLINESTERASE AS A DIAGNOSTIC TEST FOR ORGANOPHOSPHATE AND CARBAMATE POISONING

Currently red cell (erythrocyte) acetylcholinesterase activity or serum plasma cholinesterase (also known as pseudocholinesterase or butyrylcholinesterase) activity can be measured for diagnosis as well as for monitoring the progress of therapy in a patient poisoned with organophosphate or carbamate insecticides. However, such laboratory tests also have limitations. In one case report, despite the return of the plasma cholinesterase level to normal following therapy, the patient died (17). However, Addullat et al demonstrated the value of serial measurements of the plasma cholinesterase level in the management of both organophosphate and carbamate poisoning. A 23% variation in the acetylcholinesterase level among normal individuals is expected. Therefore, a baseline value is important to overcome individual variation, and such a practice is possible where occupational exposure to organophosphate or carbamate exposure is expected. A worker's baseline value can be compared with the postexposure value to determine the severity of the poisoning. However, for an accidental or intentional overdose, such an approach is not possible. In such a case the baseline sample after organophosphate or carbamate poisoning indicates severe inhibition of plasma acetylcholinesterase activity. A significant increase in activity is expected with time if the causative agent is carbamate because of the reversible spontaneous cleavage of the carbamylated cholinesterase enzyme. In contrast, with organophosphate poisoning, persistent inhibition of acetylcholinesterase with time is expected because the enzyme organophosphorus complex is formed irreversibly. In such situations a bolus dose of oxime antidotes is necessary to achieve a significant increase in enzyme activity with the progress of therapy. In the study by Addullatet et al, the lowest plasma acetylcholinesterase activity varied from 12% to 27%, and the number of serial determinations of enzyme activity among these 10 patients varied from 4 to 11 measurements. For example, in one patient who ingested methomyl insecticide, the initial plasma cholinesterase level showed moderate activity (22%) of the normal value. However, after 2–4 hours postexposure, the enzyme activities increased up to 36% of the normal level indicating carbamate poisoning (methomyl is a carbamate pesticide). The patient was successfully treated with atropine. However, in another patient where the source of poisoning was unknown, the initial plasma acetylcholinesterase activity was only 10% of the normal value, and subsequent enzyme levels drawn 2–4 hours later were also low, indicating organophosphate poisoning. After administrating the antidote oxime, the values increased to 30% of normal activity. The authors determined decreased plasma acetylcholinesterase activity by comparing the patient's value with normal individuals where plasma acetylcholinesterase activity was 100%. The procedure involves the addition of a 20-μL plasma sample to 0.1 mL of 5% solution of acetylcholine iodide to 3 mL of dithiobisnitrobenzoic acid solution followed by incubation at 25°C for 2 minutes. The difference in absorbance between 0 and 2 minutes was considered for the estimation of plasma acetylcholinesterase activity (18).

Cander et al studied the prognostic value of the Glasgow coma scale and serum acetylcholinesterase levels for evaluating patients with organophosphorus poisoning based on a study of 25 patients with a mean age of 37 years (range: 20–80 years). Three patients died. The mean score on the Glasgow coma scale was statistically lower in patients who died versus patients who survived (4 versus 11.7). Although mean acetylcholinesterase levels were lower in patients who survived, the difference between the acetylcholinesterase levels in patients who died versus patients who survived was not significant. The authors concluded that although the serum acetylcholinesterase level can aid in the diagnosis of organophosphate poisoning, the level may not be able to predict the final outcome (19). Aurbeck et al commented that serum acetylcholinesterase (also termed butyrylcholinesterase) is an inappropriate diagnostic test for the therapeutic monitoring of oxime treatment in organophosphate poisoning because oxime-induced reactivation of the enzyme is too slow to accomplish pseudo-catalytic function (20). Plasma cholinesterase activity can also be decreased for certain pathophysiological conditions. In one report, the reduced activity of plasma cholinesterase in a patient was initially suspected due to organophosphorus poisoning, but there were no characteristic symptoms. Further investigation by the author revealed that the depressed activity of the enzyme was due to pulmonary Koch's and hepatitis B with associated malnutrition (21). Abou-Hatab et al reported that activities of several esterases including plasma acetylcholinesterase were significantly lower in older patients with pneumonia (22). Patients with phenylketonuria also demonstrate reduced activity of plasma acetylcholinesterases because the high level of phenylalanine in blood inhibits acetylcholinesterase activity (23).

8.5. METHANOL USE AND METHANOL POISONING

Methanol is biodegradable and widely used in the manufacturing of a variety of chemicals including formaldehyde, a precursor for manufacturing plastics, paints, synthetic textiles, adhesives, and foam cushions. Methanol is an alternative fuel. If 5–15% methanol is added to gasoline in internal combustion engines, there will be an immediate reduction in atmospheric pollution as well as an improvement in engine performance. Methanol can also be used in electrical power plants and for heating and other fuel applications (24). Methanol is also found in many household products including windshield washer fluid, carburetor cleaner, paints, varnishes, paint thinners, and various cleaning products. In addition, methanol is used in preparing denatured alcohol because the addition of methanol to ethanol makes it toxic and undrinkable. Denatured alcohol is used as a fuel for spirit burners, camp stoves, and canned heat designed to be burned directly from its can. Denatured alcohol, which usually contains 5–10% methanol, is also referred to as "methylated spirit" in many countries in the world.

8.5.1. Human Exposure to Methanol and Methanol Toxicity

Because of the widespread use of methanol in industry and the presence of methanol in some household products, human exposure to methanol may occur through different routes including inhalation, cutaneous exposure, and ingestion. Methanol is well absorbed through all three routes and may cause toxicity regardless of the route of exposure. In one report the authors studied all cases of methanol exposure from January 2003 to May 2005 using the Texas Poison Center Network database. There were 87 cases of methanol exposure through inhalation; 81 cases were through ingestion. Carburetor cleaner was responsible for most of the inhalation cases (79 of 87); most ingestions involved windshield washer fluid (39 of 81) and carburetor cleaner (20 of 81). Most of these patients (56% of patients in the inhalation group and 46% of patients in the ingestion group) were admitted to the hospital, and some patients experienced vision loss in both groups (25). Exposure to methanol may be dangerous to pregnant mothers. In one report, a 28-year-old pregnant woman was poisoned with methanol (route of exposure not reported) and was treated aggressively including using hemodialysis. Methanol was detected in the blood of the mother as well as in the blood of her newborn. Despite aggressive therapy, both the mother and her child died from methanol poisoning (26).

Case Report

In the 1980s, methanol production was introduced at a new petrochemical complex in the port of Jubail, Saudi Arabia. A consultant supervising tank cleaning prior to methanol loading wore a positive pressure breathing apparatus but no protective clothing. After working for 2–3 hours in the tank, he came out and worked on deck, but unfortunately he wore his methanol-soaked clothing that eventually dried out because methanol is very volatile with a low boiling point of only 65°C. He developed visual disturbances, a typical symptom of methanol toxicity, 8 hours after exposure but recovered fully in the hospital (27).

8.5.2. Methanol Metabolism and Toxicity

Methanol is readily absorbed after ingestion or inhalation and enters the bloodstream. A small amount of methanol is excreted unchanged in urine as well as exhaled breath. Most methanol is metabolized by alcohol dehydrogenase producing formaldehyde that is further metabolized to acetaldehyde dehydrogenase and then to formic acid.

$$CH_3OH \rightarrow HCHO \text{ (catalyzed by alcohol dehydrogenase)}$$

$$HCHO \rightarrow HCOOH \text{ (catalyzed by acetaldehyde dehydrogenase)}$$

Methanol itself is relatively nontoxic, and methanol toxicity is a classic example of "lethal synthesis" where metabolites are responsible for the toxicity of the agent. The major complication of methanol poisoning is loss of vision including partial or total blindness. Methanol toxicity may also be fatal. The lethal dose of methanol in humans is not fully established. Although it is assumed that ingestion of anywhere from 30 to 100 mL of methanol may cause death, fatality from methanol may occur even after ingestion of 15 mL of 40% methanol. However, visual impairment and blindness are more common with methanol ingestion from as little as 4 mL (28). The mechanism of methanol-induced blindness is not fully understood. Animal experiments indicate that formic acid, the end metabolite of methanol, is responsible for most of the damage. Retinol dehydrogenase present in the human retina metabolizes methanol inside the retina, producing formic acid causing retinal toxicity (29). Shin and Uhm reported a case of a 30-year-old man who presented at the hospital with decreased visual activities in both eyes after drinking alcohol containing methanol. His initial vision was limited to only hand motions and not corrected in either eye. Initial eye examination showed optic disc swelling in both eyes. Four years later his visual activity only allowed visualization of hand motions. His brain magnetic resonance imaging showed no specific change, but on fundus examination disc pallor in both eyes was observed. In addition, disc cupping with a high cup-to-disc ratio (0.9 in the left eye and 0.5 in the right eye) was noted. Methanol poisoning may induce glaucomatous disc cupping in the late stage as well as optic atrophy. One possible mechanism of disc cupping may be ganglion cell loss due to acute demyelination of the retrobulbar optic nerve (30).

Blood methanol levels may vary widely among individuals poisoned with methanol. In general if blood methanol concentration exceeds 20 mg/dL, treatment should be initiated. In one report where authors reviewed 12 fatal cases of methanol poisoning, six individuals who were found dead and received no therapy had postmortem methanol concentrations between 84 and 543 mg/dL and postmortem formic acid levels between 64 and 110 mg/dL. Six other individuals who received therapy prior to death showed blood methanol levels between 68 and 427 mg/dL and formic acid levels between 37 and 91 mg/dL (31). Interestingly, Lushine et al described a case in which a patient with a blood methanol level of 692 mg/dL on admission was treated with 4-methylpyrazole and dialysis and survived (32). Wu et al reported an exceptionally high methanol level of 1148 mg/dL in a 5-week-old infant suffering from moderate metabolic acidosis. The child was treated with folic acid and continuous ethanol infusion and survived without any permanent damage. The authors proposed that the infant survived despite a methanol level incompatible with life due to methanol metabolism using first-order rather than zero-order kinetics as well as involvement of a different enzyme such as catalase instead of alcohol dehydrogenase (usually poorly developed in infants) in the metabolism of methanol (33).

Methanol is metabolized to formic acid, which lowers blood pH causing metabolic acidosis. Wu et al reported no correlation between blood methanol level and the clinical outcome of methanol poisoning, but a blood pH of 7 or lower was a strong predictor of death or poor outcome from methanol poisoning (33). However, Jones et al reported based on a study of 153 cases of methanol poisoning where 107 deaths were attributed to methanol poisoning that blood formate concentration was greater than 50 mg/dL (mean: 85 g/dL) except in one suicide involving methanol where serum methanol concentration was 790 mg/dL and formate concentration was 12 mg/dL (34). Another complication of methanol poisoning is lactic acidosis, which may also be life threatening.

8.5.3. Laboratory Diagnosis of Methanol Poisoning

Direct determination of methanol in serum or plasma using gas chromatography is very useful in determining the severity of methanol poisoning. In addition to methanol, gas chromatographic analysis is capable of determining other volatiles such as ethanol, acetone, and isopropyl alcohol. Determination of formic acid (formate) is also useful in the diagnosis of suspected methanol poisoning. There are many published methods for simultaneous analysis of methanol, ethanol, isopropanol, ethylene glycol, and other volatiles using gas chromatography or headspace gas chromatography. Williams et al described the detection and quantitation of diethylene glycol, ethylene glycol, methanol, isopropanol, acetone, and ethanol using capillary column gas chromatography after deproteinizing specimen using ultrafiltration. The authors used n-propanol as the internal standard for alcohols and acetone; 1,3-butanediol was used as the internal standard for the quantification of glycols (35). Gas chromatography can also be used for the determination of formate concentration in serum. Fraser and MacNeil described a headspace gas chromatographic procedure for the analysis of formate where formate present in serum or whole blood was methylated in the presence of concentrated sulfuric acid sodium propionate (internal standard) (36).

Because methanol toxicity may also cause metabolic acidosis, arterial blood gas analysis is helpful in the diagnosis of methanol poisoning. Another indirect indication of methanol poisoning is the increased anion gap in the serum. The anion gap is defined as the difference between the measured level of positively charged ions (cations) in the blood and negatively charged ions (anions) in the blood.

$$\text{Anion Gap} = \text{Concentration of sodium} - (\text{Concentration of chloride} + \text{Concentration of bicarbonate})$$

Concentration of potassium is often ignored because it is very small. The normal anion gap is 8–16 mmol/L, but in methanol poisoning this anion gap may increase significantly. Both methanol and its metabolite formate are responsible for increasing the anion gap. The anion gap may increase in many pathological conditions such as renal failure, lactic acidosis, and in common poisonings such as salicylate poisoning. Methanol poisoning also increases the serum osmolar gap. In serum, sodium, potassium, chloride, bicarbonate, urea, and glucose together make up 95% of total osmolality. Serum osmolality is measured by a principle called "freezing point depression" using an osmometer and is called measured osmolality. Osmolality can also be calculated from measured concentrations of sodium, potassium, chloride, glucose, and urea.

$$\text{Osmolar gap} = \text{Measured osmolarity} - \text{calculated osmolarity}$$

Usually measured osmolarity should be close to calculated osmolality unless compounds like methanol, ethanol, acetone, ethylene glycol, or related compounds are present in serum. These compounds would increase serum osmolality. Hovda et al measured serum formate concentration enzymatically on a Cobas Mira analyzer using formate dehydrogenase and nicotinamide adenine dinucleotide as well as anion gap and osmolar gap in 15 patients poisoned with methanol. As expected, methanol was detected in all 15 patients, and 14 patients also had an elevated formate concentration in serum. The anion gap and osmolar gap were increased in 11 patients. Four patients with no symptoms of methanol poisoning had serum concentrations of formate between 2 and 38 mg/dL. The authors concluded that formate measurement is a simple and sensitive indicator of methanol poisoning (37).

Methanol poisoning can be treated using a variety of agents such as ethanol, 4-methylpyrazole (Fomepizole), and sodium bicarbonate as well as dialysis. The outcome of methanol poisoning appears to be related more to the interval of time between exposure and initiation of treatment and to the degree of acidosis than to the initial blood methanol level. Early and aggressive therapy with bicarbonate and ethanol and subsequent initiation of hemodialysis are strongly recommended whenever methanol can be detected in blood, especially in patients who also have metabolic acidosis and a demonstrated anion gap. Another effective therapy for methanol overdose is hemodialysis. Methanol is a small molecule with a molecular weight of only 32. Methanol can be effectively removed from circulation using hemodialysis. Usually hemodialysis along with ethanol therapy for methanol poisoning should be initiated if the blood methanol level is 50 mg/dL or more. Hemodialysis may also be initiated if an individual ingested 30 mL or more of methanol or based on other clinical indications as determined by the physician treating such an overdose (38). More recently, fomepizole (4-methylpyrazole), a potent competitive inhibitor of alcohol dehydrogenase, has been used as an antidote to treat methanol poisonings. This antidote can slow down the formation of toxic formaldehyde metabolite formation from methanol, and the small formaldehyde buildup can be secreted in urine. Bicarbonate therapy can also be used with fomepizole therapy to correct metabolic acidosis caused by methanol (39). Sakuma et al described two cases of methanol poisoning. In the first patient the blood methanol concentration was 56.4 mg/dL and the serum ethanol level was 2.4 mg/dL. Serum formate was not detected initially but after ethanol infusion was started, serum formate was detected later in the serum due to the slow elimination of methanol. Both his methanol and formate levels decreased after initiation of hemodialysis. In the second patient, serum methanol and formate levels were 400 mg/dL and 13.4 mg/dL, respectively, but no ethanol was detected in the blood. He underwent hemodialysis immediately, and his methanol and formate levels decreased. Both patients were eventually discharged from the hospital without any sequelae (40).

8.6. ETHYLENE GLYCOL USE AND ETHYLENE GLYCOL POISONING

Ethylene glycol is a colorless and relatively nonvolatile liquid with a high boiling point of 194°C. It is used in the preparation of polyester products and widely used as a deicing fluid. Ethylene glycol is used as a major ingredient in automobile antifreeze. Because of its sweet taste, children and pets tend to ingest it, causing ethylene glycol toxicity. An adult may drink ethylene glycol as a substitute for ethanol or in a suicide attempt.

Because ethylene glycol is relatively nonvolatile, inhalation exposure to ethylene glycol is not generally considered as an occupational health hazard. In one study, the authors measured ethylene glycol levels in 154 breathing zone air samples and 117 urine samples from 33 aviation workers exposed to deicing fluid (basket operators, deicing truck drivers, and coordinators) during 42 workdays over a winter period of 2 months at a Montreal airport and concluded that health hazards from exposure to ethylene glycol in the form of inhalation is not significant, but other routes of exposure such as the percutaneous route may cause health hazards (41). Moreover, absorption of ethylene glycol through the skin may cause serious toxicity if any skin lesion is present. Bouattar et al reported a case of toxicity in a 38-year-old man who presented at the hospital with nausea, vomiting, abdominal pain, and worsening of his mental status. The patient suffered from renal failure, and a renal biopsy revealed the presence of calcium oxalate crystals, a characteristic of ethylene glycol poisoning. It was later discovered that the patient worked in a cement factory and handled ethylene glycol without protective gloves. In addition, the patient had cutaneous psoriasis for 10 years (42).

8.6.1. Ethylene Glycol Metabolism and Toxicology

Ethylene glycol itself is relatively nontoxic like methanol, but its metabolites, especially oxalic acid, are very toxic. Ethylene glycol is also metabolized by alcohol dehydrogenase and acetaldehyde dehydrogenase, and the half-life of ethylene glycol in blood is 3–5 hours. Ethylene glycol is first metabolized to glycoaldehyde by alcohol dehydrogenase, and then glycoaldehyde is further metabolized by aldehyde dehydrogenase into glycolic acid.

$$\underset{\text{Ethylene Glycol}}{CH_2\text{-}OH\text{–}CH_2\text{-}OH} \xrightarrow{\text{Alcohol Dehydrogenase}} \underset{\text{Glycoaldehyde}}{CH_2\text{-}OH\text{–}CHO}$$

$$CH_2\text{-}OH\text{–}CHO \xrightarrow{\text{Acetaldehyde Dehydrogenase}} \underset{\text{Glycolic Acid}}{CH_2\text{-}OH\text{–}COOH}$$

The conversion of ethylene glycol in two steps to glycolic acid occurs relatively rapidly, but then glycolic acid is further transformed through to glyoxalic acid and then finally to oxalic acid. Oxalic acid binds with calcium-forming calcium oxalate, the end toxic metabolite of ethylene glycol. The mechanism of conversion of glycolic acid to oxalic acid is not fully understood, but it has been established that lactate dehydrogenase enzymes present in hepatocytes (major cells in the liver) catalyzes this transformation (43).

Ethylene glycol toxicity in human occurs in several stages. The major complications of ethylene glycol poisoning are metabolic acidosis and renal failure, and these complications may even be fatal. The lethal dose of ethylene glycol is usually assumed as 100 mL, but there are reports of fatality from ethylene glycol poisoning even from ingestion of only 30 mL (44). Death may follow from ethylene glycol poisoning if untreated within 8–24 hours after ingestion, but the prognosis of ethylene glycol poisoning is good if treated in a timely fashion. The blood level of ethylene glycol in fatal poisoning may vary widely between different individuals. Rosano et al reviewed 12 medical examiners' cases where fatality was due to ethylene glycol poisoning and observed that the ethylene glycol concentrations ranged widely from only 5.8 to 779 mg/dL with a mean value of 183 mg/dL. The concentration of glycolic acid, a metabolite of ethylene glycol, varied from 81 mg/dL to 177 mg/dL. Calcium oxalate crystals were detected in renal tissues (45). In another case report a man died from ethylene glycol poisoning with a blood ethylene glycol level of 25 mg/dL. Acute renal failure was the cause of death, and calcium oxalate crystals were identified in renal cells (tubular epithelial cells) using confocal laser scanning microscopy (46). Garg et al reported a case in which a person who died from ethylene glycol poisoning showed a very high level of ethylene glycol in postmortem blood (2340 mg/dL) but without elevated concentration of any ethylene glycol metabolites. In addition, oxalic acid crystals were not detected in the urine (47).

Ethylene glycol poisoning often results in acute renal failure, especially if treatment is delayed. The mechanism of ethylene glycol toxicity was thought to be due to the accumulation of toxic metabolites such as glycoaldehyde, glyoxylate, and oxalic acid. However, more recent investigations reveal that the accumulation of calcium oxalate crystals, mostly calcium oxalate monohydrate, accounts for most major toxicity and acute renal failure due to ethylene glycol poisoning (48). Although if untreated ethylene glycol poisoning causes renal failure and cerebral edema, immediate treatment can save lives, and patients often experience a total recovery despite consuming a large quantity of ethylene glycol (49).

8.6.2. Laboratory Diagnosis of Ethylene Glycol Poisoning

Laboratory findings of ethylene glycol poisoning include increased anion gap, metabolic acidosis (blood pH significantly lower than 7.35), and increased osmolar gap, oxalic acid crystals in urine, and detectable ethylene glycol level in blood (50). For example, if ethylene glycol is present in the blood in an amount of 50 mg/dL, it would increase the measured serum osmolality by 8.1, thus increasing the osmolar gap. In the absence of direct measurement of ethylene glycol, elevated osmolar gap in a case of suspected ethylene glycol poisoning is diagnostically useful (51).

Blood levels of ethylene glycol are usually measured by headspace gas chromatography either singly or in combination with other volatile compounds such as methanol, acetone, and isopropyl alcohol. In addition, some enzymatic methods are available for the rapid determination of blood ethylene glycol levels using an automated analyzer in the clinical laboratory. Yao and Porter described a gas chromatographic method for the analysis of ethylene glycol and its toxic metabolite glycolic acid in human serum using gas chromatography. The authors mixed 100 μL of the patient's serum, control, or calibrator with 200 μL of internal standard (3-bromo-1-propanol) in acetonitrile to precipitate proteins. After centrifugation of 100 μL of supernatant was combined with a mixture of 2,2-dimethoxypropane, N,N-dimethylformamide, and acetic acid (78:20:2 by volume) and after incubation for 5 minutes at room temperature, the mixture was heated at 60°C to reduce the volume roughly to less than 100 μL. Then 80 μL of N-O-bis(trimethylsilyl) trifluoroacetamide (BSTFA) also containing trimethylchlorosilane (TMS) was added, and then 1 μL of the derivatized specimen was injected into the gas chromatograph. The authors achieved baseline separation between all analytes including the internal standard (52). Chromatogram of serum blank and a serum specimen containing ethylene glycol, propylene glycol, glycolic acid and internal standard are shown in Figure 8.2. However, use of propylene glycol as the internal standard may be inappropriate because some medications contain propylene glycol and diagnosis of such patients may be difficult if propylene glycol is used as the internal standard.

Case Report

A 32-year-old woman was admitted to the hospital after a suicide attempt using multiple drugs. On admission, she was found agitated. Gastric lavage

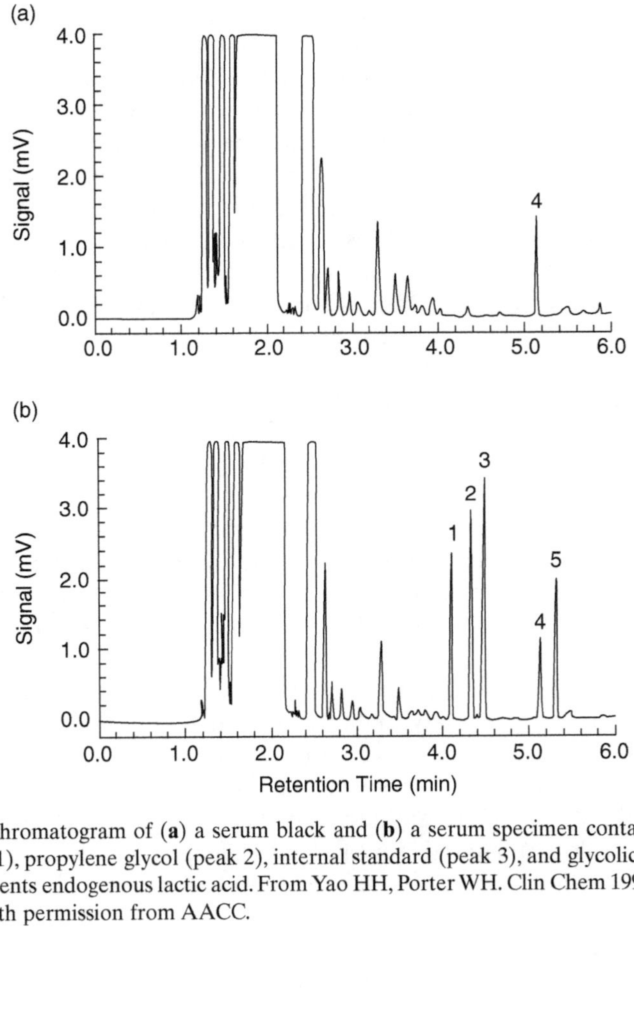

Figure 8.2. Chromatogram of (**a**) a serum black and (**b**) a serum specimen containing ethylene glycol (peak 1), propylene glycol (peak 2), internal standard (peak 3), and glycolic acid (peak 5). Peak 4 represents endogenous lactic acid. From Yao HH, Porter WH. Clin Chem 1996;42:292–297. Reprinted with permission from AACC.

was performed and activated charcoal administered. At the same time she was given intravenous diazepam and lorazepam for sedation. When ethylene glycol concentration in serum was analyzed using gas chromatography, a compound present in the serum was eluted along with the internal standard propylene glycol due to similar retention time. Propylene glycol concentration increased with time after admission (propylene glycol level was 210 mg/L at 0.5 hours after admission but peaked at 1240 mg/L 15 hours after admission). However, no ethylene glycol was found in the serum. It was later discovered that both diazepam and lorazepam preparations contained propylene glycol. In addition, lactic acid was also found at an elevated concentration 15 hours after admission probably due to propylene glycol toxicity. The use of propylene glycol is therefore inappropriate for the analysis of ethylene glycol, and a different internal standard such as 1,3-propanediol should be used (53).

Although gas chromatographic methods are widely used for the determination of ethylene glycol concentrations in serum, there is also gas chromatography combined with mass spectrometric methods for analysis of ethylene glycol. Dasgupta and Macaulay used 4-carboethpxyhexafluorobutyryl chloride for the derivatization of ethylene glycol along with the internal standard 1,2-butanediol after extraction from serum. A distinct pronated molecular ion peak using chemical ionization at m/z 563 was observed for the derivatized ethylene glycol (54). Gembus et al described a simple extraction and derivatization procedure for the analysis of eight glycol including ethylene glycol, diethylene glycol, triethylene glycol, 1,2-propanediol, 1,3-propanediol, 1,2-butanediol, 2,3-butanediol, and hexylene glycol using only 2-μL serum or whole blood. Following deproteinization with acetonitrile, glycols were converted into mono- or di-trimethylsilyl derivatives and analyzed by gas chromatography combined with mass spectrometry. γ-Hydroxybutyrate d6 was used as the internal standard (55). Meyer et al described a gas chromatography combined with mass spectrometric protocol for the simultaneous analysis of ethylene glycol, 1,2-propylene glycol, lactic acid, glycolic acid, γ-hydroxybutyric acid, diethylene glycol, triethylene glycol, and tetraethylene glycol in human plasma or urine after derivatization with bis-N,O-trimethylsilyl trifluoroacetamide and the catalyst dimethylformamide. Derivatization was assisted by microwave (56).

Ethylene glycol is known to interfere with lactic acid measurement using an ABL blood gas analyzer but usually does not interfere with lactic acid measurement using a chemistry analyzer (57). In another report when blood gas was analyzed in a 29-year-old man in the emergency department using the ABL 725 Radiometer analyzer, elevated lactate was noted. However, toxicological analysis revealed ethylene glycol poisoning. Repeated analysis using a clinical chemistry analyzer showed minimal plasma elevation of lactate (58).

Ethylene glycol poisoning is treated using bicarbonate, ethanol, fomepizole, and hemodialysis. If treatment can be started early enough after ingestion, a simple administration of ethanol intravenously may be sufficient for full recovery from ethylene glycol poisoning. Fomepizole is also useful for pediatric ethylene glycol, diethylene glycol, butoxyethanol, and methanol poisoning (59). Hemodialysis, along with ethanol or fomepizole infusion and bicarbonate therapy to correct metabolic acidosis, may also be needed in treating patients with more serious ethylene glycol toxicity. In general, patients with severe metabolic acidosis, high serum potassium values, seizure, or coma during admission show poor outcomes from ethylene glycol poisoning. Hovda et al described an interesting case in which a 26-year-old woman with a dissociative disorder was admitted to the hospital with ethylene glycol poisoning a total of 154 times. She was treated with fomepizole 99 times, ethanol 60 times, and dialysis 73 times. She was finally found dead outside the hospital with a ethylene glycol blood concentration of 506 mg/dL (60).

8.7. CONCLUSIONS

Although poisoning with organophosphate and carbamate insecticide poisoning are different in presentation than methanol and ethylene glycol poisoning, the toxicology laboratory may play an important role in the diagnosis and monitoring of progress of therapy with each of these poisons. Antidotes are available for treating organophosphate and carbamate poisoning. In addition, effective therapies are also available for treating methanol and ethylene glycol poisoning. The activity of serum acetylcholinesterase can be easily determined using a spectroscopic method for the diagnosis of organophosphate and carbamate poisoning. Although a small community hospital may not have a gas chromatograph available for measuring either methanol or ethylene glycol concentration in serum, the indirect measurement of the anion gap and osmolar gap along with arterial blood gas are also useful in the diagnosis of methanol and ethylene glycol poisoning.

REFERENCES

1. Litchfield MH. Estimate of acute pesticide poisoning in agriculture workers in less developed countries. Toxicol Rev 2005; 24: 271–278.
2. Reeves M, Sachafer KS. Greater risk fewer rights; US farm workers and pesticides. Int J Occup Environ Health 2003; 9: 30–39.
3. Calvert GM, Karnik J, Mehler L, Beckman J et al. Acute pesticide poisoning among agricultural workers in the United States, 1998–2005. Am J Ind Med 2008; 51: 883–898.
4. Liu T, Zhang CH, Zhang P, Jin MH. Situation of pesticide poisoning in Huzhou from 2006 to 2009 [in Chinese]. Zhonghua Lao Dong Wei Sheng Zhi Ye Bing Za Zhi 2011; 29: 53–55.
5. Eddleston M, Buckley NA, Eyer P, Dawson AH. Management of acute organophosphate pesticide poisoning. Lancet 2008; 371: 597–607.
6. Davies J, Roberts D, Eyer P, Buckley N et al. Hypotension in severe dimethoate self-poisoning. Clin Toxicol (Phila) 2008; 46: 880–884.
7. Lima JS, Reis CA. Poisoning due to illegal use of carbamate as a rodenticide in Rio de Janeiro. J Toxicol Clin Toxicol 1995; 33: 687–690.
8. Solomon GM, Moodley J. Acute chlorpyrifos poisoning in pregnancy: a case report. Clin Toxicol (Phila) 2007; 45: 416–419.
9. Isbister GK, Mills K, Friberg LE, Hodge M et al. Human methyl parathion poisoning. Clin Toxicol (Phila) 2007; 45: 956–960.
10. Bajgar J. Organophosphate/nerve agent poisoning: mechanism of action, diagnosis, prophylaxis and treatment. Adv Clin Chem 2004; 38: 151–216.
11. Jayawardane P, Dawson AH, Weerasinghe V, Karalliedde L et al. The spectrum of intermediate syndrome following acute organophosphate poisoning: a prospective cohort study from Sri Lanka. Plos Med 2008; 15: e147.

12. Karalliedde L, Baker D, Marrs TC. Organophosphate induced intermediate syndrome: etiology and relationship with myopathy. Toxicol Rev 2006; 25: 1–14.
13. Wang CY, Wu CL, Tsan YT, Hsu JY et al. Early onset of pneumonia in patients with cholinesterase inhibitor poisoning. Respirology 2010; 15: 961–968.
14. Jokanovic M. Medical treatment of acute poisoning with organophosphorus and carbamate poisoning. Toxicol Lett 2009; 190: 107–115.
15. Hoizey G, Canes F, Binet L, Kaltenbach ML et al. Thiodicarb and methomyl tissue distribution in a fatal multiple compound poisoning. J Forensic Sci 2008; 53: 499–502.
16. Hu YH, Yang CC, Deng JF, Wu ML. Methomyl-alphamethrin poisoning presented with cholinergic crisis, cortical blindness and delayed peripheral neuropathy. Clin Toxicol (Phila) 2010; 48: 859–862.
17. Pinakini KS, Kumar TS. Serial cholinesterase estimation in carbamate poisoning. J Clin Forensic Med 2006; 13: 274–276.
18. Abdullat IM, Battah AH, Hadidi KA. The use of serial measurement of plasma cholinesterase in the management of acute poisoning with organophosphorus and carbamates. Forensic Sci Int 2006; 162: 126–130.
19. Cander B, Dur A, Yildiz M, Koyuncu F et al. The prognostic value of the Glasgow coma scale, serum acetylcholinesterase and leukocyte levels in acute organophosphorus poisoning. Ann Saudi Med 2011; 21: 163–166.
20. Aurbek N, Thiermann H, Eyer F, Eyer P et al. Suitability of human butyrylcholinesterase as therapeutic marker and pseudo catalytic scavenger in organophosphate poisoning: a kinetic analysis. Toxicology 2009; 259: 133–139.
21. Mahadeshwara P, Gouda HS, Hallikeri VR. Plasma cholinesterase: double edged parameter in the diagnosis of acute organophosphorus poisoning. Med Sci Law 2010; 50: 159–160.
22. Abou-Hatab K, Ganeshalingam K, O'Mahony MS, Giurani F et al. The effect of community-acquired pneumonia on plasma esterases in older people. Eur J Clin Pharamcol 2001; 57: 55–60.
23. Tsakiris S, Schulpis KH, Tjamouranis J, Michelakakis H et al. Reduced acetylcholinesterase activity in erythrocyte membranes from patients with phenylketonuria. Clin Biochem 2002; 35: 615–619.
24. Reed TB, Lerner RM. Methanol: a versatile fuel for immediate use: Methanol can be made from gas, coal or wood. It is stored and used in existing equipments. Science 1973; 182: 1299–1304.
25. Givens M, Kalbfleisch K, Bryson S, Carl R. Comparison of methanol exposure routes reported to Texas poison control center. Western J Emerg Med 2003; 9: 150–153.
26. Belson M, Morgan BW. Methanol toxicity in a newborn. J Toxicol Clin Toxicol 2004; 42: 673–677.
27. Downie A, Khattab TM, Malik MI, Samara IN. A case of percutaneous industrial methanol toxicity. Occup Med (London) 1992; 42: 47–49.
28. Kruse JA. Methanol poisoning. Intensive Care Med 1992; 18: 391–397.
29. Garner CD, Lee EW, Terzo TS, Louis-Ferdinard RT. Role of retinal metabolism in methanol induced retinal toxicity. J Toxicol Environ Health 1995; 44: 43–56.

30. Shin YW, Uhm KB. A case of optic nerve atrophy with severe disc cupping after methanol poisoning. Korean J Opthalmol 2011; 25: 146–150.
31. Wallage HR, Watterson JH. Formic acid and methanol concentrations in death investigations. J Anal Toxicol 2008; 32: 241–247.
32. Lushine KA, Harris CR, Holger JS. Methanol ingestion: prevention of toxic sequelae after massive ingestion. J Emerg Med 2004; 24: 433–436.
33. Wu AH, Kelly T, McCay C, Ostheimer D et al. Definitive identification of an exceptionally high methanol concentration in an intoxication of a surviving infant: methanol metabolism by first order elimination kinetics. J Forensic Sci 1995; 40: 315–320.
34. Jones GR, Singer PP, Rittenbach K. The relationship of methanol and formate concentrations in fatalities where methanol is detected. J Forensic Sci 2007; 52: 1376–1382.
35. Williams RH, Shah SM, Maggiore JA, Erickson TB. Simultaneous detection and quantitation of diethyl glycol, ethylene glycol, and the toxic alcohols in serum using capillary column gas chromatography. J Anal Toxicol 2000; 24: 621–626.
36. Fraser AD, MacNeil W. Gas chromatographic analysis of methyl formate and application in methanol poisoning cases. J Anal Toxicol 1989; 13: 73–76.
37. Hovda KE, Urdal P, Jacobsen D. Increased serum formate in the diagnosis of methanol poisoning. J Anal Toxicol 2005; 29: 586–588.
38. Gonda A, Gault H, Churchill D, Hollomby D. Hemodialysis for methanol intoxication. Am J Med 1978; 64: 749–758.
39. Hovda KE, Anderson KS, Urdal P, Jacobsen D. Methanol and formate kinetics during treatment with fomepizole. Clin Toxicol (Phila) 2005; 43: 221–227.
40. Sakuma I, Arakawa M, Hirose Y, Sekiguchi H et al. Two cases of methanol poisoning with chronological measurements of blood concentrations of methanol, ethanol and formate [in Japanese]. Chudoka Kenkyu 2010; 23: 232–237.
41. Gerin M, Patrice S, Begin D, Goldberg MS et al. A study of ethylene glycol exposure and kidney function of aircraft de-icing workers. Int Arch Occup Environ Health 1997; 69: 255–265.
42. Bouattar T, Madani N, Hamzaqui H, Alhamany Z et al. Severe ethylene glycol intoxication by skin absorption [in French]. Nephrol Ther 2009; 53: 205–209.
43. Baker PR, Cramer SD, Kennedy M, Assimos DG et al. Glycolate and glyoxylate metabolism in HepG2 cells. Am J Physiol Cell Physiol 2004; 287: C1350–1365.
44. Walder AD, Tyler CKG. Ethylene glycol antifreeze poisoning: three case reports and a review of treatment. Anesthesia 1994; 49: 964–967.
45. Rosano TG, Swift TA, Kranick CJ, Sikirica M. Ethylene glycol and glycolic acid in postmortem blood from fatal poisonings. J Anal Toxicol 2009; 33: 508–513.
46. Pomara C, Fiore C, D'Errico S, Riezzo I et al. Calcium oxalate in acute ethylene glycol poisoning: a confocal laser scanning microscopy study in a fatal case. Clin Toxicol (Phila) 2008; 46: 322–324.
47. Garg U, Frazee C, Johnson L, Turner JW. A fatal case involving extremely high levels of ethylene glycol without elevation of its metabolites or crystalluria. Am J Forensic Med Pathol 2009; 30: 273–275.
48. McMartin K. Are calcium oxalate crystals involved in the mechanism of acute renal failure in ethylene glycol poisoning? Clin Toxicol 2009; 47: 859–869.

49. Holyoak AL, Fraser TA, Gelperowicz P. Cooling in the tropics: ethylene glycol overdose. Crit Care Resusc 2011; 13: 28–32.

50. Eder AF, McGarth CM, Dowdy YG, Tomaszewski JE et al. Ethylene glycol poisoning: toxicokinetics and analytical factors affecting laboratory diagnosis. Clin Chem 1998; 44: 168–177.

51. Zaldibar EE, Aguilera CL, Aguayo CE. Severe acute ethylene glycol poisoning: diagnostic utility of osmolar gap monitoring [in Spanish]. Rev Esp Anestesiol Reanim 2011; 58: 183–185.

52. Yao HH, Porter WH. Simultaneous determination of ethylene glycol and its major toxic metabolite glycolic acid in serum by gas chromatography. Clin Chem 1996; 42: 292–297.

53. Apple FS, Goorgins, Resen D. Propylene glycol interferes with gas-chromatographic assay of ethylene glycol [Letter to the editor]. Clin Chem 1993; 39: 167.

54. Dasgupta A, Macaulay R. A novel derivatization of ethylene glycol from human serum using 4-carboethoxyhexafluorobutyryl chloride for unambiguous gas chromatography-chemical ionization mass spectrometric identification and quantification. Am J Clin Pathol 1995; 104: 283–288.

55. Gembus V, Goulle JP, Lacroix C. Determination of glycols in biological specimens by gas chromatography-mass spectrometry. J Anal Toxicol 2002; 26: 280–285.

56. Meyer MR, Weber AA, Maurer HH. A validated GC-MS procedure for fast, simple and cost effective quantification of glycols and GHB in human plasma and their identification in urine and plasma developed for emergency toxicology. Anal Bioanal Chem 2011; 400: 411–414.

57. Meng QH, Adeli K, Zello GA, Porter WH et al. Elevated lactate in ethylene glycol poisoning: true or false? Clin Chim Acta 2010; 411: 601–604.

58. Sandberg Y, Rood PP, Russcher H, Zwaans JJ et al. Falsely elevated lactate in severe ethylene glycol intoxication. Neth J Med 2010; 68: 320–323.

59. Brent J. Fomepizole for treatment of pediatric ethylene and diethylene glycol butoxyethanol and methanol poisoning. Clin Toxcol (Phila) 2010; 48: 401–406.

60. Hovda KE, Julsurd J, Ovrebo S, Brors O et al. Studies on ethylene glycol poisoning: one patient—154 times. Clin Toxicol (Phila) 2011; 49: 478–484.

Chapter 9

Poisoning with Warfarin and Superwarfarin: What Can Laboratory Testing Do?

9.1. INTRODUCTION

The coumarin-type anticoagulants such as warfarin, phenprocoumon, and acenocoumarol have been used for a long time. Warfarin (Coumadin) is the most commonly used drug in this category, commonly prescribed for the treatment of venous and arterial thromboembolic disorders (1). Warfarin was initially marketed in 1948 as a rodent poison, but later its usefulness as an anticoagulant for humans was discovered, and since the 1950s this drug has been used in clinical practice.

Vitamin K is essential for the activation of various clotting factors (II, VII, IX, and X), and in this process vitamin K is oxidized to vitamin K epoxide. Through the action of the enzyme vitamin K epoxide reductase complex 1 (VKORC1), vitamin K epoxide is converted back into vitamin K in the liver. Warfarin and its related compounds act as vitamin K antagonists by inhibiting VKORC1, and as a result the hepatic synthesis of various blood clotting factors such as prothrombin (factor II), factor VII, factor IX, and factor X become impaired. In addition, warfarin also interferes with the action of other anticoagulant proteins such as protein C and protein S (2). Following initiation

Resolving Erroneous Reports in Toxicology and Therapeutic Drug Monitoring: A Comprehensive Guide, First Edition. Amitava Dasgupta.

of warfarin therapy, 5–7 days are needed for its anticoagulant effect because preexisting clotting factors must be degraded naturally before defective clotting factors are produced by the liver after the start of warfarin therapy. Because warfarin therapy has an induction period, patients with clots, for example patients with pulmonary embolism or deep vein thrombosis, may need short-term therapy with heparin. The three most common indications for warfarin therapy are atrial fibrillation, venous thromboembolism, and prosthetic heart valves.

Warfarin is a synthetic compound first developed at the Wisconsin Alumni Research Foundation in 1947, and hence the name warfarin (derived from WARF, standing for Wisconsin Alumni Research Foundation, plus "arin" indicating its link to the natural product coumarin). Currently warfarin is available as a racemic mixture of 50% R-warfarin and 50% S-warfarin sodium salt (3). The S-warfarin is four times more potent than R-warfarin as an anticoagulant and is also susceptible to drug interactions. Warfarin is highly water soluble, and peak plasma concentration can be achieved within 60–90 minutes after oral administration. Warfarin binds to the enzyme vitamin K 2, 3-epoxide reductase in liver microsomes stopping the cycle of vitamin K and reducing γ-carboxylation of the precursors of vitamin K–dependent pro- and anticoagulant factors. Warfarin is strongly bound to serum albumin (97–99%); it is the small free fraction of warfarin that exerts its pharmacological action. The elimination half-life of warfarin varies greatly between 35 and 45 hours, but the S-isomer has an average half-life shorter than the R-isomer (4).

9.2. DOSING AND THERAPEUTIC DRUG MONITORING OF WARFARIN: GENETIC POLYMORPHISM

Warfarin therapy can be complicated by either ineffective coagulation or risk of bleeding. In addition, due to the genetic polymorphism of CYP2C9, VKORC1, the dose requirement of warfarin among different individuals varies widely. Moreover, warfarin can interact with many drugs and food, and such interactions can be either pharmacokinetic or pharmacodynamic. Given all these factors that may affect the effectiveness and safety of warfarin therapy, pharmacodynamic monitoring of warfarin therapy is critical. The most commonly adopted monitoring of warfarin therapy is maintaining an acceptable prothrombin time expressed as the international normalized ratio (INR). The INR allows prothrombin time results to be normalized, thus making them comparable across different clinical laboratories because different clinical laboratories may use different assay methods. However, INR reflects a complex physiological end point of warfarin therapy that is affected by both vitamin K–dependent (II, VI, and X) and vitamin K–independent (I, V) clotting factors. An INR value of 1.0 is the average value for healthy individuals not on warfarin or other anticoagulant therapy. The target INR value for patients receiving warfarin is often 2.5, although the INR target range can range from 1.5 to

3 (5). The risk of hemorrhage increases at INR values greater than 4.0. Depending on the patient, it may take weeks to even a month to reach a stable INR. Flexible protocols, computer aids, and nomograms have helped to improve control, but there are still significant limitations in these approaches (6–10). Nevertheless, the major risk of bleeding from warfarin therapy has been reported to range from 1% to 3% per year (11). Warfarin serum and plasma levels are rarely determined for clinical purposes and are typically available only at large reference or academic medical center clinical laboratories.

When initiating warfarin therapy, clinicians should avoid loading doses that can raise INR excessively; instead warfarin should be initiated at a dose of 5 mg (or 2–4 mg for the elderly). With a 5-mg initial dose, the INR should not rise initially except for the few patients who require a small dose due to genetic polymorphism. The considerable variability in the warfarin dose-response relationship between individuals is explained mainly by genetic polymorphism of CYP2C9, the major enzyme that metabolizes warfarin and its target enzyme (VKORC1) (12). Unfortunately, current warfarin-dosing algorithms do not incorporate genetic factors in their calculation of warfarin dosage. In reality, age, genotype, and body surface area all affect warfarin dosage. CYP2C9 polymorphism and age markedly affect blood level of more active S-warfarin and hence dosage. Sconce et al reported that the incorporation of age, CYP2C9 and VKORC1 genotype, and height of the patient allow the best warfarin maintenance dose (13). The common polymorphism of the CYP2C9 gene includes the CYP2C9*2 isoform, which is due to the replacement of arginine at amino acid residue 144 by cysteine. This mutation reduces the catalytic activity of the enzyme to about 12% of the wild-type enzyme. CYP2C9*3 results in a substitution of leucine for isoleucine at amino acid position 359, which results in a reduction of catalytic activity to about 5% of the wild type. Therefore, as expected, both CYP2C9*2 and CYP2C9*3 genetic variants, which are common in white populations, with allele frequencies varying from 3.3% to 18% (14,15) result in significantly impaired hydroxylation of warfarin compared with the wild-type CYP2C9 enzyme. Therefore, these individuals require much lower dosage of warfarin and a longer time to achieve stable warfarin dosing, and they are at high risk of bleeding from warfarin therapy (16). In patients who are homozygous for CYP2C9*3, the clearance of S-warfarin can be reduced by 90% which can lead to a 10-fold increase in the half-life of warfarin (17). Therefore these individuals require much lower dosage of warfarin and are at an increased risk of bleeding.

Case Report

A 30-year-old Korean woman presented to the clinic with complaints of dizziness and dyspnea. The patient had a history of hypertrophic cardiomyopathy with complete atrioventricular block, and a pacemaker had been placed 13 years prior to her current clinic visit. She was taking Lasix 20 mg and spironolactone 2.5 mg along with teprenone 50 mg twice daily. Her

INR was 1.03, and cardiomegaly was found on chest radiograph. The patient was started with 4 mg of warfarin daily (as the standard dose) to prevent thromboembolism. When she revisited the outpatient clinic a week later, her prothrombin time was elevated to 88.2 seconds and her INR was highly elevated to 6.36, indicating a high risk of bleeding. She was admitted, and 10 mg of vitamin K was injected along with warfarin withdrawal. Eventually after normalization of her INR, the patient was discharged from the hospital after 16 days with her warfarin dosage reduced to 2 mg per day. The patient had a genotype of CYP2C9*1/*3, explaining why she needed a very low dosage of warfarin. The authors further commented that the CYP2C9*3 variant had a higher impact on the low clearance of warfarin compared with the CYP2C9*2 variant, and in general CYP2C9*2 and CYP2C9*3 variants require 21% and 34% lower maintenance doses of warfarin compared with wild types CYP2C9. Poor metabolizers with various variant alleles (CYP2C9*2/*2, CYP2C9*2/*3, and CYP2C9*3/*3) require a 60–90% lower dosage of warfarin than extensive metabolizers. In general, whites possess a higher frequency of CYP2C9*3 variants than Asians (6–10% whites versus 2–5% in Asians). Although there is no documented CYP2C9*2 among Asians, 8–20% of whites have the CYP2C9*2 allele. However, Asian patients with the CYP2C9*1/*3 genotype are intolerant to warfarin (18).

Moyer et al reported that polymorphism in CYP2C9 and VKORC1 have considerable effects on warfarin dosing among white patients. White individuals show considerable variability in CYP2C9 allele types, whereas Asian and African descents infrequently carry CYP2C9 allelic variants. The VKORC1AA allele associated with high warfarin sensitivity, however, predominates in Asians, whereas white and those of African descent have VKORC1BB (low warfarin sensitivity), VKORC1AB (moderate warfarin sensitivity), or VKORC1AA allele associated with high warfarin sensitivity (19). In the light of the effect of polymorphism of CYP2C9 and VKORC1 on warfarin dosage, the U.S. Food and Drug Administration (FDA) changed the label of warfarin, suggesting that clinicians consider genetic testing before initiation of warfarin therapy. Eckman et al reported that the genotype-guided dosing of warfarin resulted in a better outcome but at a relatively high cost ($400). Therefore, there is only a 10% chance that genotype-guided dosing of warfarin is going to be cost effective. Therefore, genetic testing is useful only for patients with a high risk of hemorrhage (20). However, regular INR monitoring is the standard of care of patients receiving warfarin therapy. If INR values are tightly controlled between 2 and 3, the risks of bleeding or treatment failure are significantly reduced. But if target INR is not achieved, the risk of death from bleeding, stroke, myocardial infarction, or a systemic embolism event may be related to the INR value. In one study, the authors observed that the rates of higher annual mortality (4.2%) and a major bleeding episode (3.9%) were higher in a poorly controlled INR group than in a well-controlled INR group (1.7% mortality and 1.6% major bleeding episode). Therefore, good INR control is the key to a good outcome in patients being treated with warfarin (21).

Regular INR monitoring is also useful in identifying clinically significant warfarin–drug and warfarin–herb interactions. Adjustment of warfarin dosage depends on INR values, which should be measured frequently (two to four times per week) during the dosing phase and then at 4- to 6-week interval in stable patients. Patients with elevated INR may require more frequent monitoring, and even a nonbleeding patient with an INR of 9 should be treated with vitamin K1 given orally. Patients with an excessively high INR may require immediate therapy with a clotting factor such as fresh-frozen plasma (22). In addition to INR monitoring, alternative monitoring of warfarin therapy by the measurement of clotting factors (factors II and X) has also been proposed. This is particularly useful in patients when INR values are not in accordance with patient response (23).

9.3. CLINICALLY SIGNIFICANT INTERACTIONS BETWEEN WARFARIN AND VARIOUS DRUGS

There are many significant interactions between warfarin and various drugs. Such interactions may potentiate or reduce the efficacy of warfarin. Both pharmacokinetic and pharmacodynamic interactions between warfarin and drugs have been reported. As mentioned earlier, warfarin is administered as a racemic mixture where pharmacologically more active S-warfarin is metabolized by CYP2C9, whereas the R-warfarin is metabolized by CYP3A4. Therefore, drugs that are inhibitors of CYP2C9, for example metronidazole, interact pharmacokinetically with warfarin, reducing its clearance and thus increasing INR and the risk of bleeding. However, drugs such as diltiazem, which interfere with the clearance of R-warfarin through inhibiting CYP3A4, have a modest effect on INR. However, drugs that affect clearance of both isomers have a profound effect on the efficacy of warfarin. Inducer of hepatic drug metabolizing enzymes, for example carbamazepine, may reduce the efficacy of warfarin. Many antibiotics are known to interact with warfarin. Some antibiotics interact pharmacokinetically with warfarin by interfering with the hepatic metabolism of warfarin; other antibiotics may inhibit bacterial flora in the intestine, thus reducing the amount of vitamin K produced by intestinal bacteria. Many drugs with antiplatelet effects also interact with warfarin pharmacodynamically. The most common drugs in this category are aspirin and nonsteroidal anti-inflammatory drugs. Even selective serotonin reuptake inhibitors used for treating depression have some antiplatelet effect and may interact with warfarin. Therefore in patients where there is a change in preexisting medications, a closer monitoring of INR (within 3–7 days of change of medication) is strongly recommended (24). In one paper the authors commented there are four very significant drug interactions involving warfarin that may cause a potentially life-threatening situation. These drug interactions involve interaction of warfarin with fluconazole, amiodarone, sulfamethoxazole/trimethoprim, and metronidazole. These drugs interactions are also common because these drugs have a broad spectrum of indication of use in elderly patients (25).

Sayal et al reviewed the interactions between various psychotropic drugs and warfarin and commented that fluoxetine and fluvoxamine have significant interactions with warfarin, whereas antidepressants such as citalopram, nefazodone, and sertraline have a relatively low interaction. Carbamazepine appears to reduce the anticoagulant effect of warfarin due to its ability to induce cytochrome P450 liver enzymes. In contrast, lithium, gabapentin, sulpiride, and amisulpride are predominantly cleared by renal excretion and are unlikely to interact with warfarin (26). Examples of interaction of warfarin with common drugs are listed in Table 9.1.

Case Report

A 74-year-old man receiving warfarin for atrial fibrillation experienced an abrupt increase in INR after taking acetaminophen. To investigate this effect, after stabilizing his INR, he was given 1 g of acetaminophen four times a day for 3 days, and his INR increased from 2.3 to 6.3 on the day after discontinuing acetaminophen. Warfarin therapy was suspended and his INR was reduced to 2.0. Warfarin was restarted with the same dosage without acetaminophen, and his INR was between 2.0 and 3.0 for 6 months. His plasma warfarin concentration was 1.54 μg/mL before acetaminophen therapy and 1.34 μg/mL during acetaminophen therapy, indicating that the drug interaction between the warfarin and acetaminophen was pharmacodynamic. The authors speculated that acetaminophen or a metabolite may enhance the effect of warfarin by augmenting with vitamin K antagonism (27).

Case Report

A 54-year-old white man receiving long-term warfarin therapy (past 11 years) for the prevention of ischemic stroke with a stable INR ranging from 1.9 to 2.5 for the last 5 months experienced a sharp increase in his INR (8.0). Other than warfarin, the patient was also receiving isosorbide dinitrate (20 mg/day), aspirin (325 mg/day), metoprolol (100 mg/twice a day), rosiglitazone (4 mg/day), atorvastatin (20 mg/day), gemfibrozil (600 mg/twice a day), levothyroxine (0.1 mg/day), ipratropium inhaler two puffs three times a day, multivitamins, and nitroglycerine as needed. The only finding during his elevated INR was that the patient started drinking a small amount of beer (a half can of light beer, approximately 5.4 g of alcohol per day), which is considered low consumption after 35 years of abstinence. He started drinking beer because of the perception that alcohol is protective

TABLE 9.1. Examples of Interaction of Warfarin with Common Drugs

Interacting Drug	Effect/Mechanism
Drugs Potentiating Effect of Warfarin	
Selected antibiotics (ciprofloxacin, clarithromycin, erythromycin, metronidazole, bactrim (sulfamethoxazole/trimethoprim)	Increased effect of warfarin (INR) due to inhibition of intestinal flora producing less vitamin K
Acetaminophen	Increased INR and increased risk of bleeding
NSAIDs	Increased INR and increased risk of bleeding due to antiplatelet activity of NSAIDs, direct mucosal injury by NSAIDs may also increase risk of bleeding during therapy with warfarin
Amiodarone	Increased INR and risk of bleeding because Amiodarone inhibits CYP2C9
Clofibrate	Potentiation of effect of warfarin by affecting coagulation factors
Antifungal (fluconazole, miconazole)	Increased INR and increased risk of bleeding due to inhibition of CYP2C9
Antidepressants (SSRIs: fluoxetine, fluvoxamine, etc.)	Increased INR due to some antiplatelet activity of SSRI; fluoxetine also may inhibit CYP2C9 fluoxetine, fluvoxamine, etc.)
Cimetidine, omeprazole	Potentiation of effect due to stereoselective clearance of R-isomer
Clopidogrel, ticlopidine	Increased effect of warfarin due to altered platelet function
Drugs Inhibiting Effect of Warfarin	
Rifampin	Decreased INR and decreased effect of warfarin because rifampin induces CYP2C9
Carbamazepine	Decreased effect of warfarin because carbamazepine induces liver enzymes
Barbiturates, chlordiazepoxide	Decreased effect of warfarin due to induction of enzymes

Note: These are examples of drug interactions involving warfarin; this list does not include all drugs (more than 100) known to interact with warfarin.
INR, international normalized ratio; NSAID, nonsteroidal anti-inflammatory drug; SSRI, selective serotonin reuptake inhibitor.

for heart disease. The patient's warfarin therapy was withheld for 3 days, and the INR was reduced to 2.0. The patient decided to discontinue drinking beer, and his warfarin therapy was resumed to 90 mg/week (previously it was 93.75 mg/week); after 1 week his INR was 1.9. Because the INR was reduced to 1.5 during follow-up visits after 2.5. weeks, the warfarin dosage was increased to 93.75 mg/week, the dosage before the episode (28).

It is possible that beer consumption may increase INR due to protein binding interaction and decreased metabolism of warfarin through cytochrome P450 mixed function enzymes, although concurrent administration of aspirin and other drugs that are also metabolized through cytochrome P450 enzymes may have enhanced the interaction that occurred in this patient (27). Other clinical trials investigating the effect of consumption of moderate wine (10–20 ounces [28.2–56.4 g] of alcohol) on warfarin therapy demonstrated that wine consumption has no effect (29,30).

Although aspirin potentiates the effect of warfarin, low-dose aspirin (81 mg) can be used along with warfarin in treating patients with severe intracranial stenosis. In one report, the authors observed no serious hemorrhagic complication and no incidence of myocardial infarction or sudden death in 10 men and 6 women treated with a combination of warfarin and low-dose aspirin. The authors concluded that the warfarin and low-dose aspirin combination therapy is useful in preventing recurrent stroke from symptomatic, intracranial, large arterial occlusive disease after failure of either aspirin or warfarin monotherapy (31). However, one case report indicates the potentiation of warfarin action even after the topical application of methyl salicylate. A 22-year-old white woman presented with an extremely elevated INR of 12.2 after applying a topical pain relieving gel to her knees for 8 days. Methyl salicylate is systematically absorbed through the skin in measurable amounts and may increase the pharmacological action of warfarin by affecting vitamin K metabolism or displacing warfarin from protein-binding sites. Although other investigators reported a significant interaction between warfarin with the high-dose use of methyl salicylate, these authors demonstrated that even low-dose methyl salicylate may potentiate the effect of warfarin (32).

There are several drug information compendiums available to aid clinicians in identifying the interaction between warfarin and a drug including the package insert of warfarin (Coumadin, Bristol Myer-Squibb, Princeton, NJ). Anthony et al evaluated three such compendiums (Clinical Pharmacology, ePocrates, and Micromedex) approved by the FDA for listings of interactions between warfarin and drugs, foods, and dietary supplements and observed that of 648 entries listed in these sources, only 40 were common in each source. Moreover, there was little agreement between these compendiums regarding the interactions of warfarin with drugs, food, and dietary supplements as well as the warfarin label approved by the FDA (33).

Although interaction between a vaccine and warfarin is uncommon, there is one report of fatal intracranial bleeding potentially due to the interaction of warfarin with the influenza vaccine.

Case Report

A 64-year-old white man with a history of 2 days of bleeding from his rectum was admitted to the hospital. He had no change in medication and he was receiving warfarin as a prophylaxis for stroke secondary to atrial fibrillation. The patient received an inactivated influenza vaccine 4.5 weeks prior to presentation, and during vaccination his INR was 2.0. His INR value had been relatively stable for the last 6 months (range: 1.4–4.7). On admission his INR was critically elevated to 15. A noncontrast computed tomography scan of his head showed a large parenchymal hemorrhagic infarction involving the left temporal parietal as well as occipital lobes. Despite therapy the patient died 17 hours after admission. In general receiving influenza vaccine does not affect warfarin therapy significantly, although there are some reports of increased anticoagulation after receiving the influenza vaccine. The mechanism of such interaction is not known, but it may be due to the interaction of certain components of the vaccine that is changed almost annually with warfarin. The authors concluded that the INR must be monitored more frequently for 4–6 weeks following influenza vaccination in patients receiving warfarin (34).

9.4. CLINICALLY SIGNIFICANT INTERACTIONS BETWEEN WARFARIN AND DIETARY SUPPLEMENTS/FOOD

The interaction of warfarin with various herbal supplements is of great clinical significance. Although the interaction of warfarin with St. John's wort (induces liver enzymes and reduces the effectiveness of warfarin) and milk thistle (inhibits liver enzymes and increases the effectiveness of warfarin) is pharmacokinetic, many clinically significant interactions of warfarin with herbal supplements are pharmacodynamic. In addition, some herbal supplements contain coumarin, which may exert an additive effect with warfarin and cause excessive anticoagulation.

The anticoagulant effect of warfarin increases, thus increasing the risk of bleeding, if a patient takes any coumarin-containing herbal remedies such as fenugreek and dong quai (35). In addition, herbs such as dan shen, garlic, and ginkgo biloba can potentiate the effect of warfarin because these herbs have antiplatelet activities. However, using garlic as a spice may not have any significant effect on warfarin therapy. Common herbs that may interact with warfarin are listed in Table 9.2. Paoletti et al commented that approximately 30% of patients receiving warfarin therapy also take herbal supplements. Therefore, interaction between herbal supplements and warfarin is a serious patient safety issue. Moreover, the authors reported a case of INR reduction in a patient taking warfarin and concentrated bilberry juice, although there is no reported interaction between warfarin and bilberry in the literature. Vitamin A taken in high doses with warfarin therapy may cause bleeding. The same is true for vitamin E and fish oil because these supplements affect platelet aggregation or vitamin K–dependent coagulation factors (36). Soy milk, which is a rich source of protein, may interact with warfarin, reducing its potency and causing subtherapeutic INR. Ginseng, a popular Chinese herbal supplement, also decreases the anticoagulant effect of warfarin, although another popular Chinese, herb dan shen, which is indicated for treating various symptoms of coronary heart disease such as angina pectoris, heart failure, and

TABLE 9.2. Interaction of Warfarin with Dietary Supplements

Supplements Potentiating Effect of Warfarin	
Angelica root	Bogbean
Borage seed oil	Bromelain
Chamomile	Dong Quai
Devil's claw	Fenugreek
Feverfew	Grape seed
Garlic	Ginkgo biloba
Licorice	Fish oil supplement
Horse chestnut	Royal jelly
Saw palmetto	Danshen
Supplements Reducing Effect of Warfarin	
St. John's wort	Green tea
Milk thistle	Goldenseal
Ginseng	

myocardial infarction, potentiates the effect of warfarin. Interestingly, dan shen not only can be taken orally but also can be inhaled because this herbal supplement is incorporated into some brands of Chinese cigarettes. Because of both pharmacokinetic and pharmacodynamic interaction with warfarin, dan shen should be avoided in patients receiving warfarin (37). Dietary supplement coenzyme Q10 may accelerate the metabolism of warfarin thus reducing its efficacy (38).

The interaction between cranberry juice and warfarin is controversial. Cranberry juice is effective in preventing and treating urinary tract infections (39,40). Griffiths et al reported a case of fatal internal hemorrhage in an elderly man who mostly consumed cranberry juice for 2 weeks and was receiving warfarin therapy along with phenytoin and digoxin. Three weeks before the episode he was given cephalexin for a chest infection. On admission, his INR was over 15. After his death an autopsy was performed that revealed fatal hemorrhages into the pericardial sac and gut as well as gastric ulceration. The authors speculated that possible interaction between warfarin and cranberry juice was the cause of his fatal hemorrhage (41). Although, there are other case reports of increased INR in patients receiving warfarin and drinking cranberry juice, Mellen et al using 10 male patients taking warfarin found no interaction between warfarin and cranberry juice when these subjects drank high amounts (240 mL twice daily) for 1 week (42). Zikria et al also concluded that the moderate consumption of cranberry juice should not affect the anticoagulation effect of warfarin (43). Nevertheless, a recently published case report indicated an increased INR in a patient due to the interaction between warfarin and cranberry juice (44).

Case Report

A 64-year-old man who was being followed in the anticoagulation clinic for 3 years presented with an elevated INR of 5.5. His weekly warfarin dosage of 60 mg was unchanged for a year and there was no change in drug history. The patient as taking captopril, furosemide, gabapentin, pravastatin, metoprolol, sertraline, and hydrocodone 5 mg in combination with 500 mg of acetaminophen. The patient denied taking any herbal supplement but stated that he started drinking two glasses of cranberry juice every day for the past month. He was advised to stop drinking cranberry juice, and a follow-up 2 weeks later showed an INR of 3.6. The authors concluded that health care providers should educate patients regarding the possible interaction between cranberry juice and warfarin (44).

Grapefruit juice interacts with many drugs but does not have any clinically significant interaction with warfarin. One study in which patients receiving warfarin also consumed 8 ounces of grapefruit juice daily for 1 week showed no significant change in prothrombin time (PT) during the time they consumed grapefruit juice (45).

The daily requirement for vitamin K is 90 μg/day for women and 120 μg/day for men. However, many vegetables, especially leafy vegetables, supply more than that quantity of vitamin K from a single serving, and consuming such vegetables may reduce the effectiveness of warfarin therapy. These green leafy vegetables and intestinal bacteria are major sources of vitamin K. Because many patients receiving warfarin therapy are also enthusiastic about consuming green leafy vegetables as a part of a healthy diet, a careful dietary history should be obtained in a patient receiving warfarin whenever an unexplained decrease in INR is observed (46). Vegetables with high vitamin K content are listed in Table 9.3. Green tea may also inhibit the pharmacological activity of warfarin if consumed in excess (47).

Case Report

A 44-year-old white man receiving warfarin therapy for thromboembolic prophylaxis secondary to St. Jude mechanical valve replacement in the aortic position had a stable INR. His INR was 3.79 when he entered into the authors' clinic, but 22 days later, his INR was reduced to 1.37. One month later his INR further dropped to 1.14. The authors found out that the man started drinking a half gallon to a gallon of green tea each day a week prior to his drop of INR to 1.37. On discontinuation of drinking green tea his INR increased to 2.55. Green tea can be a significant source of vitamin K, thus reducing the efficacy of warfarin (48).

TABLE 9.3. Some Common Vegetables Rich in Vitamin K

Vegetable	Single Serving/Amount	Vitamin K (μg)
Kale	1 cup (130 g)	1062.1
Spinach, cooked	1 cup (190 g)	1027.3
Turnip greens	1 cup (164 g)	851.0
Brussels sprouts	1 cup (155 g)	299.9
Broccoli, cooked	1 cup (156 g)	220.1
Onions: spring or scallion	1 cup (100 g)	218.9
Lettuce, butterhead	1 head (163 g)	166.7
Parsley, raw	10 sprigs (10 g)	164.0
Asparagus	1 cup (180 g)	144.0
Lettuce, iceberg	1 head (539 g)	129.9
Lettuce, green leaf	1 cup (56 g)	97.2
Okra	1 cup (160 g)	88.0
Cabbage	1 cup (150 g)	57.8
Coleslaw	¾ cup (99 g)	56.4

Source: U.S. Department of Agriculture: National nutrient database for standard reference, release 17.

Cigarette smoking affects the metabolism of certain drugs, but Weiner et al based on the review of the charts of 200 patients concluded that smoking history does not affect the warfarin dosage requirement (49).

9.5. CLINICAL UTILITY OF DIRECT MEASUREMENT OF WARFARIN

Although warfarin therapy is monitored by measuring INR, there are situations where direct measurement of warfarin in human serum or urine is clinically important. Pharmacogenomics play an important role in warfarin therapy because there are wide variations in the dosages of warfarin required to treat individual patients. In addition, dietary and supplemental vitamin K consumption also affects warfarin therapy (50). Determination of warfarin or related anticoagulant concentration in patients where PT or INR is very high is important to confirm accidental warfarin overdose. Therapeutic plasma concentration of warfarin is 600–2600 μg/L (51); therapeutic concentrations of acenocoumarol and phenprocoumon are 30–100 μg/L and 160–3600 μg/L, respectively (52). There was a significant correlation between warfarin dosage and plasma warfarin concentration in Korean patients with atrial fibrillation, and the authors concluded that direct monitoring of warfarin can help determine dosage adjustments and improve our understanding of individual patient response to warfarin therapy (51). In the case of suspected warfarin overdose, direct measurement of warfarin concentration in blood is useful. Cruickshank et al commented that warfarin toxicity is a commonly encountered problem in the emergency department (53). Suwanvecho and Baker reported an interesting case of a 57-year-old woman who showed a profound prolongation of prothrombin time of 56.9 seconds and an INR of 22.18. She never had an episode of bleeding or received any anticoagulant. Her coagulopathy was rapidly corrected after the administration of fresh-frozen plasma and vitamin K. It was discovered that she was given warfarin (proprietary name in Mexico: Romesa) instead of her prescribed medicine ramipril for treating hypertension, which in Mexico has the proprietary name of Ramace. Therefore, acquisition of a prescription outside the United States caused this error and potentially life-threatening situation in this patient (54).

The direct measurement of warfarin level is important for death investigations. A 73-year-old man receiving warfarin orally with a history of prosthetic heart valve replacement was struck by a truck and admitted to the hospital. The patient died 22 hours after admission. Autopsy findings revealed marked subcutaneous and intramuscular hematoma in the lumbar and gluteal region and anemic changes in the viscera. However, his postmortem serum warfarin level as determined by gas chromatography/mass spectrometry was 854 ng/mL, which was within the therapeutic range. Therefore, the person died of hemorrhagic shock caused by the force acting in the lumbar region accompanied by the influence of the anticoagulant action of warfarin (55).

For forensic investigations, the direct measurement of warfarin is crucial to establish the cause of death. Heininger reported that warfarin is one of the top 10 most commonly encountered drugs in death investigations (56). Warfarin also undergoes significant drug interactions with other drugs causing clinically confusing and life-threatening situations. Direct measurement of blood warfarin level has clinical value in such situations. In a postmortem study, the authors observed that in 33% of 328 warfarin cases, at least one interacting drug was present, and paracetamol (acetaminophen) was the most abundant in 53 cases. When paracetamol and warfarin were detected simultaneously, a fatal bleed was 4.6 and 2.7 times higher compared with paracetamol or warfarin use alone. Most of the postmortem blood samples had warfarin levels less than 500 μg/L, which is within the therapeutic range (57).

9.6. SUPERWARFARIN POISONING

Although warfarin was used in the past as a rat poison, it is no longer used because rats are resistant to warfarin. To combat this problem, superwarfarin compounds are used that are either derivatives of 4-hydoxycoumarin (brodifacoum, bromadiolone, and difenacoum) or indandione compounds (diphacinone, chlorophacinone, and pindone). These compounds possess 100 times more activity and at least 3 times the duration of action of warfarin. For example, brodifacoum has a half-life of 487 hours and an ability to inhibit 2, 3-epoxide of vitamin K for up to 3 years (58). Superwarfarin poisoning is considered a significant health problem in the United States. In 2004 there were 16,054 cases of poisoning; most were accidental ingestion of rat bait by children, but 4576 patients required hospitalization, 23 patients had major adverse effects, and 1 died. A farmer who presented with hematuria 9 days after spilling a rodenticide containing superwarfarin over himself (transcutaneous exposure) showed a PT longer than 200 seconds. He was treated with fresh-frozen plasma and vitamin K and discharged. He presented to the hospital again 2 days later with hematuria and an INR higher than 10. He required inpatient treatment with vitamin K for 1 week, and upon discharge he needed oral vitamin K and INR monitoring for a month. The superwarfarin the patient was exposed to was chlorophacinone (59).

Case Report

A 28-year-old healthy woman was transferred to the hospital with a 2-month history of recurring episodes of bleeding. Administration of vitamin K and prothrombin complex concentrate had only temporarily corrected both a highly elevated INR and prolonged activated partial thromboplastin time. Although patient had no medical reason for such bleeding episodes, a sustained deficiency of vitamin K–dependent clotting factors was detected, but again no medical explanation was found for such an observation. Analysis of serum of the patient using liquid chromatography combined with tandem mass spectrometry showed a pharmacologically active concentration of flocoumafen, a superwarfarin rodenticide. Because the half-life of this superwarfarin compound in humans has not been established, the patient management included both administration of vitamin K and monitoring of serum flocoumafen. The flucoumafen level reached the limit of quantification on day 48, and the half-life of this superwarfarin was established as 6.7 days in the woman. In contrast, the half-life of this compound is 220 days in rodents. The authors concluded that the measurement of flucoumafen in such patients may provide a rational basis for the duration of vitamin K therapy in affected patients (60).

Fatal poisoning from the ingestion of superwarfarin has also been reported. A 15-year-old girl deliberately and fatally ingested brodifacoum, a commonly used superwarfarin rodenticide. The mechanism of death was massive pulmonary hemorrhage. Brodifacoum was quantified in liver, spleen, lung, brain, bile, vitreous humor, heart blood, and femoral blood using high performance liquid chromatography with fluorescence detection. The highest brodifacoum concentrations were detected in bile (4276 ng/mL) and femoral blood (3919 ng/mL). Detection and quantification of superwarfarin are essential in forensic death investigations (61).

9.7. DETERMINATION OF WARFARIN AND RELATED COMPOUNDS BY CHROMATOGRAPHIC METHODS

Although warfarin and related compounds can be determined in blood, urine, and other biological matrix by high performance liquid chromatography combined with ultraviolet detection (62–65) or gas chromatography or gas chromatography combined with mass spectrometry (66–68), many investigators use liquid chromatography combined with mass spectrometry for the quantification of warfarin and related compounds.

Authors de Vries and Kymber analyzed coumarin anticoagulants warfarin, phenprocoumon, acenocoumarol, and dicoumarol by liquid chromatography-thermospray mass spectrometry (LC-TSP-MS) as well as liquid chromatography-electron impact mass spectrometry. The LC-TSP MS showed a single [M + 1]+ ion; the electron impact mass spectrometry showed fragment ions (69). Huang et al measured unbound concentration (free) warfarin in human plasma using liquid chromatography combined with tandem mass spectrometry. The protein-free ultrafiltrate of plasma was prepared by high-speed ultrafiltration. Warfarin and the internal standard coumachlor (p-chlorowarfarin) were extracted using liquid-liquid extraction with ethyl acetate. Chromatographic separation was achieved using the Water's Acquity ethylene-bridged C-18 column (column temperature maintained at 40°C) and a mobile phase composed of water and acetonitrile (50:50 by volume) each containing 0.1% formic acid at a flow rate of 0.2 mL/minute. The analyses were carried out by multiple reaction monitoring (MRM) using precursor to product combinations of m/z 307.3 → 161.2 for warfarin and 341.2 → 161.3 for the internal standard. The lower limit of quantitation was 0.25 ng/mL and the analytical measurement range was 0.25–16 ng/mL for free warfarin (70). In another report, the authors used two-dimensional enantioselective liquid chromatography combined with electrospray mass spectrometry for rapid quantification of enantiomers of warfarin, phenprocoumon, and acenocoumarol in human serum. For two-dimensional liquid chromatography, injected sample was online extracted with a Perfusion column Poros R2/20 (2 mm × 30 mm, Applied Biosystem); the chiral separation (second chromatographic dimension) was performed by a Chira-Grom-2-column (250 mm × 1 mm, 8 micron particle size, Grom Analytik GmbH, Herrenberg, Germany), preceded by a guard column containing the same stationary phase. Choral column was thermostated at 40°C. The injected sample was flushed through the Poros column with a high flow rate of 1.9 ml/minute of 0.5% formic acid for 1 minute. In the meantime the chiral column was flowed by 0.1 ml/minute of an aqueous solution of acetonitrile/methanol/formic acid (33/33/0.4). Subsequently, with valve switching, the analytes trapped by the Poros column were moved to the chiral column. The electrospray tandem mass spectrometer was operated on the multiple reaction mode and transitions m/z 309.3 > 163.1 was used for quantification for warfarin, transition 281.3 > 203.2 for phenprocoumon, and transition 354.3 > 163.1 for acenocoumarol, respectively (71).

Kollroser and Schober described a protocol for the determination of coumarin-type anticoagulants warfarin, acenocoumarol, and phenprocoumon in human plasma using liquid chromatography and electrospray ionization tandem mass spectrometry with ion trap detector. The authors used p-chlorowarfarin as the internal standard. To a 1-mL specimen, the authors added 20 μL orthophosphoric acid and 10 μL diluted internal standard

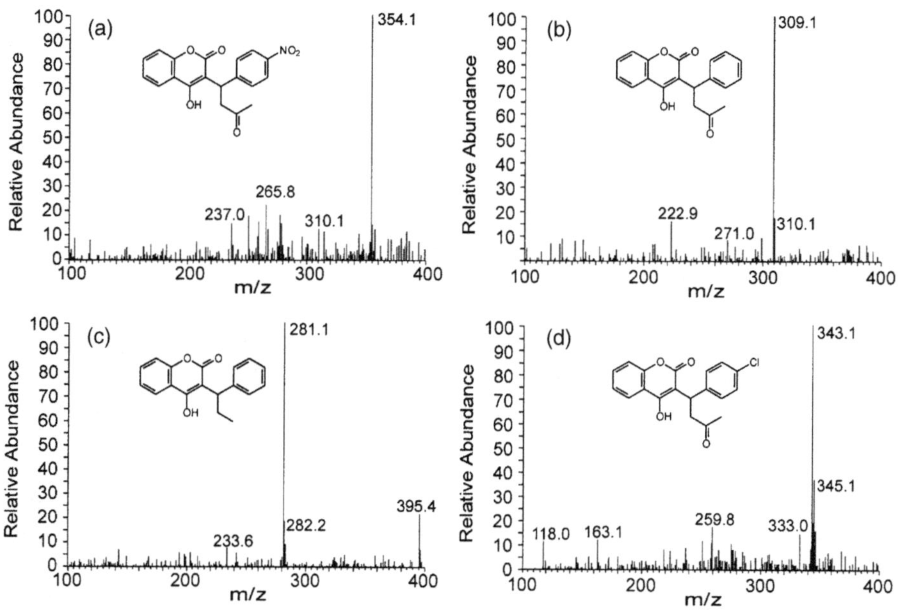

Figure 9.1. Electrospray mass spectra of (**a**) acenocoumarol; (**b**) warfarin; (**c**) phenprocoumon; and (**d**) p-chlorowarfarin as the internal standard. From Kollroser, M, Schober C. Clin Chem 2002; 48: 84–91. Reprinted with permission from AACC.

solution (10 mg/L). Then coumarin anticoagulants along with the internal standard were extracted using a solid phase extraction column and analytes were extracted using 1 mL of methanol. Then 20 μL of elute was injected into the chromatographic system and analyses were carried out using a reverse phase C-18 column (150 × 30 mm; particle size 5 micron; Waters) protected by a guard column at ambient temperature. The flow rate was 0.5 mL/minute, and the mobile phase composition was acetonitrile and 1 g/L formic acid (75:25 by volume). The mass spectrometer was operated in the positive ionization mode. The electrospray mass spectra of acenocoumarol, warfarin, phenprocoumon, and the internal standard p-chlorowarfarin are shown in Figure 9.1. In the MS/MS experiments, the protonated [M + H]+ of acenocoumarol (m/z 354.1), phenprocoumon (m/z 281.1), warfarin (m/z 309.1), and the internal standard (m/z 343.1) were selected and fragmented by the helium gas in the ion trap at a relative collision energy of 35%. The mass spectra resulted from these fragmentation were acquired in the full scan mode from m/z 100 to m/z 450. Several product ions were obtained. The most abundant product ions, m/z 163 for acenocoumarol, m/z 203 for phenprocoumon, m/z 163 for warfarin, and m/z 163 for the internal standard were extracted and selected for quantification. Extraction and separation of all three coumarin anticoagulants can be accomplished in 9 minutes. Product ion chromatograms and spectra of an extracted patient plasma sample containing 201 ng/mL acenocoumarol are shown in

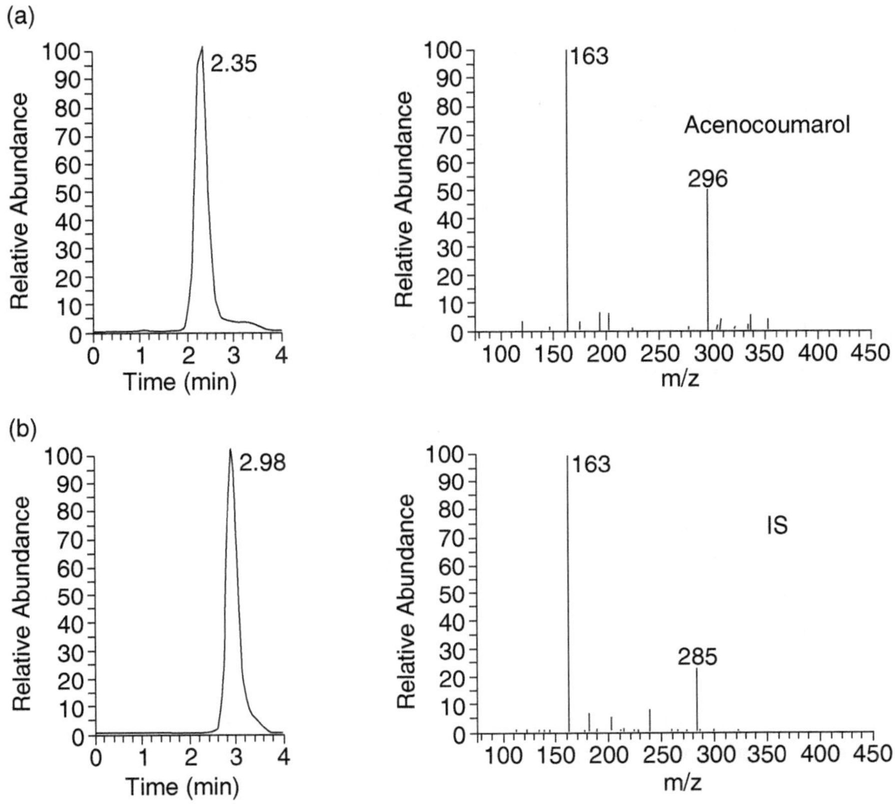

Figure 9.2. Product ion chromatograms and spectra of an extracted patient plasma sample containing 201 ng/mL acenocoumarol. (**a**) Acenocoumarol; (**b**) internal standard; p-chlorowarfarin. From Kollroser, M, Schober C. Clin Chem 2002; 48: 84–91. Reprinted with permission from AACC.

Figure 9.2. The limits of detection were 1 μg/L for phenprocoumon and warfarin, 10 μg/L for acenocoumarol. The assay was lined up to 4000 μg/L for phenprocoumon, 5000 μg/L warfarin, and 600 μg/L for acenocoumarol (72).

Hou et al described a protocol for the determination of warfarin enantiomers in human plasma using a novel polymeric surfactant for micellar electrokinetic chromatography combined with mass spectrometry. The novel polymeric chiral surfactant polysodium –N-undecenoyl-L, L-leucyl-valinate was used as a pseudo-stationary phase for the chiral separation of warfarin enantiomers. The authors used coumachlor (p-chlorowarfarin) as the internal standard. The analytical measurement range was 0.25–5.0 μg/mL, and the detection limit was 0.1 μg/mL (73). A method for the quantification of chiral isomers of phenprocoumon in human plasma using enantioselective liquid chromatography (Chira-Grom -2 chiral column and isocratic stationary phase) and electron ionization mass spectrometry was also described. The analytical range of the assay was 62.5–1000 ng/mL with a limit of detection of 12.5 ng/mL (74). In 2010, Zuo et al described a protocol for the simultaneous measurement of S-warfrain, R-warfarin, S-7-hydroxywarfarin, and R-7-hydroxywarfarin in human plasma by liquid chromatography-tandem mass spectrometry. The authors used solid phase extraction and chiral column (Chirobiotic V column) for their analysis. The mass spectrometer was operated in multiple-reaction monitoring mode. The limit of detection was 5 ng/mL for all four analytes (75).

9.8. DETERMINATION OF VARIOUS SUPERWARFARIN COMPOUNDS BY CHROMATOGRAPHIC METHODS

There are several liquid chromatography combined with mass spectrometric determination methods for determination the concentrations of various superwarfarin compounds in human plasma of serum. Grobosch et al described a protocol for the simultaneous identification and quantitation of five superwarfarin compounds (brodifacoum, bromadiolone, difenacoum, difethialone, and flucoumafen) that are used as rodenticides and five coumarin anticoagulants (warfarin, acenocoumarol, coumatetralyl, phenprocoumon, and coumachlor) using liquid chromatography combined with mass spectrometry. The method is based on acidic (pH 4.2) liquid-liquid extraction using 1-chlorobutane. The chromatographic separation was achieved by using a reverse phase C-18 column and a solvent gradient composed of methanol (solvent A) and a mixture of methanol and 0.1% formic acid (10:90 by volume; solvent B). The lower limit of detection for each analyte was 5 μg/L and the upper end of quantitation was 250 μg/L. The authors applied their protocol for diagnosis of poisoning of a 55-year-old man with bromadiolone, a superwarfarin, who showed a serum concentration of 440 μg/L (76).

Jin et al described a liquid chromatography combined with electrospray ionization tandem mass spectrometric method for the simultaneous determination of two superwarfarin compounds, flocoumafen and brodifacoum, in whole blood. The authors use warfarin as the internal standard. Superwarfarin compounds along with the internal standard were extracted from the whole blood using ethyl acetate, and the detection was performed by negative electrospray ionization in multiple reaction monitoring mode. The analytical measurement range was 0.1–100 ng/mL with a limit of detection of 0.05 ng/mL (77). In another report, the authors also used high performance liquid chromatography combined with electrospray tandem mass spectrometry for the determination of bromadiolone concentration in whole blood. The authors also used warfarin as the internal standard and liquid-liquid extraction with ethyl acetate to extract bromadiolone along with the internal standard from the whole blood. The mass spectrometer was operated in multiple reaction mode using m/z 527 → 465 for the determination of bromadiolone and m/z 307 → 161 for determination of the internal standard warfarin. The analytical measurement range was 0.5–199 ng/mL. This method can be successfully applied for the diagnosis of a poisoned human being because the method is capable of detecting even trace amounts of bromadiolone (78).

Jin et al also reported the rapid determination of three superwarfarin rodenticides bromadiolone, flocoumafen, and brodifacoum in whole blood using warfarin as the internal standard and high performance liquid chromatography combined with electrospray ionization mass spectrometry. These compounds were extracted from whole blood using ethyl acetate and analyzed using a reverse phase C-18 column. The mobile phase composition was 0.2% acetic acid/methanol (12:88 by volume) at a constant flow rate of 0.5 ml/minute. The analytes were detected using negative electrospray ionization in selected ion monitoring mode. The molecular ion [M-H]- of m/z 527, 541, 523, and 307 were selected for quantification of bromadiolone, flocoumafen, brodifacoum, and the internal standard. The analytical measurement range was 0.50–100 ng/mL (79). Measurement of indandione rodenticides is important in the diagnosis and treatment of accidental rodenticide ingestion, and liquid chromatography combined with the electrospray tandem mass spectrometric method has been described for the simultaneous determination of indandione-type rodenticides (diphacinone, chlorophacinone, valone, and pindone) in human serum using warfarin as the internal standard. After the addition of the internal standard, the sample was extracted with 10% methanol in acetonitrile and cleaned by solid-phase extraction. The analytes were separated on a C-18 column, and identification as well as quantification were achieved by using an ion trap mass spectrometer in the negative electrospray ionization mode. The multiple-reaction monitoring ion pairs were m/z 339 → 167, m/z 373 → 201, m/z 229 → 145, m/z 229 → 172, and m/z 307 → 161 for diphacinone, chlorophacinone, valone, pindone, and the internal standard, respectively. Recoveries were between 81.5% and 94.6%, and the limits of quantification were 0.2–0.5 ng/mL. The assay was linear in the range of 0.5–100.0 ng/mL (80).

9.9. CONCLUSIONS

Although oral anticoagulants warfarin and related compounds are monitored pharmacodynamically using INR, direct determination of these compounds in human serum has merit under special circumstances where INR changes are unexpected. For the diagnosis of poisoning and death investigations involving warfarin and related compounds, direct determination of warfarin is also essential. Warfarin is now rarely used as a rodenticide. Various superwarfarin compounds used as rodenticides can cause serious poisoning in humans and even death. Direct determination of these compounds is essential under such circumstances. Although there is no commercially available immunoassay for the rapid determination of these compounds in a toxicology laboratory, several chromatographic methods are available for the analysis of these compounds.

Liquid chromatography combined with mass spectrometry is the most sensitive and specific method for the analysis of these compounds.

REFERENCES

1. Hirsch J, Dalen JE, Anderson DR, Poller L et al. Oral anticoagulants: mechanisms of actions, clinical effectiveness and optimal therapeutic range. Chest 1998; 114: 445S-469S.
2. Haustein KO. Pharmacokinetic and pharmacodynamic properties of oral anticoagulants, especially phenprocoumon. Semin Thromb Hemost 1999; 25: 5–11.
3. Baselt RC, Cravey RH. *Disposition of Toxic Drugs and Chemicals in Man*. 4th ed. Foster City, CA: Chemical Toxicology Institute; 1995:777–779.
4. Palareti G, Legnani C. Warfarin withdrawal: pharmacokinetic pharmacodynamic considerations. Clin Pharmacokinet 1996; 30: 300–313.
5. Pengo V, Cucchini U, Denas G, Davidson BL et al. Lower versus standard intensity oral anticoagulant therapy (OAT) in elderly warfarin-experienced patients with non-valvular atrial fibrillation. Thromb Haemost 2010; 103: 442–449.
6. Bon Homme M, Reynolds KK, Valdes R Jr, Linder MW. Dynamic pharmacogenetic models in anticoagulation therapy. Clin Lab Med 2008; 28: 539–552.
7. Crowther MA. Oral anticoagulant initiation: rationale for the use of warfarin dosing nomograms. Semin Vasc Med 2003; 3: 255–260.
8. Lazo-Langner A, Kovacs MJ. Predicting warfarin dose. Curr Opin Pulm Med 2010; 16: 426–431.
9. Lazo-Langner A, Monkman K, Kovacs MJ. Predicting warfarin maintenance dose in patients with venous thromboembolism based on the response to a standardized warfarin initiation nomogram. J Thromb Haemost 2009; 7: 1276–1283.
10. Le Gal G, Carrier M, Tierney S, Majeed H et al. Prediction of the warfarin maintenance dose after completion of the 10 mg initiation nomogram: do we really need genotyping? J Thromb Haemost 2010; 8: 90–94.
11. Lee A, Crowther M. Practical issues with vitamin K antagonists: elevated INRs, low time in therapeutic range, and warfarin failure. J Thromb Thrombolysis 2011; 31: 249–258.
12. Lurie Y, Loebstein P, Kurnik D, Almog S et al. Warfarin and vitamin K intake in the era of pharmacogenomics. Br J Clin Pharmacol 2010; 70: 164–170.
13. Sconce EA, Khan TI, Wynne HA, Avery P et al. The impact of CYP2C9 and VKORC1 genetic polymorphism and patient characteristics upon warfarin dose requirements: proposal for a new dosing regimen. Blood 2005; 106: 2329–2333.
14. Kirchheiner J, Brockmoller J. Clinical consequences of cytochrome P450 2C9 polymorphisms. Clin Pharmacol Ther 2005; 77: 1–16.
15. Ross KA, Bigham AW, Edwards M, Gozdzik A et al. Worldwide allele frequency distribution of four polymorphisms associated with warfarin dose requirements. J Hum Genet 2010; 55: 582–589.
16. Caraco Y, Blotnick S, Muszkat M. CYP2C9 genotype-guided warfarin prescribing enhances the efficacy and safety of anticoagulation: a prospective randomized controlled study. Clin Pharmacol Ther 2008; 83: 460–470.

17. Scordo MG, Pengo V, Spina E, Dahl MI et al. Influence of CYP2C9 and CYP2C19 genetic polymorphism on warfarin maintenance dose and metabolic clearance. Clin Pharamcol Ther 2002; 72: 702–710.
18. Lee SY, Kim JS, Kim JW. A case of intolerance to warfarin dosing in an intermediate metabolizer of CYP2C9. Yonsei Med J 2005; 46: 843–846.
19. Moyer TP, O'Kane DJ, Baudhuin LM, Wiley CL et al. Warfarin sensitivity genotyping: a review of the literature and summary of patient experience. Mayo Clin Proc 2009; 84: 1079–1094.
20. Eckman MH, Rosand J, Greenberg SM, Gage BF. Cost effectiveness of using pharmacogenetic information in warfarin dosing for patients with nonvalvular atrial fibrillation. Ann Intern Med 2009; 150: 73–83.
21. White HD, Fruber M, Feyzi J, Kaatz S et al. Comparison of outcomes among patients randomized to warfarin therapy according to anticoagulant control: results from SPORTIF II and V. Arch Intern Med 2007; 167: 239–245.
22. Gage BF, Finh SD, White RH. Management and dosing of warfarin therapy. Am J Med 2000; 109: 481–488.
23. Cosat IM, Soares PJ, Afonso M, Ratado P et al. Therapeutic monitoring of warfarin: the appropriate response marker. J Pharm Pharamcol 2000; 52: 1405–1410.
24. Tadros R, Shakib S. Warfarin: indications, risk and drug interactions. Aust Fam Physician 2010; 39: 476–479.
25. Thi L, Shaw D, Bird J. Warfarin potentiation: a review of the "FAB-4" significant interactions. Consult Pharm 2009; 24: 227–230.
26. Sayal KS, Duncan-McConnell DA, McConnell HW, Taylor DM. Psychotropic interactions with warfarin. Acta Psychiatry Scand 2000; 102: 250–255.
27. Gebauer MG, Nyfort-Hansen K, Henschke PJ, Gallus AS. Warfarin and acetaminophen interaction. Pharmacotherapy 2003; 23: 109–112.
28. Havrda DE, Mai T, Chonlahan J. Enhanced antithrombotic effect of warfarin associated with low dose alcohol consumption. Pharmacotherapy 2005; 25: 303–307.
29. O'Reilly R. Lack of effect of mealtime wine on the hypoprothrombinemia of oral anticoagulant. Am J Med Sci 1979; 27: 189–194.
30. O'Reilly R. Lack of effect of fortified wine ingested during fasting and anticoagulant therapy. Arch Intern Med 1981; 141: 458–459.
31. Bekavac I, Hanna JP, Sila CA, Furland AJ. Warfarin and low dose aspirin for stroke prevention from severe intracranial stenosis. J Stroke Cerebrovasc Dis 1999; 8: 33–37.
32. Joss JD, LeBlond RF. Potentiation of warfarin anticoagulant associated with topical methyl salicylate. Ann Pharamcother 2000; 43: 729–733.
33. Anthony M, Romero K, Malone DC, Hines LE et al. Warfarin interactions with substances listed in drug information compendia and in the FDA-approved label of warfarin sodium. Clin Pharamcol Ther 2009; 86: 425–429.
34. Carroll DN, Carroll DG. Fatal intracranial bleed potentially due to a warfarin and influenza vaccine interaction. Ann Pharmacother 2009; 43: 754–760.
35. Heck AM, DeWitt BA, Lukes AL, Potential interaction between alternative therapies and warfarin. Am J Health Syst Pharm 2000; 57: 1228–1230.
36. Paoletti A, Gallo E, Benemei S, Vietri M et al. Interactions between natural health products and oral anticoagulants: spontaneous reports in the Italian surveillance

system of natural health products. Evid Based Complement Alternat Med 2011; 612150.

37. Chan TY. Interaction between warfarin and danshen (*Salvia miltiorrhiza*). Ann Pharamcother 2001; 35: 501–504.
38. Zhou Q, Zhou S, Chan E. Effect of coenzyme Q10 on warfarin hydroxylation in rat and human liver microsome. Curr Drug Metab 2005; 6: 67–71.
39. Jepson RG, Craig JC. A systematic review of the evidence for cranberries and blueberries in UTI prevention. Mol Nutr Food Res 2007; 51: 738–745.
40. Hess MJ, Hess PE, Sullivan MR, Nee M et al. Evaluation of cranberry tablets for the prevention of urinary tract infections in spinal cord injured patients with neurogenic bladder. Spinal Cord 2008; 46: 622–626.
41. Griffiths AP, Beddall A, Pegler S. Fatal hemopericardium and gastrointestinal hemorrhage due to possible interaction of cranberry juice and warfarin. J R Soc Promot Health 2008; 128: 324–326.
42. Mellen CK, Ford M, Rindone JP. Effect of high dose cranberry juice on the pharmacodynamics of warfarin in patients. Br J Clin Pathol 2010; 70: 139–142.
43. Zikria J, Goldman R, Ansell J. Cranberry juice and warfarin: when bad publicity trumps science. Am J Med 2010; 123: 384–392.
44. Roberts D, Flanagan P. Case report: cranberry juice and warfarin. Home Healthc Nurse 2011; 29: 92–97.
45. Sullivan DM, Ford MA, Boyden TW. Grapefruit juice and the response to warfarin. Am J Health Syst Pharm 1998; 55: 1581–1583.
46. Cheng TO. Warfarin interaction with herbal drugs and food [Letter to the editor]. Int J Cardiol 2007; 119: 107–108.
47. Cheng TO. Not only green tea but also green leafy vegetables inhibit warfarin [Letter to the editor]. Int J Cardiol 2008; 125: 101.
48. Taylor JR, Wilt VM. Probable antagonism of warfarin by green tea. Ann Pharmacother 1999; 33: 426–428.
49. Weiner B, Faraci PA, Fayad R, Swanson L. Warfarin dosage following prosthetic valve replacement: effect of smoking history. Drug Intell Clin Pharm 1984; 18: 904–906.
50. Lurie Y, Loebstein R, Kurnik D, Almong S et al. Warfarin and vitamin K intake in the era of pharmacogenetics. Br J Clin Pharmacol 2010; 70: 164–170.
51. Kwon MJ, Kim HJ, Kim JW, Sohn KH et al. Determination of plasma warfarin concentration in Korean patients and its potential for clinical application. Korean J Lab Med 2009; 29: 515–523.
52. Schulz M, Schmoldt A. Therapeutic and toxic blood concentrations of more than 500 drugs. Pharamazie 1997; 12: 895–911.
53. Cruickshank J, Ragg M, Eddey D. Warfarin toxicity in the emergency department: recommendations for management. Emerg Med (Fremantle) 2001; 13: 91–97.
54. Suwanvecho S, Baker JR. Accidental over-coagulation: substitution error by a foreign pharmacy. Ann Pharamcother 2000; 34: 1132–1135.
55. Hitosugi M, Maebashi K, Abe M, Takatsu A et al. Hemorrhagic shock death caused by not so severe injury during the medication of anticoagulants [in Japanese]. Nihon Hoigaku Zasshi 1998; 52: 331–335.

56. Heninger MM. Commonly encountered prescription medications in medical-legal death investigation: a guide for death investigation and medical examiners. Am J Forensic Med Pathol 2000; 21: 287–299.
57. Launiainen T, Sajantila A, Rasen I, Vuori E et al. Adverse interaction of warfarin and paracetamol: evidence from a post-mortem study. Eur J Clin Pharmacol 2010; 66: 97–103.
58. Park BK, Choonac IA, Haynes BD et al. Abnormal vitamin K metabolism in the presence of normal clotting factor activities in factory workers exposed to 4-hydroxycoumarin. Br J Clin Pharamcol 1986; 21: 289–294.
59. Binks S, Davies P. Case of the month: Oh Drat-A case of transcutaneous superwarfarin poisoning and its recurrent presentation. Emerg Med J 2007; 24: 307–308.
60. Boettcher S, Wacker A, Moerike K, Kopp HG et al. Acquired coagulopathy caused by intoxication with the superwarfarin type anticoagulant rodenticide flocoumafen. Eur J Haematol 2011; 86: 173–175.
61. Palmer RB, Alakija P, de Baca JE, Nolte KB. Fatal brodifacoum rodenticide poisoning: autopsy and toxicological findings. J Forensic Sci 1999; 44: 851–855.
62. Denooz R, Douamba Z, Charlier C. Fatal intoxication by acenocoumarol, phenprocoumon and warfarin: method validation in blood using the total error approach. J Chromatogr B 2009; 877: 2344–2348.
63. Tasker RA, Nakatsu K. Rapid, reliable and sensitive assay for warfarin using normal phase high performance liquid chromatography. J Chromatogr 1982; 228: 346–349.
64. de Wolf FA, Tetteroo-Tempelman CA, Edelbroek PM. Determination of nanogram levels of anticoagulant acenocoumarin in serum by high performance liquid chromatography. J Anal Toxicol 1980; 4: 156–159.
65. Uni T, Niloka T, Hayakari M, Sugawara K et al. Simultaneous determination of warfarin enantiomers and its metabolite in human plasma by column switching high performance liquid chromatography with chiral separation. Ther Drug Monit 2007; 29: 333–339.
66. Duffield PH, Birkett DJ, Wade DN, Duffield AM. Quantitation of plasma warfarin levels by gas chromatography chemical ionization mass spectrometry. Biomed Mass Spectrom 1979; 6: 101–104.
67. De Vries JX, Simon M, Zimmermann R, Jarenberg J. Identification of phenprocoumon metabolites in human urine by high performance liquid chromatography and gas chromatography-mass spectrometry. J Chromatogr 1985; 338: 325–334.
68. Maurer HH, Artl JW. Detection of 4-hydroxycoumarin anticoagulants and their metabolites in urine as a part of systematic toxicological analysis procedure for acidic drugs and poisons by gas chromatography-mass spectrometry after extractive methylation. J Chromatogr B 1998; 714: 181–195.
69. De Vries JX, Kymber KA. Thermospray and particle beam liquid chromatographic-mass spectrometric analysis of coumarin anticoagulants. J Chromatogr 1991; 562: 31–38.
70. Huang C, Yang J, Du Y, Miao L. Measurement of free concentrations of highly protein-bound warfarin in plasma by ultra performance high performance liquid chromatography-tandem mass spectrometry and its correlation with the international normalization ratio. Clin Chim Acta 2008; 393: 85–89.

71. Vacchione G, Casetta B, Tomaiuolo M, Gradnone E et al. A rapid method for the quantification of the enantiomers of warfarin, phenprocoumon and acenocoumarol by two dimensional enantioselective liquid chromatography/electrospray tandem mass spectrometry. J Chromatogr B 2007; 850: 507–514.
72. Kollroser M., Schober C. Determination of coumarin type anticoagulants in human plasma by HPLC-electrospray ionization tandem mass spectrometry with an ion trap detector. Clin Chem 2002; 48: 84–91.
73. Hou J, Zheng J, Shamsi SA. Separation and determination of warfarin enantiomers in human plasma using a novel polymeric surfactant for micellar electrokinetic chromatography-mass spectrometry. J Chromatogr A 2007; 1159: 208–216.
74. Kammerer B, Kahlich R, Ufer M, Laufer S et al. Determination of (R)-and (S)-phenprocoumon in human plasma by enantioselective liquid chromatography/electrospray tandem mass spectrometry. Rapid Commun Mass Spectrom 2004; 18: 458–464.
75. Zuo Z, Wo SK, Lo CM, Zhou L et al. Simultaneous measurement of S-warfrain, R-warfarin, S-7-hydroxywarfarin and R-7-hydroxywarfarin in human plasma by liquid chromatography-tandem mass spectrometry. J Pharm Biomed Appl 2010; 52: 305–310.
76. Grobosch T, Angelow B, Schonberg L, Lampe D. Acute bromadiolone intoxication. J Anal Toxicol 2006; 30: 281–286.
77. Jin MC, OuYang XK, Chen XH. High performance liquid chromatography coupled with electrospray ionization mass spectrometry for determination of flocoumafen and brodifacoum in whole blood. J Appl Toxicol 2007; 27: 18–24.
78. Jin MC, Ren YP, Xu XM, Chen XH. Determination of bromadiolone in whole blood by high performance liquid chromatography coupled with electrospray tandem mass spectrometry. Forensic Sci Int 2007; 171: 52–56.
79. Jin MC, Chen XH. Rapid determination of three anticoagulant rodenticides in whole blood by liquid chromatography coupled with electrospray ionization mass spectrometry. Rapid Commun Mass Spectrom 2006; 20: 2741–2746.
80. Jin MC, Cai MQ, Chen XH. Simultaneous measurement of indiandione type rodenticides in human serum by liquid chromatography-electrospray ionization tandem mass spectrometry. J Anal Toxicol 2009; 33: 294–300.

Chapter 10

Plant Poisoning and the Clinical Laboratory

10.1. INTRODUCTION

Among more than 276,000 species of plants that have been identified worldwide, less than 5% are considered poisonous. The exact number of poisonous plants is difficult to determine due to the use of multiple common names for a given plant, which may be known in different parts of the world by different names. Therefore, plant poisoning is an uncommon source of poisoning throughout the world. Nevertheless, accidental plant poisoning in children is relatively more common than with adults (1). Curious about their surroundings, brightly colored berries, pods, leaves, and pretty flowers are interesting, attractive, and difficult to resist for many children, and even a bitter or bad taste may not dissuade an inquisitive child from testing them. Older children may hear about the use or edibility of some plants and mistake a poisonous one for a nonpoisonous one. Adult exposures are usually accidental with most cases due to misidentification of a toxic plant as an edible plant. Mushroom poisoning is a common example of such misidentification, although mushrooms are fungi not plants. Other exposures, particularly when herbals are involved, result when adults use the plant in a manner different from its traditional or historical use.

Serious toxicity from plant poisoning is also rare, and fatality is uncommon in comparison with exposure to other toxic substances, for example alcohol

---

Resolving Erroneous Reports in Toxicology and Therapeutic Drug Monitoring: A Comprehensive Guide, First Edition. Amitava Dasgupta.

and drugs of abuse. Nevertheless, although each year more than 100,000 toxic plant exposures are reported to the poison centers throughout the United States, most of these exposures cause minimal toxicity (1). Carter and Neuspiel reported that two thirds of all calls to poison control centers reporting exposure to toxic plants involve children younger than 6 years. Most of these exposures are benign, however, and only fewer than 10% of such exposures require treatment by a health care professional. Most ingestion of plant material by young children constitutes a small quantity, and if symptoms arise, they usually resolve within a short period of time. Gastrointestinal symptoms are the most common. However, ingestion of plants by adolescents and adults may be intentional and usually involves ingestion of relatively large amounts of plant material. Such exposure may cause serious toxicity requiring treatment in a medical setting (2). In another report dealing with calls to poison informations centers in Germany, the authors observed that of 80,000 to 100,000 calls annually, plant exposures were the fourth most popular category, accounting for 22% of all pediatric exposures. Between 1998 and 2004, 58,641 cases of pediatric plant exposures (children 0–14 years of age) were reported involving 248 different plant genera. Although most exposures did not have any clinical effect at the time of the call, 9.6% of cases had noticeable effects, and 0.4% of cases were classified as moderate to major effects. The most severe poisonings were observed in children with Brugmansia (angel's trumpet), Laburnum (golden chain), Phaseolus (wild bean), and Thuja (red cedar or white cedar) (3). In addition, several plant species may cause life-threatening toxicity in adults as well as children. These plants include but are not limited to oleander (*Nerium oleander*), foxglove (*Digitalis purpurea*), castor beans (*Ricinus communis*), jequirity peas (*Arbus precatorius*), monkshood (*Aconitum napellus*), yew (*Taxus baccata*), deadly nightshade (*Atropa belladonna*), and water hemlock (*Cicuta maculata*), among many others (4,5). Common poisonous plants are listed in Table 10.1.

Identification of a poisonous plant is not a simple task because many poisonous plants are often quite similar in appearance to nonpoisonous plants at various times during development. Sometimes even botanists have mistaken poisonous plants as nontoxic. In addition, all or part of the plant may be toxic. For example, the leaves may be toxic but the stem or roots are not. Toxicity may change throughout the growing season or in dormancy, and it may differ between two identical plants due to differences in soil and growing conditions. Therefore, treating a patient with plant poisoning is challenging because species responsible for poisoning may not be easily identified, and without identifying the species it is not possible to look for the specific plant toxin in the victim's blood, urine, or other body fluid.

Another major public health concern regarding plant poisoning is the popularity of herbal medicines among the general world population. Many herbal preparations are plant-based materials, and some of these herbal supplements may cause life-threatening toxicity or even fatality. Unfortunately, the U.S. Food and Drug Administration (FDA) has no control over the sale of herbal

TABLE 10.1. Common Poisonous Plants

Plant	Toxic Part	Plant Toxin
Angel's trumpet and Jimson weed	All parts	Atropine, hyoscine, hyoscyamine
Bleeding heart	Foliage and roots	Isoquinoline
Buttercup	All parts	Ranunculin
Caladium	All parts	Calcium oxalate
Castor bean	Seeds	Ricin
Elderberry	Bark, roots, leaves	Sambunigrin
Foxglove	Leaves	Digitalis
Lantana	All parts but mostly berries	Lantadene, lantanin
Lily of the valley	All parts	Convallatoxin and glycosides
Monkshood	All parts	Aconitine
Oleander	All parts	Oleandrin
Poison hemlock	All parts	Coniine and related compounds
Pokeweed	All parts	Phylolaccagenin and oxalate
Water hemlock	All parts	Cicutoxin, citutol

TABLE 10.2. Herbal Supplements Associated with Reports of Death

Herbal Supplement	Cause of Death
Ephedra-containing herbs	Cardiac failure
Oleander tea	Cardiac failure
Chan Su	Cardiac failure
Coltsfoot	Liver failure
Comfrey	Liver failure
Kava	Liver failure
Germander	Liver failure
Pennyroyal oil	Liver failure
Thunder god vine	Cardiac shock/Damage
Aconite-containing Chinese herbs/Monkshood	Cardiac shock/Failure

supplements because they are sold under the 1994 Dietary Supplement Act as nutritional supplements. Several herbal supplements such as kava, pennyroyal oil, oleander, and comfrey have been linked with death. Table 10.2 summarizes the common toxic herbs associated with a reported fatality.

10.2. TOXINS RESPONSIBLE FOR PLANT POISONING AND POISONING WITH HERBAL SUPPLEMENTS

Plants produce toxins as a defense mechanism to discourage animals from eating them. The compounds responsible for toxicity after exposure to a poisonous plant are usually synthesized via secondary metabolic pathways inside special plant cells. Many plant toxins are alkaloids, but other classes of compounds such as cardiac glycosides, saponins, coumarin, peptides, and essential oils have also been identified as plant toxins causing toxicity in both animals and humans.

10.2.1. Toxic Plant Alkaloids

Alkaloids are the most common plant toxins; more than 10,000 have been identified to date. Indole alkaloids comprise the largest group of toxicologically important plant alkaloids. Commonly known alkaloids in this group include physostigmine, strychnine, and ergot. Gelsemine is of interest in North America because it is the primary toxin in Carolina jessamine (*Gelsemium sempervirens*). Gelsemine acts by binding to the acetylcholine receptor to interfere with cholinergic transmission at the neuromuscular junction as well as in the central nervous system (6). The atropine alkaloids comprise another large group of toxicologically important plant toxins that are found in the nightshade family (Solanaceae) of plants. Examples of these compounds include *l*-hyoscyamine, atropine (*d,l*-hyoscyamine), and scopolamine. These compounds competitively block the action of acetylcholine. Toxic effects include tachycardia, arrhythmias, delirium, hallucinations, coma, myoclonus, hyperreflexia, respiratory depression, and hypothermia. Fortunately, poisoning with the nightshade family of plants can be treated with the specific antidotes physostigmine salicylate or neostigmine methylsulfate (7).

Pyrrolizidine alkaloids are found in some herbal medicines that are known to cause toxicity, predominantly hepatotoxicity. The most commonly known pyrrolizidine alkaloids are found in the toxic herbal medicine comfrey, which has also been associated with death. As many as 14 pyrrolizidine alkaloids have been isolated from comfrey including 7-acetylintermedine and 7-acetylcopsamine (8). In additions, pyrrolizidine alkaloids are also found in many traditional Chinese medicines (9). Pyridine and piperidine alkaloids include nicotine, anabasine anagyrine, and coniine (*Conium maculatum* [poison hemlock]), along with numerous derivatives. Some of these compounds are neurotoxins that cause paralysis, tremors, weakness, and eventually death from respiratory failure. Terpenoid alkaloids such as aconite are found in monkshoods, larkspurs, and buttercups. Aconite is also commonly found in some traditional Chinese medicine and is very toxic (10).

Examples of steroidal alkaloids are glycoalkaloids, imperialin protoverine, and jervine (found in false hellebores [*Veratrum L.*]). Protoverine and similar compounds act by increasing the permeability of sodium channels of nerve cells causing them to fire continuously. As a reflex effect, blood pressure, heart, and respiratory rates are decreased (11). Isoquinoline alkaloids include medicinally important compounds such as codeine, morphine, and ipecacuanha (syrup of ipecac).

10.2.2. Toxic Plant Glycosides

Toxic plant glycosides can be classified as cyanogenic, steroid, and anthraquinone. Cyanogenic glycosides are α-hydroxynitriles linked to a sugar that are hydrolyzed to form hydrogen cyanide. Amygdalin from bitter almonds (*Prunus dulcis*) is a common example of a cyanogenic glycoside. The steroid glycosides are well known for their cardiac toxicities. More than 200 naturally occurring glycosides have been identified in plants and some amphibians. The best known of these are the digitalis glycosides from foxglove. Distinct glycosides are identified in *Digitalis purpurea* (digitoxins) and *Digitalis lanata* (lanatosides). Oleandrin, folinerin, and digitoxigenin are present in the pink oleander (*Nerium oleander*); thevetin A, thevetin B, thevetoxin neriifolin, peruviside, and ruvoside have been identified in the yellow oleander (*Thevetia peruviana*). Digitalis and oleander cardenolide act by binding to and thus inhibiting sodium potassium adenosine triphosphatase activity. This action leads to an increase in intercellular ionized calcium. Symptoms are akin to those of digoxin intoxication and include anorexia, nausea, vomiting, weakness, and confusion. Poisoning from foxglove plant or oleander may also be fatal (12).

10.2.3. Plant Coumarin and Saponins

Plant coumarin compounds are found in many plants, for example clovers. But these compounds have no anticoagulant activity until oxidized by various fungi to 4-hydroxycoumarin, which is a biologically active vitamin K antagonist. The toxicity associated with this transformation was first noted in 1924 when Schofield reported a hemorrhagic disorder in cattle fed moldy sweet clover hay (13). Bergamottin, a common furanocoumarin, is found in grapefruit juice (14).

Saponins are glycosides of terpenoid aglycone with surfactant properties. Examples of this class include medicagenic acid (alfalfa [*Medicago sativa*]), sarsaponins (*Yucca L.*), escins (buckeyes [*Aesculus L.*]), and anemonic acid (Christmas rose [*Helleborus niger*]). Some of these compounds are irritants to mucosal membranes and skin, but a few (some aglycone compounds) increase the permeability of erythrocyte membranes, rendering the cells subject to lysis.

10.2.4. Oxalates as Plant Toxins

Oxalates are present in the leaves, stems, and roots of many plants and sometimes may constitute up to 80% of the plant's dry weight. Sodium and potassium oxalates are found in rhubarb leaves; calcium oxalates are present in dieffenbachia, philodendron, and a number of other species (15). Needle-shaped elongated crystals of calcium oxalate (also called raphides) are often sequestered in the vacuoles of specialized cells, and such cells eject the needles in response to slight changes in pressure (for example, when brushing the plant or chewing on it). Therefore, touching such plants may cause contact dermatitis that can range from mild to severe. With oral contact, the needles penetrate the mucosa of the mouth and throat, causing mast cells to release histamine. Watson et al described an outbreak of foodborne illness in Chicago where 10 people became ill after eating a lunch in the office cafeteria. These subjects experienced oral burning and facial edema, and one person was admitted to the intensive care unit for potential airway obstruction secondary to severe edema. Four of the 10 subjects continued to have symptoms even 2 weeks later. Food from the cafeteria's international buffet was consumed by all 10 subjects, and there was a Chinese vegetable entree in the buffet. Analysis of plant material from a Chinese vegetable demonstrated the presence of calcium oxalate (16).

10.2.5. Amino Acids, Peptides, and Proteins

Although some amino acids, peptides, and proteins of plant origin are a good source of nutrition, other products of same class may be toxins. The lathyrogenic amino acids found in the seeds of the Lathyrus species include α-amino-β-oxalyl-aminopropionic acid, l-β-cyanoalanine and β—N-(γ-l-glutamyl)-aminopropionitrile, canavanin (jack bean [*Canavalia ensiformis*]), mimosine (*Mimosa* spp.), and willardiine (*Acacia* and *Mimosa*), and they are example of plant toxins that are also amino acids or peptides. The term *toxalbumin* is used to describe a plant toxin that may inhibit protein synthesis in animals and humans causing serious toxic symptoms. The best known toxalbumin is ricin found the seeds and other parts of the castor bean (*Ricinus communis*) (17).

10.2.6. Essential Oils

Essential oils of plant origin can be used for their medicinal benefits, but some essential oils can be toxic, for example myristicin, pulegone, and thujone. Pennyroyal oil, derived from *Hedeoma pulegiodes* or *Mentha pulegium*, has been used as an abortifacient and an emmenagogue. The active component of pennyroyal oil is pulegone, an aromatic ketone metabolized to methofuran. These compounds act by protein binding and depletion of glutathione and are toxic. If untreated, pennyroyal oil may cause severe hepatotoxicity (18). Thujone is the primary toxic component of wormwood oil (*Artemisia absinthium*). Acute symptoms of thujone toxicity include vomiting, delirium, paranoia, and seizures. Chronic toxicity includes gastrointestinal effects, mental status changes, pica, visual effects, and tremors.

10.3. ROLE OF THE CLINICAL LABORATORY IN TREATING PATIENTS WITH PLANT POISONING AND TOXICITY FROM USING HERBAL REMEDIES

Clinical laboratory tests are crucial in treating patients following plant poisoning because most cases of plant poisonings are treated with supportive care. Electrolytes and other biochemical parameters are useful in treating and monitoring such patients when supportive care is the only choice because either the specific plant or herbal supplement responsible for toxicity cannot be identified or no specific antidote is available for that particular toxin. The role of applicable clinical tests for the diagnosis of plant poisoning and toxicity from herbal supplements can be divided into several categories:

1. Use of a digoxin immunoassay in the indirect detection of plant poisoning or toxicity from an herbal remedy. However, this test can be used as a screening assay only in the case of suspected poisoning, and further confirmatory tests are needed using the appropriate chromatographic method, especially liquid chromatography combined with tandem mass spectrometry in cases involving a medicolegal scenario.
2. Use of biochemical tests such as liver function tests, renal function tests, a cardiac panel, and so on, for the diagnosis and monitoring of treatment of certain plant or herbal supplement poisonings. One of the examples is monitoring a liver function test in a patient being treated for poisoning with kava-kava, an herbal sedative.
3. Use of more sophisticated techniques such as gas chromatography/mass spectrometry (GC/MS) and high performance liquid chromatography (HPLC) combined with tandem mass spectrometry to confirm the presence of a plant toxin. Such facilities are generally available in large medical centers or reference laboratories.

Certain antidotes are useful for treating poisoning due to plant and herbal remedies. For example, several studies indicate the usefulness of Digibind (the Fab fragment of antidigoxin antibody useful for treating digoxin and digitoxin overdose) in treating poisoning due to the ingestion of oleander leaf, the Chinese medicine Chan Su, digitalis lantana poisoning, and related toxicity from cardiac glycosides (see later discussion).

Case Study

A 4-year-old girl was brought to the emergency department after her mother suspected a poisoning 15 minutes earlier. The girl had ingested multiple orange and red berries from a climbing vine located in the backyard. On admission she showed symptoms consistent with anticholinergic poisoning. She was also nonverbal and responded only to painful stimuli. On examination her heart rate was fast and regular without ectopic beat or murmur. Neurological examination revealed withdrawal to pain, but her blood chemistry was normal. The poison control center was informed, and the plant and berries were identified as nightshade based on a telephone consultation with a local botanist. Physostigmine, the antidote for an anticholinergic crisis, was administered (0.2 mg calculated based on body weight), and the dose was repeated twice over the next 50 minutes. The patient responded 20 minutes after the third dose with normalization of her heart rate and improved mental status. After 36 hours of monitoring, she was discharged in stable condition (19).

10.4. DIGOXIN IMMUNOASSAY IN DETECTING PLANT POISONING AND HERBAL TOXICITY

Digoxin immunoassays can be used for indirectly identifying oleander plant poisoning if a substantial apparent digoxin level in the blood is observed in a patient suspected of ingesting oleander leaf but not receiving digoxin. Similarly, poisoning with the Chinese medicines Chan Su and Lu-Shen-wan can be indirectly detected from observed apparent digoxin levels in a patient poisoned with such traditional Chinese medicines but again not receiving digoxin. The oleander plant contains cardiac glycoside oleandrin, which due to its structural similarity with digoxin cross-reacts with antibodies of various digoxin immunoassays. Chan Su and Lu-Shen-wan contain bufalin, which also has a structural similarity to digoxin (see Chapter 12). Poisoning from the foxglove plant and related cardiac glycoside can also be indirectly diagnosed by using digoxin immunoassays. In addition, Digibind is the Fab fragment of the polyclonal antidigoxin antibody, and this antidote for both digoxin and digitoxin poisoning can also be used in treating poisoning from the oleander plant, foxglove plant, and related poisoning involving cardioactive steroids found in the Chinese medicines Chan Su and Lu-Shen-wan. Table 10.3 lists the alternative medicines and plant poisonings that can be detected indirectly by the digoxin immunoassay.

Digibind therapy has been proposed for treating patients poisoned with oleander. Camphauser et al reported a case in which a 7-year-old girl who had ingested yellow oleander presented to the hospital with complete heart block. The patient was successfully treated with two intravenous doses of 190 mg of Digibind (20). Zylber-Katz et al reported a case in which treatment with a digoxin-specific Fab fragment was successful in treating a 24-year-old man who ingested oleander leaf extract (21). Safadi et al demonstrated the effectiveness

TABLE 10.3. Interference of Plants and Herbal Products in Therapeutic Drug Monitoring of Digoxin and Digitoxin

Herbal Product/Plant	Drug Monitoring	Comments
Lu-Shen-Wan	Digoxin	Lu-Shen-wan interferes with digoxin assay due to its active ingredient bufalin
Chan Su	Digoxin	Chan Su has active components like bufalin that cross-reacts with various digoxin assays
Uzara root (diuretic)	Digoxin	Interferes with digoxin assay
Oleander	Digoxin	Oleander plant poisoning and taking oleander-containing herbs leads to apparent digoxin level in a patient never exposed to digoxin because oleandrin present in oleander cross-reacts with digoxin immunoassays
Foxglove	Digitoxin	Foxglove plant contains digitoxin and other cardiac glycosides

of Digibind in treating a patient who ingested oleander leaf (22). Another article indicated that in the absence of Digibind therapy, fatalities from oleander poisoning increased threefold in Sri Lanka (23).

Case Study

A 37-year-old man presented to the emergency department after ingesting a handful of oleander leaves he harvested from a highway median. He ingested oleander leaves to commit suicide because he was taught as a child that oleander was poisonous. After ingesting the leaves he walked for a few hours and then decided to seek medical treatment when he developed dry mouth, cramping abdominal pain, nausea, vomiting, dizziness, and an irregular heartbeat. He denied taking any digitalis or drugs but occasionally took alprazolam. The patient had a history of chronic depression and suicide attempts that included hospitalization 2 months earlier than this episode for attempted self-drowning. His electrocardiogram showed marked bradycardia. His electrolytes were normal and urine drug screen was negative. However, his serum digoxin level was 1.5 ng/mL (RIA assay, Becton Dickinson Company, Orangeburg, NY). The apparent digoxin level in the patient's serum was due to the cross-reactivity of cardiac glycosides found in oleander leaf with digoxin immunoassay. He was given ipecac and vomited without any return of plant material. He was also given activated charcoal followed by an empirical dose of five vials (40 mg each; 200 mg total dose) of Digibind administered intravenously. After 30 minutes of receiving Digibind, the patient showed improvement in his heart rate and conduction. After monitoring for 4 days, the patient was discharged to an inpatient psychiatric unit (24).

In vitro studies indicate that in the presence of Digibind there is a significant reduction in free apparent digoxin concentrations (measured in the protein-free ultrafiltrate by the fluorescence polarization immunoassay [FPIA] digoxin assay) indicating successful binding of free oleandrin and related cardiac glycosides by Digibind. Therefore, monitoring free apparent digoxin using a digoxin immunoassay with a high cross-reactivity with oleander may be used to monitor progress of therapy in patients poisoned with oleander and being treated with Digibind (25).

Poisoning from foxglove (*Digitalis purpurea*) may be accidental, intentional, or due to the drinking of herbal tea. One report indicates that a man poisoned with foxglove misidentified a foxglove plant as a comfrey plant (26). Another case of foxglove poisoning due to the ingestion of herbal tea in an 85-year-old man produced a serum digoxin level of 1.8 ng/mL and a plasma potassium of 5.4 mmol/L. The serum digitoxin concentration on the second day was 59.0 ng/mL (27). Glycoside composition of foxglove extract contains digitoxin, purpurea A, lanatoside, gitoxin, purpurea B, and may be a trace amount of digoxin.

Rich et al presented a case of a 22-year-old man who presented to the emergency department after an intentional overdose of a homemade foxglove extract. The patients showed a serum digoxin level of 7.82 ng/mL (FPIA) and a serum digitoxin level of 172.5 ng/mL (FPIA). The patient was treated initially with 10 vials of Digibind (800 mg) and then 24 hours later with an additional 400 mg of Digibind. The patient was eventually discharged indicating that Digibind can be used as an antidote in treating foxglove poisoning. The presence of apparent digoxin in patients poisoned with foxglove was due to the cross-reactivity of various glycosides with the FPIA digoxin assay (28). Lacassie et al, using liquid chromatography combined with electrospray mass spectrometry, demonstrated a gitoxin concentration of 13.1 ng/mL, digitoxin; 112.6 ng/mL, digitoxigenin; 3.3 ng/mL, digitoxigenin mono-digitoxoside, and 8.9 ng/mL the first day in a 36-year-old female patient who ingested a concoction of foxglove. A peak gitaloxin level of 112.6 ng/mL was also observed on the fifth day (29).

Plantain (*Plantago major*), a green weedy plant native to Europe and Asia, now grows everywhere in the world, and its leaf is used as an herbal remedy to treat bronchitis, cough, peptic ulcer, and urinary tract infection. Plantain should not be confused with the banana-like vegetable of the same name. Slifman et al reported two cases in which one patient showed a serum digoxin level of 3.66 ng/mL and another patient showed a digoxin level of 0.9 ng/mL. The cause of positive digoxin was contamination of plantain, one of the products both patients received with digitalis lantana (30). Miller also reported that plantain may interfere with the therapeutic drug monitoring of digoxin (31). More recently, Cole and Fetrow commented on the contamination of plantain with digitalis lantana (32). Our experience indicates that contamination of plantain with digitalis may be isolated incidences. Recently we tested three different plantain preparations and found no digoxin-like immunoreactivity using the FPIA assay (33).

Barrueto et al describe a case report in which a 36-year-old woman took an herbal internal cleansing agent. Her serum digoxin level was 1.7 ng/mL (as measured by the FPIA digoxin assay) and 0.34 ng/mL (as measured by Tina-Quant, Roche). Analysis of serum by HPLC showed the presence of digoxin (0.3 ng/mL) and active digitoxin metabolites (no digitoxin). The patient was treated successfully using Digibind (34).

Thurmann et al reported that glycosides from Uzara roots may interfere with serum digoxin measurement by immunoassays. The authors investigated digoxin and digitoxin concentrations after four healthy volunteers ingested 1.5 mL (approximately 22 drops) of Uzara. Maximum digoxin concentrations of 1.4–6.34 μg/L (1.1–4.9 ng/mL) were observed 6 hours after dosing (35). Uzara roots are not usually found in herbal stores in the United States but are a popular remedy in Germany.

In vitro studies show that Digibind can neutralize digoxin-like immunoreactive components of the Chinese medicines Chan Su, Danshen, and Asian ginsengs, and such an effect can be monitored by measuring apparent free digoxin concentrations in the protein-free ultrafiltrates (36). Brubacher et al reported successful treatment with Digibind of two patients poisoned with Chan Su (37).

10.5. ABNORMAL LIVER FUNCTION TESTS AS AN INDICATION FOR PLANT POISONING AND HERBAL TOXICITY

Some plant toxins as well as herbal remedies are known to cause liver damage, and abnormal liver function tests in an otherwise healthy individual during a routine checkup are a good indication of such toxicity. Hepatotoxicity has been reported from consuming plants containing pyrrolizidine alkaloid such as germander (*Teucrium chamaedrys*), chaparral (*Larrea tridentata*), *Atractylis gummifera, Callilepsis laureola,* as well as certain Chinese herbal medicines such as Jin Bu-Huan, ma huang, and Sko-Saiko-to (38). Lantana (*Lantana camara Linn*) is a noxious weed that grows in many tropical and subtropical parts of the world. This weed contains hepatotoxins such as pentacycline triterpenoids called lantadenes. Green unripe fruits of this plant are toxic to humans (39).

Another dangerous herbal supplement containing pyrrolizidine alkaloid is coltsfoot, which is often found in combination with other products and used as a cough remedy. There are several reports of death in the medical literature due to the use of a coltsfoot supplement. Three women were admitted to the hospital due to severe liver toxicity. They were taking an herbal tea, and symptoms developed 19–45 days afterward. Two patients stopped taking the tea, but the third patient continued taking the tea against medical advice and eventually died. It was determined that the tea contained coltsfoot. Death in an infant was reported because her mother drank an herbal tea that contained coltsfoot and breastfed the infant (40).

Kava-kava is an herbal remedy taken for anxiety. However, the use of this remedy can cause severe hepatotoxicity. Heavy consumption of kava has been associated with increased concentrations of γ-glutamyltransferase (GGT), suggesting potential hepatotoxicity. Escher et al described a case in which severe hepatitis was associated with kava use. A 50-year-old man took three to four kava capsules daily for 2 months (maximum recommended dose: three capsules). Liver function tests showed 60- to 70-fold increases in aspartate aminotransferase (AST) and alanine aminotransferase (ALT). Tests for viral hepatitis (hepatitis A, B, and C) were all negative as were tests for cytomegalovirus (CMV) and human immunodeficiency virus. The patient eventually received a liver transplant (41). Humberston et al also reported a case of acute hepatitis induced by kava-kava (42). Other cases of hepatotoxicity due to the use of kava have been documented (43). In January 2003, kava extracts were banned in the entire European Union and Canada and also in the United States. The FDA strongly cautioned against using kava. There are at least 11 cases of serious hepatic failure and 4 deaths directly linked to kava extract

TABLE 10.4. Herbal Supplements and Plant Poisoning That May Cause Abnormal Liver Function Tests

Specimen	Toxicity	Abnormal Laboratory Tests
Kava-kava	Hepatotoxicity	Elevated ALT, AST, GGT, ALP, bilirubin
Comfrey	Hepatotoxicity	Elevated ALT, AST, ALP, bilirubin
Germander	Hepatotoxicity	Elevated ALT, AST, ALP, bilirubin
Chaparral	Hepatotoxicity Carcinogenic	Elevated ALT, AST, ALP, bilirubin
Mistletoe	Hepatotoxicity	Elevated ALT, AST, LDH
Breynia (Chi R Yun)	Hepatotoxicity	Elevated ALT, AST, ALP, GGT
Jimsonweed	Hepatotoxicity	Elevated ALT, AST, CK, LDH

ALT, alanine aminotransferase; AST, aspartate aminotransferase; GGT, γ-glutamyltransferase; ALP, alkaline phosphatase; LDH, lactate dehydrogenase; CK, creatinine kinase.

consumption, and there are also 23 reports indirectly linking kava with hepatotoxicity (44). Anke and Ramzan reviewed hepatotoxicity due to the use of kava (45). Common herbal supplements that may cause abnormal liver function tests are listed in Table 10.4.

Chaparral, found in health food stores as capsules and tablets, is used as an antioxidant and anticancer herbal product. Leaves, stems, and bark in bulk are also available for brewing tea. However, this product can cause severe hepatotoxicity. Several reports of chaparral-associated hepatitis have been reported. Gordon et al reported a case in which a 60-year-old woman took chaparral for 10 months and developed severe hepatitis for which no other cause was found. On admission her bilirubin was 12.4 mg/dL, ALT 341 U/L, AST 1191 U/L, and alkaline phosphatase 186 U/L. All tests for viral hepatitis were negative. Eventually she received a liver transplant (46).

Case Study

A 45-year-old woman who took chaparral 160 mg/day for 10 weeks presented with jaundice, anorexia, fatigue, nausea, and vomiting. Liver enzymes and other liver function tests showed abnormally high values (ALT 1611 U/L, AST 957 U/L, alkaline phosphatase 265 U/L, GGT 993 U/L, and bilirubin 11.6 mg/dL). Viral hepatitis, CMV, and Epstein-Barr virus were ruled out. Liver biopsy showed acute inflammation with neutrophil and lymphoplasmacytic infiltration, hepatic disarray, and necrosis. The diagnosis was drug-induced cholestatic hepatitis, which in this case was due to the use of chaparral (47).

Germander has been used as a remedy for weight loss and as a general tonic. It is an aromatic plant in the "mint family," and Germander tea made from the aerial parts of the plant has been used for many centuries. Several cases of liver toxicity were reported in Europe due to the ingestion of germander (48). A 55-year-old woman taking 1600 mg/day of germander became jaundiced after 6 months. Her bilirubin was 13.9 mg/dL, AST 1180 U/L, ALT 1500 U/L, and alkaline phosphatase (ALP) 164 U/L. Serological tests for all types of hepatitis were negative. Liver biopsy suggested drug-induced hepatitis. Germander therapy was discontinued, and the hepatitis resolved in 2 months (49). Bosisio et al described a HPLC method for the detection of teucrin A, the active component of germander in beverages (50).

Mistletoe is a parasitic evergreen plant that lives on trees such as oaks, elms, firs, pines, apple, and elms. Mistletoe was used in folk medicine as a digestive aid, heart tonic, and a sedative. It was also used in treating arthritis, hysteria, and other mental disturbances, as well as to treat cancer. Leaf of mistletoe is usually used as an herbal remedy. A 49-year-old woman presented with nausea, general malaise, and a dull pain in her abdomen. Liver tests suggested hepatitis: ALT, 123 U/L, lactate dehydrogenase (LDH) 395 U/L, and AST 250 U/L. Liver biopsy also suggested hepatitis. However, all tests for viral hepatitis were negative. The patient was diagnosed with a drug-induced hepatitis due to the use of mistletoe (51).

Breynia officinalis is a Chinese plant from which the proprietary Chinese medicine Chi R Yun is prepared. However, this plant is also mistaken for a similar plant, *Securinega suffruticosa*, and cooked in a soup for the treatment of muscle soreness. Lin et al reported cases of 19 patients (11 males and 8 females) who consumed an average of 130 mL of a soup containing *Breynia officinalis*. Fourteen patients developed diarrhea, 10 experienced nausea and a chilly sensation, and 7 had vomiting. Hepatocellular liver injury was observed in some patients where ALT levels ranged from 89 to 9440 U/L, AST levels ranged from 47 to 67756 U/L, ALP levels ranged from 224 to 278 U/L, and GGT levels ranged from 84 to 313 U/L (52). *Breynia officinalis* contains toxic phenolic glycosides.

Ingestion of jimsonweed (*Dutra stramonium*) may cause severe toxicity because it contains a variety of alkaloids including atropine, hyoscyamine, and scopolamine. Other species of Dutra are also toxic, and patients usually present with the symptoms of atropine intoxication. Elevated levels of liver enzyme may be observed in these patients through an atropine-induced hepatotoxicity. In one patient, AST, ALT, creatinine kinase (CK), and LDH were 1829 U/L, 2052 U/L, 6970 U/L, and 1087 U/L at the second day of admission where initial concentrations were 370 U/L, 304 U/L, 5739 U/L, and 1014 U/L, respectively. The mechanism of elevation in CK concentration is not clear (53). For unambiguous diagnosis, atropine and scopolamine can be confirmed in a biological matrix using liquid chromatography combined with mass spectrometry.

10.6. OTHER ABNORMAL LABORATORY TESTS AS INDICATIONS FOR PLANT POISONING AND HERBAL TOXICITY

Kelp (seaweed) tablets are available in health food stores and used as a thyroid tonic, anti-inflammatory and metabolic tonic, and also as a dietary supplement.

Kelp tablets are rich in vitamins and minerals but also contain substantial amounts of iodine. Teas et al reported that the iodine content varies widely among various commercially available seaweed preparations, and some Asian seaweed dishes may exceed the tolerable upper iodine intake of 1100 μg/day (54). Clark et al studied the effect of kelp supplementation on thyroid function in euthyroid subjects and concluded that short-term dietary supplementation with kelp significantly increases both basal and poststimulation thyroid-stimulating hormone (TSH) (55).

Case Study

A 72-year-old woman with no history of thyroid disease presented with typical symptoms of hyperthyroidism in the clinic. She was taking four to six kelp tablets a day for 1 year. Her TSH was low (1.3 mIU/L), total T4 was 185.3 nmol/L (normal: up to 160 nmol/L), and total T3 was 4.38 nmol/L (normal: 1.07–3.38 nmol/L). After cessation of ingestion of kelp tablets, her hyperthyroidism resolved and thyroid function test returned to normal (TSH, 3.1 mIU/L; total T4, 108.1 nmol/L; total T3, 2.15 nmol/L) (56).

Several cases of licorice-induced hypokalemic myopathy cases have been reported. Laboratory findings include mean serum potassium of 1.98 mmol/L, mean total CK of 5383 U/L, plasma aldosterone activity 2.92 ng/dL, and mean plasma rennin activity of 0.17 ng/mL per hour (57). Cheng et al reported a case of hypokalemia leading to paralysis in a patient with prostate cancer. On admission his potassium was 1.7 mmol/L, and he also had metabolic alkalosis (bicarbonate: 42.6 mmol/L). Other abnormal findings were low plasma rennin activity and a low aldosterone level but a normal cortisol level indicating a state of pseudohyperaldosteronism. The patient was consuming eight packs of Korean herbal tonic (100 mL per pack) daily to treat his prostate cancer for 2 months. A significant amount of glycyrrhizic acid (0.23 mg/mL), an active ingredient of licorice, was detected in the tonic (58).

10.7. CHROMATOGRAPHIC METHODS FOR THE DIAGNOSIS OF PLANT POISONING AND HERBAL TOXICITY

Chromatographic methods are the gold standard for identifying a plant toxin responsible for plant poisoning or poisoning from an herbal supplement (Table 10.5). Unfortunately, only large reference laboratories, academic medical centers, and forensic toxicology laboratories have such capability; most hospital-based clinical laboratories do not offer such testing. Nevertheless, for forensic death investigations, the identification of plant toxin by a chromatographic method is essential to establish the cause of death.

TABLE 10.5. Commonly Reported Chromatographic Methods for Identifying Various Toxic Substances Found in Herbal Remedies and Plants

Herbal/Plant	Toxic Substance	Chromatographic Method
Ma huang	Ephedra alkaloids	GC/MS, HPLC
Bitter orange	Synephrine	GC/MS
Aconite/Monkshood	Aconitine and related compounds	LC/MS/MS
Castor bean	Ricin	GC/MS, HPLC
Angel's trumpet (Datura)	Atropine, Scopolamine	GC/MS
Nutmeg	Myristicin	HPLC

GC/MS, gas chromatography/mass spectrometry; HLPC, high performance liquid chromatography; LC/MS/MS, liquid chromatography combined with tandem mass spectrometry.

10.7.1. Chromatographic Analysis of Ma Huang and Other Weight Loss Products

Ephedra is a small perennial shrub with thin stems that rarely grows taller than 1 foot. Some of the better known species include *Ephedra sinic*a and *Ephedra equisetina* (collectively called ma huang) from China. Ephedrine is the predominant active compound found in ephedra plants, although other compounds such as pseudoephedrine, norephedrine, and phenylpropanolamine are also found. Ma huang contains approximately 1% ephedrine, which has the potential to stimulate the central nervous system and is marketed as weight loss product. Other promoted purposes include bodybuilding and enhancement of athletic performance. "Herbal ecstasy" is also an ephedrine-containing product that can induce a euphoric state. Haller and Benowitz evaluated 140 reports of ephedra-related toxicity that were submitted to the FDA between June 1997 and March 31, 1999. The authors conclude that 31% of the cases were definitely related to ephedra toxicity, and another 31% were possibly related. A total of 47% of the reports of ephedra toxicity involved cardiovascular problems; 18% involved problems with the central nervous system. Hypertension was the single most frequent adverse reaction followed by palpitation, tachycardia, stroke, and seizure. Ten events resulted in death, and 13 events caused permanent disability. The authors conclude that use of a dietary supplement that contains ephedra, a sympathomimetic agent, may pose a serious health risk (59). On April 12, 2004, the FDA prohibited the sale of ephedra containing dietary weight loss supplements in the United States. Inchiosa et al recently commented that experience with sympathomimetic agents for weight loss is mostly negative due to its cardiovascular toxicity (strokes and cardiac arrhythmias) (60).

A urine drug screen in a patient suspected of an herbal weight loss product containing ma huang or ephedra should be positive for amphetamine due to the cross-reactivity of ephedrine and related compounds with amphetamine immunoassays. Nishiguchi et al observed two false-positive amphetamine/methamphetamine tests out of 187 normal subjects screened by a Triage drugs-of-abuse panel who had a history of ingesting herbal drug ma huang. To investigate this positive reaction further, the authors collected urine specimens from seven volunteers who ingested ma huang. All seven specimens tested positive for amphetamine/methamphetamine, but neither amphetamine nor methamphetamine can be confirmed by HPLC or GC/MS. The authors concluded that ma huang causes a false-positive response with the amphetamine immunoassay (61). Ephedrine alkaloids in a dietary supplement can be analyzed by liquid chromatography combined with tandem mass spectrometry. In one report, the authors confirmed the presence of ephedrine, pseudoephedrine, norephedrine, norpseudoephedrine, methyl-ephedrine, and methylpseudoephedrine in dietary supplements using a phenyl liquid chromatography column and a mobile phase composed of ammonium acetate, glacial acetic acid, acetonitrile, and water (62). Alternatively, ephedrine alkaloids can also be determined using GC/MS. Ranieri and Ciolino determined ephedra alkaloids in ma huang and related herbs after extracting them using ammonical chloroform followed by a two-step derivatization that forms O-trimethylsilyl, N-trifluoroacetyl derivatives (63).

Bitter orange or Seville orange has been used mainly as a weight loss product. Topically, bitter orange is also used for treating ringworm and athlete's foot. Following withdrawal of ephedrine from the marketplace of dietary supplements, weight loss products containing bitter orange are gaining popularity, but at this point there is little evidence that bitter orange may promote weight loss. However, synephrine and other structurally related compounds are present in the bitter orange, and these compounds are structurally similar to ephedra. Health Canada reported that from January 1, 1998, to February 28, 2004, it received 16 reports in which products containing bitter orange or synephrine were suspected of being associated with cardiovascular-adverse effects such as blackout, transient collapse, cardiac arrest, tachycardia, and ventricular fibrillation (64). A 55-year-old man who took 300 mg of bitter orange extract every day had an acute myocardial infarction (65).

Case Study

A 33-year-old man presented to the emergency department with a recent onset of dizziness, memory loss, and difficulty concentrating. He had no past medical history of any serious illness. He was a light smoker but denied alcohol or drug abuse. One week before his symptoms he took one or two capsules per day for weight loss (Stacker 2 ephedra-free weight loss product). According to the manufacturer (container label) each capsule contained 6 mg synephrine and 200 mg of caffeine alkaloids (kola nut extract). For the 4 days before presenting to the hospital, he also experienced episodes of dizziness. On admission his blood pressure was normal (120/70 mm Hg), his heart rate was 56 beats/minute, respiratory rate 16/minute, and temperature of 36°C, and his electrocardiogram showed a normal sinus rhythm. However, his physical examination showed a mildly ataxic gait and a substantial impairment of both memory and concentration. His routine laboratory test results were unremarkable, but computed tomography of the brain and magnetic resonance imaging showed subacute infarction in the left thalamus and left cerebellum in the distribution of the vertebrobasilar circulation. Various causes of ischemic stroke were considered, and a vasospastic origin was suspected probably due to synephrine in the weight loss formula. The patient was discharged after 1 week with a nearly complete clinical recovery (66).

Although synephrine has some structural similarity with amphetamine, in one report the authors demonstrated that the cloned donor enzyme immunoassay (CEDIA) for amphetamine showed a negative response when urine specimens were tested from six healthy volunteers after each of them ingested 900 mg dietary supplement capsule (Nature's Way, Bitter Orange) standardized to 6% synephrine (67). Toxicity due to the use of synephrine-containing herbal weight loss products can be confirmed by using GC/MS (68).

Guarana, a popular herbal weight loss product (and also used as one of the ingredients of many herbal weight loss products), is prepared from the seeds of guarana containing 2.5–7% caffeine (200 mg/dose). In contrast, one cup of coffee contains 100 mg or less of caffeine. At the recommended dosage for weight loss, a person may take up to 1800 mg of caffeine per day. Such high caffeine intake may produce adverse effects such as increasing blood pressure to a dangerous level in a person who already has hypertension. In addition to taking such supplement, if a person also drinks coffee or take any medicine such as pseudoephedrine that may also increase blood pressure, there may be an additive effect on blood pressure that may cause cardiac problems. A 25-year-old woman with preexisting mitral valve prolapse died from using an herbal supplement containing guarana and ginseng due to the high caffeine content of the herbal preparation (69).

10.7.2. Chromatographic Methods for the Diagnosis of Aconite Poisoning

Aconite is an herb native to China and Europe commonly known as monkshood. Although the toxicity of aconite was known for centuries, it is still found as an active ingredient in Chinese herbal remedies and homeopathic remedies. Because aconite is very toxic, its use can cause irregular heartbeat, heart block, heart failure, and even death. Aconite is even dangerous to use in the form of a cream for topical application because it can be absorbed from the skin. Even touching this plant may cause an allergic reaction in allergy-prone individuals. Patients usually present with a combination of neurological, cardiovascular, and gastrointestinal features, and the main cause of death is usually ventricular arrhythmias and asystole. The overall hospital mortality is estimated to be 5.5%. There is no antidote for aconite poisoning; treatment is supportive (70).

The Chinese medicines Chuanwu and Caowu are prepared from the roots of various species of aconite plants, and they are believed to have anti-inflammatory, analgesic, and cardiotonic effects. Although soaking and boiling roots may hydrolyze some aconite alkaloids to less toxic compounds, these products still contain highly toxic compounds such as aconitine, mesaconitine, and hypaconitine. Death may occur from ventricular arrhythmia usually within 24 hours of the ingestion of herbal supplements. In one report, three patients who consumed an aconite-containing herbal remedy died from a very abnormal heart rhythm (71). Another patient also was very ill 90 minutes after taking an aconite-containing Chinese herbal preparation and survived, although his symptoms lasted for 2 days (72). There are other reports of death in the medical literature due to the ingestion of aconite-containing herbal remedies. There is no readily available test in clinical laboratories for the quick diagnosis of aconite poisoning. Usually the urine toxicology screen is negative. Pullela et al presented a case in which a 25-year-old man died suddenly following a recreational outing with his friends where he ingested a number of wild berries and plants. One of the plants was later identified as monkshood. All routine blood and urine toxicology tests were negative for this patient, but a high level of aconitine was found in his postmortem blood (3.6 μg/mL) and urine (149 μg/mL), confirming the cause of death as monkshood plant poisoning. The authors used liquid chromatography combined with tandem mass spectrometry (LC/MS/MS) for identification of aconitine in postmortem body fluids (73). In another report, a 54-year-old man died from refractory ventricular fibrillation after ingestion of a plant for suicide. He showed a whole blood aconitine level of 24 μg/mL, a much higher level than reported in other aconitine-related deaths (74).

10.7.3. Castor Bean (Ricin) Poisoning and Chromatographic Methods for Confirming Such Poisoning

The U.S. Centers for Disease Control and Prevention (CDC) lists ricin, the toxin extracted from castor beans, on its second highest priority list as a potential agent for bioterrorism. Ricin is a glycoprotein (molecular weight: 60–65 kDa) that can be isolated from castor beans (*Ricinus communis*). Castor oil produced from the castor bean is used in paints, varnishes, and lubricating oil for jet engines. Castor oil is also used as a purgative, and ricin is contained in the bean part following separation of oil from the bean pulp. No ricin is present in the castor oil, and ricin is also inactivated during the extraction process if heated conditions are used. *Ricinus communis* grows in the southwestern parts of the United States where it is considered a common weed. The castor beans are glossy, oval, and speckled with white black or white spots, and intact beans can pass through the digestive system without producing any significant toxicity. However, if ricin is extracted from the castor beans, it is a deadly poison. Inhalation is the most lethal route, and illness can occur within 8 hours. Symptoms include cough, respiratory distress, fever, hypotension, liver and renal dysfunction, and possible death. Treatment is mostly supportive with vasopressor drugs such as dopamine. There is no antidote, and ricin is not dialyzable. Laboratory findings may include increased ALT, AST, bilirubin, total CK, renal dysfunction, and anemia. The measurement of ricin in biological fluid is difficult (0.1 ng/ml sensitivity), and only mass spectrometric methods are available. Reference laboratories like the CDC and U.S. Army Medical Research Facility have the capability to analyze ricin (75,76). Although castor beans may pass through the digestive system, castor bean extract may contain ricin. Coopman et al described a 49-year-old man who committed suicide by intravenous and subcutaneous injection of castor bean extract. The man died after 9 hours from multiple organ failure. The authors identified ricinine in blood, urine, vitreous humor, and castor bean extract by solid phase extraction followed by full scan GC/MS, HPLC with photodiode combination, as well as LC/MS (77).

10.7.4. Chromatographic Methods for the Diagnosis of Angel's Trumpet Poisoning

Angel's trumpet (*Datura stramonium*) poisoning is common in North America, especially in children. Sometimes adolescents and adults abuse this plant to achieve hallucination. Weibe et al described cases of four adolescent patients who were hospitalized after the intentional ingestion of angel's trumpet. All four patients had a decreased level of consciousness measured by the Glasgow coma scale, visual hallucinations, dilated pupils, and agitation. All four had a history of substance abuse. Use of sedation and restrain was sufficient, and all patients recovered with an average hospital stay of 2 days (78). A 5-year-old healthy boy presented with typical findings of Guillain-Barré syndrome and unilateral tonic pupil after ingestion of angel's trumpet. He was placed on mechanical ventilation for 35 days and experienced stepwise recovery from his poisoning (79). Popoutsis et al described the cases of seven individuals who were admitted to two different hospitals after eating cooked vegetables. All patients showed the anticholinergic syndrome. The investigation revealed that the vegetable was *Datura innoxia* (thorn apple). Urine and plasma samples of seven patients were analyzed by GC/MS, and both atropine and scopolamine were confirmed. Urinary concentration of atropine ranged from 67.1 ng/mL to 691.7 ng/mL; urinary scopolamine concentrations ranged from 32.4 ng/mL to 186.4 ng/mL. The concentrations of atropine and scopolamine were 0.8 μg/g and 1.2 μg/g, respectively, in cooked vegetables. All patients recovered, although some of them required mechanical ventilation (80).

10.8. MISCELLANEOUS PLANT AND HERBAL MEDICINE POISONINGS

There are other plant poisoning and toxicity from using certain herbal supplements that may even cause fatality. However, there is no readily available laboratory tests for the diagnosis of such poisoning, and again only chromatographic methods can be used for identifying the toxic agent responsible for poisoning.

Poison hemlock produced respiratory failure in a 2-year-old boy after he was exposed to this common weed that contains the toxin compound coniine that stimulates nicotinic receptors and causes a syndrome of rapidly progressive muscle weakness and paralysis. Fortunately, the boy recovered within 24 hours (81). Beyer et al developed a method for the quantitative determination of several plant toxins (aconitine, atropine, colchicine, coniine, cystine, nicotine, its major metabolite cotinine, physostigmine, and scopolamine) in human plasma using LC/MS (82). Water hemlock poisoning is due to cicutoxin. Ingestion of water hemlock can lead to serious complications that may even be fatal. Treatment is supportive (83).

Pokeweed (American nightshade), a large perennial herb reaching a height of 10–12 feet during the summer months, is found in the eastern parts of North America, California, Hawaii, and Canada as well as other parts of the world. The berries and dried roots are used for preparing herbal remedies and traditionally were used by Native Americans for treating a variety of conditions including skin disease, syphilis, cancer, infections, emetics, and narcotics. Unfortunately, all parts of the pokeweed are toxic. In the nineteenth century, poisonings from pokeweed were common in the United States because pokeroot tinctures were used to treat rheumatism (84). Eating uncooked berries also cause pokeweed poisoning. Although cooked young leaves (poke salad made by cooking twice and discarding water) and cooked berries (after cooking twice and discarding the water) are eaten by some people, there is no guarantee that such cooked foods are safe for human consumption. The toxicity of pokeweed increases with its maturity, but green berries are more toxic than red berries. There are several cases of pokeweed poisoning reported in the literature. Because of the toxicity, it is advisable to avoid pokeroot tea and herbal supplements containing pokeweed.

Thunder god vine (*Triptergium wilfordii*) has been used in traditional Chinese medicine for more than 2000 years for the local treatment of arthritis and inflammatory tissue swelling. This supplement has also been indicated for treating rheumatoid arthritis. Unfortunately, thunder god vine can cause severe adverse reactions and also is poisonous if it is not carefully extracted from the skinned roots. Other parts of the plant such as the leaves, flowers, and skin of the root are very toxic to humans and may cause death if ingested. A 36-year-old man was admitted to the hospital with severe diarrhea and vomiting for 3 days. Three days before his admission to the hospital he consumed an herbal supplement. The person died 15 hours after admission to the hospital due to shock, hypotension, and cardiac damage. The herbal supplement the patient was taking was identified as thunder god vine (85).

Seeds of the nutmeg are used as a spice, and nutmeg oil has many benefits including its antibacterial effect. Historically, it was used as a stimulant and abortifacient as well as for promoting menstruation. Nutmeg contains volatile oil composed of several active compounds including myristicin. Nutmeg is also abused in large quantities to achieve its hallucinogenic effects because myristin is metabolized to an active compound with an effect likened to lysergic acid diethyl amide (LSD). Due to its euphoric and hallucinogenic effects, nutmeg has long been used as a low-cost substitute for other drugs of abuse. To get this effect, the consumption of 15–20 g of nutmeg is needed, which can produce severe toxicity. From 1998 to 2004, 17 calls were reported to the Texas Poison Center Network involving nutmeg poisoning, and 64.7% of the cases of poisoning (all males) were due to the intentional abuse of nutmeg (86). Stein et al described a fatal case in which a 55-year-old woman died from nutmeg abuse as confirmed by the presence of myristicin (4 µg/mL in postmortem serum). Identification of myristicin was achieved by using ultraviolet-visible spectroscopy and thin layer chromatography; quantitation in serum was performed by HPLC (87). Toxicity may occur with ingestion of approximately 5 g of nutmeg, but it is very unlikely that the intake of nutmeg as a spice would cause any toxicity because the amount of ingested myristicin and other active compounds would be very low (88).

Case Study

A healthy 18-year-old student presented with complains of palpitations, drowsiness, dizziness, nausea, thirst, and dry mouth. She was anxious, restless, and described as being in a trance state. However, she was not experiencing hallucinations. Her medical history was unremarkable. Her blood pressure was 106/68 and cardiopulmonary examination unremarkable. Her serum and urine toxicology screens were negative, and other laboratory tests were within normal limits; electrocardiogram showed a fast sinus arrhythmia (rate: 95–110/minute). She was admitted and after 5 hours confessed that she had taken a large dose of nutmeg to get high. Some 50 g of commercially available nutmeg was blended into a milkshake, and she drank three quarters of the amount. Her symptoms appeared 30 minutes after drinking the milkshake. Her symptoms resolved with time, and she was eventually discharged from the hospital 10 hours after admission (89).

Lobelia (*Lobelia inflata*), also called Indian tobacco, has been used traditionally for treating asthma and bronchitis. The leaves and the seed of lobelia are used in making herbal remedies. Unfortunately, lobelia is a potentially toxic herb, and ingestion of a lobelia-containing herb may cause nausea, vomiting, rapid heartbeat, low blood pressure, and possibly coma. Therefore, it is advisable to avoid lobelia-containing herbals.

Yohimbe bark was traditionally used in Africa to increase sexual desire and now is used as a dietary supplement for treating sexual dysfunction including erectile dysfunction in men. Yohimbe bark can be brewed in tea and consumed, or the extract of the bark, which is available commercially, is consumed. Although yohimbe is effective in treating erectile dysfunction, its benefits seems to be outweighed by the risk of taking this supplement (90). Yohimbe use has been associated with high blood pressure, increased heart rate, dizziness, and other symptoms, and if taken in large doses for a long time it can be dangerous. A 43-year-old man died in a hotel room during sexual relations with a colleague. He was taking Viagra, and pill boxes found in the room had several drugs including yohimbe (91). Therefore, yohimbe does not have a good safety record and probably should be avoided.

10.9. CONCLUSIONS

Although plant poisoning is a rare cause of death in the United States, toxicity from various herbal supplements is an important public health issue. Unfortunately, there is no antidote for most plant poisoning and toxicity after using certain herbal supplements, and there are only a few routine laboratory tests available for the diagnosis of such poisoning. Nevertheless, abnormal liver function tests in a healthy person may indicate toxicity due to the use of certain herbal supplements such as kava, germander, chaparral, and mistletoe. Hypokalemia may due to an overdose with licorice. An apparent digoxin level in a patient not taking digoxin may be related to the use of the Chinese medicine Chan Su, an oleander-containing herb, or oleander plant poisoning. Digibind, a specific antidote for treating digoxin and digitoxin overdose, may be beneficial in treating oleander plant poisoning. Although chromatographic techniques, especially GC/MS and liquid chromatography combined with mass spectrometry or tandem mass spectrometry, are capable of detecting various plant toxins as well as toxic ingredients of an herbal supplement, such methods are not available in many hospital-based small clinical laboratories.

REFERENCES

1. Froberg B, Ibrahim D, Furbee RB. Plant poisoning. Emerg Med Clin North Am 2007; 25: 375–433.
2. Carter K, Neuspiel DR. Toxic plants. Pediatr Rev 2010; 31: 174–175.
3. Pietsch J, Koch I, Hermanns-Clausen M, Huller G et al. Pediatric plant exposures in Germany, 1998–2004. Clin Toxicol (Phila) 2008; 46: 686–691.
4. Krenzelok EP, Jacobsen TD. Plant exposures. A national profile of the most common plant genera. Vet Hum Toxicol 1997; 39:248–249.
5. Caksen H, Odabas D, Akbayram S, Cesur Y et al. Deadly nightshade (*Atropa belladonna*) intoxication: an analysis of 49 children. Hum Exp Toxicol 2003; 22: 665–668.
6. Blaw ME, Adkisson MA, Levin D, Garriott JC, Tindall RSA. Poisoning with Carolina Jessamine (*Gelsemium sempervirens* (L) Ait). J Pediatr 1979; 94: 998–1001.
7. Parisi P, Francia A. A female with ventral anticholinergic syndrome responsive to neostigmine. Pediatr Neurol 2000; 23: 185–187.

8. Mei N, Guo L, Fu PP, Fuscoe JC et al. Metabolism, genotoxicity and carcinogenicity of comfrey. J Toxicol Environ Health B Crit Rev 2010; 13: 509–526.
9. Yu L, Xu Y, Feng H, Li SF. Separation and determination of toxic pyrrolizidine alkaloids in traditional Chinese herbal medicines by micellar electrokinetic chromatography with organic modifier. Electrophoresis 2005; 26: 3397–3404.
10. Chan TY. Cause and prevention of herb induced aconite poisoning in Asia. Hum Exp Toxicol 2011; 30: 2023–2026.
11. Schep LJ, Schmierer DM, Fountain JS. Veratrum poisoning. Toxicol Rev 2006; 25: 73–8.
12. Ramlakhan SL, Fletcher AK. It could have happened to van Gogh: a case of fatal purple foxglove poisoning. Eur J Emerg Med 2007; 14: 356–359.
13. Schofield FW. A brief account of a disease in cattle simulating hemorrhagic septicemia due to feeding sweet clover. Can Vet J 1984; 25: 453–55.
14. De Castro WV, Mertens-Talcott S, Derendorf H, Butterweck V. Effect of grapefruit juice, naringin, narigenin and bergamottin on the intestinal carrier-mediated transport of talinolol in rats. J Agri Food Chem 2008; 56: 4840–4845.
15. Franceschi VR, Nakata PA. Calcium oxalate in plants: formation and function. Annu Rev Plant Biol 2005; 56: 41–71.
16. Watson JT, Jones RC, Siston AM, Diaz PS et al. Outbreak of food borne illness associated with plant material containing raphides. Clin Toxicol (Phila) 2005; 43: 17–21.
17. Audi J, Belson M, Patel M, Schier J et al. Ricin poisoning: a comprehensive review. JAMA 2005; 294: 2342–2351.
18. Sullivan JB, Rumack BH, Thomas H Jr, Peterson RG et al. Pennyroyal oil poisoning and hepatotoxicity. JAMA 1979; 242: 2873–2874.
19. Ceha LJ, Presperin C, Young E, Allswede M et al. Anticholinergic toxicity from nightshade berry poisoning responding to physostigmine. J Emerg Med 1997; 15: 65–69.
20. Camphauser C, Haas NA, Mattke AC. Successful treatment of oleander intoxication (cardiac glycosides) with digoxin specific fab antibody fragment in a 7-year old child: case report and review of literature. Z Kardiol 2005; 94: 817–823.
21. Zylber-Katz E, Safadi R, Kevy I, Caraco Y. Digoxin specific fab fragment in the treatment of oleander poisoning in man. Ther Drug Monit [abstract] 1995; 17: 419.
22. Safadi R, Levy I, Amitai Y, Caraco Y. Beneficial effect of digoxin-specific fab antibody fragments in oleander intoxication. Arch Intern Med 1995; 23: 2121–2125.
23. Eddelson M, Senarathna L, Mohamed F, Buckley N et al. Deaths due to absence of an affordable antitoxin for plant poisoning. Lancet 2003; 362: 1041–1044.
24. Shumanik GM, Wu AW, Ping AC. Oleander poisoning: treatment with digoxin-specific Fab antibody fragments. Ann Emerg Med 1988; 17: 732–735.
25. Dasgupta A, Datta P. Rapid detection of oleander poisoning by using digoxin immunoassays: comparison of five assays. Ther Drug Monit 2004; 26: 658–663.
26. Bain RJ. Accidental digitalis poisoning due to drinking herbal tea [case report]. Br Med J (Clin Res Ed) 290: 1624.
27. Dickstein E, Kunkel FW. Foxglove tea poisoning. Am J Med 1980; 69: 167–169.

28. Rich SA, Libera JM, Locke RJ. Treatment of foxglove extract poisoning with digoxin-specific fab fragment. Ann Emerg Med 1993; 22: 1904–1907.

29. Lacassie E, Marquet P, Martin-Dupont S, Gaulier JM et al. A non-fatal case of intoxication with foxglove documented by means of liquid chromatography-electrospray mass spectrometry. J Forensic Sci 2000; 45: 1154–1158.

30. Slifman NR, Obermeyer WR, Aloi BK, Musser SM et al. Contamination of botanical dietary supplements by digitalis lantana. N Eng J Med 1998; 339: 806–812.

31. Miller LG. Herbal medicinal: selected considerations focusing on known potential drug-herb interactions. Arch Intern Med 1998; 158: 2200–2211.

32. Cole MR, Fetrow CW. Adulteration of dietary supplements. Am J Health-Syst Pharm 2003; 60: 1576–1580.

33. Dasgupta A, Davis B, Wells A. Effect of plantain on therapeutic drug monitoring of digoxin and thirteen other common drugs. Ann Clin Biochem 2006; 43: 223–225.

34. Barrueto F, Jortani SA, Valdes R, Hoffman RS et al. Cardioactive steroid poisoning from an herbal cleansing preparation. Ann Emerg Med 2003; 41: 396–399.

35. Thurmann PA, Neff A, Fleisch J. Interference of Uzara glycosides in assays of digitalis glycosides. Int J Clin Pharmacol Ther 2004; 42: 281–284.

36. Dasgupta A, Szelei-Stevens K. Neutralization of free digoxin-like immunoreactive components of Oriental medicines DanShen and Lu-Shen-Wan by the fab fragment of antidigoxin antibody (Digibind). Am J Clin Pathol 2004; 121: 276–281.

37. Brubacher JR, Ravikuman PR, Bania T, Heller MB et al. Treatment of toad venom poisoning with digoxin-specific Fab fragment. Chest 1996; 110: 1282–1288.

38. Stickel F, Egerer G, Seitz HK. Hepatotoxicity of botanicals. Public Health Nutr 2000; 3: 113–124.

39. Sharama OP, Sharma S, Pattabhi V, Mahato SB et al. A review of the hepatotoxic plant Lantana camara. Crit Rev Toxicol 2007; 37: 313–352.

40. Dailey A, Johns Cupp M. Coltsfoot. In: Johns Cupp M, ed. *Toxicology and Clinical Pharmacology of Herbal Products*. Totowa, NJ: Humana Press; 2000.

41. Escher M, Desmeules J. Hepatitis associated with kava, a herbal remedy. Br Med J 2001; 322: 139.

42. Humberston CL, Akhtar J, Krenzelok EP. Acute hepatitis induced by kava-kava. J Toxicol Clin Toxicol 2003; 41: 109–113.

43. Stickel F, Baumuller HM, Seitz K, Vasilakis D et al. Hepatitis induced by kava (*Piper methysticum rhizoma*). J Hepatol 2003; 39: 62–67.

44. Clouatre DL. Kava kava: examining new reports of toxicity. Toxicol Lett 2004; 150: 85–96.

45. Anke J, Ramzan I. Kava hepatotoxicity: are we close to the truth? Planta Med 2004; 70: 193–196.

46. Gordon DW, Rosenthal G, Hart J, Sirota R et al. Chaparral ingestion. The broadening spectrum of liver injury caused by herbal medications. JAMA 1995; 273: 489–490.

47. Alderman S, Kailas S, Goldfarb S et al. Cholestatic hepatitis after ingestion of chaparral leaves: confirmation by endoscopic retrograde cholangiopancreatography and liver biopsy. J Clin Gastroenterol 1994; 19: 242–247.

48. Perez Alvarez J, Saez-Royuela F, Gento Pena E, Lopez Morante A et al. Acute hepatitis due to ingestion of *Teucrium chamaedrys* infusion [in Spanish]. Gasteroenterol Hepatol 2001; 24: 240–243.
49. Laliberte L, Villeneuve JP. Hepatitis after use of germander, a herbal remedy. CMAJ 1996; 154: 1689–1692.
50. Bosisio E, Givarini F, Dell'Agli M, Galli G, Galli CL. Analysis of high-performance liquid chromatography of teucrin A in beverages flavored with an extract of *Teucrium chamaedrys L.* Food Addit Contam 2004; 21: 407–414.
51. Harvey J, Colin-Jones DG. Mistletoe hepatitis. Br Med J (Clin Res Ed) 1981; 282: 186–187.
52. Lin TJ, Su CC, Lan CK, Jiang DD et al. Acute poisoning with *Breynia officinalis*—an outbreak of hepatotoxicity. J Toxicol Clin Toxicol 2003; 41: 591–594.
53. Ertekin V, Selimogly MA, Altinkaynak S. A combination of unusual presentation of *Datura stramonium* intoxication in a child: rhabdomyolysis and fulminant hepatitis [letter]. J Emerg Med 2005; 28: 227–228.
54. Teas J, Pino S, Critchley A, Braverman LE. Variability of iodine content in common commercially available edible seaweeds. Thyroid 2004; 14: 836–841.
55. Clark CD, Bassett B, Burge MR. Effect of kelp supplementation on thyroid function in euthyroid subjects. Endocr Pract 2003; 9: 363–369.
56. Shilo S, Hirsch HJ. Iodine-induced hyperthyroidism in a patient with a normal thyroid gland. Postgrad Med J 1986; 62: 661–662.
57. Shintani S, Murase H, Tsukagoshi H, Shiigai T. Glycyrrhizin (licorice) induced hypokalemic myopathy: report of 2 cases and review of literature. Eur Neurol 1992; 32: 44–51.
58. Cheng CJ, Chen YH, Chau T, Lin SH. A hidden cause of hypokalemic paralysis with prostate cancer. Support Care Cancer 2004; 12: 810–812.
59. Haller CA, Benowitz NL. Adverse and central nervous system events associated with dietary supplements containing ephedra alkaloids. N Eng J Med 2000; 343: 1833–1838.
60. Inchiosa MA. Experience (mostly negative) with the use of sympathomimetic agents for weight loss. J Obes 2011; 2011 pii 764584.
61. Nishiguchi M, Kinoshita H, Higasa K, Taniguchi T et al. The false positive reaction of the Triage panel drugs-of-abuse by herbal drug ma huang (Ephedra sinica (Ephedraceae) [in Japanese]. Nihon Hoigaku Zasshi 2001; 55: 331–338.
62. Trujillo WA, Sorenson WR. Determination of ephedra alkaloids in dietary supplements and botanicals by liquid chromatography/tandem mass spectrometry: collaborative study. J AOAC Int 2003; 86: 657–668.
63. Ranieri TL, Ciolino LA. Rapid selective screening and determination of ephedra alkaloids by GC-MS footnote mark. Phytochem Anal 2008; 19: 127–135.
64. Haaz S, Fontaine KB, Cutter G, Limdi N et al. Citrus aurantium and synephrine alkaloids in the treatment of overweight and obesity: an update. Obes Res 2006; 7: 79–88.
65. Hess AM, Sullivan DL. Potential for toxicity with use of bitter orange extract and guarana for weight loss. Ann Pharmacother 2005; 39: 574–575.

66. Bouchard NC, Howland MA, Greller HA, Hoffman RS et al. Ischemic stroke associated with use of an ephedra-free dietary supplement containing synephrine. Mayo Clin Proc 2005; 80: 541–545.
67. Nguyen DT, Bui LT, Ambrose PJ. Response to CEDIA amphetamines assay after a single dose of bitter orange. Ther Drug Monit 2006; 28: 252–254.
68. Rossato LG, de Pinho PG, Silva R, Carmo H et al. Development and validation of a GC/IT-MS method for simultaneous quantitation of para and meta synephrine in biological samples. J Pharm Biomed Anal 2010; 52: 721–726.
69. Hess AM, Sullivan DL. Potential for toxicity with use of bitter orange extract and guarana for weight loss. Ann Pharmacother 2005; 39: 574–575.
70. Chan TY. Aconite poisoning. Clin Toxicol (Phila) 2009; 47: 279–285.
71. But PP, Tai YT, Young K. Three fatal cases of herbal aconite poisoning. Vet Hum Toxicol 1994; 36: 212–215.
72. Cahn TY, Tomlinson B, Chan WW, Yeung VT et al. A case of acute aconite poisoning caused by chunanwu and caowu. J Trop Med Hyg 1993; 96: 62–63.
73. Pullela R, Young L, Gallagher B, Avis SP et al. A case of fatal aconitine poisoning by monkshood ingestion. J Forensic Sci 2008; 53: 491–494.
74. Strzelecki A, Pichon N, Gaulier JM, Amiel JB et al. Acute toxic herbal intake in a suicide attempt and fatal refractory ventricular arrhythmia. Basic Clin Pharamacol Toxicol 2010; 107: 689–699.
75. Doan LG. Ricin: mechanism of toxicity, clinical manifestations and vaccine development. A review. J Toxicol Clin Toxicol 2004; 42: 201–208.
76. Audi J, Belson M, Patel M, Schier J et al. Ricin poisoning: a comprehensive review. JAMA 2005; 294: 2342–2351.
77. Coopman V, De Leeuw M, Cordonnier J, Jacobs W. Suicidal death after injection of a castor bean extracts (*Ricinus communis*). Forensic Sci Int 2009; 189: e13–20.
78. Weibe TH, Sigurdson ES, Katz LY. Angel's trumpet (*Datura stramonium*) poisoning and delirium in adolescents in Winnipeg, Manitoba: summer 2006. Paediatr Child Health 2008; 13: 193–196.
79. Sevketoglu E, Tatli B, Tugcu B, Demirelli Y et al. An unusual cause of fulminant Guillain-Barré syndrome: angel's trumpet. Pediatr Neurol 2010; 43: 368–370.
80. Papoutsis I, Nikolaou P, Athanaselis S, Stefanidou M et al. Mass intoxication with Dutra innoxia—case series and confirmation by analytical toxicology. Clin Toxicol (Phila) 2010; 48: 143–145.
81. West PL, Horowitz BZ, Montanaro MT, Lindsay JN. Poison hemlock-induced respiratory failure in a toddler. Pediatr Emerg Care 2009; 25: 761–763.
82. Beyer J, Peters FT, Kraemer T, Maurer HH. Detection and validated quantification of toxic alkaloids in human plasma-comparison of LC-APCI-MS with LC-ESI-MS/MS. J Mass Spectrom 2007; 42: 621–633.
83. Schep LJ, Slaughter RJ, Becket G, Beasley DM. Poisoning due to water hemlock. Clin Toxicol (Phila) 2009; 47: 270–278.
84. Lewis WH, Smith PR. Pokeroot herbal tea poisoning. JAMA 1979; 242: 2759–2760.
85. Chou WC, Wu CC, Yang PC, Lee YT. Hypovolemic shock and mortality after ingestion of *Tripterygium wilfordii* hook F: a case report. Int J Cardiol 1995; 49: 173–177.

86. Forrester MB. Nutmeg intoxication in Texas: 1998–2004. Hum Exp Toxicol 2005; 563–566.
87. Stein U, Greyer H, Hentschel H. Nutmeg poisoning—report on a fatal and a series of cases recorded by a poison information center. Forensic Sci Int 2001; 118: 87–90.
88. Hallstrom H, Thuvander A. Toxicological evaluation of myristicin. Nat Toxin 1997; 5: 186–192.
89. Demetriades AK, Wallman PD, McGuiness A, Gavalas MC. Low cost high risk: accidental nutmeg intoxication. Emerg Med J 2005; 22: 223–225.
90. Ernst E, Pittler MH. Yohimbine for erectile dysfunction: a systematic review and meta-analysis of randomized clinical trials. J Urol 1998; 159: 433–436.
91. Dumestre-Toulet V, Cirimele V, Gromb S, Belooussoff T et al. Last performance with Viagra: post-mortem identification of sildenafil and its metabolites in biological specimens including hair sample. Forensic Sci Int 2002; 126: 71–76.

Chapter 11

Sources of Erroneous Results in Therapeutic Drug Monitoring Due to Preanalytical Errors, High Bilirubin, Hemolysis, and Lipids

11.1. INTRODUCTION

Many analytical techniques are used in the clinical laboratory for therapeutic drug monitoring including immunoassays, gas chromatography, and liquid chromatography as well as chromatographic techniques coupled with mass spectrometry. In general, immunoassays are more susceptible to interferences than chromatography techniques, but blood collected in tubes with the wrong anticoagulant may also cause erroneous results regardless of the method of analysis. In addition, one of the most common causes of unexpected test results in therapeutic drug monitoring is collecting a specimen in a serum separator tube where the barrier gel may absorb certain drugs. High bilirubin or hemolyzed specimens also cause erroneous results in therapeutic drug monitoring.

Resolving Erroneous Reports in Toxicology and Therapeutic Drug Monitoring: A Comprehensive Guide, First Edition. Amitava Dasgupta.

11.2. SERUM VERSUS PLASMA AND GLASS VERSUS PLASTIC TUBE

Although serum and heparinized plasma specimens are considered equivalent for many assays performed in clinical laboratories, differences in results may be obtained for several analytes. Significant differences have been reported for albumin, alkaline phosphatase, calcium, carbon dioxide, chloride, creatinine kinase, glucose, inorganic phosphorus, potassium, and total protein between serum and heparinized plasma. In general, potassium, lactate dehydrogenase, calcium, and magnesium values can be higher in serum than plasma. For potassium, there are differences between serum and plasma reference values (1).

For therapeutic drug monitoring, a green-top tube containing heparin is widely used, although a serum specimen (collected in a red-top tube without barrier gel) can provide accurate results for most therapeutic drugs except for immunosuppressant drugs such as cyclosporine, tacrolimus, sirolimus, and everolimus where whole blood is the required specimen. It is usually recommended that lavender-top tubes (contains ethylenediamine tetra-acetic acid [EDTA] as the coagulant) must be used for collecting cyclosporine, tacrolimus, sirolimus, and everolimus. For cyclosporine, EDTA is preferred over heparin as the anticoagulant of choice. The stability of cyclosporine in EDTA has been demonstrated. The differences in cyclosporine concentrations collected in glass (K_3EDTA) and plastic (K_2EDTA) tubes are less than 10% (2). The small differences observed could be attributed to the dilution effects by the liquid K_3EDTA compared with the dry K_2EDTA. Seger et al described simultaneous quantification of cyclosporine, tacrolimus, sirolimus, and everolimus using whole blood specimens collected in a tube containing EDTA as an anticoagulant and high performance liquid chromatography combined with mass spectrometry for analysis (3). However, Korte et al reported that heparinized whole blood is equivalent to EDTA whole blood for the analysis of cyclosporine using fluorescence polarization immunoassay (FPIA) and cloned enzyme donor immunoassay assay and Beckman Coulter DXC analyzer (Brea, CA) (4). Interestingly, another immunosuppressant drug, mycophenolic acid, is monitored in a heparinized plasma specimen (5). Alternatively, serum can also be used for the determination of mycophenolic acid concentration because heparinized plasma specimens showed a mean bias of only negative 1.6% with serum samples for measuring mycophenolic acid levels (6).

Green-top tubes contain heparin as an anticoagulant, but heparin can be used in such tubes as lithium, ammonium, or sodium salt. Although lithium heparin can be used in collecting specimens for most therapeutic drugs, it is unsuitable for therapeutic drug monitoring of lithium for obvious reasons. Lithium is used in the treatment of bipolar affective disorder, and due to a narrow therapeutic range (0.6–1.2 mEq/L), the therapeutic drug monitoring of lithium is essential. Lithium toxicity is usually encountered at 1.5 mEq/L (or mmol/L) level or higher, and lithium overdose has been associated with fatality. However, in one report the authors observed severe neurotoxicity at a therapeutic lithium level of 0.8 mEq/L in one patient (7). Fatality has even been reported from a therapeutic lithium serum level (8). The preferred specimen for lithium determination is serum (collected in a red-top tube or a tube containing serum separator gel) to avoid confusion, although a sodium heparin tube is also acceptable. Wills et al reported that lithium levels can be falsely elevated up to 4 mmol/L if lithium heparin-containing green-top tubes are used for specimen collection (9).

Case Study

A 37-year-old woman delivered a female infant at full term (birthweight: 3.1 kg). The mother was on 900 mg/day of lithium throughout the pregnancy, and her serum lithium level was 0.9 mmol/L at the time of delivery and 0.7 mmol/L 11 days later. The infant was breastfed, and her serum lithium level was undetectable at 3 days after birth, but the infant's blood lithium level increased to 0.7 mmol/L at day 6 and 1.1 mmol/L at day 10 and day 18. The authors questioned the validity of the serum lithium level in the infant because she showed no signs of lithium toxicity or any renal insufficiency. Later it was found that the infant's blood was wrongly collected in a tube containing lithium heparin as an anticoagulant. A correctly collected blood specimen later from the infant showed an undetectable level of lithium (10).

Matzke et al reported that heparin interferes with tobramycin determination by the enzyme multiplied immunoassay technique (EMIT) (Syva, Palo Alto, CA; also marketed by Siemens Diagnostics, Deerfield, IL) and measured values can be significantly lower compared (up to 82%) with a radioimmunoassay. The magnitude of interference depended on the heparin concentration of the tube (11). In another report, the authors observed significant negative bias in the measurement of the aminoglycosides gentamicin, netilmicin, and tobramycin using EMIT assays if specimens were collected in heparinized tubes. However, radioimmunoassay and FPIA (Abbott Laboratories, Abbott Park, IL) were not affected by heparin (12). Therefore, it is important to check the package insert for appropriate specimen collection because requirements may change from manufacturer to manufacturer. However, the presence of heparin is not usually a factor with therapeutic drug monitoring using automated analyzers. For example, the Vitros, Dimension, and Centaur suggest either serum or plasma for their drug assays, except digoxin on the Centaur is restricted to serum (information from respective package inserts). However, for a serum specimen, insufficiently clotted blood may contain small fibrin clots, creating the potential for inaccurate results or instrument error. However, if blood is collected in a heparinized tube, it must be inverted 8–15 times to ensure proper mixing of the blood with the anticoagulant, but specimens can be centrifuged without the 30-minute waiting period.

TABLE 11.1. Appropriate Collection Tube for Therapeutic Drug Monitoring

Tube	Anticoagulant	Drugs
Red top	None	All drugs including free drugs.
Green top	Lithium/sodium/ ammonium heparin	Lithium cannot be analyzed collected in green-top tube containing lithium heparin but for most drug heparinize plasma can be used. However any aminoglycoside should not be analyzed using heparinized plasma if EMIT method is used. Digoxin assay using Centaur analyzer prefers serum specimens.
Lavender top	EDTA	Recommended for analysis of cyclosporine, tacrolimus, sirolimus, and everolimus using whole blood containing EDTA as anticoagulant. However, mycophenolic acid is usually analyzed using plasma rather than whole blood.
Royal blue top	EDTA/Acid washed	Should be used for all heavy metal analysis.
Serum separator tubes (SSTs)	May or may not contain anticoagulant	Unsuitable for analysis of certain drugs (see Table 11.2).

Analysis of heavy metals such as lead, mercury, arsenic, and cadmium are often carried out in a toxicology laboratory. Heavy metal can be analyzed in whole blood, serum, or urine specimens. However, to avoid contamination, a specially designed blood collection tube or urine collection bottle must be used because standard red-top tubes are contaminated with zinc and other heavy metals (lead, mercury, cadmium, nickel, chromium, etc.), and rubber stoppers are often contaminated with aluminium. Royal blue–top tubes are especially designed for blood collection for heavy metal analysis, and in such tubes EDTA is used as the anticoagulant. Heavy metals can be analyzed in whole blood or serum. For urine collection, clean plastic containers with no metal caps or glued inserts should be used. In addition, urine must not be collected from metal urinals. Colored containers must not be used for urine collection because metals may be found in the dye. Table 11.1 summarizes the recommended tubes for blood collection for therapeutic drug monitoring.

Because plastic blood collection tubes are not breakable, such tubes are widely used for blood collection for various laboratory tests. Correlation studies between glass and plastic blood collection tubes for common analytes and some hormones and tumor markers have been documented. Dasgupta et al examined the stability of 13 therapeutic drugs stored in plastic tubes compared with glass tubes using quality controlled materials. No significant reduction in concentrations of caffeine, primidone, procainamide, N-acetylprocainamide, acetaminophen, salicylate, amikacin, valproic acid, methotrexate, or cyclosporine was observed. When comparing volume of serum in the tube, the authors observed significant reductions in the concentration of phenytoin, phenobarbital, carbamazepine, quinidine, and lidocaine after storage of 500 μL of serum versus 1 mL in both glass and plastic gel tubes. However, the lower recovery of these drugs was attributed to the gel in the tubes rather than the tube material (13). Faynor and Robinson examined the suitability of plastic blood collection tubes for cyclosporine measurement. Specimens from renal transplant patients were collected in glass and plastic Becton-Dickinson Vacutainer EDTA tubes. Tubes were stored at room temperature and 4°C after collection, and cyclosporine levels were determined between testing intervals of 0, 1, 4, and 7 days. The drug levels in the plastic tubes were slightly higher than those from the glass tubes at both storage temperatures, but time had no effect on the cyclosporine concentration. All of the differences for individual pairs of samples were within 10%. Therefore, the authors concluded that cyclosporine levels are stable in plastic tubes over 7 days at room temperature or refrigerated (14). Boeynaems et al compared Terumo's Venoject glass tubes with Venosafe PET tubes with clot activator and heparin for several therapeutic drugs at 2 and 24 hours postcollection. Blood was spiked with the parent drug to achieve low, mid, and high therapeutic levels. The authors found no consistent significant differences among the tube types for the panel of drugs tested (15).

However, plastic glass collection tubes are unsuitable for the collection of blood for the analysis of tetrahydrocannabinol (THC). As expected, THC concentrations remained unchanged when blood collected in glass tubes were stored for 4 weeks at −20°C or 4 days at room temperature, but THC concentrations were reduced by 60–100% when blood was collected and stored in plastic tubes (16). In urine specimens, the major metabolite of marijuana (11-nor-delta-9-tetrahydrocannibinol -9-carboxylic acid [THC-COOH]) is analyzed as a proof of marijuana abuse. Rapid loss of THC-COOH at 4°C for urine specimens stored in both polypropylene (14% maximum loss) and polyethylene (17% maximum loss) plastic bottles was observed, but significantly less loss was observed when specimens were stored at 25°C using both types of plastic containers (loss less than 5%). Interestingly, all reductions in concentrations of THC-COOH were observed within 1 hour, and no further losses were observed up to 1 week. The authors concluded that polypropylene and polyethylene plastic containers can be used for the collection of urine for the analysis of the marijuana metabolite (17). In another report the authors observed that adsorptive loss of THC-COOH is a phenomenon observed in the acidic urinary pH (4.6) but relatively absent near neutral or basic pH. Moreover, both borosilicate glass containers and polypropylene containers showed similar absorptive loss of THC-COOH when urine specimens were stored in such containers for a prolonged time and depended on the solution to container exposure time (18). Romberg and Past studied the stability of THC-COOH and benzoylecgonine (metabolite of cocaine) in urine specimens stored for a prolonged time and observed that in 85 THC-COOH-positive urine specimens stored frozen for 1–10 months (average: 2.3 months), the average decline of THC-COOH was 25%; based on 61 benzoylecgonine specimens studied, the average decline was 19%. Because drug testing laboratories performing legal drug testing are required to save positive urine specimens for at least a year in the case of requested reanalysis of specimens during legal challenge, their reports indicated that values of reanalyzed specimen may not match the original drug metabolite levels determined in a relatively fresh urine specimen (19).

11.3. SUITABILITY OF SERUM SEPARATOR GEL TUBES FOR BLOOD COLLECTION FOR THERAPEUTIC DRUG MONITORING

Serum separator tubes (SSTs) are widely used for blood collection for clinical laboratory tests including therapeutic drug monitoring because barrier gel facilitates the rapid separation of serum from cellular constituents of blood, thus reducing time-consuming preanalytical steps because serum can be stored over the barrier gel without the possibility of hemolysis. If analysis of the serum is not performed shortly after processing, collecting serum in separator tubes provides a closed system that allows for collection, transport, processing, sampling, and storage of specimens. Different barrier materials are available among tube manufacturers, but all are thixotropic materials that facilitate the separation of serum or plasma from the cells and prevent hemolysis on prolonged storage, and the base material for preparing the gel is acrylic, silicone, or a polyester polymer. The stability of various analytes when blood is collected in gel tubes has been studied extensively. Significant absorption of certain drugs in specimens collected in gel barrier tubes was first reported in the 1980s. Although improvements have been made to the gel composition over time to prevent such passive absorption, certain drugs are still absorbed by the gel material of various commercially available gel tubes. Therefore, using a blood collection tube without gel is the best way to avoid such a problem. In addition to specific therapeutic drugs, certain steroid hormones are also absorbed by the gel. Several factors affect the stability of therapeutic drugs in sera when stored in serum separator gel tubes with serum in contact with the gel including the following:

1. The chemical nature of the drug and the gel is the primary factor that determines absorption of the drug by the gel. Thus, if the gel is hydrophobic, hydrophobic drugs tend to be absorbed and hydrophilic drugs remain in the aqueous fraction.
2. Specimen volume and time of storage are also important factors that determine the stability of therapeutic drugs in sera stored in gel tubes. In general, if analysis is done within 2–3 hours of blood collection, even certain drugs that are absorbed by the barrier gel did not show clinically significant decline in concentrations. When blood is collected into a gel tube, the total volume of blood is exposed to a constant surface area of gel. Drug absorption in whole blood is typically less than 1% of the total concentration as shown by comparing the nongel tube with the gel tube drug levels at the initial time after processing. Additionally, the surface

TABLE 11.2. Effect of Collecting Specimen in Gel Separator Tube on Therapeutic Drug Monitoring Results

Type of Gel Tube/Drugs Affected	Comments
Becton-Dickinson SST Tube	
Phenytoin Phenobarbital Carbamazepine Lidocaine Quinidine	Decreased value of therapeutic drugs depends on time of storage and volume. Fill the tube completely and analyze within 4 hours of collection to avoid error.
Becton-Dickinson SST II Tube	
Phenytoin	Decreased value of phenytoin less significant than carbamazepine.
Carbamazepine	Fill the tube completely and analyze within 24 hours of collection to avoid error.
Corvac Tube	
Lidocaine	Decrease in lidocaine concentration is volume and time dependent. Fill the tube completely and analyze within 4 hours of collection to avoid error.
Greiner Vacuette Tube	
Tricyclic antidepressant	Decreased value of therapeutic drug with prolonged storage.
Carbamazepine	Fill the tube completely and analyze within 4 hours of collection to avoid error.

area of the gel centrifuged in a fixed-angle centrifuge is greater than that when centrifuged in a swing-bucket centrifuge. A higher gel surface area can lead to greater drug absorption into the gel. If the serum separator tube is filled almost to its capacity, declines in serum drug levels with time is less significant if the tube is half filled or less than half filled.

3. The temperature of storage also influences drug stability. The rate of absorption increases with increasing temperatures until equilibrium is achieved. The stability of various drugs stored in serum separator gel tubes is summarized in Table 11.2.

As early as 1983, Quattrocchi et al studied the stability of 12 commonly monitored drugs in sera stored in gel tubes and observed significant decreases in the concentrations of lidocaine, pentobarbital, and phenytoin upon storage. The authors concluded that this problem can be circumvented by processing at least 2 mL of the specimen when analyzed within 1 hour (20). Landt et al reported that when blood was processed and analyzed up to 24 hours after collection, none of the separator tubes had any effect on the seven drugs studied (theophylline, digoxin, phenytoin, phenobarbital, gentamicin, and ethanol). However, phenobarbital concentration was reduced after 24 hours of storage. The authors also examined the effect of partially filling various barrier tubes with blood instead of full draw on these seven drugs. Only phenytoin, at a 1-mL fill volume in a 7-mL polyester separator tube, showed a modestly lower recovery (92%) (21). Bailey et al studied the stability of 11 drugs (amikacin, carbamazepine, digoxin, gentamicin, lithium, methotrexate, phenobarbital, phenytoin, quinidine, theophylline, and tobramycin) as well as two heavy metals (copper and zinc) in plasma specimens stored in a Corvac gel septation tube over a week of storage in the refrigerator and found no significant absorption of any drugs by the barrier gel. The authors used 10 mL of serum supplemented with appropriate drugs for their study (22). Devine also did not observe any decline in drug concentration either over time or with respect to sample size when aliquots of heparinized blood (3–9 mL) were supplemented with lidocaine or phenytoin and stored in Corvac serum separator gel tubes for 24 hours (23). In contrast, Koch and Platoff demonstrated that both Corvac and Becton-Dickinson Vacutainer SST serum separator blood collection tubes were unsuitable for the storage of specimens containing lidocaine due to a 25–30% decline in concentrations over a 72-hour period. In addition, for phenytoin, SST serum separator blood collection tubes may extract phenytoin even with the immediate separation of serum. The authors concluded that SST serum separator blood collection tubes were unsuitable for quinidine (storage greater than 24 hours), lidocaine (greater than 6 hours), and phenytoin. Partially filled tubes caused additional errors (24). In another report the authors observed a decline in free and total phenytoin concentrations as well as free and total carbamazepine concentrations when specimens were collected in Becton-Dickinson SST serum separator blood collection tubes, but no decline was observed in specimens collected in Terumo Autosep tubes also containing barrier gels (25). Dasgupta et al studied the stability of several therapeutic drugs in serum following storage in Becton-Dickinson SST serum separator blood collection tubes and Corvac serum separator blood collection tubes and observed significant reductions in drug concentrations (ranging from 5.9% to 64.5%) for phenytoin, phenobarbital, lidocaine, quinidine, and carbamazepine. In contrast, concentrations of theophylline and salicylate did not change under identical specimen storage conditions. The reduction in drug concentration depended on the time of storage (more pronounced effect with storage over 2–6 hours) as well as specimen volume with more significant changes observed in specimens where sample volumes were low. No significant change in the concentration of phenytoin, phenobarbital, carbamazepine, theophylline, quinidine, and salicylate was observed when serum was stored in Corvac serum separator blood collection tubes, whereas concentration of lidocaine declined (range: 31.5–72.6%) after storage in Corvac tubes, especially if specimen volume was low (200–500 μL). The decline in drug concentration due to storage of specimens in a gel separator tube was due to slow passive absorption of certain drugs by the gel as evidenced by recovery of drugs from the gel following chemical extraction with methanol. For phenytoin and phenobarbital, the reduction in total drug concentration also resulted in a proportional reduction in free drug concentration, and such a reduction in free drug level also depended on the extent of the protein binding of the drug. As expected, specimens stored in standard red-top Vacutainer tubes without any separator gel showed no reduction in concentration of any drugs (26).

Case Study

A 6-day-old boy weighing 3.6 kg was hospitalized for possible sepsis and meningitis after experiencing two seizures at home. During the first 4 days of hospitalization, seizures continued despite phenobarbital and diazepam therapy. Because of the seizures, phenytoin therapy was initiated on day 6. The infant received a wrong dose of phenytoin (360 mg) and suffered cardiac arrest. Shortly after resuscitation he was transferred to the University of Chicago Hospital and admitted to the pediatric intensive care unit. Over the next 9 days his phenytoin concentrations were monitored closely; still they declined to less than 10 μg/mL. During the first 2–3 days at the University of Chicago Hospital, several phenytoin concentrations were reported to be substantially below the expected values (up to 40%), and all such specimens involved drawing blood in SST serum separator blood collection tubes with small sample volumes (less than 400 μL). All these specimens were submitted as routine specimens and were analyzed 2–3 hours after receiving such specimens. When such specimens were reanalyzed the following day, a further reduction in phenytoin concentration was observed. It was determined that lower values were due to the absorption of phenytoin by the barrier gel (26).

The most extensive study on the suitability of gel barrier tubes for therapeutic drug monitoring was published by Karppi et al. The authors studied the stability of 41 drugs, including tricyclic antidepressants, benzodiazepines, antiepileptics, asthma drugs, aminoglycosides, other antibiotics, and cardioactive drugs, when specimens were stored in three different gel tubes (BD SST tubes, Terumo Autosep, and Sarstedt Microvette gel tubes). After 24-hour storage time, absorbed drugs ranged from 5% to 20% the concentration at the initial time for all analyzed drugs. The authors concluded that the studied gel tubes were satisfactory for blood collection for antiepileptic, antibiotics, asthma, and cardioactive drug measurement because absorptions were slight (0–5%). Antidepressant drugs can also be collected in gel separation tubes if the separation step is performed within 3 hours of blood clotting because absorption was less than 10%. However, gel tubes are not suitable for the collection of benzodiazepines because the absorption can be as high as 30% (27). In another study, the authors investigated the stability of 15 common therapeutic drugs (amikacin, gentamicin, tobramycin, vancomycin, digoxin, quinidine, theophylline, carbamazepine, phenobarbital, phenytoin, valproic acid, tricyclic antidepressants, salicylate, acetaminophen, and ethanol) in specimens stored in Greiner Vacuette blood collection tubes containing serum separator gel and observed that most drugs did not show any significant decline in concentration except for tricyclic antidepressants and carbamazepine. Interestingly, the concentration of amitriptyline declined more than its metabolite nortriptyline, and the concentration of imipramine also decreased more significantly than its metabolite desipramine. Concentration of carbamazepine also declined statistically over prolonged storage, but decline may not be clinically significant (28).

To circumvent the problem of absorption of phenytoin, carbamazepine, and phenobarbital by the Becton-Dickinson Vacutainer serum separator SST gels, a new gel tube was formulated and marketed by Becton-Dickinson using clear gel (SST II tubes) rather than yellow gel. Based on a study, the new gel was superior to the old because concentration of carbamazepine declined only by 10%, and the concentration of phenytoin only declined by 4% on prolonged storage. This was a significant improvement over the old gel formulation. These new tubes are effective for collecting blood for therapeutic drug monitoring if phenytoin and carbamazepine are analyzed within a day of collection (29). Schouwers et al studied the effect of serum separator gel in Sarstedt S-Monovette serum tubes on various therapeutic drugs, hormones, and proteins by using sera from patients. The authors tested four therapeutic drugs (amikacin, vancomycin, valproic acid, and acetaminophen) as well as cortisol, free thyroxine, thyroid-stimulating hormone, transferrin, prealbumin, and carcinoembryonic antigen. The authors observed no statistically significant difference in values of any analytes on the day of phlebotomy. Although after 1 day, statistically significant differences were observed, except for free T4, the differences were not clinically significant (30). In another study the authors demonstrated that Becton-Dickinson Vacutainer PST II blood collection tube provide equivalent results to serum and plasma nongel tubes for free thyroxine, total thyroxine, follicle-stimulating hormone, luteinizing hormone, ferritin, vitamin B_{12}, folate, and testosterone (31). Li et al reported that Becton-Dickinson Vacutainer SST II Plus tubes are suitable for blood collection and storage for analysis of tumor markers (32).

11.4. IMPACT OF DOCUMENTING DOSING TIME AND OTHER FACTORS THAT AFFECT RESULTS OF THERAPEUTIC DRUG MONITORING

Most specimens for therapeutic drug monitoring are collected 15–20 minutes before the next dosage except aminoglycosides and vancomycin where both peak and trough concentrations are monitored due to the extreme toxic nature of these drugs. A therapeutic drug monitoring specimen must be collected after five half-lives of the drug to estimate drug concentration at the steady state. In addition, collection of specimen for digoxin monitoring is difficult because digoxin exhibits a long distribution phase as well as a lag in the time between dosing and pharmacological response. Therefore, specimens for digoxin measurement should be collected no earlier than 6–8 hours after dosing and preferably after 12 hours. Specimens collected before the distribution phase is complete are often elevated, and such elevated concentrations are confusing because digoxin toxicity may not be experienced due to such elevated levels. In one extensive study involving 666 institutions and 18,675 toxic digoxin results, the authors demonstrated that for every five specimens for which toxic concentrations were obtained, three were inappropriately collected with respect to the dosing interval (33). Sidwell et al reported that only 29% of specimens for therapeutic drug monitoring of digoxin were collected appropriately with indication, sample timing, and subsequent dosage adjustment. The authors also demonstrated that in 19% of cases, the specimens for digoxin analysis did not reflect a steady state condition, and in 5% of the cases, dosage adjustment of digoxin was made inappropriately (34).

Therapeutic drug monitoring of vancomycin and aminoglycosides pose additional challenges because accurate time of dosage is important to determine if the specimen represents peak or trough drug concentrations. Blaser et al studied the relevance of proper timing of specimen collection for therapeutic drug monitoring of gentamicin, amikacin, and netilmicin using 58 patients and 139 dosage intervals. Serum levels of the appropriate drug were measured after 30 minutes of infusion of the drug and compared with 2-hour postdosage values. However, considerable variability was observed between 30 minutes and 2 hours postdosage concentrations of gentamicin, netilmicin, and amikacin that cannot be explained by interindividual differences, renal function, or drug half-life. The authors concluded that timing of peak drug level for these three drugs is important to interpret data from therapeutic drug monitoring (35). Samples are also collected at other points in the dosing cycle for aminoglycosides and vancomycin, for example as with the once-daily dosing protocols. In these protocols, samples are collected at defined points within the dosing cycle. Some laboratories call these "random" samples because they are neither peak nor trough, and the time of collection may vary, but this is an inaccurate use of the term. For the result to be useful, the collection time must be controlled and known, not random. The once-daily protocols for aminoglycosides differ from other protocols in another manner, and peak samples are used to assess toxicity, but in these protocols, trough concentrations are used because at this time the patient should have completely cleared the drug. The finding that the aminoglycoside remains in the circulation at this time indicates reduced clearance and hence the likelihood of toxicity. Roberts et al recently reviewed the therapeutic drug monitoring of aminoglycosides in detail (36). Traugott et al reported that there was a significant increase in the number of appropriately ordered and drawn specimens for vancomycin after implementation of computerized prescriber-order entry (37).

In an effort to improve efficiency, prepare patients for discharge earlier in the day, and minimize the use of blood, many hospitals have consolidated phlebotomy collections and begin routine collection rounds often as early as 4 A.M. This has led to some therapeutic drug monitoring specimens being collected as part of these very early routine chemistry and hematology collections, and such specimens may not represent true troughs levels of drugs. In contrast to therapeutic drug monitoring, specimens for toxicology testing are typically random. For monitoring toxicity from salicylate or acetaminophen, it may be useful to obtain multiple specimens because the victim may not remember the time of ingestion of such drugs. Therefore, it may be possible to assess whether the drug is in the absorption phase or elimination phase to determine the proper therapy to treat the patient.

11.5. OTHER PREANALYTICAL FACTORS THAT MAY AFFECT TEST RESULTS

In addition to the use of the wrong anticoagulant and storage for certain drugs prior to analysis collected in serum separator gel tubes, there are other common errors in preanalytical steps that not only affect the results in toxicology tests but potentially all clinical laboratory tests. It has been estimated that preanalytical errors account for more than two thirds of all laboratory errors; errors in the analytical phase and postanalytical phase account for one third of all laboratory errors. Carraro and Plebani reported that among 51,746 clinical laboratory analyses performed in a 3-month period in the authors' laboratory (7615 laboratory orders, 17,514 blood collection tubes), clinicians notified the laboratory regarding 393 questionable results of which 160 results were confirmed due to laboratory errors. Of 160 confirmed laboratory errors, 61.9% were determined to be preanalytical errors, 15% were analytical errors, and 23.1% were postanalytical errors. However, the overall frequency of the laboratory errors was only 0.309%. The preanalytical phase showed the highest percentage of errors, the most frequent problems arising from mistakes in tube filling with incorrect blood-to-anticoagulant ratio for coagulations tests and empty or inadequately filled tubes. Other common errors in the preanalytical steps include wrong types of blood collection tube used, errors in requested test procedures, wrong patient identification, contradictory demographic data from different information system, missing tubes, samples diluted with intravenous infusion solution, and other problems. The authors also identified 24 errors in the analytical phase; in particular, some results were released in the presence of unacceptable internal quality control data due to a problem in the calibration verification process as well as random errors due to pipetting difficulty during the analytical phase due because of fibrin clots and short specimen volume. The postanalytical stage accounted for only 37 of the 160 laboratory errors. Most of the laboratory errors fortunately had no impact on patient care (121 errors of 160), but 1 error caused inappropriate intensive

TABLE 11.3. Laboratory Errors Encountered in STAT Laboratory Testing

Type of Error	Frequency, %
Preanalytical Errors	
Tube filling error	13.1
Patient identification error	8.8
Inappropriate container	8.1
Request procedure error	7.5
Empty tube	6.9
Data communication conflict	3.8
Missing tube	3.1
Check-in not performed in laboratory information system	2.5
Physician's request order missed	1.9
Specimen collected wrongly from an infusion line	1.9
Specimen not stored/refrigerated properly	1.9
Missing tube	0.6
Specimen contaminated	0.6
Digoxin test timing error	0.6
All errors in preanalytical phase	61.9
Analytical Errors	
Analytical inaccuracy not recognized before release of result	13.1
Instrument caused random error	1.9
All errors in analytical phase	15
Postanalytical Errors	
Result of communication error	20
Lack of communication with the laboratory	1.9
Excessive turnaround time	1.3
All errors in postanalytical phase	23.1

Source: Data from Howanitz and Steindel (33).

care admission (0.6%), 2 errors caused inappropriate transfusion (1.3%), 9 errors (5.6%) resulted in inappropriate investigation, and 27 errors (16.9%) required repeated laboratory tests (38). Types of laboratory errors (preanalytical, analytical, and postanalytical) are summarized in Table 11.3.

The worst preanalytical error is incorrect patient identification where a physician may act on test results from the wrong patient. Another common error is blood collection from an intravenous line that may falsely increase the test results of glucose, electrolytes, or a therapeutic drug due to contamination with infusion fluid. Incorrect orders of draw can cause laboratory errors. This is the correct order of draw for blood specimens:

Blood culture tubes

Citrate tube (light blue)

Serum tube (red top) or tube with gel separator/clot activator (gold top, tiger top)

Heparin tube (green top)

EDTA tube (purple top)
Fluoride tube (gray top)

Hemolysis during blood draw may cause problems in the analytical phase. Fortunately, many newer versions of automated analyzers can detect hemolyzed specimens and flag the results. In addition, the specimen after collection must be mixed properly by inverting tubes so anticoagulant can mix properly with the specimen. Because EDTA is the least soluble anticoagulant, blood collected in a purple top tube must be mixed very carefully to avoid micro-clot formation.

The effect of monitoring the drug concentrations through lines used to administer the drugs has been studied extensively. Wanwimolruk and Murphy performed an in vitro study in which digoxin, aminophylline, and phenytoin were administered through central catheters in concentrations stimulating those used clinically. Drug solutions were allowed to stay in the lumen for a time similar to that encountered clinically and then either remained in the tubing or were flushed out with 5 mL of normal saline. After 6 hours, a 5-mL sample was withdrawn for drug analysis. With undiluted concentration of drug, only flushing the line after the dose and discarding a sample prior to final sampling prevented an artifactual rise in drug concentration (39). Boodhan et al the studied extent of agreement between peripheral and single-lumen venous centrally drawn serum samples for determining gentamicin concentrations and observed a lack of agreement between gentamicin concentrations between these specimens (40).

Case Study

A 59-year-old woman was admitted to the hospital due to a transient ischemic heart attack. During the first day of hospitalization she experienced a generalized tonic-clonic seizure, and a 1000-mg intravenous phenytoin loading dose was administered followed by an oral dose of 100 mg of phenytoin every 3 hours for a total of three doses. For the following 6 days, the patient received 100 mg phenytoin intravenously or orally every 8 hours. On the evening of day 5, she was administered two additional 300-mg doses of phenytoin intravenously 2 hours apart due to recurrent episodes of seizure. Beginning on day 7, the dose was 100 mg intravenously every 6 hours. On day 5, a serum specimen was drawn, and the phenytoin concentration was determined to be 17.0 μg/mL. On day 7, another specimen drawn showed the phenytoin concentration of 13.4 μg/mL; the next day, the phenytoin level was highly toxic (80.7 μg/mL). However, the patient did not show any symptoms of phenytoin toxicity. Another sample drawn 7 hours later showed a phenytoin concentration of 12.4 μg/mL. The authors suspected that the falsely elevated serum phenytoin level was due to the drawing of the specimen from the same line through which the intravenous phenytoin was administered (41).

11.6. EFFECT OF HIGH BILIRUBIN, HEMOLYSIS, AND HIGH LIPID ON THERAPEUTIC DRUG MONITORING

Most of the interferences from bilirubin, hemoglobin, or lipids are due to interference during the optical detection process of the assay. Bilirubin absorbs around 450–460 nm; hemoglobin absorbs from 340 nm to 560 nm, and the absorbance peak is observed at 541 nm (oxyhemoglobin). Therefore, hemolysis affects assays that use the absorbance properties of nicotinamide adenine dinucleotide (NADH) or nicotinamide adenine dinucleotide phosphate (NADPH) at 340 nm (42). For example, selecting a longer wavelength (more than 650 nm) for detection may minimize the interference from bilirubin or hemolysis. Bilirubin and hemoglobin may also interfere in assays through unintended side reactions. The package inserts from most commercial assay kits normally list the effect of these interfering substances. Most manufacturers follow the EP7-P protocol from the National Committee for Clinical Laboratory Standardization (currently called the Clinical Laboratory Standards Institute). The protocol recommends testing interference of bilirubin at 20 mg/dL, hemoglobin at 500 mg/dL, and lipid at 1000 mg/dL levels. Urine is the most commonly used specimen for drugs of abuse testing. Urine samples less frequently contain significant amounts of hemoglobin bilirubin to cause significant interference with an assay. Moreover, high bilirubin or hemolyzed specimens have a minimal impact on therapeutic drug monitoring or toxicology tests performed by chromatographic techniques such as gas chromatography, gas chromatography combined with mass spectrometry, and liquid chromatography combined with mass spectrometry because during specimen processing, such interfering substances are eliminated.

11.6.1. Effect of Bilirubin on Assays

Bilirubin is derived from the hemoglobin of aged or damaged red blood cells. Bilirubin does not have iron and is rather a derivative of the heme group. Some part of serum bilirubin is conjugated as glucuronides ("direct" bilirubin); the unconjugated bilirubin is also referred to as indirect bilirubin. In normal adults, bilirubin concentrations in serum are from 0.3 to1.2 mg/dL (total) and less than 0.2 mg/dL (conjugated). If bilirubin is not elevated, no interference is likely in any assay, and usually the total bilirubin concentration up to 20 mg/dL is not a problem for most assays used for therapeutic drug monitoring. In different forms of jaundice, total bilirubin may increase over 20 mg/dL, and the ratios of direct to indirect bilirubin also vary. In obstructive jaundice, the increase in total bilirubin is contributed mainly by direct bilirubin. In hemolytic and neonatal jaundice, the increase is mostly in indirect bilirubin. Elevated bilirubin may cause interference with certain assays. Elevated bilirubin causes interference proportional to its concentration. The interference of bilirubin in therapeutic drug monitoring assays is mainly caused by bilirubin absorbance at 454 or 461 nm. In addition, bilirubin may interfere in a colorimetric enzyme-linked immunosorbent assay that uses alkaline phosphatase label and p-nitro phenol phosphate substrate (405 nm). However, if the assay is enzymatic or colorimetric, bilirubin may also interfere by reacting chemically to the reagents (43).

Interference of high bilirubin in colorimetric assay for acetaminophen has been reported, but immunoassays are free from such interference. In one case study, a severely jaundiced 17-year-old male patient (total bilirubin: 19.8 mg/dL) with abdominal pain and increased serum transaminase results was suspected of an acetaminophen overdose, although the patient denied using any medications containing acetaminophen in the previous week. The apparent plasma acetaminophen concentration by an enzyme method was found to be 3.4 mg/dL. In this method, acetaminophen is enzymatically (by arylacylamidase) hydrolyzed to p-aminophenol, which is condensed with o-cresol in the presence of periodate to form the blue indophenol chromophore. The method was run on the Roche Modular chemistry analyzer, with absorbance measurement at 600 nm (2-point rate) with a background correction at 800 nm. To investigate false-positive results from elevated bilirubin, the authors investigated 12 specimens from patients with hyperbilirubinemia without a history of acetaminophen exposure and detected measurable amounts of acetaminophen, but when such specimens were diluted to achieve bilirubin concentrations less than 5 mg/dL, no acetaminophen was detected. The authors concluded that high bilirubin may cause a false acetaminophen level in serum (44). Another example of bilirubin interference was noted in the acetaminophen assay but using a different method. When 15 serum samples, none containing acetaminophen but with total bilirubin concentrations between 2.2 and 16.7 mg/dL, were tested in an acetaminophen assay involving the reaction of the analyte with ferric-2,4,6-tripyridyl-s-triazine, false-positive acetaminophen between 0.7 and 13.6 mg/dL was observed (45). Polson et al observed false-positive acetaminophen results in serum specimens using colorimetric assays when bilirubin concentrations exceeded 10 mg/dL. However, immunoassays (fluorescence polarization immunoassay and enzyme multiplied immunoassays), as well as the more sophisticated gas chromatography combined with mass spectrometric method, were free from such interferences (46). Both ultrafiltration and serial dilution of specimen can be used to investigate such interferences (47).

Case Study

A 31-year-old woman was admitted to the hospital with abdominal pain, decreased appetite, confusion, and tea-colored urine. Her liver enzymes and bilirubin were elevated despite a history of liver disease. Her condition worsened, and she was transferred to the authors' hospital. On admission, her alanine aminotransferase was 6170 U/L, aspartate aminotransferase 5080 U/L, alkaline phosphates 150 U/L, lactate dehydrogenase 6830 U/L, and bilirubin 70.7 mg/dL. Her plasma acetaminophen level was 18.3 µg/mL (121 µmol/L). Although her general condition improved, her plasma acetaminophen level stayed above 100 µmol/L. The acetaminophen levels were determined by the Vitos acetaminophen method (Johnson & Johnson, Rochester, NY), which is based on enzymatic conversion of acetaminophen to para-aminophenol and subsequent reaction with ortho-cresol to form blue-colored indophenol, which is measured by a change of absorption at 600 nm. An increase in background to 600 nm caused by high bilirubin (more than 20 mg/dL) may cause interference. The authors suspected falsely elevated acetaminophen in their patient due to the presence of high bilirubin in specimens, and no acetaminophen was detected in the protein-free ultrafiltrate of such specimens. The patient was discharged in 3 weeks with supportive therapy (48).

Wood et al reported a case in which increased bilirubin (22.6 mg/dL), especially consisting of a high percentage of conjugated fraction (82%), caused negative interference in a fluorescence polarization immunoassay (FPIA) for vancomycin. In their study, the authors first compared 28 plasma samples with total bilirubin less than 5.9 mg/dL between two different Abbott's vancomycin assays using a TDx analyzer and an AxSYM analyzer. The method used in the TDx analyzer is a homogeneous FPIA, using a polyclonal sheep antibody and fluorescein-labeled antigen. The assay on the AxSYM analyzer also uses the same assay principle but utilizes a different monoclonal mouse antibody. The vancomycin results from these 28 samples, ranging from 2.0 to 34.5 µg/mL, were in close agreement between the assays performed using two different analyzers (correlation coefficient: $r^2 = 0.996$). When the authors analyzed plasma specimens containing abnormal bilirubin, they observed discordant results between the two vancomycin assays. For example, in a specimen containing 22.6 mg/dL of total bilirubin, the vancomycin concentration observed by using the TDx analyzer was 2.6 µg/mL, but with the AxSYM analyzer it was 8.0 µg/mL. The authors concluded that although the assay on the AxSYM analyzer somehow was not affected by high direct bilirubin, the assay on the TDx analyzer demonstrated false-negative results for such samples (49). However, in 2010 the manufacturer (Abbott Laboratories, Abbott Park, IL) discontinued the TDx vancomycin assay.

Bilirubin also interferes with the colorimetric salicylate assay (Trinder salicylate assay) based on the ability of salicylate to form a colored complex with ferric ion. The colored complex can be measured at 560 nm. However, bilirubin also forms a colored complex with the reagent that absorbs at 600 nm. Salicylate assay on the Beckman SYNCHRON LX analyzer utilizes a primary wavelength at 560 nm and a secondary wavelength at 700 nm. Therefore, a high concentration of bilirubin causes positive interference with serum salicylate determination using this assay. Trinder salicylate reagent is used in the salicylate assay for application in both the Hitachi 917 and SYNCHRON analyzer, but high bilirubin caused negative interference in the salicylate determination using a primary wavelength of 546 nm and no secondary wavelength in the Hitachi 917 analyzer. Interestingly, fluorescence polarization immunoassay for salicylate for application on the AxSYM analyzer (Abbott Laboratories) is free from bilirubin interference (50).

11.6.2. Effect of Hemoglobin on Therapeutic Drug Monitoring

Whole blood specimens must be used for most of the immunosuppressant drugs. False-positive tacrolimus results were reported in a microparticle enzyme immunoassay (MEIA) for patients with low hematocrit values and high imprecision at tacrolimus concentrations less than 9 ng/ml. The EMIT assay for tacrolimus was not affected. When the authors divided the study specimens into three groups by hematocrit percentage (less than 25%, 25–35%, and more than 35%), the difference between the MEIA and EMIT assays increased as the hematocrit percentage decreased. Moreover, false-positive results were reported in 63% of specimens with MEIA where patients did not receive any tacrolimus but only 2.2% of specimens used EMIT. Such false-positive values in the MEIA and EMIT methods ranged up to 3.7 ng/mL and 1.3 ng/mL, respectively (51).

Hemolysis can occur in vivo during venipuncture and blood collection or during the processing of the sample. Hemoglobin interference depends on its concentration in the sample. Serum appears hemolyzed when the hemoglobin concentration exceeds 20 mg/dL (52). However, icteric serum may contain a higher concentration of hemoglobin before hemolysis can be noticed. Hemoglobin interference is caused not only by the spectrophotometric properties of hemoglobin but also by its participation in chemical reaction with sample or reactant components (53). The absorbance maxima of the heme moiety in hemoglobin are at 540–580 nm wavelengths. However, hemoglobin begins to absorb around 340 nm, absorbance increasing at 400–430 nm as well. The iron atom in the center of the heme group is the source of such absorbances. Of the many variants of hemoglobin, methemoglobin and cyanmethemoglobin (cyanide complex of hemoglobin) also absorb at 500 and 480 nm, respectively. Methods that use the absorbance properties of NADH or NADH (340 nm) may thus be affected by hemolysis. When hemoglobin is oxidized to methemoglobin, the absorbance at 340 nm decreases (54).

11.6.3. Effect of High Lipid Content on Therapeutic Drug Monitoring

Lipids in serum or plasma exist as complexed with proteins called lipoproteins. Lipoproteins, consisting of various proportions of lipids, range from 10 nm to 1000 nm in size (the higher the percentage of the lipid, the lower the density of the resulting lipoprotein and the larger the particle size). Chylomicrons (diameter: 70–1000 nm, density less than 0.95 g/mL) are present in plasma after a person ingests a meal but should disappear within 2 hours of the meal unless the person has the genetic disorder of chylomicron metabolism. Chylomicrons are absorbed by the adipose tissue and liver. The liver secretes lipoprotein particles called very low density lipoproteins (VLDLs; density less than 1.006 g/mL), low-density lipoproteins (density: 1.006–1.063 g/mL), and high-density lipoproteins (density: 1.063–1.21 g/mL) containing decreasing amounts of lipids in that order. The lipoprotein particles with high lipid contents are micellar and the main source of assay interference. Unlike bilirubin and hemoglobin, lipids normally do not participate in chemical reactions and mostly cause interference in assays due to their turbidity by scattering light. Because scattered light does not follow the Lambert-Beer law of absorbance, scattering normally reduces absorbance producing false results (positive or negative, depending on the reaction principle). Among the plasma lipoproteins, chylomicrons and VLDL particles only scatter light. VLDLs exist in three size classes: small (27–35 nm), intermediate (35–60 nm), and large (60–200 nm). Only the latter two sizes of VLDLs scatter light. Lipemic interference is most pronounced with spectrophotometric assays, less important with fluorometric methods, and rarely interferes with chemiluminescent methods. Thus assays that use turbidimetry for signal are the ones most affected by lipid interference (55). Lipemia may also interfere with assays for fat-soluble analytes such as steroids and their derivatives. In such cases interference arises from solvent partitioning and solute exclusion of the analyte between the lipid and aqueous phases. Like bilirubin and hemolysis, package inserts do report the extent of lipid interference in a commercial assay. Lipids, however, present a special problem due to the lack of readily available standardized materials. Most manufacturers use Intralipid, a synthetically produced emulsion containing soybean oil and egg phospholipids, for intravenous administration, to spike specimens to simulate lipemic samples. However, samples with Intralipid do not perfectly mimic lipemic samples (56).

Ji and Meng recently performed an extensive study to evaluate the potential interference of high bilirubin (icterus), hemolysis, and lipid on various assays performed using Cobas 600 analyzers (Roche Diagnostics, Indianapolis, IN) and commented that hemolysis, icterus, and lipemia commonly interfere with the spectrophotometric method, with hemolysis the most common interfering substance. Although direct spectral interference of a chemistry analyzer has been significantly reduced by using bichromatic and kinetic analysis, constituent from red blood cells during hemolysis certainly falsely elevates serum potassium and lactate dehydrogenase concentrations. Although the authors' data for most analytes regarding the interferences of bilirubin, hemolysis, and lipemia agreed with Roche reagent package inserts for these analytes, some assays were affected with lower concentrations of these interfering substances than stated in the package inserts. These assays included ceruloplasmin, high-density lipoprotein cholesterol, phosphorus, rheumatoid factors, total bilirubin, unsaturated iron binding capacity, and uric acid. In addition, the authors' positive interference of hemoglobin on alanine aminotransferase, lipase, total protein, ferrous iron, and negative interference on calcium and creatinine kinase assay were not described in the respective package inserts. Acetaminophen levels were false increased by both hemoglobin and bilirubin. Hemolysis caused negative interference (falsely lower values) with digoxin, valproic acid, and vancomycin assays. Bilirubin caused positive interference (falsely increased values) with gentamicin, phenobarbital, and theophylline but caused negative interference with tobramycin assay. Blood levels of phenytoin, salicylate, theophylline, and vancomycin were significantly decreased due to the presence of high lipids in specimens (57).

11.7. CONCLUSIONS

Because preanalytical errors are responsible for most laboratory errors, it is an important consideration for improving patient safety. Wrong anticoagulant use, wrong specimen collection tube, wrong timing recorded for dosage for specimen collection for therapeutic drug monitoring, filling a tube partially rather than completely, and incorrect patient identification on the tube can be eliminated by surveillance and are integral parts of improving laboratory practice. If care is taken during blood draw, in vitro hemolysis can be avoided. Ashakiran et al commented that improper request, incorrect timing of the specimen, wrong tube collection, and in vitro hemolysis accounted for the major proportion of errors in the authors' clinical laboratory (58). Proper education of staff is needed to improve the quality of a clinical laboratory by avoiding such errors (59).

REFERENCES

1. Miles RR, Roberts RF, Putnam AR, Roberts WL. Comparison of serum and heparinized plasma for measurement of chemistry analyzers [letter to the editor]. Clin Chem 2004; 50: 1704–1705.
2. Faynor SM, Robinson R. Suitability of plastic collection tubes for cyclosporine measurements. Clin Chem 1998; 44: 2220–21.
3. Seger C, Tentschert K, Stoggl W, Griesmacher A et al. A rapid HPLC-MS/MS method for the simultaneous quantification of cyclosporine A, tacrolimus, sirolimus and everolimus in human blood samples. Nat Protoc 2009; 4: 526–534.
4. Korte WC, Engler H, Aldebert E, Riesen WF. Heparinized blood provides equivalent results to EDTA in the CEDIA and FPIA cyclosporine immunoassays, thus facilitating routine cyclosporine determination. Clin Chim Acta 2008; 390: 138–140.
5. Millan O, Oppenheimer F, Brunet M, Vilardeli J et al. Assessment of mycophenolic acid induced immunosuppression: a new approach. Clin Chem 2000; 46: 1376–1383.
6. Bunch DR, Wang S. Measurement of mycophenolic acid in plasma or serum by a commercial enzyme inhibition technique in comparison with a high performance liquid chromatography. Clin Chem Lab Med 2008; 46: 1281–1284.

7. Gruenberger EC, Maria-Rountree E, Baron-Shoty E, Kahn DA. Neurotoxicity with therapeutic lithium levels: a case report. J Psychiatr Prac 2009; 15: 60–63.
8. Venkatarathnamma PN, Patil AR, Nanjundaiah N. Fetal lithium toxicity with therapeutic levels—a case report. Int J Clin Pharamcol Ther 2011; 49: 336–338.
9. Wills BK, Mycyk MB, Mazor S, Zell-Kanter M et al. Factitious lithium toxicity secondary to lithium heparin containing blood tubes. J Med Toxicol 2006; 2: 61–63.
10. Tanaka T, Moretti ME, Verjee ZH, Shupak M et al. A pitfall of measuring lithium levels in neonates. Ther Drug Monit 2008; 30: 752–754.
11. Matzke GR, Piveral K, Halstenson CE, Abraham PA. Heparin interferes with tobramycin serum concentration determination by EMIT. Drug Intell Clin Pharm 1984; 18: 517–519.
12. O'Connell ME, Heim KL, Halstenson CE, Matzke GR. Analytical accuracy of determination of aminoglycoside concentrations by enzyme multiplied immunoassay, fluorescence polarization immunoassay, and radioimmunoassay in the presence of heparin. J Clin Microbiol 1984; 20: 1980–1082.
13. Dasgupta A, Blackwell W, Bard D. Stability of therapeutic drug measurement in specimens collected in VACUTAINER plastic blood-collection tubes. Ther Drug Monit 1996; 18: 306–309.
14. Faynor SM, Robinson R. Suitability of plastic collection tubes for cyclosporine measurements [letter to the editor]. Clin Chem 1998; 44: 2220–2221.
15. Boeynaems JM, De Leener A, Dessars B, Villa-Lobos HR et al. Evaluation of a new generation of plastic evacuated blood collection tubes in clinical chemistry, therapeutic drug monitoring, hormone and trace metal analysis. Clin Chem Lab Med 2004; 42: 67–71.
16. Christophersen AS. Tetrahydrocannabinol stability in whole blood: plastic versus glass containers. J Anal Toxicol 1986; 10: 129–131.
17. Stout PR, Horn CK, Lesser DR. Loss of THCCOOH from urine specimens stored in polypropylene and polyethylene containers at different temperatures. J Anal Toxicol 2000; 24: 567–571.
18. Jamerson MH, McCue JJ, Klette KL. Urine pH, container composition, and exposure time influence adsorptive loss of 11-nor-delta9-tetrahydrocannbinol-9-carboxylic acid. J Anal Toxicol 2005; 29: 627–631.
19. Romberg RW, Past MR. Reanalysis of forensic urine specimens containing benzoylecgonine and THC-COOH. J Forensic Sci 1994; 39: 479–485.
20. Quattrocchi F, Karnes HT, Robinson JD, Hendeles L. Effect of serum separator blood collection tubes on drug concentrations. Ther Drug Monit 1983; 5: 259–362.
21. Landt M, Norling LL, Steelman M, Smith CH. Monoject Samplette capillary blood container with serum separator evaluated for collection of specimens for therapeutic drug assays and common clinical chemical tests. Clin Chem 1986; 32: 523–526.
22. Bailey DN, Coffe JJ, Briggs JR. Stability of drug concentrations in plasma stored in serum separator blood collection tubes. Ther Drug Monit 1988; 10: 352–354.
23. Devine JE. An assessment of the Corvac blood collection tube for drug specimen processing. Clin Chem 1984; 30: 1034.
24. Koch TR, Platogg G. Suitability of collection tubes with serum separator gels for therapeutic drug monitoring. Ther Drug Monit 1990; 12: 277–280.

25. Mauro LS, Mauro VF. Effect of serum separator tubes on free and total phenytoin and carbamazepine serum concentrations. Ther Drug Monit 1991; 13: 240–243.
26. Dasgupta A, Dean R, Saldana S, Kinnaman G, McLawhon RW. Absorption of therapeutic drugs by barrier-gel in serum separator blood collection tubes. Volume and time dependent reduction in total and free drug concentrations. Am J Clin Pathol 1994; 101: 456–461.
27. Karppi J, Akerman K, Parviainen M. Suitability of collection tubes with separator gels for collecting and storing blood samples for therapeutic drug monitoring. Clin Chem Lab Med 2000; 38: 313–320.
28. Dasgupta A, Yared MA. Wells A. Time dependent absorption of therapeutic drugs by the gel of the Greiner Vacuette blood collection tube. Ther Drug Monit 2000; 22: 427–431.
29. Nush V, Blennerhasset J, Wells A, Dasgupta A. Stability of therapeutic drugs in serum collected in a Vacutainer serum separator tubes containing a new gel (SST II). Ther Drug Monit 2001; 23: 259–262.
30. Schouwers S, Brandt I, Willemse J, Van Rogenmortel N et al. Influence of separator gel in Sarstedt S-Monovette serum tubes on various therapeutic drugs, hormones and proteins. Clin Chim Acta 2012; 413: 100–104.
31. Chance J, Berube J, Vandersmissen M, Blanckaert N. Evaluation of the BD Vacutainer PST II blood collection tube for special chemistry analysis. Clin Chem Lab Med 2009; 47: 358–361.
32. Li Z, Feng Z, Yan C, Tan R. Evaluation of BD Vacutainer SSTII Plus tubes for common tumor marker tests by Roche Diagnostics Modular E170 analyzer. J Clin Lab Anal 2010; 24: 418–421.
33. Howanitz PJ, Steindel SJ. Digoxin therapeutic drug monitoring practices. A College of American Pathologists Q-Probes study of 666 institutions and 18,679 toxic levels. Arch Pathol Lab Med 1993; 117: 684–690.
34. Sidwell A, Barclay M, Begg E, Moore G. Digoxin therapeutic drug monitoring: an audit and review. N Z Med J 2003; 116: U708.
35. Blaser J, Simmen HP, Gonzenbach HR, Sonnabend W et al. Aminoglycoside monitoring: timing of peak level is critical. Ther Drug Monit 1985; 7: 303–307.
36. Roberts JA, Norris R, Paterson DL, Martin JH. Therapeutic drug monitoring of antimicrobials. Br J Clin Pharmacol 2012; 73: 27–36.
37. Traugott KA, Maxwell PR, Green K, Frie C et al. Effect of therapeutic drug monitoring criteria in a computerized prescriber-order-entry system on appropriateness of vancomycin level order. Am J Health Systs Pharm 2011; 68: 347–352.
38. Carraro P, Plebani M. Errors in STAT laboratory; types and frequency 10 years later. Clin Chem 2007: 53: 1338–1342.
39. Wanwimolruk S, Murphy JE. Effect of monitoring drug concentrations through lines used to administer the drugs: an in vitro study. Ther Drug Monit 1991; 13: 443–447.
40. Boodhan S, Maloney AM, Dupuis LL. Extent of agreement in gentamicin concentration between serum that is drawn peripherally from central venous catheters. Pediatrics 2006; 118: e1650–1656.
41. Murphy JE, Ward ES. Elevated phenytoin concentration caused by sampling through the drug-administered line. Pharmacotherapy 1991; 11: 348–350.

42. Fonseca-Wolheim FD. Hemoglobin interference in the bichromatic spectrophotometry of NAD(P)H at 340/380 nm. Eur J Clin Chem Clin Biochem 1993; 31: 595–601.
43. Perlstein MT, Thibert RJ, Watkins RJ, Zak B. Spectrophotometric study of bilirubin and hemoglobin interactions in several hydrogen peroxide generating procedures [abstract]. Clin Chem 1977; 23: 1133.
44. Bertholf RL, Johannsen LM, Bazooband A, Mansouri V. False-positive acetaminophen results in a hyperbilirubinemic patient. Clin Chem 2003; 49: 695–698.
45. Kellmeyer K, Yates C, Parker S, Hilligoss D. Bilirubin interference with kit determination of acetaminophen. Clin Chem 1982; 28: 554–555.
46. Polson J, Wians FH, Orsulak P, Fuller D et al. False positive acetaminophen concentrations in patients with liver injury. Clin Chim Acta 2008; 391: 24–30.
47. Zeidler J, Kavsak PA. An alternative approach for detecting interferences in enzymatic acetaminophen assay [letter to the editor]. Clin Chem 2011; 57; 1203–1204.
48. Fong MB, Siu TS, Tam S. Persistently increased acetaminophen concentration in a patient with acute liver failure. Clin Chem 2011; 57: 9–13.
49. Wood FL, Earl JW, Nath C, Coakley JC. Falsely low vancomycin results using the Abbott TDx. Ann Clin Biochem 2000; 37: 411–413.
50. Dasgupta A, Zaldi S, Johnson M, Chow L et al. Use of fluorescence polarization immunoassay for salicylate to avoid positive/negative interference by bilirubin in the Trinder salicylate assay. Ann Clin Biochem 2003; 40: 684–688.
51. Armedariz Y, Garcia S, Lopez R et al. Hematocrit influences immunoassay performance for the measurement of tacrolimus in whole blood. Ther Drug Monit 2005; 27: 766–769.
52. Sonntag O, Glick MR. Serum-index und interferogram-ein neuer weg zur prufung und darstellung von interferengen durch serumchromogene. Lab Med 1989; 13: 77–82.
53. Wenk RE. Mechanism of interference by hemolysis in immunoassays and requirements for sample quality. Clin Chem 1998; 44: 2554.
54. Fonseca-Wolheim FD. Hemoglobin interference in the bichromatic spectrophotometry of NAD(P)H at 340/380 nm. Eur J Clin Chem Clin Biochem 1993; 31: 595–601.
55. Weber TH, Kaoyho KI, Tanner P. Endogenous interference in immunoassays in clinical chemistry. Scand J Clin Lab Invest Suppl 1990; 201: 77–82.
56. Bornhorst JA, Roberts RF, Roberts WL. Assay-specific differences in lipemic interference in native and Intralipid-supplemented samples. Clin Chem 2004; 50: 2197–2201.
57. Ji JZ, Meng QH. Evaluation of the interference of hemoglobin, bilirubin, and lipids on Roche Cobas 6000 assays. Clin Chim Acta 2011; 412: 1550–1553.
58. Ashakiran S, Sumati ME, Murthy NK. A study of per-analytical variables in clinical biochemistry laboratory. Clin Biochem 2011; 44: 944–945.
59. Simundic AM, Nikolac N, Vukasovic I, Vrkic N. The prevalence of preanalytical errors in a Croatian ISO 151189 accredited laboratory. Clin Chem Lab Med 2010; 48: 1009–1014.

Chapter 12

Challenges in Therapeutic Drug Monitoring of Digoxin Using Immunoassays

12.1. INTRODUCTION

Digoxin, a cardiac glycoside, is an old drug, but despite the introduction of many relatively new cardioactive drugs, it is still regularly used for treating patients with cardiac illnesses. Digitalis glycosides derived from foxglove plant were used for therapeutic purposes as early as the sixteenth century, and foxglove was listed in the London Pharmacopeia in 1661. However, due to the lack of knowledge of its proper use, reports of toxicities resulted in the removal of digitalis from the Pharmacopeia in 1745. Later, William Withering's classical description of the effects of digitalis was published in 1785 because he treated many patients with congestive heart failure successfully with digitalis and also recorded digitalis toxicity systematically. Today, digoxin isolated from a species of the foxglove plant, *Digitalis lantana,* is the only cardiac glycoside prescribed in the United States (1).

Digoxin can be administered acutely or long term, both orally and intravenously. Digoxin is safe and beneficial in treating both acute and chronic heart failure because it increases the strength and vigor of heart contractions (2). Heart failure is a growing public concern. Approximately 5 million patients in

Resolving Erroneous Reports in Toxicology and Therapeutic Drug Monitoring: A Comprehensive Guide, First Edition. Amitava Dasgupta.

the United States have heart failure, and despite therapy approximately 53,000 patients die each year primarily from heart failure. Digoxin is an inexpensive drug that improves symptoms in patients suffering from heart failure and reduces hospitalization, thus reducing health care expenditure related to hospitalization and the treatment for patients with heart failure. Currently digoxin accounts for approximately 20 million prescriptions each year (3). Heart failure is also more common among elderly patients. Rate of heart failure is approximately 15.2 per 1000 population between the ages of 65 and 74, but the rate of heart failure is significantly more in older populations (31.7 per 1000 population between the ages of 75 and 84 and 65.2 per 1000 people older than 85). Fortunately, elderly patients gain a comparable benefit from digoxin therapy as do younger patients in the management of heart failure (4).

The American College of Cardiology/American Heart Association consensus guidelines recommend that digoxin be considered for the outpatient treatment of all patients with persistent signs of heart failure as well as after emergent treatment of heart failure. The dosage of digoxin should be 0.125 to 0.25 mg in most patients, and the lowest dosage should be used in patients older than 70 years. Digoxin has a narrow therapeutic range, but digoxin toxicity may occur even at the therapeutic digoxin level if hypokalemia, hypomagnesemia, or hypothyroidism coexists. In addition, drugs such as quinidine, verapamil, spironolactone, flecainide, and amiodarone may increase serum digoxin levels causing toxicity (5). Other than treating heart failure, digoxin is effective in treating various heart diseases including atrial fibrillation, atrial flutter, and paroxysmal atrial tachycardia (6).

Digoxin is absorbed well after oral administration with bioavailability between 60% and 80% after the administration of Lanoxin tablets. However, bioavailability was higher (90–100%) when digoxin in solution in the form of capsules (Lanoxicaps) was administered to healthy volunteers (7). The intestinal absorption of digoxin is mostly a passive nonsaturable diffusion process, although a carrier-mediated component may play some part. The presence of food may reduce peak digoxin concentration but does not affect the total amount of digoxin absorbed. Digoxin is extensively bound to the myocardium as well as renal and skeletal muscle but does not significantly bind to adipose tissue. Plasma protein binding is relatively low, varying between 20% and 30%. Therefore, displacement of digoxin by other strongly protein-bound drugs does not cause clinically significant interactions. In addition, hemodialysis or exchange transfusions do not significantly alter the body load of digoxin. The apparent volume of distribution of digoxin is between 5 and 3.3 L/kg, and elimination half-life is approximately 36 hours. Digoxin is metabolized minimally with 60–80% excreted in urine unchanged by both passive glomerular filtration and active tubular secretion. Therefore, renal function plays an important role and must be considered during digoxin dosage, especially in elderly patients who may already have diminished renal function. Careful monitoring of serum digoxin concentration is an important therapeutic option (8).

12.2. MECHANISM OF ACTION OF DIGOXIN

Digoxin acts directly on the cells of the myocardium by inhibiting the membrane-bound alpha unit of the sodium potassium adenosine triphosphatase (Na, K+ ATPase) pump. The amino acids that constitute ouabain binding site are highly conserved across the evolutionary spectrum probably because of their role in maintaining the essential physiological functions of the body (9). As a result of inhibiting the Na, K+ ATPase pump, digoxin produces a positive inotropic effect due to the increased intracellular calcium that can be used by contractile proteins of both normal and failing hearts. However, in failing hearts, digoxin-mediated increase in myocardial contractility leads to a reduction in peripheral resistance. This process improves cardiac output in a failing heart. In addition, digoxin also improves the performance of a failing heart by attenuating the activation of the neurohormonal system (10,11).

Digoxin also increases the refractory period and decreases impulse velocity in certain myocardial tissue (such as the AV node). The electrophysiological properties of digitalis are reflected in the electrocardiogram (ECG) by shortening of the QT interval. A wide variety of placebo-controlled trials have unequivocally confirmed that therapy with digoxin can improve symptoms, exercise tolerance, and quality of life in patients with mild to severe heart failure. Unlike some other newer cardioactive agents with positive inotropic properties, digoxin does not increase all-cause mortality from heart failure. Digoxin therapy is also associated with a reduction in the rate of hospitalization due to heart failure (5).

12.3. THERAPEUTIC DRUG MONITORING OF DIGOXIN

Due to its narrow therapeutic range, the monitoring of serum or plasma digoxin level is essential for patient management to avoid digoxin toxicity. Usually trough blood samples are used for the monitoring of digoxin, and specimens must be drawn at least 6 hours after the last digoxin dose. In addition, digoxin levels must be monitored after reaching a steady state, which is usually at least after four half-lives of digoxin (digoxin half-life: 1.6 days), that is, 6 days after the initiation of digoxin therapy. Inappropriate sampling may lead to unexpected digoxin results and may cause confusion in interpreting serum digoxin concentration by the clinician or pharmacist, which may lead to dosage error. Unfortunately, inappropriate digoxin level monitoring is a problem. Mordasini et al, based on a study of assessing 210 digoxin levels, observed that 125 of the results (59%) were drawn inappropriately, 81 (39%) were drawn appropriately, and 4 case evaluations could not be performed. Of 125 specimens drawn inappropriately, most (79%) were performed as routine digoxin monitoring. Most of the errors in digoxin monitoring were due to the wrong timing of the specimen (32%) including 17 specimens (8%) drawn during the distribution phase of digoxin that may result in a noninterpretable

TABLE 12.1. Indications for Therapeutic Drug Monitoring of Digoxin

Indication	Comments
Initiation/Restart of digoxin therapy	Monitoring must be initiated after serum digoxin is level and reaches steady state. Trough level (at least 6 hours after last dose) should be monitored.
After change of dosage	Therapeutic drug monitoring is essential to ensure digoxin level is within the therapeutic range.
High-risk patients	Monitoring is essential in patients with renal impairment or with hypokalemia or hypomagnesemia.
Suspected overdose	Digoxin level is crucial in the investigation of acute or chronic overdose. In the case of acute overdose, serum digoxin level is useful in the administration of Digibind or DigiFab therapy.
Noncompliance	Treatment failure due to noncompliance can be easily identified by determining the digoxin serum level.
Lack of effect	Subtherapeutic digoxin level despite adequate dosage is an indication of malabsorption.
Digoxin-drug interaction	Serum digoxin monitoring is essential in a suspected digoxin–drug interaction.
Pharmacokinetic study	For selected patients with high risk of digoxin toxicity, pharmacokinetic study (half-life, area under the curve, etc.) pharmacokinetic study based on at least two digoxin-level determinations may be helpful.

result due to a clinically irrelevant increased serum digoxin level (12). In another study, the authors observed that the mean serum digoxin level was 0.78 ng/mL in 118 patients when the digoxin sample was collected at the right time, but in 54 patients with premature sampling, the mean digoxin concentration was significantly elevated to 1.25 ng/mL. The authors concluded that premature digoxin sample collections were common in their hospital, resulting in higher digoxin levels (13). Sidwewll et al, based on a study of 100 consecutive requests for therapeutic drug monitoring of digoxin, concluded that at their hospital only 29% of requests were appropriate. In 19% of cases, the digoxin sample did not reflect the steady sate condition, and in 5% of occasions, the authors believed the subsequent decision regarding dosage adjustment for digoxin was clearly inappropriate. The authors concluded that in their hospital, the practice of therapeutic drug monitoring of digoxin was often inappropriate and medical staff education is needed to improve such practices (14). The indications for therapeutic drug monitoring of digoxin are listed in Table 12.1.

12.3.1. Controversies in Therapeutic and Toxic Ranges of Digoxin

Although it is traditionally assumed that digoxin has a therapeutic range of 0.8–2.0 ng/mL while toxicities are encountered at levels higher than 2.0 mg/mL, there are conflicting reports in the literature regarding the therapeutic and toxic range of digoxin. Sameri et al commented that although the therapeutic range of digoxin is frequently cited as 0.8–2.0 ng/mL or 0.5–2.0 ng/mL, studies in the past decade suggest an upper level of the therapeutic range of 1.0 ng/mL for treating patients with heart failure. In addition, this upper limit is also applicable for patients with both heart failure and atrial fibrillation with rapid ventricular response. The authors recommend that reducing the upper limit of therapeutic range of digoxin from 2.0 ng/mL to 1.0 ng/mL on computerized and paper laboratory report forms may guide clinicians to avoid unnecessary high serum digoxin concentrations in their patients, thus minimizing digoxin toxicity without any compromise of the therapeutic benefits of digoxin in treating patients with heart failure (15). Another report also indicated that the therapeutic range of digoxin must be lower than 0.8–2.0 ng/mL. In an analysis of digoxin concentrations in both men and women, those with digoxin levels between 0.5 and 0.8 ng/mL had a 6.3% lower all-cause mortality and a 5.9% lower hospitalization rate compared with placebo, but patients with a digoxin serum level exceeding 1.2 ng/mL had an 11.8% higher mortality rate than placebo. Therefore, there is a recommendation of lower serum therapeutic digoxin concentrations of 0.5–0.9 ng/mL or less than 1.0 ng/mL in patients receiving digoxin for symptomatic heart failure. Because the new therapeutic window of digoxin is associated with improved outcome, more intensive dosage refinement based on new dosage guidelines and a nomogram should be considered in patients with heart failure (16).

There is usually a consensus that a digoxin level higher than 2.0 ng/mL is associated with toxicity. Based on a study population of 125 hospitalized patients, Jitapunkul et al reported that 43 patients (33.6%) were identified as having definite digoxin toxicity, 9 patients (7.2%) were classified as having a probable digoxin toxicity, and 74 patients (59.2%) did not experience any digoxin toxicity. The most common symptom of digoxin toxicity was atrial fibrillation with block. The authors observed no significant differences in digoxin dosage in patients experiencing toxicity versus patients not experiencing digoxin toxicity. However, the mean serum digoxin level in patients experience toxicity was 2.28 ng/mL, which was significantly higher than the mean digoxin serum level of 1.05 ng/mL in patients not experiencing any digoxin toxicity. The authors concluded that the best cut-off of serum digoxin for the diagnosis of digoxin toxicity is 2 ng/mL (17). Ong et al, based on a study of 67 patients (24 clinically toxic with digoxin and 43 asymptomatic), observed that the mean serum digoxin level (2.09 ng/mL) in toxic group was significantly higher than the nontoxic group (1.20 ng/mL), but the serum digoxin level cannot be the sole criteria in the diagnosis of digoxin toxicity due to overlapping levels between the toxic and nontoxic groups. However, the serum digoxin level above 3 ng/mL in the presence of suggestive clinical features is a strong indicator of digoxin toxicity (18).

Abad-Santos et al, based on a study of 109 patients, observed significant overlap in serum digoxin concentrations in patients experiencing digoxin toxicity versus asymptomatic patients. Intoxicated patients were more frequently women, older, with a lower body mass or declining renal function. No difference was found in serum electrolytes levels between these two groups. Although serum digoxin levels were significantly higher in toxic patients (3.08 ± 0.20 ng/mL) than in nontoxic ones (1.58 ± 0.10 ng/mL), there was considerable overlap between the two groups. Digoxin level was below 2 ng/mL in 10 intoxicated patients and higher than 2 ng/mL in 16 asymptomatic subjects. For toxicities that developed in 10 patients with serum digoxin levels lower than 2 ng/mL, 3 of them had hypokalemia and another 3 had hypomagnesemia. Interestingly, these 10 patients in general showed lower serum potassium concentrations than the other toxic patients with digoxin levels higher than 2 ng/mL (19). Cauffield et al commented that although serum digoxin concentrations of 0.8–2.0 ng/mL can be considered the therapeutic range of digoxin, renal function plays a major role in the pharmacokinetics of digoxin and must be considered while making digoxin dosage decisions. Many medications including quinidine, amiodarone, and verapamil alter digoxin pharmacokinetics and may result in two- to threefold increases in the serum digoxin level. Therefore, these factors must be considered during the interpretation of serum digoxin concentrations (20). In another report the authors concluded that digoxin use among patients who were receiving dialysis treatment was associated with higher mortality, especially among patients with low predialysis potassium levels (21). Although creatinine is usually used as a marker for renal failure, Hallberg et al reported that cystatin C, which has been suggested to be superior to creatinine in measuring glomerular filtration rate, is a better marker of assessing renal function in patients receiving digoxin (22). Table 12.2 summarizes the factors that increase the risk of digoxin toxicity.

TABLE 12.2. Factors That Increase the Risk of Digoxin Toxicity

Factors	Comments
Impaired renal function	Decreased renal clearance of digoxin leading to higher serum digoxin level
Electrolyte imbalance	Hypokalemia, hypomagnesemia, or hypercalcemia alone may predispose to digoxin toxicity
Elderly population	Patients older than 70 years are more susceptible to digoxin toxicity
Drug–drug interaction	
Digoxin–antibiotic interaction	Clarithromycin, erythromycin, roxithromycin, tetracycline, gentamicin
Digoxin–antiarrhythmic interaction	Quinidine, amiodarone, verapamil
Digoxin–thiazide diuretics	Hydrochlorothiazide, metolazone, bendroflumethiazide, indapamide, trichlormethiazide
Digoxin–loop diuretics	Furosemide, bumetanide
Digoxin–potassium-sparing diuretics	Spironolactone

12.3.2. Significant Digoxin–Drug Interaction: Role of Therapeutic Drug Monitoring

A number of drugs interact with digoxin pharmacokinetically. Quinidine, amiodarone, and verapamil cause an increase in the serum digoxin level by reducing its clearance. Clarithromycin, erythromycin, and tetracycline are also known to increase the serum digoxin level by altering gut flora that is important for digoxin metabolism (23). Mordel et al., based on a study of 141 patients, reported that the digoxin toxicity rate was 4.9% in patients receiving digoxin alone, 5.0% in patients receiving digoxin along with amiodarone or verapamil, and 50% of patients receiving a combination of digoxin and quinidine experienced digoxin toxicity, indicating the severity of the digoxin–quinidine interaction. The authors observed no toxicity when digoxin levels were below 1 ng/mL. However, for digoxin concentrations between 1.0 and 2.0 ng/mL, the presumed therapeutic concentration of digoxin, only 1 of 41 patients (2.4%) experienced digoxin toxicity, whereas 4 of 15 patients (26%) receiving both digoxin and quinidine experienced toxicity with serum digoxin concentrations between 1.0 and 2.0 ng/mL. The authors concluded that the digoxin–quinidine interaction significantly increases digoxin toxicity even in the therapeutic range of digoxin (24). Although clarithromycin and roxithromycin exhibited significant interaction with digoxin, another macrolide antibiotic, azithromycin, demonstrated a minimal interaction with digoxin. Therefore, azithromycin is relatively safe to treat an infection in a patient also receiving digoxin (25).

Case Study

A 28-year-old man presented to the emergency department with a 3-day history of anorexia and vomiting. He was receiving digoxin, clarithromycin, captopril, and frusemide. Physical examination of the patient revealed decreased breath sounds bilaterally and an irregular pulse, but the patient was stable hemodynamically. His ECG showed a paced rhythm with frequent premature ventricular beats. In addition, episodes of ventricular couplets and nonsustained ventricular tachycardia were also observed during continuous monitoring. His serum potassium was within normal range (4.8 mmol/L), but his serum digoxin concentration was at the toxic level of 5.0 ng/mL. He took his last digoxin tablet 20 hours prior to his presentation to the emergency department. His toxic digoxin was due to the interaction between digoxin and clarithromycin, and both medications were withheld. The day following his admission, the digoxin level was reduced to 3.4 ng/mL, and his serum digoxin concentration was reduced further to 1.2 ng/mL 6 days after admission. The patient was treated with sulfamethoxazole/trimethoprim for pneumonia and discharged from the hospital on day 6 of his admission with digoxin, captopril, and frusemide as the only drugs to take. A subsequent follow-up visit by the patient revealed normal serum digoxin levels (0.5–1.2 ng/mL) (26). Interestingly, in the author's hospital, the therapeutic range of digoxin is 0.5–1.2 ng/mL, which is substantially lower than the traditionally used serum digoxin therapeutic range of 0.8–2.0 ng/mL.

Case Study

A 72-year-old woman with chronic atrial fibrillation was receiving 0.25 mg of digoxin per day and warfarin 22.5 mg/week. To eradicate *Helicobacter pylori* infection, 500 mg clarithromycin three times a day was prescribed. The patient presented to the emergency department with gastrointestinal symptoms, weakness, dizziness, and visual changes 12 days after initiation of clarithromycin therapy. Her serum digoxin level was toxic (4.6 ng/mL), and her international normalization ratio was also significantly elevated to 7.3 (normal range: 2.0–3.0) indicating overcoagulation from warfarin therapy. Digoxin, warfarin, and clarithromycin were all discontinued, and the patient was admitted to the hospital. Reduced dosages of digoxin (0.125 mg/day) and warfarin (17.5 mg/week) were restarted on day 7 of her hospital stay, and the patient was discharged in a stable condition on day 11. In this case the elevated digoxin concentration was due to alteration of digoxin metabolizing gut flora by clarithromycin; the warfarin toxicity was due to an inhibition of the cytochrome P450 drug metabolizing enzyme in the liver by clarithromycin (27).

Gentamicin, an aminoglycoside, also increases the serum digoxin level. Alkadi et al reported that gentamicin significantly increased serum digoxin levels in diabetic patients and patients with congestive heart failure and concluded that an increased digoxin level when combined with gentamicin should be considered as a risk factor for digoxin toxicity (28). Digoxin also interacts with a variety of antiarrhythmic drugs.

Case Study

A 31-year-old man with dilated cardiomyopathy was hospitalized for the new onset of atrial fibrillation, and oral amiodarone 600 mg/day was prescribed to control his arrhythmia. The patient continued with his daily digoxin dose of 0.125 mg. The plasma digoxin level peaked to 2.93 ng/mL on his third day of hospital stay. Over the next 3 days, his plasma digoxin levels fluctuated, but he developed no symptoms of digoxin toxicity. His increased digoxin level correlated with the timing of the oral administration of amiodarone. His renal function was also stable during the course of amiodarone therapy. The authors concluded that when a loading dose of oral amiodarone is administered in a patient receiving digoxin, the digoxin dosage first should be reduced and digoxin therapy adjusted based on symptoms of digoxin toxicity due to the interaction between digoxin and amiodarone (29).

Digoxin also interacts with various diuretics. In general, a combined therapy of digoxin with any diuretic is associated with a 3.08-fold increase in the risk of digoxin intoxication. Regarding various classes of diuretics, interaction of digoxin with loop diuretics is greater than its interaction with thiazide or potassium-sparing diuretics (Table 12.2). In addition, digoxin–diuretic interaction also increases the risk of hospitalization due to digoxin toxicity (30). Although most interactions between digoxin and another drug result in an increased serum digoxin level, rifampin reduces the serum digoxin level. Gault et al reported that digoxin dosages were increased in two dialysis patients when rifampin was added to their drug regime. When rifampin was discounted, digoxin doses were reduced to the level prior to rifampin therapy. The authors concluded that serum digoxin level may fall to an ineffective level with rifampin therapy (31). Therapeutic drug monitoring of digoxin is helpful in identifying the interaction of various drugs with digoxin and the readjustment of digoxin dosage based on serum digoxin levels.

12.4. ANALYTICAL METHODS FOR THERAPEUTIC DRUG MONITORING OF DIGOXIN

Although chromatographic methods for the therapeutic drug monitoring of digoxin have been reported in the literature, most hospital laboratories adopt various digoxin immunoassays on automated analyzers for routine therapeutic drug monitoring of digoxin. In general, chromatography-based methods for the therapeutic drug monitoring of digoxin are superior compared with immunoassays because these methods are mostly free from interferences.

Kanno et al described a method for the simultaneous analysis of digoxin, digitoxin, and related compounds using liquid chromatography combined with tandem mass spectrometry (LC-MS-MS) (32). Oiestad et al also described a LC-MS-MS method for digoxin and digitoxin in whole blood for autopsy cases. Samples were prepared by liquid-liquid extraction using ethyl acetate/heptane/dichloromethane (3:1:1 by volume); chromatographic separations were achieved by a C-18 reverse phase column. Mass detection was performed by positive ion mode electrospray on ammonium adducts, and deuterated digoxin (digoxin-d3) was used as the internal standard (33). Recently, Xue et al also described a LC-MS-MS protocol for the simultaneous analysis of digoxin along with midazolam and its metabolite 1′-hydroxy midazolam. Digitoxin was used as the internal standard. After a single step of liquid-liquid extraction with tert-butyl methyl ether/dichloromethane (75:25 by volume), chromatographic separation was achieved using a C-18 reverse phase column. The mass spectrometer was operated using positive electrospray ionization under selected reaction monitoring mode. The transits used for determination were 798.33 → 651.36 for digoxin and 782.67 → 635.34 for the internal standard (34).

One of the major limitations of using immunoassays for serum digoxin monitoring in clinical laboratories is the vulnerability of digoxin immunoassays to various endogenous and exogenous interfering substances. The main reason why digoxin immunoassays are more susceptible to interferences compared with immunoassays for any other drug is the requirement of digoxin assays to be highly sensitive in order to reveal a very low level of digoxin in serum. Compared with other drugs, digoxin serum drug concentrations are a hundredfold to a thousandfold lower in therapeutic range because the therapeutic range of digoxin is 0.8–2.0 ng/mL, whereas most other drugs show therapeutic range in the range of micrograms. For example, the therapeutic range of phenytoin, a widely monitored classical anticonvulsant, is 10–20 μg/mL. For example, let's assume that 100 ng/mL of a endogenous cross-reactive substance with only 1% cross-reactivity with digoxin immunoassay is present in a patient's specimen submitted for therapeutic drug monitoring of digoxin. Further assume that this cross-reactant falsely increases the serum digoxin level and such interference has an additive effect on the serum digoxin measurement. Therefore, if the true digoxin concentration in a patient is 1 ng/mL which is therapeutic, the observed digoxin value in the presence of this cross-reactant would be 2 ng/mL because 100 ng/mL of this cross-reactant would contribute to a 1 ng/mL false digoxin reading. In contrast, if this cross-reactive substance has even 10% cross-reactivity with the phenytoin immunoassay and assuming again 100 ng/mL of this substance is present in a specimen, this cross-reactant would falsely increase phenytoin concentration by 10 ng/mL. If true procainamide concentration is 10 μg/mL, this value would be falsely increased to only 10. 010 μg/mL. Such an increase is nonsignificant.

Digoxin immunoassays may be affected by digoxin metabolites, endogenous digoxin-like immunoreactive substances, or by a variety of drugs and metabolites including spironolactone, potassium canrenoate, and their common metabolite canrenone. In addition, certain Chinese medicines such as Chan Su, Lu-Shen-Wan, and oleander-containing herbal supplements may interfere with serum digoxin measurements using various immunoassays. Various factors that affect digoxin immunoassays are listed in Table 12.3.

12.5. EFFECT OF ENDOGENOUS FACTORS ON SERUM DIGOXIN MEASUREMENTS BY IMMUNOASSAYS

Usually endogenous factors such as high bilirubin, triglycerides, blood urea nitrogen, and total protein do not significantly affect the digoxin immunoassay (35). Moderate hemolysis is usually tolerated by most digoxin immunoassays. However, elevated concentrations of endogenous digoxin-like immunoreactive substances (DLIS) and the presence of a heterophilic antibody may affect serum digoxin measurement by certain digoxin immunoassays.

TABLE 12.3. Various Endogenous and Exogenous Factors That Affect Serum Digoxin Measurements by Immunoassays

Factors	Comments
Endogenous Factors	
Digoxin-like immunoreactive substances	These substances are elevated in volume- expanded patients and affect polyclonal antibody-based digoxin immunoassays more than monoclonal antibody-based digoxin immunoassays
Endogenous Factors	
Digoxin metabolites	Although digoxin is minimally metabolized, metabolites cross-react with various digoxin immunoassays
Potassium-sparing diuretics	Spironolactone minimally affects various digoxin immunoassays, but another drug, potassium canrenoate (not used in the United States but used in Europe) may have significant interferences with various digoxin immunoassays. Eplerenone also cross-reacts with some antidigoxin antibodies.
Digibind/DigiFab	Digibind/DigiFab used in treating severe digoxin overdose interferes with most digoxin immunoassays. Free digoxin monitoring is recommended in patients overdosed with digoxin and being treated with Digibind or DigiFab.
Herbal supplements	Chinese medicine Chan Su, Lu-Shen-Wan, and oleander-containing herbal supplements significantly affect serum digoxin measurement by immunoassays. Ingestion of the hellebore plant may also cause a false-positive digoxin result.

12.5.1. Endogenous DLIS and Serum Digoxin Measurements

The presence of DLIS was first described in a volume-expanded dog in 1980 (36). After publication of that initial report, many investigators confirmed the presence of endogenous DLIS in serum and other biological fluids in volume-expanded patients not limited to patients with uremia, liver disease, essential hypertension, transplant recipients, eclampsia, pregnant women, and preterm infants (37,38). Interestingly, in one report, the authors found that when six healthy subjects underwent a 20-day exposure to high altitude (16,174 feet [4930 meters]), significant increases in plasma and serum DLIS were observed in these subjects. Moreover, a trend to a greater loss of sodium through urinary excretion correlated with urinary DLIS values was noted (39). Bisordi and Holt postulated that DLIS may represent a class of volume-sensitive natriuretic factors because chronic renal failure patients are known to have elevated concentrations of natriuretic factors in their circulation and also have detectable DLIS. The authors measured DLIS levels in 15 stable hemodialysis patients and observed that mean predialysis DLIS concentration was 109.3 pg/mL digoxin equivalents and postdialysis mean DLIS was 97.5 pg/mL digoxin equivalents. The degree of extracellular fluid volume expansion predialysis also correlated with predialysis DLIS. As expected, the mean DLIS in 50 healthy subjects (30.0 pg/mL digoxin equivalents) was significantly lower than the DLIS level in hemodialysis patients. More interestingly, changes in DLIS with dialysis were paralleled by similar changes in alpha-atrial natriuretic peptide levels (40). In addition, DLIS can inhibit Na, K+ ATPase, the same enzyme that is inhibited by digoxin (41). Ebara et al also reported that DLIS isolated from cord blood has natriuretic activity and plays a role in water and sodium hemostasis in perinatal life (42). Therefore, in comparison with endorphins, the endogenous equivalent of opiates, DLIS can also be considered an endogenous equivalent of digoxin. DLIS is not a single compound but may be a complex mixture of several compounds that have a common steroid-like structure (43). Averina et al reported elevated concentrations of digitalis-like cardiotonic steroids in patients with preeclampsia. Interestingly, antidigoxin antibody Digibind can neutralize such factors (44). Another published paper also documented the capability of Digibind to neutralize such endogenous cardiotonic steroids, and immunoneutralization of these factors in patients with preeclampsia may have therapeutic benefits (45). Ihenetu et al reported that endogenous digoxin-like immunoreactive factors can inhibit the release of proinflammatory cytokines from human peripheral blood mononuclear cells, suggesting such compounds may also have an anti-inflammatory effect (46).

The highest magnitude of interferences with serum digoxin measurement in the presence of elevated concentrations of DLIS were reported in digoxin radioimmunoassays (RIAs) and florescence polarization immunoassay (FPIA, Abbott Laboratories, Abbot Park, IL) (47). Saccois et al in 1996 reported that the EMIT 2000 immunoassay had minimal interference with DLIS compared with a RIA digoxin assay (Amerlex 125I assay) as well as the FPIA digoxin assay as assessed by analyzing cord blood samples of neonates, which presumably contain high amounts of DLIS (48). However, RIA digoxin assays were discontinued in the 1990s. Because the FPIA digoxin assay also showed very high cross-reactivity with DLIS, many investigators also used this assay to measure DLIS concentrations in various patient groups. For example, Berendes et al used the FPIA digoxin assay for measuring DLIS in 401 critically ill patients and 62 healthy volunteers, and they found that DLIS levels were significantly elevated in critically ill patients compared with healthy volunteers (49). The Abbott Laboratories recently discontinued the manufacturing of the FPIA digoxin assay for application on the TDx analyzer. In contrast to digoxin, which is poorly protein bound (20–30%), DLIS are strongly bound to serum proteins. Valdes and Graves first reported the strong protein binding of DLIS (50).

Although DLIS causes positive interference with most digoxin immunoassays, the microparticle enzyme immunoassay (MEIA) digoxin assay marketed by the Abbott Laboratories for application on the AxSYM analyzer showed negative interference. However, taking advantage of the strong protein binding of DLIS and the only 25% protein binding of digoxin, interference of DLIS in MEIA assay can be eliminated by monitoring the free digoxin concentration in the protein-free ultrafiltrate. Due to the high concentration of digoxin in the protein-free ultrafiltrate (approximately 75% of total digoxin concentration), immunoassays designed for total digoxin can also be used for measuring free drug digoxin, and there is an insignificant matrix effect from the lack of presence of proteins in the protein-free ultrafiltrate used for measuring free digoxin concentrations. Protein-free ultrafiltrate can be easily prepared by centrifuging serum for approximately 20–25 minutes in a Centrifree Micropartition filter with a molecular cut-off of 30,000 Da (51). Other investigators also reported that interference of DLIS in serum digoxin measurement can be eliminated by monitoring the free digoxin concentration (52).

Newer, more specific monoclonal antibody-based digoxin assays are virtually free from DLIS interferences. Nevertheless, due to narrow therapeutic range, discordant digoxin values measured using two different assays may occur. Jones and Morris analyzed digoxin values in 36 plasma samples by sending aliquots to two different laboratory using different digoxin immunoassays (CEDIA [cloned enzyme donor immunoassay] DRI digoxin, Microgenics Corporation, Fremont, CA, and DGNA digoxin assay, Dade Behring, now Siemens Diagnostics, Deerfield, CA). The authors observed clinically significant discordance in 39% of these samples and commented that DLIS interference may explain only some of the discordance (53).

12.5.2. Heterophilic Antibody and Digoxin Measurement

The presence of human anti-animal antibody, especially those directed against mouse in serum (HAMA), may cause interference with certain immunoassays. The clinical use of mouse monoclonal antibody for radioimaging and treatment for certain cancers may cause an accumulation of HAMA. Anti-animal antibodies are also found among veterinarians, farm workers, or pet owners due to exposure to animals, and these antibodies are broadly classified as heterophilic antibodies. In one study Koshida et al investigated the prevalence of HAMA in patients undergoing routine examination by developing an enzyme-linked immunosorbent assay to detect HAMA in human serum. Based on a study of 290 patients, the authors detected the presence of HAMA in 34 patients (11.7%) (54).

Usually the presence of heterophilic antibodies in serum may interfere with sandwich assays designed for measuring relatively large molecules such as human chorionic gonadotropin (hCG). Rotmensch and Cole reported that 12 women were diagnosed with postgestational choriocarcinoma on the basis of persistently positive hCG test results in the absence of pregnancy, and despite therapy (extirpative surgery, chemotherapy, or both), no decline in serum hCG values was observed. The authors identified false-positive hCG results in all these women by observing a discrepancy of serum hCG values determined by two different assays as well as by not detecting hCG or its degradation products in urine specimens. Based on these results, further therapies were suspended in these women. The authors concluded that when a serum specimen is positive for hCG, a corresponding urine specimen must also be investigated to avoid false-positive test results in serum hCG testing (55). Because heterophilic antibodies are absent in urine due to their large molecular weight, determination of β-hCG in urine is free from such interferences. In a recent report, the authors observed that 5 of 378 patients showed interference of the heterophilic antibody in serum calcitonin measurement, and four values were clinically relevant false-positive calcitonin results (56). Bonetti et al described a case report in which the serum myoglobin concentration was falsely increased using an immunoassay (Beckman Access assay) due to the presence of a heterophilic antibody (57).

Usually heterophilic antibodies do not interfere with antibody-based competitive assays used for determining the concentrations of small molecules such as many therapeutic drugs. Nevertheless, Liendo et al described a case report of a patient with cirrhotic liver disease and atrial fibrillation who was being treated with spironolactone and digoxin and showed an elevated digoxin concentration of 4.2 ng/mL. Despite the toxic digoxin level, the patient was asymptomatic, and after discontinuation of both drugs, digoxin values higher than 3.0 ng/mL were observed for approximately 5 weeks in the patient's serum. The authors ruled out DLIS, digoxin antibodies, and spironolactone as potential sources of the observed interference. Because both ultrafiltration and treating with protein A markedly lowered the digoxin value, the authors concluded that the interference was due to a heterophilic antibody, especially an anti-mouse antibody in the serum, because protein A can selectively remove heterophilic antibody (immunoglobulin G class) by forming complex. In addition, due to it high molecular weight (approximately 150,000 Da), heterophilic antibodies are absent in the protein-free ultrafiltrate that usually utilizes a filter with a molecular weight cut-off of 30,000 Da (58).

12.6. EFFECT OF EXOGENOUS FACTORS ON SERUM DIGOXIN MEASUREMENTS BY IMMUNOASSAYS

A large number of exogenous factors including digoxin metabolites, antidigoxin antibodies used for treating a life-threatening digoxin overdose, various structurally related drugs, and active components of some Chinese and herbal medicines also interfere with serum digoxin measurement using various immunoassays. In general, polyclonal antibody-based digoxin immunoassays are more affected that monoclonal-based digoxin immunoassays.

12.6.1. Effect of Digoxin Metabolites on Digoxin Immunoassays

The major metabolites of digoxin are digoxigenin, digoxigenin monodigitoxoside, digoxigenin bisdigitoxoside, and dihydrodigoxin. These metabolites exhibit significantly different cross-reactivities against various antidigoxin antibodies used in digoxin immunoassays. In general, digoxin concentrations in serum obtained by a more specific analytical technique such as high performance liquid chromatography (HPLC) tend to be lower than the corresponding digoxin concentrations obtained by immunoassay due to metabolite cross-reactivity. Miller et al reported that digoxigenin has 0.7% cross-reactivity with the ACS assay (Ciba-Corning Diagnostics, now Siemens Medical Diagnostic Solutions), 103% cross-reactivity with Stratus assay (Baxter Corporation, now Dade-Behring, Deerfield, IL), and 153% cross-reactivity with Magic digoxin assay (Ciba-Corning) (59). However, due to the relatively low levels of digoxin metabolites in serum in comparison with digoxin, effects of metabolite cross-reactivities are minimal on serum digoxin measurements by immunoassays in patients with normal renal function. Vlasses et al reported highly elevated digoxin levels of 8.6 ng/mL in serum as measured by both radioimmunoassay and fluorescence polarization immunoassay (FPIA) in a patient with combined hepatic and renal dysfunction. Despite elevated digoxin, no evidence of digoxin toxicity was found based on ECG. Further investigations using a chromatographic method reveled that the false-positive digoxin level was due to an accumulation of an inactive digoxin metabolite (conjugates of digoxigenin monodigitoxoside) (60). In another report, the authors compared a chromatographic method for measuring serum digoxin concentration with a commercially available immunoassay (ACA digoxin assay, DuPont) and concluded that digoxin values obtained by the immunoassay compared well with values obtained by the chromatographic method in patients without renal disease, liver disease, or diabetes. However, for patients with renal disease, liver disease, or diabetes, the immunoassay significantly overestimated digoxin values (in the range of 0.3 to 1.1 ng/ml) compared with the chromatographic method (61).

12.6.2. Effect of Digibind and DigiFab on Serum Digoxin Measurement

Digibind and DigiFab are Fab fragments of the antidigoxin antibody used in treating life-threatening acute digoxin overdose. Digibind is the first antidote available for treating digoxin overdose in the United States that was marketed in 1986 by Glaxo Wellcome Company. In 2001 the U.S. Food and Drug Administration approved DigiFab, the second product for treating digoxin overdose. Digibind is produced by immunizing sheep with digoxin-human albumin conjugate followed by isolation of a digoxin-specific antibody from the blood. Then papain digestion of the isolated antibody finally generates the Fab fragment of the antibody. The Fab fragment of the ant-digoxin antibody is further purified prior to human administration. The DigiFab is prepared by injecting sheep with digoxin-dicarboxymethylamine, a digoxin analog followed by isolation of the antidigoxin antibody from the blood, papain digestion, and then purification of the Fab fragment using affinity chromatography. The purpose of papain cleavage is to produce the Fab fragment, which is a region of the antibody responsible for antigen recognition and binding. Moreover, the Fab fragment is substantially smaller in size compared with the original antibody and can be easily filtered through glomerular filtration process and excreted in the urine. The molecular weight of DigiFab (46,000 Da) is similar to the molecular weight of Digibind (46,200 Da). Usually Digibind dosage is 80 times the digoxin body burden or if neither the dose ingested nor the plasma digoxin concentration is known, then 380 mg of Digibind is usually injected into the patient. The half-life of the Fab fragment in humans is 12–20 hours but may be prolonged in patients with renal failure (62). Digibind and DigiFab are equivalent in their digoxin-binding capacity (63).

Both Digibind and DigiFab interfere with serum digoxin measurement using immunoassays. The magnitude of interference depends on the assay design and the specificity of the antidigoxin antibody. Valid determination of digoxin concentration in serum after imitation of immunotherapy with Fab fragment of antidigoxin antibody is not possible using digoxin immunoassays because such immunoassays not only measure pharmacologically active free digoxin but also inactive Fab-bound digoxin. After initiation of immunotherapy, unbound digoxin in serum readily binds with the Fab fragment, disturbing the equilibrium between the myocardium digoxin and serum digoxin. Then digoxin bound to myocardium may release back to circulation and binds with Fab, thus increasing total digoxin due to the increased concentration of Fab-bound inactive digoxin. McMillin et al studied the effect of Digibind and DigiFab on 13 different digoxin immunoassays. Significant interference was observed with both Digibind and DigiFab, although the magnitude of interference was somewhat less with DigiFab. The magnitude of interference varied significantly with each method; IMMULITE, Vitros, Dimension, and Access digoxin methods showed the highest interference. Minimal interferences were observed with FPIA, MEIA, SYNCHRON, and CEDIA methods. The authors also commented that monitoring free digoxin (in the protein-free ultrafiltrate) eliminates this interference because both Digibind and DigiFab due to approximate the molecular weight of 46,000 Da are absent in protein-free ultrafiltrate (64).

Case Study

A 35-year-old woman intentionally swallowed 100 Lanitop tablets (0.1 mg methyldigoxin per tablet, 10 mg of methyldigoxin) in a suicide attempt. Methyldigoxin is a semisynthetic cardiac glycoside that has a similar effect as digoxin, but the onset of effect is faster. Methyldigoxin is rapidly converted into digoxin after oral administration. On admission, approximately 19 hours after ingestion, her serum digoxin level was 7.4 ng/mL, which is in the toxic range. The patient was treated immediately with 80 mg of Fab fragment of antidigoxin antibody followed by continuous infusion at a rate of 30 mg/hour. A total of 395 mg of Fab was administered. The total serum digoxin level increased significantly and peaked at 125 ng/mL, 23 hours after ingestion and decreased according to a two-phase exponential decay function with a calculated terminal half-life of 34 hours. However, the free serum digoxin level immediately decreased to a nontoxic level immediately after the initiation of Fab therapy, indicating that the free digoxin level must be monitored in patients receiving immunotherapy for a digoxin overdose. Her symptoms of digoxin toxicity (ECG) as well as nausea and vomiting resolved within 3 hours of initiation of therapy, and the patient was discharged from the hospital after 3 days (65).

12.6.3. Interference of Potassium-Sparing Diuretics with Digoxin Immunoassays

Although various diuretics interact with digoxin and may increase the serum digoxin level, potassium-sparing diuretics such as spironolactone, potassium canrenoate, and eplerenone not only interact with digoxin pharmacokinetically but also may interfere with serum digoxin measurements using various digoxin immunoassays. Spironolactone is a competitive aldosterone antagonist that blocks the binding of aldosterone to the renal receptor, causing sodium loss and potassium retention and acts as a diuretic. This drug is used clinically in treating primary aldosteronism, essential hypertension, congestive heart failure and edema, and may be used along with digoxin therapy. After oral administration, spironolactone is rapidly and extensively metabolized to several metabolites including canrenone, which is an active metabolite. Potassium canrenoate is also metabolized to canrenone, which is not approved for clinical use in the United States but used clinically in Europe and other countries throughout the world. Because of the structural similarity between spironolactone and related compounds with digoxin, these substances interfere with serum digoxin assays, especially assays utilizing polyclonal antibody against digoxin. Chemical structure of spironolactone and digoxin is depicted in Figure 12.1.

The first report of the positive interference of spironolactone and canrenone in the radioimmunoassay for digoxin was published in 1974 (66). Morris et al reported the positive interference of spironolactone in digoxin measurement using FPIA (Abbott Laboratories) (67), but this assay is no longer commercially available. In another report, the authors discussed two cases in which the cross-reactivity of potassium canrenoate with the digoxin assay caused a clinical problem and recommended the use of the OPUS digoxin assay that showed minimum cross-reactivity with potassium canrenoate (68). Steimer et al reported for the first time the negative interference of canrenone in serum digoxin measurement using the microparticle enzyme immunoassay (MEIA, Abbott Laboratories, Abbott Park, IL). Misleading subtherapeutic concentrations of digoxin as measured on several occasions led to falsely guided digoxin dosing, resulting in serious digoxin toxicity in the patients (69). In a follow-up study, Steimer et al reported that spironolactone, potassium canrenoate, and

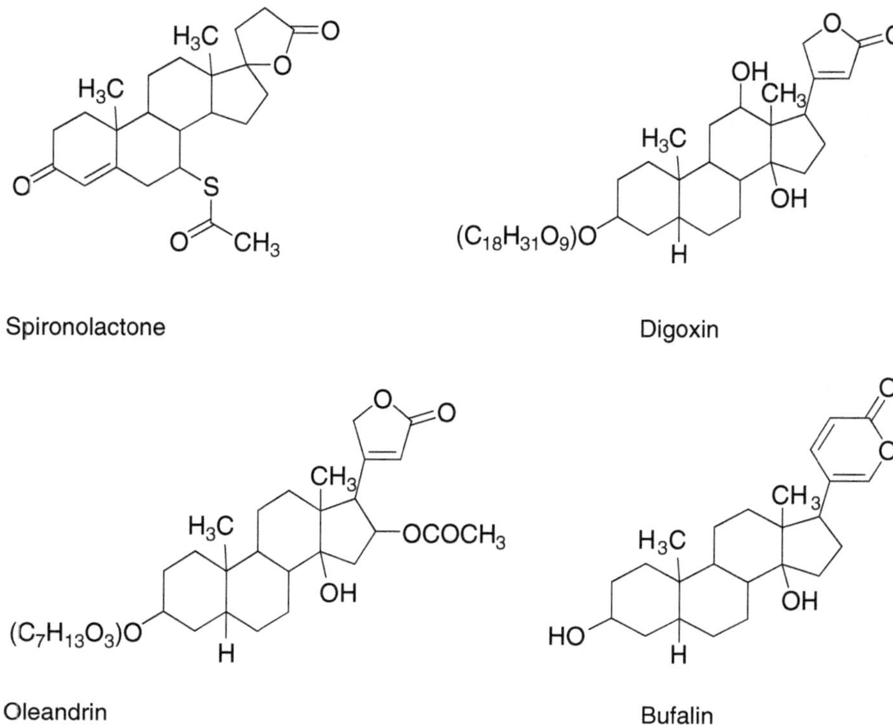

Figure 12.1. Chemical structure of spironolactone, digoxin, bufalin, and oleandrin.

their common metabolite canrenone can cause both positive interference and negative interference in serum digoxin measurement using immunoassays. Positive interference was observed using FPIA, aca (Dade Behring, Deerfield, IL), or Elecsys (Roche Diagnostics, Indianapolis, IN). Digoxin values are falsely lower (negative interference) if measured by MEIA, IMx (both from Abbott Laboratories), and Dimension digoxin assays (Dade Behring, Deerfield, IL). The magnitude of interference is more significant with potassium canrenoate where concentration of its metabolite canrenone can be significantly higher compared with spironolactone therapy. The authors observed a 42% decline in the expected value of serum digoxin in the presence of 3125 ng/mL of canrenoate using MEIA, 78% decline in using Dimension, and 51% decrease using IMx. A positive bias was observed with the aca (0.7 ng/mL), TDx (0.62 ng/mL), and Elecsys (0.58 ng/mL). EMIT 2000, Tina-Quant (Roche Diagnostics, Indianapolis, IN), and the Vitros digoxin assay are free from such interference (70).

Howard et al demonstrated that low-dose spironolactone (up to 25 mg per day) as used for oral therapy does not cause clinically significant negative interference in the MEIA digoxin assay on the AxSYM analyzer by comparing results with the EMIT assay that is free from spironolactone interference (71).

However, with a higher spironolactone dose, such as 200 mg/day, a significant interference may be observed with the MEIA assay (72). Fortunately, spironolactone and its metabolite canrenone are strongly bound to serum protein and virtually absent in the protein-free ultrafiltrate. Therefore, interference of spironolactone in the MEIA assay can be eliminated by monitoring free digoxin concentration (73). In addition, the Tina-Quant digoxin assay manufactured by Roche Diagnostics (Indianapolis, IN) is free from the interference of spironolactone potassium canrenoate and its common metabolite canrenone, although another digoxin assay, Digoxin III, for application on the AxSYM platform (Abbott Laboratories, Abbott Park, IL), was affected significantly by these compounds, and digoxin values were falsely elevated when digoxin pools prepared from patients receiving digoxin were further supplemented with these compounds (74). However, the relatively new digoxin assays manufactured by the Abbott Laboratories for application on ARCHITECT clinical chemistry platforms (*c*Dig, particle-enhanced turbidimetric inhibition immunoassay, PETINIA) and ARCHITECT immunoassay platforms (*i*Dig, chemiluminescent microparticle immunoassay, CMIA) are completely free from such interferences (75). Yamada et al reported that another potassium-sparing diuretic, eplerenone, interferes with FPIA and affinity column mediated immunoassays (ACMIA; on Dimension analyzer, Siemens Diagnostics, Deerfield, IL) but has no cross-reactivity with the MEIA digoxin assay (Abbott Laboratories, Abbott Park, IL). The cross-reactivities of eplerenone with these two digoxin assays were smaller in magnitude compared with spironolactone or canrenone (76).

12.6.4. Effect of Chinese Medicines and Herbal Supplements on Digoxin Immunoassays

In the United States, complementary and alternative medicines are classified as dietary supplements and sold pursuant to the Dietary Supplement Health and Education Act of 1994. Complementary and alternative medicines including Chinese medicines are becoming increasingly popular in the United States, Europe, and other parts of the world. Interestingly, herbal supplements only interfere with digoxin immunoassays, and the magnitude of interference depends on the antibody specificity of the assay. As expected, polyclonal antibody-based digoxin immunoassays are more affected by these supplements more than specific monoclonal antibody-based digoxin immunoassays. Significant interferences of Chinese medicines Chan Su, Lu-Shen-Wan, and oleander-containing herbal products with various digoxin assays (both monoclonal and polyclonal antibody based) have been reported. Asian ginseng and related products only demonstrated modest interference with the FPIA assay, but as mentioned earlier, Abbott Laboratories discontinued manufacturing this assay in 2009.

Chan Su is prepared from the dried white secretion of the auricular glands and the skin glands of Chinese toads (*Bufo melanostictus* Schneider or *Bufo*

bufo gargarzinas Gantor) and used for treating various heart diseases in traditional Chinese medicine. Bufalin, the active component of Chan Su, is also responsible for interference with various digoxin assays due to it structural similarity with digoxin (see Fig. 12.1 for structure of bufalin). Lu-Shen-Wan is also a Chinese medicine that contains multiple components including toad venom and may cause irritant contact dermatitis (77).

The interference of Chan Su and Lu-Shen-wan to serum digoxin measurement can be positive (falsely elevated digoxin concentrations) or negative (falsely lowered digoxin concentration) depending on the assay design. Although the Beckman assay (SYNCHRON LX system, Beckman Coulter) and Roche assay (Tina-Quant) showed positive interference in the presence of Chan Su, the MEIA digoxin assay on the AxSYM platform (Abbott Laboratories) showed negative interference of Chan Su in serum digoxin measurement. However, the components of Chan Su responsible for digoxin-like immunoreactivity are significantly bound to serum proteins (more than 90%) and are virtually absent in the protein-free ultrafiltrate. Therefore measuring free digoxin concentrations in the protein-free ultrafiltrate could be used to mostly eliminate the interference of Chan Su in serum digoxin measurements, but complete elimination of such interference is not possible especially if relatively higher amounts of Chan Su or Lu-Shen-Wan were consumed (78). Digoxin III immunoassay on the AxSYM platform is also significantly affected by Chan Su and Lu-Shen-Wan (79).

The oleanders are evergreen ornamental shrubs that grow in the southern parts of the United States from Florida to California and in Australia, India, Sri Lanka, China, and other parts of the world. There are two major varieties of oleander tree. The pink oleander plant (*Nerium oleander)* grows widely in the southern United States with beautiful pink flowers; the yellow oleander tree (*Thevetia peruviana*) is common through much of the tropics and subtropics. All parts of both types of oleander plants are toxic, and the toxic effect may occur after exposure from a small amount of the plant. Even ingestion of a single leaf may be fatal, especially in children. Boiling or drying the plant does not inactivate the toxins. Wasfi et al described a case of fatal oleander poisoning in a diabetic man who had gastrointestinal distress and cardiac symptoms 1 hour after ingestion of oleander leaves. Oleandrin was detected in his blood by gas chromatography combined with mass spectrometry, confirming the cause of death as oleander poisoning (80).

Many cardenolides have been isolated from yellow oleander including thevetin A, thevetin B (major components), peruvoside, neriifolin, thevetoxin, ruvoside, and theveridoside. These cardenolides have a structural similarity with digoxin and cross-react with the antidigoxin antibodies used in digoxin immunoassays. Oleandrin, the active component of pink oleander, also has structural similarity with digoxin and interferes with digoxin immunoassays. Eddleston et al reported a mean apparent serum digoxin concentration of 1.49 nmol/L (1.16 ng/ml) in patients who were poisoned with yellow oleander but eventually discharged from the hospital. Severe toxicity from oleander resulted in a mean apparent serum digoxin concentration of 2.83 nmol/L (2.21 ng/mL) as measured by the FPIA digoxin assay (81). Although the FPIA digoxin assay demonstrated the highest cross-reactivity with pink oleander extract as well as oleandrin, this assay is no longer commercially available. The Beckman digoxin assay on SYNCHRON LX as well as the turbidimetric assay on the ADVIA 1650 analyzer (Bayer Diagnostics) also showed significant interference with oleander, although the magnitude of interference was approximately 65% less with both the Beckman assay and the turbidimetric assay. The chemiluminescent assay (Bayer Diagnostics, now Siemens) is virtually free from interference of oleander. Although oleander causes positive interference with most digoxin assays, the MEIA digoxin assays (Abbott Laboratories) demonstrated negative interference (82). However, the Digoxin III assay on the AxSYM platform showed significant positive interference with oleander. This assay can be used for rapid indirect diagnosis of oleander poisoning in a patient not receiving digoxin (83).

There is one case report of interference of Siberian ginseng in serum digoxin measurement in the literature. A 74-year-old man had a steady serum digoxin level of 0.9–2.2 ng/ml for 10 years. His serum digoxin increased to 5.2 ng/mL on one occasion after taking Siberian ginseng. Although the level was toxic, the patient did not experience any signs or symptoms of digoxin toxicity. The patient stopped taking Siberian ginseng, and his digoxin level returned to normal (84). However, in our experience Siberian ginseng only has a very modest interference with the FPIA and has no effect on the other digoxin assays investigated. Therefore, it is possible that the Siberian ginseng ingested by this patient was mislabeled; the plant was probably from a digitalis group of plants. A recent article indicates that false hellebore plant (*Veratrum viride*) can cause a positive digoxin level (0.38 ng/mL) after ingestion using a turbidimetric immunoassay on the Abbott Architect chemistry platform (85).

12.7. CONCLUSIONS

Digoxin immunoassays, although widely used for routine therapeutic drug monitoring of digoxin in clinical laboratories, are subjected to interferences from both endogenous and exogenous substances including digoxin metabolites. However, taking advantage of poor protein binding of digoxin (20–30%) and strong protein binding (more than 90%) of endogenous DLIS and various interfering potassium-sparing diuretics interferences of these compounds can be partly or fully eliminated by monitoring free digoxin concentrations in the protein-free ultrafiltrate instead of convention digoxin monitoring in human serum or plasma. In addition, interferences of certain herbal supplements can also be partly or completely eliminated by monitoring free digoxin concentrations (Table 12.4). As mentioned earlier, chromatographic methods are mostly free from interferences but are rarely used in therapeutic drug monitoring of digoxin. In 2010 Li et al developed a rapid liquid chromatography combined

TABLE 12.4. Eliminating Interferences in Digoxin Immunoassays by Monitoring Free Digoxin Concentration

Interfering Substance	Comment
Endogenous DLIS	DLIS are very strongly protein bound; monitoring free digoxin concentration completely eliminates this interference.
Spironolactone	Spironolactone and its metabolite canrenone are strongly protein bound; monitoring free digoxin eliminates this interference.
Potassium canrenoate	Potassium canrenoate and its metabolite canrenone are strongly protein bound; monitoring free digoxin may completely or partially eliminate this interference depending on the dosage of potassium canrenoate.
Chan Su and Lu-Shen-Wan	Bufalin, the digoxin-like immunoreactive component, is strongly protein bound; monitoring free digoxin may completely or partially eliminate such interference depending on dosage.
Oleandrin	Oleandrin, the digoxin like immunoreactive component is strongly protein bound and monitoring free digoxin may completely of partially eliminate such interference depending on dosage.

DLIS, digoxin-like immunoreactive substances.

with tandem mass spectrometric method for the therapeutic drug monitoring of digoxin where chromatographic separation can be achieved within 3 minutes after liquid-liquid extraction of digoxin along with digoxin-d3, the internal standard using methyl-tert-butyl ether (86).

REFERENCES

1. Bessen HA. Therapeutic and toxic effects of digitalis: William Withering 1785. J Emerg Med 1986; 4: 243–248.
2. Gheorghiade M, Harinstein ME, Filippatos GS. Digoxin for the treatment of chronic and acute heart failure. Acute Card Care 2009; 11: 83–87.
3. O'Connell JB, Bristow MR. Economic impact of heart failure in the United States: time foe a different approach. J Am Coll Cardiol 2009; E1–E90.
4. Cheng JW, Rybak I. Use of digoxin for heart failure and atrial fibrillation in elderly patients. Am J Geriatr Pharmacother 2010; 8: 419–427.
5. Dec GW. Digoxin remains useful in the management of chronic heart failure. Med Clin North Am 2003; 87: 317–337.
6. Mittal MK, Chockalingan P, Chockalingam A. Contemporary indications and therapeutic implications for digoxin use. Am J Ther 2011; 18: 280–287.
7. Pouwels MJ, Hooymans PM, van der Aa GC, Gribnau FW. Comparison of steady state serum concentration of digoxin in tablets (Lanoxin) and capsules (Lanoxicaps) in the elderly. DICP 1992; 25: 1043–1046.

8. Mooradian AD. Digitalis: an update of clinical phramacokinetics, therapeutic monitoring techniques and treatment recommendations. Clin Pharamcokinet 1988; 15: 165–179.
9. Lingrel JB. The physiological significance of the cardiotonic steroid/ouabain binding site of the Na, K-ATPase. Annu Rev Physiol 2010; 72: 395–412.
10. Gheorghiade M, Adams KF, Colucci WS. Digoxin in the management of cardiovascular disorders. Circulation 2004; 109: 2959–2964.
11. Smith TW. Pharmacokinetics, bioavailability and serum levels of cardiac glycosides. J Am Coll Cardiol 1985; 5 (Suppl A): 43A-50A.
12. Mordasini MR, Krahenbuhl S, Schilenger RG. Appropriateness of digoxin level monitoring. Swiss Med Wkly 2002; 123: 506–512.
13. KuMana CR, Chan YM, Kou M. Audit exposed flawed sampling for digoxin levels. Ther Drug Monit 1992; 14: 155–158.
14. Sidwell A, Barclay M, Begg E, Moore G. Digoxin therapeutic drug monitoring: an audit and review. N Z Med J 2003; 116: U708.
15. Sameri RM, Soberman JE, Finch CK, Self TH. Lower serum digoxin concentrations in heart failure and reassessment of laboratory report forms. Am J Med Sci 2002; 324: 10–13.
16. Bauman JL, DiDomenico RJ, Viana M, Fitch M. A method for determining the dose of digoxin for ehart failure in the modern era. Arch Intern Med 2006; 166: 2539–2545.
17. Jitapunkul S, Kongswat V, Sutheparak S. Digoxin toxicity in Thai medical patients: clinical manifestations and an appropriate diagnostic serum level. Southeast Asian J Trop Med Public Health 2002; 33: 608–612.
18. Ong HT, Chang SL, Masduki A, Chandrasekharan N. Digoxin toxicity: clinical and laboratory assessment. Med J Malaysia 1989; 44: 296–301.
19. Abad-Santos F, Carcas AJ, Ibanez C, Frias J. Digoxin level in clinical manifestations as determined in the diagnosis of digoxin toxicity. Ther Drug Monit 2000; 22: 163–168.
20. Cauffield JS, Gums JG, Grauer K. The serum digoxin concentration: ten questions to ask. Am Fam Physician 1997; 56: 495–503.
21. Chan KE, Lazarus JM, Hakim RM. Digoxin associates with mortality in ESRD. J Am Soc Nephrol 2010; 21: 1550–1559.
22. Hallberg P, Melhus H, Hansson LO, Larsson A. Cystatin C vs creatinine as markers of renal function in patients on digoxin treatment. Ups J Med Sci 2004; 109: 247–253.
23. Prybys KM. Deadly drug interactions in emergency medicine. Emerg Med Clin North Am 2004; 22: 845–863.
24. Mordel A, Halkin H, Zulty L, Almog S et al. Quinidine enhance digitalis toxicity at therapeutic serum digoxin levels. Clin Pharmacol Ther 1993; 53: 457–462.
25. Hughes J, Crowe A. Inhibition of P-glycoprotein mediated efflux of digoxin and its metabolites by macrolide antibiotics. J Pharmacol Sci 2010; 113: 315–324.
26. Nordt SP, Williams SR, Manoguerra AS, Clark RF. Clarithromycin induced digoxin toxicity. J Accid Emerg Med 1998; 15: 194–195.

27. Gooderham MJ, Bolli P, Fernandez PG. Concomitant digoxin toxicity and warfarin interaction in a patient receiving clarithromycin. Ann Pharmacother 1999; 33: 796–799.
28. Alkadi HO, Nooman MA, Raja's YA. Effect of gentamicin on serum digoxin level in patients with congestive heart failure. Pharm World Sci 2004; 26: 107–109.
29. DeVore KJ, Hobbs RA. Plasma digoxin concentration fluctuations associated with timing of plasma sampling and amiodarone administration. Pharmacotherapy 2007; 27: 472–475.
30. Wang MT, Su CY, Chan AL, Lian PW et al. Risk of digoxin toxicity in heart failure patients exposed to digoxin-diuretic interactions: a population based study. Br J Clin Pharamcol 2010; 70: 258–267.
31. Gault H, Longerich L, Dawe M, Fine A. Digoxin-rifampin interaction. Clin Pharmacol Ther 1984; 35: 750–754.
32. Kanno S, Watanabe K, Yamagishi I, Hirano S et al. Simultaneous analysis of cardiac glycoside in blood and urine by thermo-responsive LC-MS-MS. Anal Bioanal Chem 2011; 399: 1141–1149.
33. Oiestad EL, Johansen U, Stokke Opdal M, Bergan S et al. Determination of digoxin and digitoxin in whole blood. J Anal Toxicol 2009; 33: 372–378.
34. Xue X, Huang M, Xiao H, Qin X et al. Rapid and simultaneous measurement of midazolam, 1′-hydroxymidazolam, and digoxin by liquid chromatography/tandem mass spectrometry: application to an in vivo study to simultaneously measure P-glycoprotein and cytochrome P 450 3A activity. J Pharm Biomed Anal 2011; 55: 187–193.
35. Brustolin D, Sirtoli M, Tarenghi G. An enzyme labeled immunometric assay for quantitation of digoxin in serum or plasma. Ther Drug Monit 1992; 14: 72–77.
36. Gruber KA, Whitaker JM, Buckalew VM. Endogenous digitalis-like substances in plasma of volume expanded dogs. Nature 1980; 287: 743–745.
37. Craver JL, Valdes R. Anomalous serum digoxin concentration in uremia. Ann Intern Med 1983; 98: 483–484.
38. Jortani SA, Valdes R Jr. Digoxin and its related endogenous factors. Crit Rev Clin Lab Sci 1997; 34: 225–274.
39. De Angelis C, Farrace S, Urbani L, Porcu S et al. Effects of high altitude exposure on plasma and urinary digoxin-like immunoreactive substance. Am J Hypertens 1992; 5: 600–607.
40. Bisordi JE, Holt S. Digitalis like immunoreactive substances and extracellular fluid volume status in chronic hemodialysis patients. Am J Kidney Dis 1989; 13: 396–403.
41. Dasgupta A, Yeo K, Malik S, Sandu P et al. Two novel endogenous digoxin-like immunoreactive substances isolated from human plasma ultrafiltrate. Biochem Biophys Res Comm 1987; 148: 623–628.
42. Qazzaz HM, Valdes R. Simultaneous isolation of endogenous digoxin-like immunoreactive factor, ouabain-like factor, and deglycosylated congeners from mammalian tissue. Arch Biochem Biophys 1996; 328: 193–200.
43. Ebara H, Suzuki S, Nagashima K, Kuroume T. Natriuretic activity of digoxin like immunoreactive substance extracted from cord blood. Life Sci 1988; 42: 303–309.

44. Averina IV, Tapilskaya NI, Reznik VA, Frolova EV et al. Endogenous Na/K-ATPase inhibitors in patients with preeclampsia. Cell Moll Biol (Noisy-le-grand) 2006; 52: 19–23.
45. Fedorova OV, Tapilskaya NI, Bzhelyansky AM, Frolova EV et al. Interaction of Digibind with endogenous cardiotonic steroids from preeclamptic placenta. J Hyperten 2010; 28: 361–366.
46. Ihenetu K, Espinosa R, de Leon R, Planas G et al. Digoxin and digitoxin-like immunoreactive factors (DLIF) modulate the release of pro-inflammatory cytokines. Inflamm Res 2008; 57: 519–523.
47. Dasgupta A. Therapeutic drug monitoring of digoxin: impact of endogenous and exogenous digoxin-like immunoreactive factors. Toxicol Rev 2006; 25: 273–281.
48. Saccoia NC, Hackett LP, Morris RG, Ilett KF. Enzyme-multiplied immunoassay (EMIT 2000) digoxin assay compared with fluorescence polarization immunoassays and amerlex 125I radioimmunoassay at two Australian centers. Ther Drug Monit 1996; 18: 672–677.
49. Berendes E, Cullen P, Van Aken H, Zidek W et al. Endogenous glycosides in critically ill patients. Crit Care Med 2003; 31: 131–1337.
50. Valdes R, Graves SW. Protein binding of endogenous digoxin-like immunoreactive factors in human serum and its variation with clinical condition. J Clin Endocrinol Metab 1985; 60: 1135–1143.
51. Dasgupta A, Trejo O. Suppression of total digoxin concentration by digoxin-like immunoreactive substances in the MEIA digoxin assay: elimination of interference by monitoring free digoxin concentrations. Am J Clin Pathol 1999; 111: 406–410.
52. Christenson RH, Studenberg SD, Beck-Davis SS, Sedor FA. Digoxin-like immunoreactivity eliminated from serum by centrifugal ultrafiltration before fluorescence polarization immunoassay of digoxin. Clin Chem 1987; 33: 606–608.
53. Jones TE, Morris RG. Discordant results from "real-world" patient samples assayed for digoxin. Ann Pharmacother 2008; 42: 1797–1803.
54. Koshida S, Asanuma K, Kuribayashi K, Goto M et al. Prevlence of human anti-mouse antibodies (HAMA) in routine examination. Clin Chim Acta 2010; 411: 391–394.
55. Rotmensch S, Cole LA. False positive diagnosis and needless therapy of presumed malignant disease in women with false positive human chorionic gonadotropin concentrations. Lancet 2000; 355: 712–725.
56. Giovanella L, Suriano S. Spurious hypercalcitoninemia and heterophilic antibodies in patients with thyroid nodules. Head Neck 2011; 33: 95–97.
57. Bonetti A, Monica C, Bonaguri C, Gnocchi C et al. Interference by heterophilic antibodies in immunoassays: wrong increase of myoglobin values. Acta Biomed 2008; 79: 140–143.
58. Liendo C, Ghali JK, Graves SW. A new interference in some digoxin assays: anti-murine heterophilic antibodies. Clin Pharmacol Ther 1996; 60: 593–598.
59. Vlasses PH, Bersarb A, Lottes SR, Conner DP et al. False positive digoxin measurements due to conjugated metabolite accumulation in combined renal and hepatic dysfunction. Am J Nephrol 1987; 7: 355–359.

60. Miller JJ, Straub RW, Valdes R. Digoxin immunoassays with cross-reactivity of digoxin metabolites proportional to their biological activities. Clin Chem 1994; 40: 1898–1903.
61. Tzou MC, Reuning RH, Sams RA. Quantitation of interference in digoxin immunoassay in renal, hepatic and diabetic disease. Clin Pharm Ther 1997; 61: 429–441.
62. Flanagan RJ, Jones AL. Fab antibody fragments: some applications in clinical toxicology. Drug Saf 2004; 27: 1115–1133.
63. Pullen MA, Harpel MR, Danoff TM, Brooks DR. Comparison of non-digitalis binding properties of digoxin specific Fab using direct binding methods. J Immunol Methods 2008; 336: 235–241.
64. McMillin GA, Qwen W, Lambert TL, De B et al. Comparable effects of DIGIBIND and DigiFab in thirteen digoxin immunoassays. Clin Chem 2002; 48: 1580–1584.
65. Fyer F, Steimer W, Muller C, Zilker T. Free and total digoxin in serum during treatment of acute digoxin poisoning with Fab fragments: case study. Am J Crit Care 2010; 19: 387–391.
66. Huffman DH. The effects of spironolactone and canrenone in digoxin radioimmunoassay. Res Comm Chem Pathol Pharmacol 1974; 9: 787–790.
67. Morris RG, Frewin DB, Taylor WB, Glistak ML et al. The effect of renal and hepatic impairment and of spironolactone on serum digoxin assay. Eur J Clin Pharmacol 1998; 34: 233–239.
68. Okazaki M, Tanigawara Y, Kita T, Komada F, Okumura K. Cross-reactivity of TDx and OPUS immunoassay system for serum digoxin determination. Ther Drug Monit 1997; 19: 657–662.
69. Steimer W, Muller C, Eber B, Emmanuilidis K. Intoxication due to negative canrenone interference in digoxin drug monitoring [letter]. Lancet 1999; 354: 1176–1177.
70. Steimer W, Muller C, Eber B. Digoxin assays: frequent, substantial and potentially dangerous interference by spironolactone, canrenone and other steroids. Clin Chem 2002; 48: 507–516.
71. Howard G, Barclay M, Florkowski C, Moore G et al. Lack of clinically significant interference by spironolactone with the AxSYM digoxin II assay. Ther Drug Monit 2003; 25: 112–113.
72. Steimer W. Lack of critically significant interference by spironolactone with the AxSYM digoxin II assay only applies to low dose therapy with spironolactone [letter]. Ther Drug Monit 2003; 25: 484–485.
73. Dasgupta A, Saffer H, Wells A, Datta P. Bidirectional (positive/negative) interference of spironolactone, canrenone and potassium canrenoate on serum digoxin measurement: elimination of interference by measuring free digoxin or using a chemiluminescent assay for digoxin. J Clin Lab Anal 2002; 16: 172–177.
74. Dasgupta A, Tso G, Wells A. Effect of spironolactone, potassium canrenoate and their common metabolite canrenone on serum digoxin measurement by Digoxin III, a new digoxin immunoassay. Ther Drug Monit 2008; 30: 744–747.
75. DeFrance A, Armbruster D, Petty D, Cooper K et al. Abbott ARCHITECT clinical chemistry and immunoassay systems: digoxin assays are free of interferences from spironolactone, potassium canrenoate and their common metabolite canrenone. Ther Drug Monit 2011; 33: 128–131.

76. Yamada T, Suzuki K, Kanada Y, Kato R et al. Interference between eplerenone and digoxin in fluorescence polarization immunoassay, and affinity column mediated immunoassay. Ther Drug Monit 2010; 32: 774–777.
77. Lee TY, Lam TH. Irritant contact dermatitis due to Chinese herbal medicine lu-shen-wan. Contact Dermatitis 1988; 18: 213–218.
78. Chow L, Johnson M, Wells A, Dasgupta A. Effect of the traditional Chinese medicine Chan Su, Lu-Shen-Wan, Dan Shen, and Asian ginseng on serum digoxin measurement by Tina-Quant (Roche) and Synchron LX system (Beckman) digoxin immunoassays. J Clin Lab Anal 2003; 17: 22–27.
79. Reyes MA, Actor JK, Risin SA, Dasgupta A. Effect of Chinese medicines Chan Su and Lu-Shen-Wan on serum digoxin measurement by Digoxin III, a new digoxin immunoassay. Ther Drug Monit 2008; 30: 95–99.
80. Wasfi IA, Zorob O, Al Katheeri NA, Al Awadhi AM. A fatal case of oleandrin poisoning. Forensic Sci Int 2008; 178: e31–36.
81. Eddleston M, Ariaratnam CA, Sjostrom L, Jayalath S, et al. Acute yellow oleander (*Thevetia peruvica*) poisoning: cardiac arrhythmias, electrolyte disturbances, and serum cardiac glycoside concentrations on presentation to hospital. Heart 2000; 83: 310–306.
82. Dasgupta A, Datta P. Rapid detection of oleander poisoning by using digoxin immunoassays: comparison of five assays. Ther Drug Monit 2004; 26: 658–663.
83. Dasgupta A, Risin SA, Reyes M, Actor JK. Rapid detection of oleander poisoning by Digoxin III, a new digoxin assay: impact on serum digoxin measurement. Am J Clin Pathol 2008; 129: 548–553.
84. McRae S. Elevated serum digoxin levels in a patient taking digoxin and Siberian ginseng. Can Med Assoc J 1996; 155: 293–295.
85. Bechtel LK, Lawrence DT, Haverstick D, Powers JS et al. Ingestion of false hellebore plants can cross-react with a digoxin clinical chemistry assay. Clin Toxicol (Phila) 2010; 48: 435–442.
86. Li S, Liu G, Jia J, Miao Y et al. Therapeutic monitoring of serum digoxin for patients with heart failure using a rapid LCF-MS/MS method. Clin Biochem 2010; 43: 307–313.

Chapter 13

Interference in Immunoassays Used for Monitoring Anticonvulsants and the Usefulness of Monitoring Free Anticonvulsants

13.1. INTRODUCTION

Usually digoxin immunoassays are mostly affected by various interfering substances, but immunoassays used for monitoring other therapeutic drugs are less affected by interferences. Monitoring concentrations of classical anticonvulsants such as phenytoin, phenobarbital, carbamazepine, and valproic acid is essential for good medical practice. Although total drug concentrations are usually monitored for classical anticonvulsants, under certain pathophysiological conditions, total phenytoin concentrations cannot predict the pharmacologically active free phenytoin concentrations. For these patients, monitoring free phenytoin, valproic acid, or carbamazepine concentrations is necessary. Another classical anticonvulsant, phenobarbital, is only 40% protein bound, and monitoring free phenobarbital concentration is not necessary.

Resolving Erroneous Reports in Toxicology and Therapeutic Drug Monitoring: A Comprehensive Guide, First Edition. Amitava Dasgupta.

Although phenytoin immunoassays are usually free from interferences from other drugs, carbamazepine immunoassays are affected by carbamazepine 10, 11-epoxide, the active metabolite of carbamazepine. The magnitude of interference may vary widely from almost no cross-reactivity to 96% cross-reactivity from epoxide, and the interference of epoxide in the carbamazepine immunoassay is clinically important.

In the last 20 years, several new anticonvulsants have been approved for clinical use by the U.S. Food and Drug Administration. These drugs include felbamate, gabapentin, lamotrigine, levetiracetam, oxcarbazepine, pregabalin, tiagabine, topiramate, vigabatrin, and zonisamide. In contrast to classical anticonvulsants, newer anticonvulsants have a wider gap between therapeutic and toxic serum concentrations, requiring relatively less or no therapeutic drug monitoring for certain newer anticonvulsants.

13.2. INTERFERENCE IN THERAPEUTIC DRUG MONITORING OF CLASSICAL ANTICONVULSANTS

In the1970s, when therapeutic drug monitoring of classical anticonvulsants was first introduced, gas liquid chromatography (also known as gas chromatography [GC]) was used exclusively to determine serum or plasma drug concentrations. Pippenger published the first paper in 1969 on therapeutic drug monitoring of classical anticonvulsants using GC (1). Since then, assays have been developed for all four classical anticonvulsants based on GC, high performance liquid chromatography (HPLC), and mass spectrometry. Later in the 1980s, immunoassays became commercially available for the determination of serum or plasma concentrations of classical anticonvulsants using automated analyzers; thus routine therapeutic drug monitoring of these drugs also became available in smaller hospital laboratories. Usually chromatographic methods for the analysis of anticonvulsants are robust and relatively free from interferences. However, most clinical laboratories today use immunoassays for routine therapeutic drug monitoring of classical anticonvulsants due to their ease of operation, automation, and capability of batch analysis of specimens. Unfortunately, immunoassays are subjected to interferences mainly due to the cross-reactivity of the metabolites. However, other structurally related drugs may also interfere with various immunoassays used for the therapeutic drug monitoring of these anticonvulsants. The major interfering substances in immunoassays for phenytoin, carbamazepine, phenobarbital, and valproic acid are listed in Table 13.1.

13.2.1. Interference in Phenytoin Immunoassays

Phenytoin (diphenylhydantoin) (Fig. 13.1) was first introduced as an anticonvulsant agent in 1938, and it is one of the most widely used anticonvulsant drugs. Phenytoin is a low-cost drug with a long history of safe use and therefore

TABLE 13.1. Major Interferences in Immunoassays for Classical Anticonvulsants

Anticonvulsant	Interfering Substance	Comment
Carbamazepine	Carbamazepine 10, 11-epoxide	Cross-reactivity varies from 0% (Vitros) to 96% (PETINIA).
	Hydroxyzine, cetirizine	Affects PETINIA assay but other assays are not affected.
	10-hydroxy-10,11-dihydro-carbamazepine metabolite of oxcarbazepine	Affects EMIT assay.
Phenytoin	(HPPH) and its glucuronide conjugate	May cross-react with phenytoin immunoassays.
	Fosphenytoin	Cross-reacts with various phenytoin immunoassays, but treating serum with 10 μL alkaline phosphatase eliminates this interference.
		In uremia an unusual metabolite of fosphenytoin is formed that interferes with phenytoin immunoassay.
Phenobarbital	Amobarbital, butabarbital, secobarbital, and phenytoin	May interfere with immunoassay if present in very high amounts (more than observed in overdose).
		Not a concern in laboratory.
Valproic acid		No significant interference reported in immunoassays.

HPPH, 5-(p-hydroxyphenyl)-5-phenylhydantoin; EMIT, enzyme multiplied immunoassay technique; PETINIA, particle-enhanced turbidimetric inhibition immunoassay.

often used as a first-line drug to treat seizure disorders that respond to phenytoin therapy. Therapeutic drug monitoring of phenytoin is essential, however, to select the proper dosage and also to avoid phenytoin toxicity. The therapeutic range is usually considered 10–20 μg/mL; toxicity is encountered at a phenytoin level exceeding 40 μg/mL. Phenytoin is highly protein bound (about 90%), so in cases where toxicity is suspected but total serum phenytoin is within the optimal therapeutic interval, it is important to measure free phenytoin concentration directly because unbound (free) phenytoin is pharmacologically active, whereas the fraction of phenytoin bound to serum protein, mostly albumin, is inactive.

Common immunoassays commercially available in clinical laboratories for the determination of phenytoin concentration include enzyme-labeled immunoassays such as enzyme multiplied immunoassay technique (EMIT) and cloned enzyme donor immunoassay (CEDIA), fluorescence polarization immunoassay (FPIA), and turbidimetric and chemiluminescent immunoassays. One of the potential interferences in immunochemical measurements of phenytoin is cross-reactivity of the major phenytoin metabolite, 5-(p-hydroxyphenyl)-5-phenylhydantoin (HPPH), and its glucuronide conjugate with phenytoin immunoassays. HPPH is the primary metabolite of phenytoin, and it is readily conjugated to glucuronide (HPPG), which is excreted

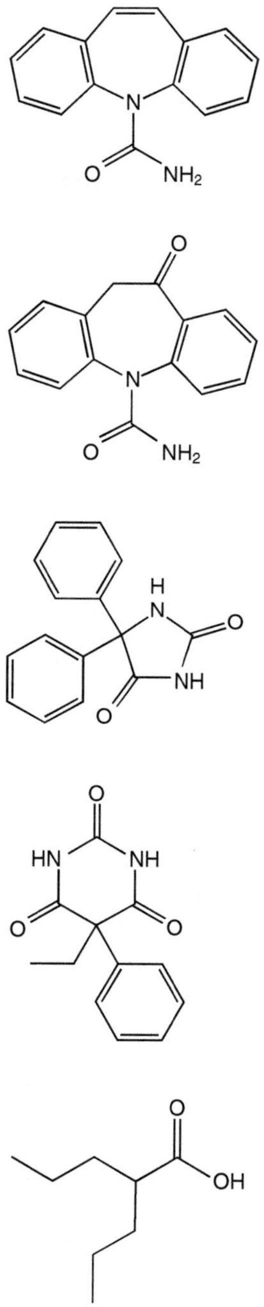

Carbamazepine

Oxacarbazepine

Phenytoin

Phenobarbital

Valproic acid

Figure 13.1. Chemical structures of classical anticonvulsants and oxcarbazepine.

in urine. It is estimated that 60–90% of the administered dose of phenytoin can by recovered in the urine as HPPG (2). This cross-reactivity becomes particularly important in patients with renal insufficiency because of the increased concentration of metabolites. An earlier study in 1981 showed a significant interference of metabolites with phenytoin immunoassays (3). However, newer monoclonal antibody-based immunoassays are less affected by phenytoin metabolites. Tutor-Crespo et al compared phenytoin concentrations determined by two immunoassays (FPIA and EMIT) with values established by the HPLC method for the determination of phenytoin using species from 36 patients receiving phenytoin. These patients had renal functions ranging from normal to severe renal insufficiency (glomerular filtration rate 10–102 mL/minute of creatinine clearance). The authors used a deviation of 15% in value as clinically significant. The mean phenytoin concentrations determined by FPIA, EMIT, and HPLC were 10.8 μg/mL, 10.8 μg/mL, and 10.5 μg/mL, respectively, indicating a negligible effect of phenytoin metabolites on the two immunoassays studied. An FPIA value deviated by more than 15% from the value determined by HPLC in only one patient with very low renal function (creatinine clearance less than 20 mL/minute). The authors concluded that both immunoassays provided accurate results in the therapeutic monitoring of phenytoin in patients with renal insufficiency (4). Datta et al studied the analytical performance of a turbidimetric assay on the ADVIA 1650 analyzer (Siemens Diagnostics) and reported that the phenytoin assay had very low cross-reactivity (5–8%) with the HPPH metabolite and virtually no cross-reactivity with oxaprozin. In addition, this assay was free from the interference of bilirubin (free bilirubin tested 34.5 mg/dL; conjugated bilirubin tested 38.5 mg/dL), triglycerides (level tested 750 mg/dL), and hemolysis (hemoglobin level tested 120 mg/dL) (5).

Another drug that must be considered for cross-reactivity with phenytoin assays is fosphenytoin (5,5-diphenyl-3-[(phosphonooxy)methyl]-2,4-imidazolidine-dione disodium salt). Fosphenytoin is a prodrug of phenytoin that is rapidly converted into phenytoin after administration. Fosphenytoin, unlike phenytoin, is readily water soluble and can be administered via intravenous (IV) or intramuscular (IM) routes. However, phenytoin is poorly soluble in an aqueous solution, and it may crystallize in commonly used IV fluids or at the site of an IM injection. Fosphenytoin is not typically monitored clinically because of its short half-life and lack of pharmacological activity. However, phenytoin is monitored in a patient after administration of fosphenytoin, but in this case monitoring of phenytoin must be initiated after the complete conversion of fosphenytoin into phenytoin because fosphenytoin may cross-react with the phenytoin immunoassay. Significant cross-reactivity of fosphenytoin in various degrees was found on the TDx phenytoin, AxSYM phenytoin II, Vitros, IMMULITE, and EMIT 2000 assays (6–9). Based on this cross-reactivity, it is recommended that specimens to determine phenytoin concentrations should not be obtained for patients on fosphenytoin until at least 2 hours after IV infusion or 4 hours after IM injection. Also incubating a 1-mL specimen with 10 μL of alkaline phosphatase enzyme (Sigma Chemical Company), it converts any fosphenytoin present in the specimen to phenytoin within 5 minutes at room temperature. This procedure eliminates the interference of fosphenytoin in phenytoin immunoassays. The authors observed a complete conversion of fosphenytoin to phenytoin by alkaline phosphatase in heparin, ethylenediamine tetra-acetic acid, and citrated plasma (10).

Roberts et al studied in detail falsely elevated phenytoin values when measured by immunoassays compared with HPLC in patients with renal failure. The authors observed falsely increased phenytoin results up to 20 times higher than the HPLC results using AxSYM, TDx Phenytoin II (Abbott Laboratories, Abbott Park, IL), ACS:180 (Bayer Diagnostics, Tarrytown, NY), and Vitros assays (currently TDx Phenytoin II and ACS:180 phenytoin assays are not commercially available). Interestingly, no fosphenytoin was detected in any of these specimens by HPLC. For example, in renal failure patient number 3 on day 9 of the hospital stay (300 mg of fosphenytoin dosage), the phenytoin concentration as measured by the HPLC was 5.3 μg/mL. The corresponding phenytoin concentrations measured by immunoassays were 12.7 (AxSYM) and 28.0 μg/mL (TDxII), respectively. Based on their study with several patients, the authors proposed the presence of a novel metabolite of fosphenytoin, which has a very high cross-reactivity with antibodies, used in several immunoassays for phenytoin (9). Annesley et al identified a unique immunoreactive oxymethylglucuronide metabolite derived from fosphenytoin in the sera of uremic patients and demonstrated that this unusual metabolite was responsible for the cross-reactivity (11).

13.2.2. Interference in Carbamazepine Immunoassays

Carbamazepine (Fig. 13.1) is an anticonvulsant drug structurally similar to tricyclic antidepressants used in the treatment of generalized tonic-clonic, partial, and partial-complex seizures. It was approved for the treatment of epileptic patients in the United States in 1974 and subsequently approved for use in children older than 6 years in 1979. Similar to phenytoin, carbamazepine is one of the most widely used anticonvulsant drugs. In addition to treating epilepsy, carbamazepine is also used in combination therapy with tricyclic antidepressants in the treatment of neuropathic pain. Like many other anticonvulsant drugs, the efficacy of carbamazepine is better correlated with serum or plasma concentrations rather than dosage. Therefore, routine therapeutic drug monitoring of carbamazepine is essential in patient management. Early methods for the monitoring of carbamazepine included GC, HPLC, or GC coupled with mass spectrometry (GC-MS) (12–14). However, most clinical laboratories currently use immunochemical methods to measure the concentrations of carbamazepine serum or plasma. In chromatographic-based methods, no significant interferences have been reported in the scientific literature with carbamazepine metabolites, but cross-reactivity of carbamaze-

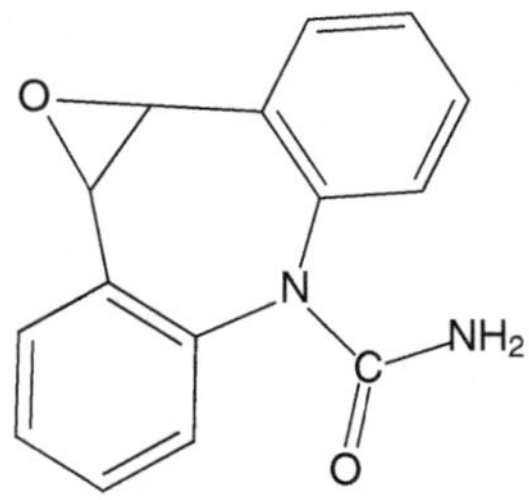

Figure 13.2. Structure of carbamazepine 10, 11-epoxide, the active metabolite of carbamazepine.

pine metabolites and structurally similar compounds may pose a problem in carbamazepine immunoassays.

The active as well as the main metabolite of carbamazepine is carbamazepine 10,11-epoxide (Fig. 13.2), which represents 10–20% of the parent drug concentration at the steady state. Many factors including renal insufficiency, polytherapy, genetic predisposition, and drug–drug interactions may disproportionately increase the concentration of the active epoxide metabolite causing carbamazepine toxicity even with a therapeutic concentration of carbamazepine determined by an immunoassay with poor cross-reactivity with the epoxide metabolite. Tutor-Crespo commented that in patients with moderate to severe renal insufficiency, the relative proportion of epoxide with respect to carbamazepine is significantly increased, and in such patients carbamazepine concentration obtained by EMIT or other assays with low cross-reactivity with the epoxide may have inadequate diagnostic efficiency because pharmacological activities of carbamazepine and the epoxide are the same (15). Potter and Donnelly suggested that the routine monitoring of the epoxide metabolite may have clinical benefits (16). The active epoxide metabolite is transformed by the microsomal epoxide hydroxylase, an enzyme that is highly polymorphic. There is a poor correlation between serum carbamazepine concentration and dosage in epileptic elderly patients due to age-related alterations in the pharmacokinetics of carbamazepine, and most elderly patients may require lower dosages than those based on median population kinetics (17). Lamotrigine is often used along with carbamazepine in treating epilepsy. In one study with eight patients (five male, four female; ages 19–31), the authors observed that after the introduction of lamotrigine (median daily dosage: 200 mg), the mean serum epoxide concentration was increased by 45% compared with the pre-lamotrigine level. In four patients, such increases in the epoxide concentration caused clinically significant toxicity (18). Quetiapine also interacts with carbamazepine and increases the concentration of the epoxide metabolite, possibly by inhibiting the epoxide hydroxylase enzyme (19). It has been well documented that valproic acid increases carbamazepine toxicity due to the accumulation of epoxide metabolite in the serum, and serum carbamazepine levels may be within the reference range in these patients (15).Valpromide, valnoctamide, and progabide also inhibit epoxide hydrolase, thus increasing the concentrations of carbamazepine 10, 11-epoxide. Inhibition of carbamazepine metabolism and the elevation of plasma carbamazepine to potential toxic concentrations can also be due to cotherapy with stiripentol, remacemide, acetazolamide, macrolide antibiotics, isoniazid, metronidazole, verapamil, diltiazem, cimetidine, danazol, or propoxyphene (20).

Many investigators have studied the cross-reactivity of carbamazepine epoxide in carbamazepine immunoassays. The cross-reactivity of carbamazepine 10, 11-epoxide with different immunoassays for carbamazepine may vary between 0% (Vitros) and 94% (Dade Dimension) (21). Parant et al also reported high cross-reactivity of the particle-enhanced turbidimetric inhibition immunoassay [PETINIA]; Siemens Diagnostics) carbamazepine assay with carbamazepine 10, 11-epoxide and negligible cross-reactivity with the EMIT 2000 assay (22). Many other carbamazepine immunoassays showed low to moderate cross-reactivity with the epoxide metabolite, and currently the PETINIA assay has the highest known cross-reactivity with the epoxide metabolite. Cross-reactivities of some carbamazepine immunoassays with the epoxide metabolite are summarized in Table 13.2. In general, the PETINIA assay actually measures the total concentration of the parent drug and the epoxide metabolite; carbamazepine assays with low cross-reactivities with epoxide measure the actual carbamazepine value. McMillin et al recently demonstrated a significant discordance in carbamazepine values obtained by two immunoassays, one with low cross-reactivity with epoxide (ADVIA Centaur carbamazepine assay) and another with high cross-reactivity with epoxide (PETINIA). The PETINIA assay is based on monoclonal antibody and can be used on the Dimension Vista analyzer. The ADVIA Centaur carbamazepine assay is a competitive immunoassay using direct chemiluminescent technology. This assay also uses a monoclonal antibody. The authors observed that cross-reactivity of epoxide in the PETINIA assay varied from 93% to 101% with an average cross-reactivity of 95.7%; the cross-reactivity of epoxide varied from 5% to 7.6% in the ADVIA Centaur assay with an average cross-reactivity of only 6.3%. In one specimen collected from a patient taking carbamazepine, the

TABLE 13.2. Cross-Reactivity of Carbamazepine 10, 11-Epoxide with Various Commercially Available Immunoassays

Carbamazepine Immunoassay	Cross-Reactivity of 10, 11-Epoxide, %
Abbott AxSYM	22.0
Beckman SYNCHRON	7.6
Syva EMIT	0.4
Vitros	0.0
CEDIA (Roche)	10.5
ADVIA Centaur	6.3
PETINIA assay	96.2

discordance was 115%, which is clinically very significant. The carbamazepine concentration using the ADVIA Centaur assay was 6.0 μg/mL; the carbamazepine concentration observed by the PETINIA assay was 12.9 μg/mL. The true carbamazepine concentration determined by liquid chromatography combined with tandem mass spectrometry (LC/MS/MS) was 6.4 μg/mL. The discordance between the carbamazepine values determined by the two assays was due to the presence of a high amount of epoxide in the specimen (epoxide concentration 8.1 μg/mL as determined by LC/MS/MS). The authors also demonstrated that the carbamazepine concentration as determined by the ADVIA Centaur assay correlated better with the values obtained by LC/MS/MS than the values obtained by the PETINIA assay. In general, PETINIA overestimates carbamazepine concentration by 12% (23).

Currently, there is no commercially available immunoassay for measuring carbamazepine 10, 11-epoxide concentration. However, both HPLC and HPLC combined with mass spectrometric methods have been reported in the literature for the simultaneous determination of both carbamazepine and its active metabolites. McMillin et al also described a mathematical approach to calculate epoxide concentration by measuring carbamazepine with two different immunoassays (PETINIA and ADVIA Centaur) with one assay showing low cross-reactivity to epoxide metabolite and another assay demonstrating high cross-reactivity to the epoxide metabolite. The average cross-reactivity of epoxide with the PETINIA assay was 96.2%. Therefore, carbamazepine concentrations observed by the PETINIA assay can be expressed as $x + 0.962\ y = a$, where a represents observed carbamazepine concentration by the PETINIA assay, x represents the true concentration of carbamazepine, and y represents the true concentration of epoxide. Similarly, if the observed carbamazepine concentration by the Centaur assay is considered as b, then $x + 0.065\ y = b$ because the average cross-reactivity of the epoxide with the Centaur carbamazepine assay is 6.5%. Therefore, true carbamazepine as well as the epoxide concentration can be calculated (carbamazepine concentration is represented as x and epoxide concentration is represented as y) by two simultaneous equations:

$$x + 0.962\ y = a \text{ (value obtained by the PETINIA assay)}$$

$$x + 0.065\ y = b \text{ (value obtained by the Centaur assay)}$$

Solving these two equations:

$$0.962\ y - 0.065\ y = a - b$$

$$\text{or } 0.897\ y = a - b$$

This can be further simplified as $0.9\ y = a - b$, or $y = a - b/0.9$. After calculating the value of y, the value of x can be easily determined by subtracting $0.96\ y$ from the value of a. Carbamazepine values obtained by the PETINIA assay and CEDIA assay can also be used to calculate epoxide as well as true carbamazepine concentration. The CEDIA carbamazepine assay was run on the Modular P800 analyzer also obtained from Roche Diagnostics (Indianapolis, IN), and the authors observed an average of 10.5% cross-reactivity with the epoxide metabolite.

$$x + 0.962\ y = a$$

$$x + 0.105\ y = c \text{ (value obtained by CEDIA assay)}$$

Therefore,

$$0.962\ y - 0.105\ y = a - c$$

Or

$$0.857\ y = a - c$$

Simplifying this, we can consider 0.86 $y = a - c$, or $y = a - c/0.86$. After calculating the value of y, value of x can be easily determined. The authors observed a good correlation between epoxide concentrations obtained by using mathematical equations with true epode concentrations determined by LC/MS/MS (24).

Oxcarbazepine (Fig. 13.1) is structurally similar to carbamazepine. This drug is also used in the treatment of epilepsy. In some cases, both drugs and their metabolites may be present in patients who are transitioning from one therapeutic regimen to the other. In a study of whether oxcarbazepine or its metabolites cross-reacted with an EMIT carbamazepine assay, it was shown that from a clinical perspective, only the 10-hydroxy-10,11-dihydro-carbamazepine metabolite of oxcarbazepine had any significant cross-reactivity with the assay; there was no significant interference from oxcarbazepine (21). Hydroxyzine is a commonly prescribed first-generation antihistamine with sedative properties. Hydroxyzine is also one of the drugs used in first-line therapy for the treatment of allergic rhinitis and chronic idiopathic urticaria. Cetirizine is a metabolite of hydroxyzine but also available as a drug used to treat allergies. Although hydroxyzine and cetirizine are structurally unrelated to carbamazepine, Parant et al documented two cases where hydroxyzine in serum caused false-positive carbamazepine levels using the PETINIA assay. A 22-year-old woman with a hydroxyzine concentration of 1.77 μg/mL and a cetirizine concentration of 2.1 μg/mL showed an apparent carbamazepine level of 5.3 μg/mL. Another patient with a hydroxyzine level of 520 ng/mL and cetirizine level of 2.18 μg/mL demonstrated a carbamazepine level of 25.4 μg/mL. The authors of this study also demonstrated the cross-reactivity of the PETINIA assay with cetirizine, oxatomide, and other benzhydryl piperazine drugs.

However, the EMIT 2000 assay for carbamazepine showed no cross-reactivity (25). The ADVIA centaur carbamazepine assay is also free from the interferences of hydroxyzine and cetirizine (26).

13.2.3. Interference in Phenobarbital Immunoassays

Phenobarbital (Fig. 13.1) is a member of the barbiturate family of drugs that was first introduced as a sedative and hypnotic drug in 1912. A few years later its utility as an anticonvulsant drug was observed, and later the drug began to be used in treating epilepsy. Phenobarbital is indicated for all types of seizures, with the exception of absence seizures, and it is the oldest anticonvulsant drug still available for use today. Phenobarbital is primarily metabolized by CYP2C19 and is a known inducer of most of the enzymes in the CYP450 family. When serum levels are above the optimal therapeutic interval, toxic effects include decreased consciousness, slowing of heart rate, shallow breathing, and even edema and renal failure in a severe overdose.

Although chromatographic methods were used initially for the therapeutic drug monitoring of phenobarbital, currently most clinical laboratories use commercially available immunoassays for the therapeutic drug monitoring of phenobarbital. In general, phenobarbital immunoassays are robust with relatively few reports of interferences. One report demonstrated cross-reactivity for amobarbital, butabarbital, secobarbital, and phenytoin, but these interferences occurred only at concentrations exceeding toxic levels and are not a concern for the therapeutic drug monitoring of phenobarbital using immunoassays (27). One case study demonstrated a possible cross-reactivity of butalbital at therapeutic concentrations, but no other report confirmed that finding (28).

13.2.4. Interference in Valproic Acid Immunoassays

Valproic acid is an 8-carbon branched chain fatty acid (Fig. 13.1) commonly used in the therapy for epileptic patients. This drug is indicated for patients with absence, tonic-clonic, and complex partial seizures. The proposed mechanism of action for valproic acid is interference with the neurotransmitter γ-aminobutyric acid. Valproic acid is highly metabolized in the liver by either glucuronidation or β-oxidation. Valproic acid is not an enzyme inducer, but it may cause clinically relevant drug interactions by inhibiting the metabolism of selected substrates such as phenobarbital and lamotrigine. Side effects of valproic acid include nausea or dyspepsia, sedation, headaches, dizziness, and tremors.

The early method of measuring valproic acid in patient specimens was by gas liquid chromatography, but currently it is mostly measured by immunoassay. There are no currently documented interferences with immunoassays for the measurement of valproic acid.

13.3. MONITORING FREE CONCENTRATIONS OF CLASSICAL ANTICONVULSANTS

The protein binding of a drug can vary widely between 0% and 99%, but it is the free fraction of the drug that demonstrates pharmacological activity. Classical anticonvulsants except phenobarbital are strongly bound to serum proteins. Free drug concentration can be estimated from total drug concentration. For example, if total phenytoin concentration is 10 μg/mL, it can be assumed that free phenytoin concentration is 1 μg/mL because phenytoin is 90% protein bound. However, protein binding may be decreased under certain pathophysiological conditions such as uremia, liver disease, and hypoalbuminemia, and for these patients free phenytoin concentrations may not be accurately predicted from total phenytoin concentration. Therefore, for such patients, directly determining a free drug level is necessary to avoid a toxic drug event. Protein binding of classical anticonvulsants and the recommended total and free therapeutic ranges are given in Table 13.3.

Drugs exist in the peripheral circulation as free (unbound) and bound to protein forms following the principle of reversible equilibrium and the law of mass action. Only free drugs can cross the plasma membrane and bind with the receptor for pharmacological action (29).

In general, there is equilibrium between a free drug and protein-bound drug:

$$[\mathrm{D}]+[\mathrm{P}]=[\mathrm{DP}]$$

$$\mathrm{K}=[\mathrm{DP}]/[\mathrm{D}][\mathrm{P}]$$

Where [D] is the unbound drug concentration; [P], the binding protein concentration, [DP], the drug/protein complex; and K, the association constant (liters/mole). The greater the affinity of the protein for the drug, the higher the K value. The free fraction of a drug represents the relationship between bound and free drug concentration and is often referred as α.

$$a=\frac{\text{Free Drug Concentration}}{\text{Total drug Concentration (Bound + Free)}}$$

TABLE 13.3. Protein-Binding, Total, and Free Suggested Therapeutic Ranges for Classical Anticonvulsants

		Therapeutic Range	
Anticonvulsant	Protein Binding, %	Total Drug	Free Drug
Carbamazepine	75–80	4–12 μg/mL	1–3 μg/mL
Phenytoin	90	10–20 μg/mL	1–2 μg/mL
Phenobarbital	40	15–40 μg/mL	Not required
Valproic acid	90–95	50–100 μg/mL	6–22 μg/mL

Free fraction does not vary with total drug concentration because protein-binding sites usually exceed the number of drug molecules present, but for certain drugs, the number of protein-binding sites may approach or be less than the number of drug molecules. Valproic acid exhibits saturable protein binding at the upper end of the therapeutic range, and as a result the free fraction of valproic acid is subject to more variation than other highly protein-bound antiepileptic drugs such as phenytoin and carbamazepine (30). For example, an albumin concentration of 3.5 g/dL is equivalent to an albumin concentration of 522 μmol/L (molecular weight of albumin: 66,500 Da). The therapeutic range of valproic acid is 50–100 μg/mL, or 347–693 μmol/L. Therefore, at the upper end of the therapeutic range, for example at a valproic acid concentration of 90 μg/mL, or 624.6 μmol/L, not enough albumin molecules are available for binding valproic acid, leading to a disproportionate increase in free valproic acid concentration. Therefore, if the patient showed albumin level at the lower end of normal (3.5 g/dL) and has a total valproic acid concentration at the upper end of normal, for example 90 μg/mL, then there would not be enough albumin binding sites available for valproic acid, and a disproportionate increase in free valproic acid level could be observed. For this patient, monitoring the free valproic acid level is necessary. In addition, in certain pathophysiological conditions, the displacement of a strongly protein-bound drug by another strongly protein-bound drug competing for the same protein-binding sites may lead to a disproportionate increase in free drug concentration. Usually in clinical laboratories, the free concentrations of strongly protein-bound classical anticonvulsants (phenytoin, valproic acid, and carbamazepine) are monitored. Phenobarbital is not strongly protein bound. Therefore, there is no need to monitor free phenobarbital concentration.

Monitoring the free concentrations of phenytoin is not needed for all patients, although Burt et al commented based on their study of 189 phenytoin requests from 139 patients over a 6-week period that monitoring total phenytoin is not as reliable as free phenytoin as a clinical indicator for therapeutic and nontherapeutic concentrations. They recommended free phenytoin monitoring for all patients. The authors observed a discrepancy between total and free phenytoin in 30% of the patients studied with the largest discordance (53%) occurring in the patient group with free phenytoin concentrations of less than 8% or more than 12% of total phenytoin concentration. The authors further observed that the free fraction of phenytoin varied widely among their patients ranging from 6.8% to 35.3% (31). Banh et al reported in their study using 163 patients that the mean free phenytoin fraction was 15%, but the free fraction varied widely from as low as 4% to as high as 61% (32). However, total phenytoin is still the most commonly measured parameter for therapeutic drug monitoring, and free phenytoin monitoring is restricted in patients with uremia, liver disease, hypoalbuminemia, and if toxicity is observed in a patient even though total phenytoin is within the therapeutic range such as a disproportionate increase in free phenytoin concentration due to displacement of phenytoin from the protein-binding site by another strongly protein-bound drug. Von Winckelmann et al commented that hypoalbuminemia, renal failure, and interactions with another highly protein-bound drug such as valproic acid can alter the protein binding of phenytoin, and when these conditions are present, the free phenytoin concentration cannot be predicted solely from the total phenytoin concentration (33). Such wide variations in protein binding are also observed with other strongly protein-bound classical anticonvulsants such as valproic acid and carbamazepine. In one study using 17 patients, the authors observed that the total valproic acid concentrations varied from 39 to 154 μg/mL; the free valproic acid concentrations varied from 3.3 to 25.6 μg/mL, and the free fraction varied from 5.9% to 24% (34). For carbamazepine, the free fraction may vary from 20% to 40% (35).

13.3.1. Monitoring Free Anticonvulsant Concentrations Patients with Uremic and Liver Disease

The unexpected elevated concentrations of free valproic acid, free phenytoin, and free carbamazepine are encountered in uremic patients, and the free fraction of valproic acid can be as high as 20–30% compared with 8.45% as observed in healthy volunteers. In uremia, the free fraction of phenytoin can be as high as 30%; in normal volunteers the free fraction is usually 10%. Uremia may also alter the hepatic metabolism of these drugs (36). Significantly elevated free fractions of highly protein-bound anticonvulsants in uremia can only be partially explained by hypoalbuminemia. It has been speculated that many uremic toxins are responsible for the displacement of strongly bound classical anticonvulsants from protein-binding sites causing elevated concentrations of free drugs. Hippuric acid and indoxyl sulfate, the two compounds that are present in elevated concentrations in uremia, can cause displacement of strongly protein-bound drugs (37). Takamura identified 3-carboxy-4-methyl-5-propyl-2-furanpropionate (CMPF) as the major uremic toxin that causes impaired protein binding of furosemide (38). Another study indicated that several endogenous compounds with low molecular weights (less than 500 Da) play significant roles in the displacement of strongly protein-bound drugs, but mid-molecular uremic toxins do not displace drugs (39). Conformational change to human albumin is also possible in uremia. Uremic toxins with indole ring and hippuric acid primarily bind to site 2, whereas the location of the CMPF binding site is in site 1 of human serum albumin (40).

Elevated free phenytoin concentration occurs in patients with hepatic disease mainly because of hypoalbuminemia (41). Hepatic disease can alter the pharmacokinetic parameters of valproic acid. Klotz et al reported that alcoholic cirrhosis and viral hepatitis decreased valproic acid protein binding from 88.7% to 70.3% and 78.1%, respectively, with a significant increase in the volume of distribution. The elimination half-life was also prolonged (42). An increase in unbound concentration of carbamazepine has been also reported in patients with hepatic disease (43).

13.3.2. Monitoring Free Anticonvulsant Concentrations in the Critically Ill, Patients with Hypoalbuminemia, Pregnant Women, and the Elderly

Critically ill patients also exhibit elevated concentrations of free anticonvulsants. Kemper et al reported severe phenytoin intoxication in two patients with hypoalbuminemia. One 35-year-old woman and a 60-year-old man developed severe phenytoin toxicity, but both patients showed total phenytoin concentrations within the recommended therapeutic range. However, the free phenytoin level was 4 μg/mL in the first patient and 8 μg/mL in the second, explaining the severe phenytoin toxicity experienced by these two patients. The authors concluded that despite total phenytoin concentrations within the therapeutic range, free phenytoin levels were disproportionately elevated in both patients due to severe hypoalbuminemia (albumin below 2.5 mg/dL in both patients). Although the first patient (the woman) survived after treatment with phenytoin was stopped, the second patient died from a phenytoin overdose (44). In one report the authors investigated free phenytoin fractions in 60 critically ill pediatric patients and concluded that total phenytoin concentrations did not accurately reflect free phenytoin concentrations in these patients. The authors reported that 10% of their patients demonstrated toxic concentrations of free phenytoin (more than 2 μg/mL) even though total phenytoin concentrations were within the recommended therapeutic range of less than 20 μg/mL. Although the mean free phenytoin fraction was 13% (10% free fraction is expected based on 90% protein binding of phenytoin), the free fraction varied widely from 6% to 42%. The free phenytoin concentrations were particularly elevated in patients with severe hypoalbuminemia (albumin less than 2.5 g/dL) (45). Zielmann et al studied the protein binding of phenytoin in 39 critically ill patients and observed that in 76% of these patients, free phenytoin fractions were increased (over 10% up to 24%) due to hypoalbuminemia or hepatic or renal failure. The authors concluded that free phenytoin monitoring is essential in critically ill patients (46).

Valproic acid is extensively bound to serum proteins, mainly albumin. Other than uremia and liver disease, hypoalbuminemia is a major cause of the disproportionate increase in free valproic acid concentration. Gidal et al reported a case in which markedly elevated plasma-free valproic acid in a hypoalbuminemic patient contributed to neurotoxicity. The total valproic acid concentration was 103 μg/mL, but the free valproic acid concentration was 26.8 μg/mL. This unexpected elevation was due to the low albumin level (3.3 g/dL) of the patient (47).

Case Study

A 53-year-old white woman with declining cardiac function was admitted to the cardiac intensive care unit. She had a heart transplant 7 years prior to this hospital admission. She then developed a chronic seizure disorder that was well controlled (seizure free for 2 years) with valproic acid and gabapentin. Her typical valproic acid serum level was 100 μg/mL; her free valproic acid level was 20 μg/mL. On hospital day 5, her total valproic acid level was 64 μg/mL; her free valproic acid level was 44 μg/mL (free fraction: 68%). Her albumin was only 2.9 g/dL, explaining such a high degree of free fraction as well as valproic acid toxicity. Her valproic acid dosage was reduced, and on day 7, her serum albumin level increased to 3.3 g/dL. Her total valproic acid was 44 μg/mL and free valproic cid was 18 μg/mL (free fraction: 44%). There was marked improvement in the patient with a reduction of unbound valproic acid concentration. Over the next month, her total valproic acid concentrations ranged from 50 to 60 μg/mL; her free valproic acid concentrations ranged from 10 to 20 μg/mL (free fraction: 13–47%), and the patient did not exhibit any symptoms of valproic acid toxicity (48).

Case Study

A 61-year-old previously healthy woman was admitted to the hospital due to urosepsis, cavitating pulmonary infiltrations, nonketonic hyperglycemia, and acute chronic renal failure. On admission her serum creatinine was 1.4 mg/dL, blood urea nitrogen 28.6 mg/dL, glucose 467 mg/dL, and albumin 2.4 g/dL. On hospital day 12 she developed multiple generalized tonic-clonic seizures along with respiratory failure and was put on mechanical ventilation. An intravenous loading dose of valproic acid (1500 mg) was administered. On day 14 she developed another generalized tonic-clonic seizure, and a single dose of phenytoin (1500 mg) was administered. On day 15 she was dialyzed due to progressing renal failure. A routine serum valproic acid concentration on hospital day 20 (on valproic acid therapy 8 days) showed no detectable level, but 2 days later her valproic acid level in serum was 1.4 μg/mL, which was severely subtherapeutic. But the patient experienced no seizure and no dosage adjustment was made. On hospital day 25 (13 days after initiation of valproic cid therapy), she was switched to valproic acid oral solution (900 mg in three divided doses). Two days later her serum valproic acid concentration was 9.8 μg/mL. Although the patient experienced no seizure due to extremely low total valproic acid concentration, a full pharmacokinetic curve was determined by measuring both total and free valproic acid concentrations. On day 28 (16 days on valproic acid therapy), 8 hours after her valproic acid dose, her total valproic acid concentration was 8.4 μg/mL and free valproic acid concentration was 7.2 μg/mL. Her free valproic acid fraction was 86% due to an extremely low albumin of 1.2 g/dL. Eventually 28 days after admission, the patient was extubated. On day 50 of her hospital admission (day 40 on valproic acid therapy), her albumin concentration was increased to 2.6 g/dL. Her total valproic acid concentration 7 hours after dosage was 26.0 μg/mL; her free valproic acid concentration was 7.5 μg/mL. Her free fraction was reduced to 29% due to an elevated concentration of albumin (49). This is the highest fraction of free valproic acid concentration for a critically ill patient with extremely low albumin reported in the literature so far.

The pharmacokinetics of many anticonvulsants undergo important changes in pregnancy due to modification in body weight, hemodynamic alteration, hormonal influence, and the contribution of the fetoplacental unit to drug distribution and disposition. Pregnancy thus affects absorption of drugs, binding to plasma protein, distribution, metabolism, and elimination on many drugs. At constant dosages plasma concentrations of anticonvulsants such as phenytoin, valproic acid, carbamazepine, and phenobarbital tend to decrease during pregnancy and then return to normal within the first or second month after delivery. Marked decreases in total phenytoin concentrations (about 40% of prepregnancy level) have been reported; the free phenytoin level decreased to a much lesser extent (50). Reports on the decline of total and free carbamazepine during pregnancy are conflicting. One study reports a 42% decline in total carbamazepine concentration and a 22% decrease in free carbamazepine concentration from pregnancy to delivery in 22 patients (51); another study reported only a slight decrease in total carbamazepine concentration and an insignificant decrease in free carbamazepine concentration from pregnancy to delivery (52). For valproate no significant change in free concentration was observed despite the reduction in total valproic acid concentration.

Clearances of many drugs are reduced in elderly people primarily due to the reduced capability of liver enzymes responsible for drug metabolism. An apparent clearance of free phenytoin tends to decrease in the elderly compared with younger people (53). Studies have shown that clearance of free valproic acid can be reduced significantly (39% according to one study and 65% according to another study) in the elderly compared with young adults (54). Elderly people also tend to have a lower albumin concentration that may cause disproportionate increases in free furcations of strongly protein-bound anticonvulsants. Therefore, monitoring free anticonvulsant concentrations is more appropriate for elderly people, especially patients older than 75 years. In addition, elderly patients also often take several medications, and drug–drug interactions may also cause elevation of free levels of classical anticonvulsants.

13.3.3. Monitoring Free Anticonvulsant Concentrations in Patients with Acquired Immunodeficiency Syndrome

Seizures are a common manifestation of central nervous system disease in patients with human immunodeficiency virus (HIV) infection. The incidence is approximately 10% in a population of hospitalized patients with an advanced stage of disease (55). Phenytoin is widely prescribed in the treatment of tonic-clonic seizures and other forms of epilepsy. Burger et al investigated serum concentrations of phenytoin in 21 patients with acquired immunodeficiency syndrome (AIDS). The total phenytoin concentrations were significantly lower in patients with AIDS compared with the control population, although phenytoin dosages were significantly higher in patients with AIDS. Calculation of Michaelis-Menten parameters demonstrated that Vmax values were similar in both patients with AIDS and the control group, but a nonsignificant trend of lower Km values was observed in patients with HIV. The authors demonstrated that unbound phenytoin concentrations were significantly higher in patients with HIV infection and concluded that the lower protein binding of phenytoin in patients with AIDS could be related to hypoalbuminemia. Because unbound phenytoin is the pharmacologically active fraction, the authors recommended monitoring unbound phenytoin concentrations for patients with HIV infection receiving phenytoin (56). In vitro experiments also confirmed the findings of Burger et al. Concentrations of free phenytoin and free valproic acid were significantly elevated in serum pools prepared from patients with AIDS and supplemented with phenytoin or valproic acid compared with serum pools prepared from normal subjects and also supplemented with the same amount of phenytoin or valproic acid. Hypoalbuminemia alone did not explain the elevation of free phenytoin or free valproic acid. Drug–drug interactions probably play a major role because an average patient with AIDS is receiving more than 10 medicines per day (57). Toler et al also described severe phenytoin toxicity as a result of decreased protein binding of phenytoin in a patient with AIDS leading to an elevated free phenytoin concentration of 4.9 μg/mL. The toxicity was due to high concentration of free phenytoin of (58).

13.3.4. Drug–Drug Interactions and Elevated Free Anticonvulsant Concentrations

Many strongly protein-bound drugs are capable of displacing phenytoin, valproic acid, and carbamazepine from protein-binding sites causing increases in free drug levels. Sandyk reported a case in which phenytoin toxicity was induced due to the coadministration of ibuprofen because of displacement of phenytoin from albumin binding by ibuprofen that is also strongly bound to albumin (59). Other nonsteroidal anti-inflammatory drugs such as salicylate, tolmetin, naproxen, mefenamic acid, and fenoprofen can displace phenytoin, valproic acid, and carbamazepine from protein-binding sites (60). Blum et al reported that tenidap sodium 120 mg/day at steady state increased the percentage of protein binding of phenytoin in plasma by 25%. The authors concluded that because tenidap increases the percentage of unbound phenytoin in plasma, when monitoring plasma phenytoin concentration, free phenytoin concentrations should also be monitored (61).

Displacement of phenytoin by another strongly albumin-bound antiepileptic drug valproic acid has been well documented in the literature. Tsanaclis et al studied plasma protein binding of phenytoin in nine epileptic patients before and during addition of sodium valproate to the drug therapy and observed that the mean free fraction of phenytoin increased from 13.5% to 18.2%. The total phenytoin concentrations were reduced due to increased concentration of free phenytoin (62). In another report, the authors demonstrated in vivo significant decreases in phenytoin protein binding due to the presence of valproic acid or primidone (63).

Penicillins including oxacillin and dicloxacillin were effective in displacing phenytoin from its binding sites. In vivo, the total phenytoin concentration in serum decreased during penicillin administration and the free phenytoin concentrations were increased (64). However, phenytoin–oxacillin interaction is not significant at the lower dose of oxacillin usually prescribed in oral therapy but is significant at higher oxacillin doses, especially in patients with hypoalbuminemia (65). In vitro and in vivo displacement of phenytoin by antibiotics ceftriaxone, nafcillin, and sulfamethoxazole have also been reported (66). Antoniou et al recently performed an extensive study to demonstrate trimethoprim/sulfamethoxazole-induced phenytoin toxicity in the elderly based on a study with 58,429 patients. There were 796 cases of patients hospitalized with phenytoin toxicity, and following multivariable adjustment for potential confounders, the authors observed a more than doubling of the risk of phenytoin toxicity after receiving trimethoprim/sulfamethoxazole. However, no such risk was observed after introducing amoxicillin (67).

13.4. MISCELLANEOUS FACTORS AFFECTING THERAPEUTIC DRUG MONITORING OF TOTAL AND FREE CLASSICAL ANTICONVULSANTS

Genetic polymorphism and thyroid dysfunction may affect phenytoin metabolism. Therapeutic drug monitoring is essential to identify such an effect. Phenytoin toxicity in thyroid dysfunction has been well documented. A 48-year-old woman presented to the emergency department with a toxic phenytoin concentration of 42.5 µg/mL. During admission it was found that the patient was profoundly hypothyroid, explaining her phenytoin toxicity. She was given thyroid replacement therapy, and her phenytoin level returned to normal when she returned to euthyroidism (68).

A rapid metabolizer may require a higher phenytoin dosage. Hawkins et al described a case of a 48-year-old man who was a rapid metabolizer and required 650 mg of phenytoin daily and experienced no phenytoin side effects (69). However, genetic polymorphism may also precipitate phenytoin toxicity.

Case Study

A 53-year-old Chinese woman was admitted to the neuroscience intensive care unit due to a traumatic brain injury following an assault. She had no significant past medical history. On admission the patient scored 7 on the Glasgow coma scale. On day 3, the patient received 100 mg phenytoin intravenously followed by a maintenance dosage of 300 mg per day. On day 7, phenytoin toxicity was suspected and her free phenytoin concentration was 4.4 µg/mL (normal free phenytoin: 1–2 µg/mL) and phenytoin was discontinued. Genetic testing indicated the CYP2C9*3 allele, which is associated with less than 5% metabolic activity. Therefore, the phenytoin toxicity in the patient was due to her poor metabolism of phenytoin even after giving a standard dosage (70).

Phenytoin toxicity secondary to an oxcarbazepine-phenytoin 2C19 interaction is also possible. Coadministration of oxcarbazepine and phenytoin resulted in a toxic concentration of phenytoin because oxcarbazepine is an inhibitor of CYP2C19 but an inducer of the CYP3A4 enzyme. Because phenytoin is also metabolized by CYP2C19, phenytoin metabolism is reduced in the presence of oxcarbazepine. Therefore, phenytoin dosage reduction is needed if oxcarbazepine is introduced in the treatment regime (71).Oxcarbazepine also interacts with valproic acid. Carbamazepine toxicity can be induced by protease inhibitors.

Case Study

A 51-year-old woman with a long history of schizoaffective disorder, bipolar type, catatonia, neuroepileptic syndrome, and diabetes was admitted to the hospital. She was on valproic acid, and her total valproic acid level was 105.1 μg/mL on admission. Oxcarbazepine was added after admission at a dosage of 600 mg twice daily. Later she developed valproic acid toxicity and her free valproic acid was highly elevated to 47.8 μg/mL. Her platelets also dropped to 80,000/mL. Oxcarbazepine was discontinued and her total valproic acid was 108.5 μg/mL and her free valproic acid was reduced to 26.8 μg/mL. Unsure of oxcarbazepine–valproic acid interaction, oxcarbazepine was introduced again at 300 mg twice daily, and 5 days later her total valproic acid was 119.6 μg/mL, free valproic acid was 39.1 μg/mL, and albumin was 3.9 g/dL. Her platelet count again dropped to 130,000/mL. Oxcarbazepine was discontinued again, and her free valproic acid and platelet count normalized after 7 days. Quetiapine was introduced later, and the patient was discharged in stable condition after 2 weeks. Oxcarbazepine is only 40% bound to serum protein, whereas valproic acid is 90% bound to serum protein. However, the authors observed an increased free valproic acid concentration while the total valproic acid level was not changed. The mechanism may be reduced free valproic acid clearance because of reduced protein-mediated valproic acid transport to the hepatocytes for clearance. In addition, metabolic inhibition may also play a role, although the exact metabolic pathway involved was unclear (72).

Case Study

A 59-year-old HIV-positive white man receiving highly active antiretroviral therapy (HAART) was changed to tenofovir (300 mg daily), lamivudine 150 mg twice daily, and lopinavir 133 mg/ritonavir 33 mg three capsules twice daily. The patient had been stable on carbamazepine 400 mg three times daily for the preceding 7 months. His serum carbamazepine level increased from 10.3 μg/mL, but after a change in HAART therapy with a new drug regime, his serum carbamazepine level was increased to 15 μg/mL, which was toxic 7 days later. The patient felt drowsy and experienced other symptoms of carbamazepine toxicity. His dosage was reduced to 400 mg twice daily. A repeat carbamazepine concentration on day 11 showed a value of 7.4 μg/mL, and his carbamazepine toxicity was resolved. Later he was admitted to the hospital with a skin rash, and he also discontinued HAART therapy due to side effects. The dermatologist thought the rash was due to HAART therapy. On admission to the hospital he was started on carbamazepine 400 mg three times a day, and on day 17, HAART therapy was reintroduced and his carbamazepine level returned to a toxic concentration of 15 μg/mL with symptoms of carbamazepine toxicity. On the reduction of carbamazepine dosage to 400 mg twice daily, carbamazepine toxicity was resolved, and the patient was discharged on day 22 with a carbamazepine therapeutic level of 9.3 μg/mL. Carbamazepine is extensively metabolized, and only 3% is excreted unchanged. Carbamazepine is metabolized by CYP3A4 and to a lesser extent with CYP2C8 to the active metabolite carbamazepine 10, 11-epoxide. Three other metabolic pathways of carbamazepine involve hydroxylation of the six-membered ring, glucuronidation of the carbamoyl side chain, and substitution of the six-membered ring with the sulfur-containing group. Carbamazepine also induces metabolism of other drugs and its own metabolism. The HIV protease inhibitor ritonavir inhibits CYP3A4. In addition, indinavir, amprenavir, nelfinavir, and atazanavir are moderate inhibitors of the CYP3A4 enzyme. Carbamazepine toxicity in this patient was most likely due to the inhibition of carbamazepine metabolism by lopinavir/ritonavir and nelfinavir (73).

13.5. THERAPEUTIC DRUG MONITORING OF NEWER ANTICONVULSANTS

Few immunoassays are available for newer anticonvulsants, and reference ranges for many of these newer anticonvulsants has not been clearly established. Although there is lesser need for therapeutic drug monitoring of newer anticonvulsants, some newer anticonvulsants such as lamotrigine, zonisamide, and topiramate may be useful. In general, gabapentin, pregabalin, tiagabine, and vigabatrin are not good candidates for therapeutic drug monitoring. Therapeutic drug monitoring of levetiracetam and pregabalin is justified in patients with renal impairment. Monitoring the active metabolite of oxcarbazepine (10-hydroxycarbazepine) has some justification. Usually chromatographic techniques are used for therapeutic drug monitoring of these newer anticonvulsants. These methods are usually free from interferences. However, there are commercially available immunoassays for zonisamide and topiramate. There are no reports of cross-reactivity or assay interference in the literature for the topiramate assay, but the package insert reports a cross-reactivity of approximately 10% for the metabolite 9-hydroxytopiramate. The package insert for zonisamide reports negligible cross-reactivity for the metabolite N-acetyl zonisamide, and a small degree (but clinically insignificant) of cross-reactivity with the metabolite 2-sulfamoylacetyl phenol. In addition, immunoassay is also available for lamotrigine, but no interference has been reported so far in the assay. However, lamotrigine metabolites showed low cross-reactivities (5–12%) with the assay according to the package insert.

13.6. CONCLUSIONS

Despite the introduction of many new anticonvulsants, classical anticonvulsants are still widely used today as first-line therapy to treat convulsions. Monitoring classical anticonvulsants and under certain circumstances free drug monitoring is essential to avoid toxicity in patients receiving these drugs. For carbamazepine, cross-reactivity of carbamazepine 10, 11-epoxide with the carbamazepine immunoassay is vital because cross-reactivity may vary from 0% to 94% depending on the assay. Therefore, discordant carbamazepine values may be observed if a patient is transferred from one small hospital to a bigger facility and if two different carbamazepine immunoassays with widely different cross-reactivities toward epoxide metabolite are used in two different hospitals. For resolving such discrepancies, the direct determination of epoxide metabolite may be necessary. Currently, there is no immunoassay for the determination of epoxide metabolite, and only chromatographic techniques are available in reference laboratories. However, it is possible to estimate epoxide concentration by measuring carbamazepine concentrations using two different immunoassays, one with low cross-reactivity with epoxide and another with high cross-reactivity. The cross-reactivity of phenobarbital metabolites with phenobarbital immunoassays is not a concern.

In addition to measuring total drug concentrations, monitoring free drug concentrations for phenytoin, carbamazepine, and valproic acid is important in patients with renal insufficiency, hepatic disease, critically ill patients, patients with hypoalbuminemia, pregnant women, and the elderly because free drug levels may be increased disproportionately due to the impaired binding capacity of albumin.

For newer anticonvulsants, however, therapeutic drug monitoring is less important except for a few drugs such as lamotrigine, zonisamide, and topiramate.

REFERENCES

1. Pippenger CE, Gillen HW. Gas chromatographic analysis for anticonvulsant drugs in biologic fluids. Clin Chem 1969; 15: 582–590.

2. Rainey PM, Rogers KE, Roberts WL. Metabolite and matrix interference in phenytoin immunoassays. Clin Chem 1996; 42: 1645–1653.

3. Matzke GR, Sawchuk RJ. Elevated serum phenytoin concentrations in uremic patient measured by enzyme multiplied immunoassay. Drug Intell Clin Pharm 1981; 15: 386–388.

4. Tutor-Crespo MJ, Hermida J, Tutor JC. Phenytoin immunoassay measurements in serum from patients with renal insufficiency: comparison with high-performance liquid chromatography. J Clin Lab Anal 2007; 21: 119–123.

5. Datta P, Scurlock D, Dasgupta A. Analytic performance evaluation of a new turbidimetric immunoassay for phenytoin on the ADVIA 1650 analyzer: effect of phenytoin metabolite and analogue. Ther Drug Monit 2005; 27: 305–308.

6. Frank EL, Schwarz EL, Juenke J, Annesley TM et al. Performance characteristics of four immunoassays for antiepileptic drugs on the IMMULITE 2000 automated analyzer. Am J Clin Pathol 2002; 118: 124–131.

7. Datta P, Dasgupta A. Cross-reactivity of fosphenytoin in four phenytoin immunoassays. Clin Chem 1998; 44: 696–697.

8. Kugler AR, Annesley TM, Nordblom GD, Koup JR et al. Cross-reactivity of fosphenytoin in two human plasma phenytoin immunoassays. Clin Chem 1998; 44: 1474–1480.

9. Roberts WL, De BK, Coleman JP, Annesley TM. Falsely increased immunoassay measurements of total and unbound phenytoin in critically ill uremic patients receiving fosphenytoin. Clin Chem 1999; 45: 829–837.

10. Dasgupta A, Warner B, Datta P. Use of alkaline phosphatase to correct underestimation of fosphenytoin concentrations in serum measured by phenytoin immunoassays. Am J Clin Pathol 1999; 111: 557–562.

11. Annesley T, Kurzyniec S, Nordblom G et al. Glucuronidation of prodrug reactive site: isolation and characterization of oxymethylglucuronide metabolite of fosphenytoin. Clin Chem 2001; 46: 910–918.

12. Hallbach J, Vogel H, Guder WG. Determination of lamotrigine, carbamazepine and carbamazepine epoxide in human serum by gas chromatography mass spectrometry. Eur J Clin Chem Clin Biochem 1997; 35: 755–759.

13. Jurgens U, May T, Hillenkotter K, Rambeck B. Systematic comparison of three basic methods of sample pretreatment for high-performance liquid chromatographic analysis of antiepileptic drugs using gas chromatography as a reference method. Ther Drug Monit 1984; 6: 334–343.

14. Lensmeyer GL. Isothermal gas chromatographic method for the rapid determination of carbamazepine ("Tegretol") as its TMS derivative. Clin Toxicol 1977; 11: 443–454.

15. Tutor-Crespo MJ, Hermida J, Tutor JC. Relative proportions of serum carbamazepine and its pharmacologically active 10, 11-epoxide derivative: effect of polytherapy and renal insufficiency. Ups J Med Sci 2008; 113: 171–180.

16. Potter JM, Donnelly A. Carbamazepine 10, 11-epoxide in therapeutic drug monitoring. Ther Drug Monit 1998; 20: 652–657.

17. Klotz U. The role of pharmacogenomics in the metabolism of antiepileptic drugs: pharmacokinetics and therapeutic implications. Clin Pharmacokinet 2007; 46: 271–279.

18. Warner T, Patsalos PN, Prevett M, Elyas AA et al. Lamotrigine induced carbamazepine toxicity: an interaction with carbamazepine 10, 11-epoxide. Epilepsy Res 1992; 11: 147–150.
19. Fitzgerald BJ, Okas AJ. Elevation of carbamazepine 10, 11-epoxide by quetiapine. Pharmacotherapy 2002; 22: 1500–1503.
20. Spina E, Pisani F, Perucca E. Clinically significant pharmacokinetic drug interactions with carbamazepine: an update. Clin Pharmacokinetic 1996; 31: 198–214.
21. Hermida J, Tutor JC. How suitable are currently used carbamazepine immunoassays for quantifying carbamazepine 10, 11-epoxide in serum samples? Ther Drug Monit 2003; 25: 384–388.
22. Parat F, Bossu H, Gagnieu MC, Lardet G et al. Cross reactivity assessment of carbamazepine 10, 11-epoxide, oxcarbazepine, and 10 hydroxy carbazepine in two automated carbamazepine immunoassay: PENTINA and EMIT 2000. Ther Drug Monit 2003; 25: 41–45.
23. McMillin GA, Juenke JM, Johnson M, Dasgupta A. Discordant carbamazepine values between two immunoassays: carbamazepine values determined by ADVIA Centaur correlate better with those determined by PETINIA assay. J Clin Lab Anal 2011; 25: 212–216.
24. McMillin GA, Juenke JM, Tso G, Dasgupta A. Estimation of carbamazepine and carbamazepine 10, 11–epoxide concentrations in plasma using mathematical equations generated with two carbamazepine immunoassays. Am J Clin Pathol 2010; 133; 728–736.
25. Parant F, Moulsma M, Gagnieu MC, Lardet G. Hydroxyzine and metabolites as a source of interference in carbamazepine particle-enhanced turbidimetric inhibition immunoassay (PETINIA). Ther Drug Monit 2005; 27: 457–462.
26. Dasgupta A, Tso G, Johnson M, Chow L. Hydroxyzine and cetirizine interfere with the PETINIA carbamazepine assay but not with the ADVIA CENTEUR carbamazepine assay. Ther Drug Monit 2010; 32: 112–115.
27. Ammann H, Vinet B. Accuracy, precision, and interferences of three modified EMIT procedures for determining serum phenobarbital, urine morphine, and urine cocaine metabolite with a Cobas-Fara. Clin Chem 1991; 37: 2139–2141.
28. Nordt SP. Butalbital cross-reactivity to an EMIT assay for phenobarbital. Ann Pharmacother 1997; 31: 254–255.
29. Chan S, Gerson B. Free drug monitoring. Clinics Lab Med 1987; 7: 279–287.
30. Meinardi H, Vander Kleijn E, Meijer JWA. Absorption and distribution of antiepileptic drugs. Epilepsia 1982; 23: 23–26.
31. Burt M, Anderson DC, Kloss J, Apple FS. Evidence based implementation of free phenytoin therapeutic drug monitoring. Clin Chem 2000; 46: 1132–1135.
32. Bahn HL, Burton ME, Sperling MR. Interpatient and intrapatient variability in phenytoin protein binding. Ther Drug Monit 2002; 24: 379–385.
33. Von Winckelmann SL, Spriet I, Willems L. Therapeutic drug monitoring of phenytoin in critically ill patients. Pharmacotherapy 2008; 28: 1391–1400.
34. Rapeport WG, Mendelow AD, French G, MacPherson P et al. Plasma protein-binding and CSF concentrations of valproic acid in man following acute oral dosing. Br J Clin Pharamcol 1983; 16: 365–369.

35. Levy RH, Friel PN, Johno I, Linthicum LM et al. Filtration for free drug level monitoring: carbamazepine and valproic acid. Ther Drug Monit 1984; 6: 67–76.
36. Yuan R, Venitz J. Effect of chronic renal failure on the disposition of highly hepatically metabolized drugs. Int J Clin Pharmacol 2000; 38: 245–253.
37. Gulyassy PF, Jarrard E, Stanfel L. Roles of hippurate and indoxyl sulfate in the impaired ligand binding by azotemic plasma. Adv Exp Med Biol 1987; 223: 55–58.
38. Takamura N, Maruyama T, Otagiri M. Effects of uremic toxins and fatty acids on serum protein binding of furosemide: possible mechanism of the binding defect in uremia. Clin Chem 1997; 43: 2274–2280.
39. Dasgupta A, Malik S. Fast atom bombardment mass spectrometric determination of the molecular weight range of uremic compounds that displace phenytoin from protein binding: absence of midmolecular uremic toxins. Am J Nephrol 1994; 14: 162–168.
40. Ghuman J, Zunszain P, Petitpas I, Bhattacharya A et al. Structural basis of drug-binding specificity of human serum albumin. J Mol Biol 2005; 353: 38–52.
41. Reidenberg MM, Affirme M. Influence of disease on binding of drugs to plasma proteins. Ann NY Acad Sci 1973; 226: 115–126.
42. Klotz U, Rapp T, Muller WA. Disposition of VPA in patients with liver disease. Eur J Clin Pharmacol 1978; 13: 55–60.
43. Hooper W, Dubetz D, Bochner F et al. Plasma protein binding of carbamazepine. Clin Pharmacol Ther 1975; 17: 433–440.
44. Kemper EM, van Khan HJ, Speelman P, de Gans K et al. Severe phenytoin intoxication in patients with hypoalbuminemia [in Dutch]. Ned Tjdschr Geneeskd 2007; 151: 138–141.
45. Wolf GK, McClain CD, Zurakoski D, Dodson B et al. Total phenytoin concentrations do not accurately predict free phenytoin concentrations in critically ill children. Pediatr Crit Care Med 2006; 7: 434–439.
46. Zielmann S, Mielck F, Kahl R, Kazmaier S et al. A rational basis for the measurement of free phenytoin concentration in critically ill trauma patients. Ther Drug Monit 1994; 16: 139–144.
47. Gidal BE, Collins DM, Beinlich BR. Apparent valproic acid neurotoxicity in a hypoalbuminemic patient. Ann Pharmacother 1993; 27: 32–35.
48. Haroldson JA, Kramer LE, Wolff DL, Lake KD. Elevated free fractions of valproic acid in a heart transplant patient with hypoalbuminemia. Ann Pharmacother 2000; 34: 183–187.
49. de Matt M, van Leeuwen HJ, Edelbroek PM. High unbound fraction of valproic acid in a hypoalbuminemic critically ill patient on renal replacement therapy. Ann Pharmacother 2011; 45: e18.
50. Pennell PB. Antiepileptic drug pharmacokinetics during pregnancy and lactation. Neurology 2003; 61 (Suppl 2): S35–S42.
51. Tomson T, Lindbom U, Ekqvist B et al. Epilepsy and pregnancy: a prospective study on seizure control in relation to free and total concentrations of carbamazepine and phenytoin. Epilepsia 1994; 35: 122–130.
52. Yerby MS, Friel PN, McCormick K. Antiepileptic drug disposition during pregnancy. Neurology 1992; 42 (Suppl 5): 12–16.

53. Wright DF, Begg EJ. The apparent clearance of free phenytoin in elderly vs younger adults. Br J Clin Pharmacol 2010; 70: 132–138.
54. Butler JM, Begg EJ. Free drug metabolic clearance in elderly people. Clin Pharmacokinetic 2008; 47: 297–321.
55. Wong MC, Suite NDA, Labar DR. Seizures in human immunodeficiency virus infection. Arch Neurol 1990; 47: 640–642.
56. Burger D, Meenhorst PL, Mulder JW et al. Therapeutic drug monitoring of phenytoin in patients with the acquired immunodeficiency syndrome. Ther Drug Monit 1994; 16: 616–620.
57. Dasgupta A, McLemore J. Elevated free phenytoin and free valproic acid concentrations in sera of patients infected with human immunodeficiency virus. Ther Drug Monit 1998; 20: 63–67.
58. Toler SM, Wilkerson MA, Porter WH, Smith AJ et al. Severe phenytoin intoxication as a result of altered protein binding in AIDS. DICP 1990; 24: 698–700.
59. Sandyk R. Phenytoin toxicity induced by interaction with ibuprofen. Afr Med J 1982; 62: 592.
60. Dasgupta A, Emerson L. Interaction of valproic acid with nonsteroidal anti-inflammatory drugs mefenamic acid and fenoprofen in normal and uremic sera: lack of interaction in uremic sera due to the presence of endogenous factors. Ther Drug Monit 1996; 18: 654–659.
61. Blum RA, Schentag JJ, Gardner MJ, Wilner KD. The effect of tenidap sodium on the disposition and plasma protein binding of phenytoin in healthy male volunteers. Br J Clin Pharmacol 1995; 39 (Suppl I): 35S–38S.
62. Tsanaclis LM, Allen J, Perucca E, Routledge PA et al. Effect of valproate on free plasma phenytoin concentrations. Br J Clin Pharmacol 1984; 18: 17–20.
63. Pospisil J, Perlik F. Binding parameters of phenytoin during monotherapy and polytherapy. Int J Clin Pharmacol Ther Toxicol 1992; 30: 24–28.
64. Arimori K, Nanko M, Otagiri M, Uekama K. Effect of penicillins on binding of phenytoin to plasma proteins in vitro and in vivo. Biochem Drug Dispos 1984; 5: 219–227.
65. Dasgupta A, Sperelakis A, Mason A, Dean R. Phenytoin-oxacillin interactions in normal and uremic sera. Pharmacotherapy 1997; 17: 375–378.
66. Dasgupta A, Dennen DA, Dean R, McLawhon RW. Displacement of phenytoin from serum protein carriers by antibiotics: studies with ceftriaxone, nafcillin and sulfamethoxazole. Clin Chem 1991; 37: 98–100.
67. Antoniou T, Gomes T, Mamdani MM, Juurlink DN. Trimethoprim/sulfamethoxazole induced phenytoin toxicity in the elderly: a population based study. Br J Clin Pharmacol 2011; 71: 544–549.
68. Betteridge T, Fink J. Phenytoin toxicity and thyroid dysfunction. N Z Med J 2009; 122: 102–104.
69. Hawkins DW, Ludden TM, Hoffman SF, Allen JP. Rapid metabolism of phenytoin: a method of calculating phenytoin dosage. Arch Neurol 1979; 36: 109–110.
70. McCluggage LK, Volis SA, Bullock MR. Phenytoin toxicity due to genetic polymorphism. Neurocrit Care 2009; 10: 222–224.

71. Soskin DP, Kane AJ, Stern TA. Phenytoin toxicity secondary to an oxcarbazepine-phenytoin 2C19 interaction. Psychosomatics 2010; 51: 532–535.
72. Xiong GL, Ferrranti J, Leamon MH. Toxic interaction between valproate and oxcarbazepine: a case detected by the free valproate level [letter to the editor]. J Clin Psychopharmacol 2008; 28: 472–473.
73. Bates DE, Herman RJ. Carbamazepine toxicity induced by lopinavir/ritonavir and nelfinavir. Ann Pharmacother 2006; 40: 1190–1194.

Chapter 14

Interference in Immunoassays Used to Monitor Tricyclic Antidepressants

14.1. INTRODUCTION

Tricyclic antidepressants (TCAs) are a class of drugs that were first introduced in the 1950s for the treatment of major depression and were named after their chemical structures. This class of compounds is heterocyclic containing three cyclic rings (two six carbon containing aromatic rings and one seven carbon containing aromatic ring) as the integral part of the structure. Chemical structures of common TCAs are given in Figure 14.1. The tetracyclic antidepressants have four rings in their structure and are closely related to TCAs. TCAs can be further classified as primary and secondary amines and due to their narrow therapeutic range require routine therapeutic drug monitoring. TCAs currently used in the United States along with their suggested therapeutic ranges are listed in Table 14.1. Because tertiary amines are metabolized to secondary amines that are also biologically active, for therapeutic drug monitoring of tertiary amines, concentrations of their active metabolites must also be included.

TCAs were the major antidepressants available in the United States until the 1980s for treating clinical depression. However, therapeutic use of TCAs

Resolving Erroneous Reports in Toxicology and Therapeutic Drug Monitoring: A Comprehensive Guide, First Edition. Amitava Dasgupta.

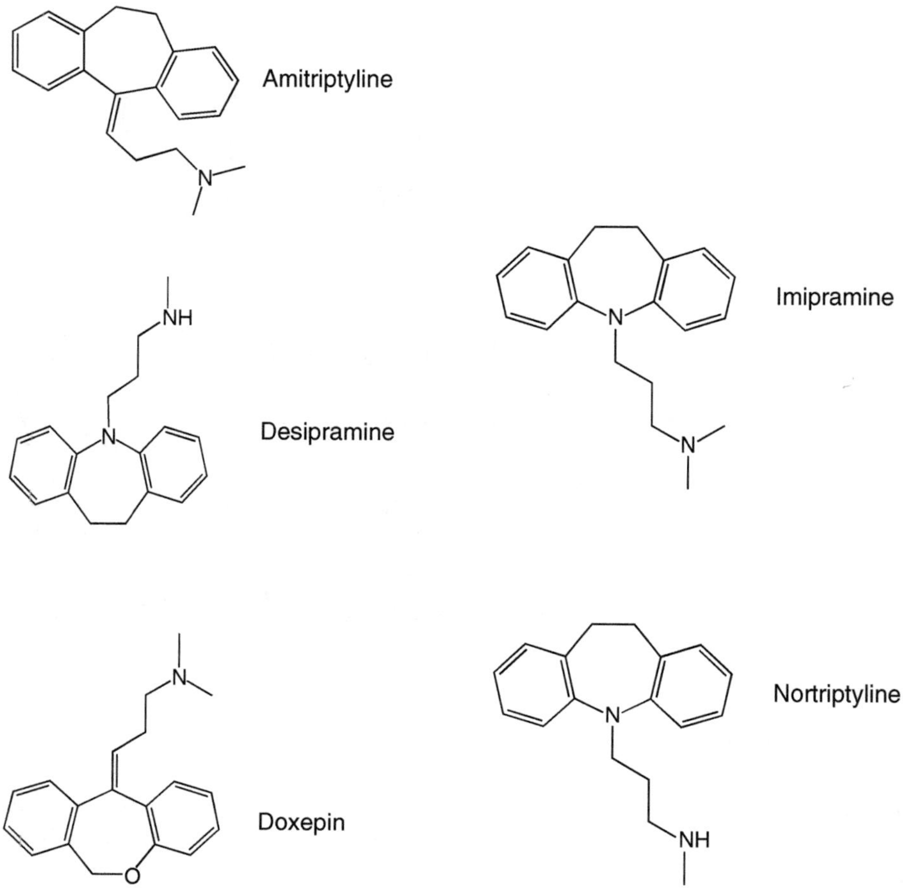

Figure 14.1. Chemical structures of common TCAs.

TABLE 14.1. Suggested Therapeutic Range for Common Tricyclic Antidepressants

Drug Class/Drug	Active Metabolite	Suggested Therapeutic Range, ng/mL
Tertiary Amines		
Amitriptyline	Nortriptyline	Amitriptyline + nortriptyline: 150–250
Imipramine	Desipramine	Imipramine + desipramine: 200–300
Doxepin	Desmethyldoxepin	Doxepin + desmethyldoxepin: 100–300
Trimipramine		Trimipramine: 59–205
Clomipramine	Norclomipramine	Clomipramine + nor clomipramine: 220–500
Secondary Amines		
Desipramine	2-Hydroxydesipramine	Desipramine: 150–300
Nortriptyline	10-Hydroxynortriptyline	Nortriptyline: 50–150
Protriptyline		Protriptyline: 70–240

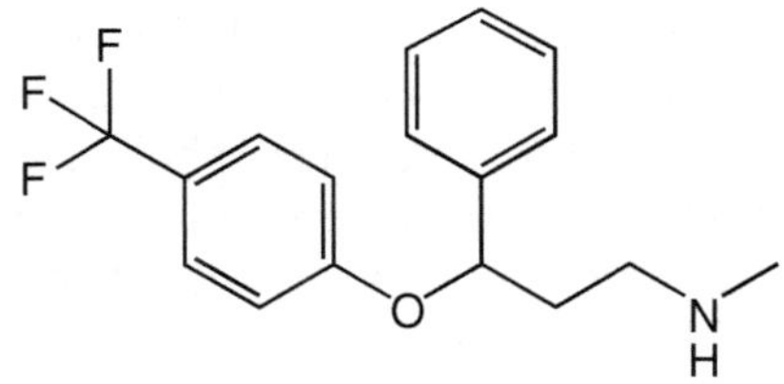

Fluoxetine

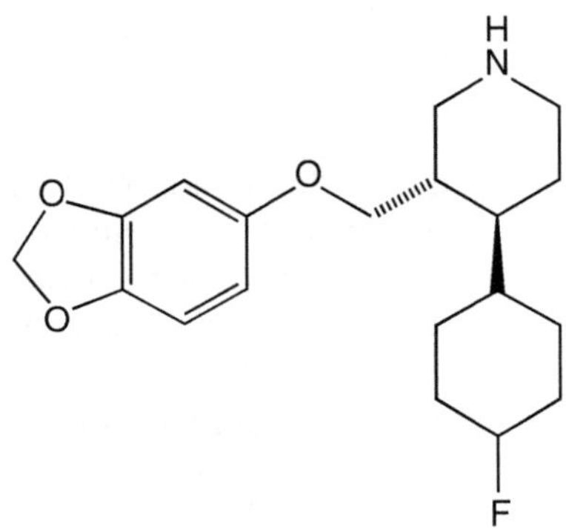

Paroxetine

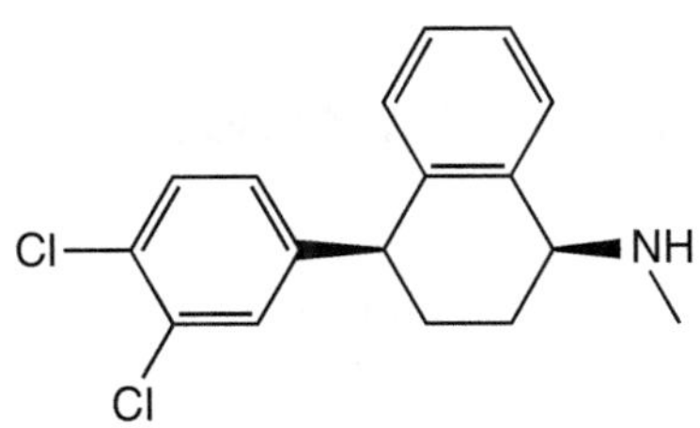

Sertraline

Figure 14.2. Chemical structures of some SSRIs.

for the treatment of anxiety disorders and major depression has decreased significantly with the introduction of other newer classes of antidepressants such as selective serotonin reuptake inhibitors (SSRIs) and serotonin-norepinephrine reuptake inhibitors (SNRIs). Chemical structures of common SSRIs are given in Figure 14.2. Nevertheless, TCAs are still used today for treating depression that fails to respond to the newer antidepressants. TCAs are not addictive and are superior to monoamine oxidase inhibitors for treating depression (1). Kok et al, based on data from eight double-blind clinical trials involving 925 elderly patients, concluded that efficacy and tolerability during the long-term treatment of depression in the elderly does not differ between TCAs and SSRIs (2).

Anxiety disorders are the most common mental health disorder. Although SSRIs due to their safety and greater tolerability are more commonly used today for treating anxiety disorders, both SSRIs and TCAs are effective in treating various anxiety disorders including panic disorders, obsessive compulsive disorder, posttraumatic stress disorder, social phobia, and generalized anxiety disorder (3). Interestingly, TCAs such as imipramine and amitriptyline with dual serotonergic-noradrenergic effects are effective across various anxiety disorders, whereas SSRIs are particularly effective in panic disorders and obsessive compulsive disorders (4). Another important therapeutic use of TCAs is the treatment of chronic pain (especially amitriptyline) (5). Jackson et al, based on a meta-analysis of 37 studies, concluded that TCAs are more effective than SSRIs in preventing migraine and tension-type headaches, although TCAs have a greater adverse effect including dry mouth, drowsiness, and weight gain (6).

One of the major limitations of treating depression with antidepressants is that there is a time lag between the initiation of therapy and a clinical response. Another limitation is that not all depressed patients respond to therapy. However, an early onset of efficacy is an indicator of success of therapy and better outcome. In one study, the authors observed that on average 70% of patients who were treated with various TCAs and showed improvement of symptoms within 14 days became responders. Moreover, the differences between the placebo and treatment groups reached a maximum point on day 14 (7).

14.2. MECHANISM OF ACTION, PHARMACOLOGY, AND TOXICOLOGY

Although the mechanism of action of TCAs for treatment of major depression has not been completely elucidated, it has been recognized that the therapeutic action is attributed to the inhibition of norepinephrine and serotonin transporters. This action serves to increase the amount of serotonin and norepinephrine in the synapses of neurons that release these neurotransmitters, thus enhancing neurotransmission (8). Interestingly, TCAs have a negligible affinity for dopamine transporters. However, serotonin and norephedrine play important roles in the pathophysiology of depression and anxiety. Therefore, facilitation of activities of these neurotransmitters is highly effective in treating depression and anxiety (9). Other actions of TCAs include inhibition of muscarinic acetylcholine receptors (anticholinergic effect) and α-adrenergic receptors (antiadrenergic effect) that are more or less contribute to the unwanted side effects of TCAs.

TCAs after oral administration are readily absorbed from the gastrointestinal tract and are subjected to first-pass metabolism. In the fasting state, absorption is almost complete, and peak plasma concentration can be achieved between 2 and 6 hours. However due to first pass, metabolism bioavailabilities of TCAs are between 30% and 70%. Usually after ingestion of the same amounts, doxepin produces the lowest plasma drug level; protriptyline due to a relatively low first-pass metabolism compared with other TCAs produces the highest plasma drug level (10,11). TCAs are metabolized by liver enzymes to more polar metabolites, and some of the metabolites are also active, for example a tertiary amine such as amitriptyline is metabolized by the liver to secondary amine metabolite nortriptyline, which is also pharmacologically active. The half-life of a TCA ranges from 24 to 76 hours at the therapeutic concentration but may be prolonged significantly in the case of an overdose (12). TCAs are mainly excreted renally. Volume of distribution of TCAs is large (10–20 L/kg) due to the lipophilic nature of these drugs (13).

Although the therapeutic effectiveness of TCAs in treating major depression is well established, the occurrence of unpleasant side effects due to anticholinergic effects (e.g., dry mouth, sedation, blurred vision or light sensitivity from mydriasis, urinary retention) or antiadrenergic effects (e.g., orthostatic hypotension) limits therapy with TCAs, especially with the availability of newer antidepressants that have substantially fewer side effects. In TCA overdose, particularly with amitriptyline, the anticholinergic symptoms can resemble severe atropine poisoning with dilated pupils, flushed skin, hyperthermia, tachycardia, and cardiac arrhythmias. Cardiac toxicity of TCAs is responsible for mortality from these agents. Cardiovascular toxicity is manifested by electrocardiogram (ECG) abnormalities, arrhythmia, and hypotension. Again, amitriptyline appears to be particularly toxic in overdose situations. ECG changes include prolongation of QRS, QT, and PR intervals along with nonspecific ST segment and T wave change. In addition, atrioventricular block, right axis deviation of the terminal 40 ms vector of the QRS complex in the frontal plane (T 40 ms axis) and the Brugada pattern (downsloping ST segment elevation in leads V1–V3 in association with right bundle branch block) may also be present in TCA overdose. Sinus tachycardia is the most commonly encountered arrhythmia associated with TCA overdose due to anticholinergic activity and the inhibition of norephedrine uptake by TCAs. Although bradyarrhythmia and tachyarrhythmia may also be encountered in TCA overdose, torsade de pointes occurs only rarely. Hypotension from TCAs is related to a combination of reduced myocardial contractility and decreased systematic vascular resistance related to α-adrenergic blockage by TCAs. Life-threatening arrhythmia and death usually occur within 24 hours of TCA overdose and QRS duration over 100 msec appears to be a good predictor of cardiovascular toxicity compared with serum TCA concentration (14). For treatment of a TCA overdose, 30–50 g of activated charcoal is administered orally or through a nasogastric tube. Sodium bicarbonate is indicated if QRS prolongation is over 100 msec or terminal right axis deviation is more than 120 degrees. Hypertonic saline appears to be an alternative to sodium bicarbonate therapy in treating severe TCA overdose (15).

Overdose of a several weeks' supply of TCAs can lead to life-threatening toxicity, with cardiac arrhythmias especially difficult to manage clinically after ingestion of 1 g of a TCA. Such a massive overdose usually results in a plasma level of the TCA (alone or in addition to its active metabolite) around

1000 ng/mL, which could be fatal; severe toxicity may be encountered with a plasma level above 500 ng/mL for the drug and the active metabolite if applicable. Fujitani et al reported a fatal hyperthermia after a severe amitriptyline overdose. A 49-year-old woman with a history of depression died of hyperthermia secondary to an amitriptyline overdose. Her rectal temperature was 41.5°C approximately 3 hours after death, and her postmortem plasma level of amitriptyline was 510 ng/mL. The concentration of active metabolite nortriptyline was 740 ng/mL (16). Chaturvedi et al described postmortem levels of amitriptyline, nortriptyline, and desipramine in two fatal overdose cases. The first case was a 20-year-old man who was found dead in his residence. His postmortem blood level of amitriptyline was 1700 ng/mL, and the concentration of active metabolite nortriptyline was 660 ng/mL. Therefore, the combined amitriptyline and nortriptyline concentration was 2300 ng/mL. The second case involved the suicide of a 19-year-old man receiving desipramine. His desipramine blood level was 1420 ng/mL. The concentrations of TCAs in blood and other biological fluids were measured by the high performance liquid chromatographic (HPLC) method (17). Death may occur with lower amounts of amitriptyline ingested if the victim also consumed alcohol (18).

Case Study

A 44-year-old woman with a history of depression was found by her husband lying in bed while several empty pill boxes were found around her. Most remarkably were 10 empty boxes of amitriptyline with 420 pills missing and one box of diazepam with 7 pills missing. She was pronounced dead on arrival at the hospital. There was no previous history of a suicide attempt. Very high amounts of amitriptyline (85,900 ng/mL in femoral blood and 379,600 ng/mL in gastric content) and nortriptyline (580 ng/mL in femoral blood and none detected in gastric content) were detected in the postmortem analysis of body fluids. Usually a combined concentration of amitriptyline and nortriptyline more than 1000 ng/mL is considered very toxic, and the victim's death was determined to be caused by severe amitriptyline toxicity (19).

14.3. THE PHARMACOGENOMICS ASPECT OF TRICYCLIC ANTIDEPRESSANT METABOLISM

Plasma concentrations and response to TCAs vary considerably among patients treated with similar dosages due to the polymorphism of cytochrome P450 isoenzymes responsible for the metabolism of TCAs. These drugs are hydroxylated by CYP2D6, but tertiary amines such as amitriptyline, clomipramine, and imipramine are also N-demethylated to the corresponding active metabolites nortriptyline, N-desmethylclomipramine, and desipramine by CYP2C19, CYP1A2, and CYP3A4 (20). Approximately 7% of whites are poor metabolizers due to lower activity of CYP2D6 and are more susceptible to toxicities from therapy with TCAs, whereas ultrarapid metabolizers with multiple CYP2D6 genes might require a higher dosage of TCAs for the optimal clinical response. The activity of CYP2D6 is in general lower in Asian population than whites due to a frequent mutation causing decreased enzyme activities. Therefore, Asians are at a higher risk of toxicity from TCA therapy (21). It has been documented that CYP2C9 poor metabolizers convert amitriptyline slowly, potentially leading to excessively high amitriptyline concentrations and making them prone to adverse drug effects. Nortriptyline is converted to 10-hydroxynortriptyline by CYP2D6, an enzyme that also shows significant genetic variation. Like CYP2C9, CYP2D6-poor metabolizers of nortriptyline exist and are prone to toxic side effects when administered standard doses of nortriptyline (22,23). Jiang et al studied the role of CYP2C19 polymorphism in N-demethylation of amitriptyline in Chinese subjects and observed that the mean area under the curve (AUC) of CYP2C19 poor metabolizers was significantly higher than CYP2C19 extensive metabolizers. However, no difference was observed between poor metabolizers and extensive metabolizers in other pharmacokinetic parameters such as maximum plasma concentration and half-life. The authors concluded that the genetic defect of CYP2C19 in the Chinese population has a significant effect on amitriptyline pharmacokinetics due to the major role played by CYP2C19 in N-demethylation of amitriptyline at a clinically therapeutic dose (24). In another study, Koshi et al noted a very high amitriptyline level of 60 mg/L (60,000 ng/mL) in one suicide victim who had no functional CYP2D6 gene (25).

The gene encoding CYP2D6 can be also be duplicated or even multiplied, leading to more than two functional copies of the gene and high CYP2D6 activity relative to the average population. These CYP2D6 ultrarapid metabolizers have been observed to metabolize nortriptyline much more rapidly than the average population. In the most extreme case reported, a subject found to have 13 functional copies of the gene for CYP2D6 demonstrated a very short half-life (19 hours compared with 54 hours in poor metabolizer) of nortriptyline (26).

14.4. THERAPEUTIC DRUG MONITORING OF TRICYCLIC ANTIDEPRESSANTS

Therapeutic drug monitoring (TDM) of TCAs has two main possible advantages. First, for some TCAs, most notably nortriptyline, therapeutic serum concentrations have been well defined (Table 14.1). Second, possible benefits of TDM are that TCA serum concentration may be affected by genetic variation of enzymes that metabolize TCAs or by concomitant medications, herbal products, or foods that alter TCA metabolism, or by interference with transporter-mediated efflux. Such effects can be detected by unexpected variation in TCA levels in blood. However, for the TDM of TCAs, it is important to monitor levels of active metabolites along with the parent drug. For example, amitriptyline is converted to nortriptyline and imipramine is converted to desipramine. It is important to recognize that for tertiary amines, therapeutic ranges are based on the combined concentrations of the parent drug and the active metabolite.

Muller et al studied the TDM of TCAs under clinical conditions by randomly assigning patients in a TDM group and a non-TDM group while receiving TCAs. Although serum levels of TCAs were determined in both cohorts, feedback and dose recommendations were only provided to the TDM group. Serum levels of TCAs were determined by a HPLC method. The study duration was 3 weeks plus or minus 8 days, and the outcome of therapy was determined by the Hamilton Depression (HAMD) Scale and other methods. Although both groups showed improvements (decreased HAMD scores), therapeutic levels of TCA in weeks 1–3 in the TDM group were associated with a significantly better outcome at the end of the study period. Moreover, considerable side effects were encountered where serum concentrations of TCAs were above therapeutic levels. The authors concluded that treating depression with TCAs can be optimized using TDM, which is superior to clinical judgment alone (27). Preskorn commented that the TDM of TCAs is a cost-effective tool that can significantly improve both the clinical outcome and safety of pharmacotherapy with TCAs. TDM is effective in improving efficacy of TCAs while avoiding cardiac and brain toxicity (28). Nelson et al determined the relationship between desipramine plasma concentration and antidepressant response in 30 depressed inpatients and observed that 89% of patients with plasma desipramine concentration above 115 ng/mL responded to the therapy. In addition, 10 nonresponder were converted into responders when the dosage of desipramine was increased to elevate plasma desipramine levels to 125 ng/mL or above (29).

Therapeutic drug monitoring can also identify genetic variations in TCA metabolism, and dose adjustment can be easily done based on the serum level. Therefore, TDM of TCA is a phenotype approach in identifying genetic variations in drug metabolism. The usual dose of amitriptyline or nortriptyline is between 75 mg and 200 mg daily. A 61-year-old woman with depression was treated with 150 mg of nortriptyline daily for 8 weeks by her general practice physician, but despite this dose her serum nortriptyline level was only 100 nmol/L (26.3 ng/mL), which was substantially below the therapeutic range (225–600 nmol/L, of 59–158 ng/mL, the reference cited by the authors). Based on the sparteine test, it was determined that the patient was an ultrarapid metabolizer. Her nortriptyline dosage was increased to 250 mg per day and her serum nortriptyline level was increased to therapeutic level (528 nmol/L; 139 ng/mL) (30). A 42-year-old woman with depression was treated with 100 mg of nortriptyline but developed a toxic serum level of 2100 nmol/L (552.6 ng/mL). Nortriptyline was withdrawn due to severe toxicity, but she experienced depression during her medication-free period. The sparteine test indicated that she was a poor metabolizer. She was started on a low dose of nortriptyline (25 mg/day), and her depression resolved completely without any side effects (31). These case reports indicate the value of TDM in identifying genetic variations among patients in metabolizing nortriptyline and the adjustment of the dosage-based serum drug level to improve the outcome of therapy.

In the United States from 1970 through 2005, approximately two to four TCAs have been among the top 200 most prescribed medications. The last 10 years, however, have seen markedly steady declines in prescriptions for TCAs. The exception to this trend is amitriptyline, which continues to be among the top 60 most prescribed medications, probably due its effectiveness as an adjunct medication in treating chronic pain (5). Therefore, TDM of TCAs continues to be an important aspect of TDM service, and most hospital laboratories are capable of providing TDM of TCAs, at least screening of TCAs in serum and urine to identify an overdose involving TCAs. Suicide attempts involving antidepressants are common in the United States, and between 2000 and 2004, 82,803 suicidal single-agent ingestions of identifiable antidepressants cases were reported to the major U.S poison control centers. Overdoses occur more predominantly in females than males, and there were 40 major or fatal outcomes per 1000 cases. Overdoses with TCAs occur more frequently in the 30s; overdoses with SSRIs peak in the teens. In addition, TCAs had higher hazard ratios than SSRIs (32). Response to antidepressants and the occurrence of adverse drug events are marked by interindividual variability that may be partly due to genetic differences (33).

Although therapeutic drug monitoring of TCAs can be performed using immunoassays or chromatographic methods, chromatographic methods are recommended for routine TDM of TCAs because immunoassays do not determine how much parent drug versus metabolite is present in the serum. In contrast, concentrations of individual TCAs and their metabolites in serum/plasma can be more specifically determined by a variety of chromatographic techniques including gas chromatography (GC) and HPLC, either alone or in combination with mass spectrometry.

Immunoassays for TCAs were first developed as toxicology screening tools starting with radioimmunoassays in the 1970s and 1980s. Homogeneous immunoassays for TCAs in serum were later developed, and today multiple commercial assays are available on the market in the United States. Homogeneous immunoassay techniques for measuring TCA serum concentrations include the fluorescence polarization immunoassay (FPIA; Abbott Diagnostics, Abbott Park, IL) and enzyme-multiplied immunoassay techniques (EMIT; Dade-Behring, now Siemens Diagnostics, Deerfield, IL). Immunoassays for TCAs generate either qualitative results (positive or negative) or semiquantitative results (34–37). Qualitative assays for TCAs have also been developed for urine (38).

14.5. INTERFERENCE IN IMMUNOASSAYS FOR TRICYCLIC ANTIDEPRESSANTS

TCAs share a structural similarity with a number of other drugs, and, as expected, these drugs may interfere with serum or urine level measurement

TABLE 14.2. Common Interferences in Immunoassays for TCAs

Interfering Substance	Comments
Metabolites of TCAs	TCAs, which are tertiary amines, are metabolized to active secondary amines that have almost 100% cross-reactivity with various immunoassays.
Phenothiazines	Bothe parent drugs and their metabolites may interfere with various immunoassays. Thioridazine may even interfere with some HPLC assays.
Carbamazepine	Carbamazepine overdose may be falsely identified due to TCAs because of interference of carbamazepine with immunoassays. The active metabolite of carbamazepine, carbamazepine 10, 11-epoxide, has lower cross-reactivity with immunoassays for TCAs compared with carbamazepine.
Cyproheptadine	The metabolite, not the parent drug, usually interferes with immunoassays for TCAs.
Quetiapine	Interference is dose dependent.
Hydroxyzine	Interference in immunoassays for TCAs may be observed after a severe overdose, but no interference is expected after a therapeutic dosage.
Cetirizine	Interference in immunoassays for TCAs may be observed after a severe overdose, but no interference is expected after a therapeutic dosage.
Diphenhydramine	A common antihistamine not structurally related to TCAs may interfere with immunoassays with TCA-only patients overdosed with diphenhydramine.
Cyclobenzaprine	Both parent drug and metabolite may interfere with both immunoassays and certain HPLC methods.

TCA, tricyclic antidepressant; HPLC, high performance liquid chromatography.

of TCAs using various immunoassays. Cyclobenzaprine is used as a skeletal muscle relaxant and structurally related to amitriptyline differing by only one double bond. Phenothiazine antipsychotics, carbamazepine, and oxcarbazepine are examples of three-ringed molecules that also share structural similarity to TCAs. In addition, the class of tetracyclic antidepressants including amoxapine, maprotiline, mianserin, and mirtazapine also has structural similarities with TCAs. Metabolites of TCAs also interfere with immunoassays (Table 14.2). Chemical structures of common drugs that interfere with immunoassays for TCAs are given in Figure 14.3.

14.5.1. Limitations of Immunoassays for Detecting TCAs in Serum and Urine

In general, a tertiary amine TCA is metabolized to a secondary amine, and the metabolite usually has almost 100% cross-reactivity with an antibody used for an immunoassay for TCAs because immunoassays for TCAs in serum usually utilize antibodies that raised imipramine or desipramine as the hapten (35). Therefore, for monitoring tertiary amines, immunoassays in general indicate

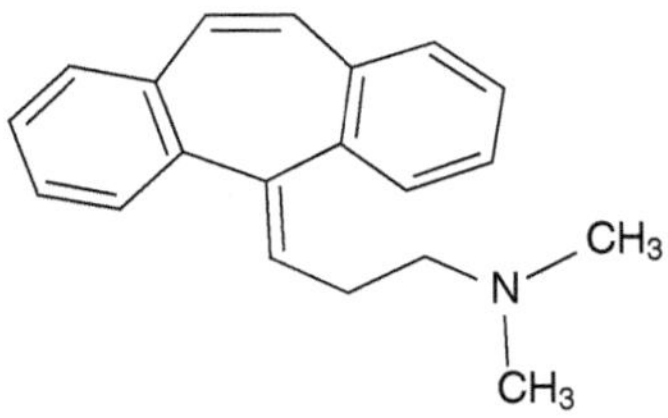

Cyclobenzaprine

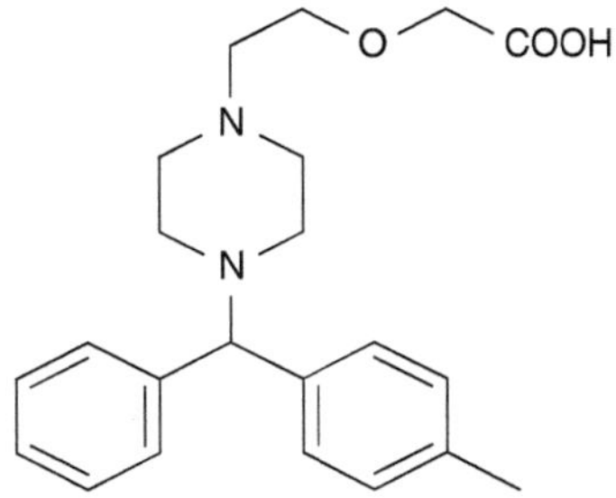

Cetirizine

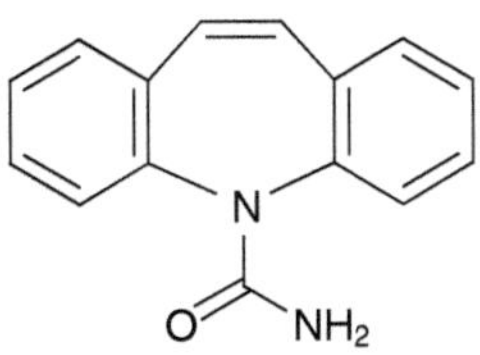

Carbamazepine

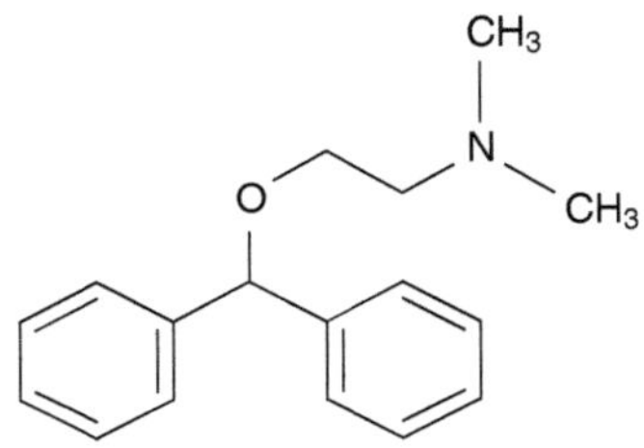

Diphenhydramine

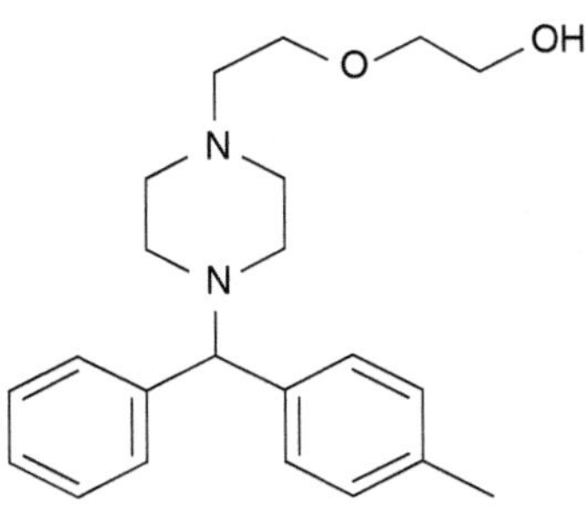

Hydroxyzine

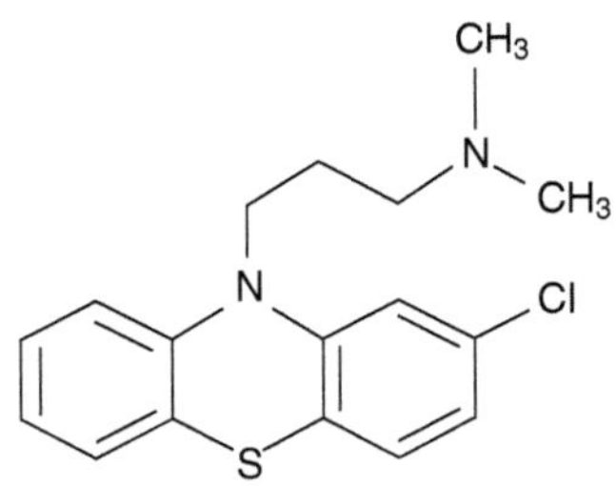

Chlorpromazine (Phenothiazine class)

Figure 14.3. Structures of common drugs that interfere with immunoassays for TCAs.

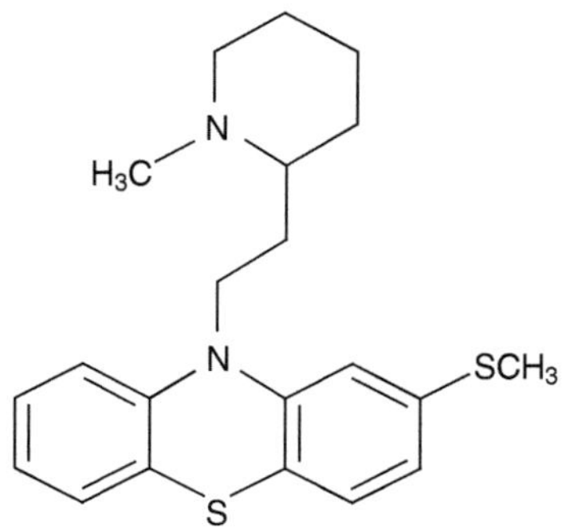

Thioridazine

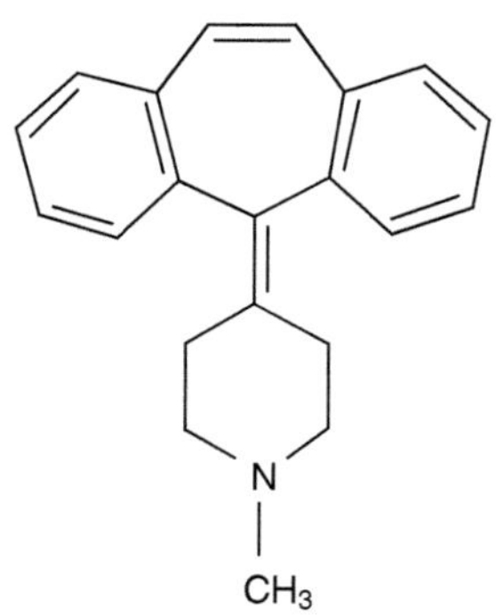

Cyproheptadine

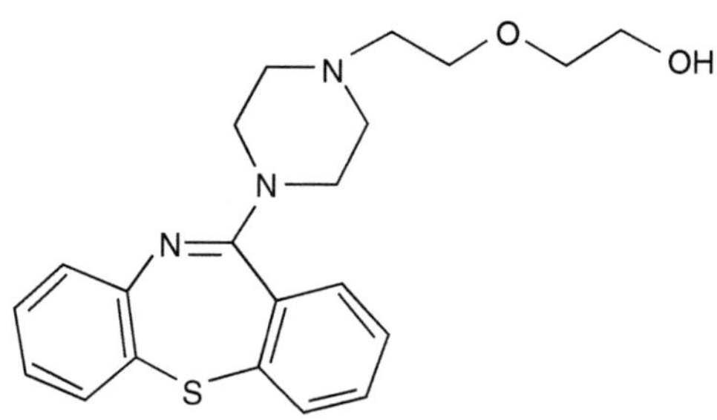

Quetiapine

Figure 14.3. (*Continued*)

total concentration of the parent drug along with the active metabolite. Poklis et al evaluated the Abbott ADX total serum tricyclic immunoassay that provides semiquantitation of imipramine or amitriptyline and respective N-demethylated active metabolites in cases of overdose and observed good correlation between results obtained by the ADX total serum tricyclic immunoassay assay and GC liquid chromatography (LC) when the patient's sera contained imipramine and desipramine. However, correlation between the reference GC method and ADX total serum tricyclic immunoassay was poor when the patient's sera contained amitriptyline and nortriptyline or doxepin and desmethyldoxepine. The ADx assay flagged "HI" serum TCAS when total imipramine and desipramine concentrations exceeded 300 ng/mL by the GC method but only flagged when serum amitriptyline and nortriptyline concentrations exceeded 360 ng/mL by the GC method. The ADX total serum tricyclic immunoassay is designed to flag all serum TCAs specimens if the total value exceeds 300 ng/mL (39). Therefore, manufacturers of such immunoassays recommend the use of these assays for screening in case of a suspected overdose involving TCAs. Such assays should not be used for routine therapeutic drug monitoring of TCAs. Although the manufacturer of the FPIA for TCAs (Abbott Laboratories) only recommends its use for the diagnosis of overdose, Hackett et al commented that this assay is widely used for therapeutic drug monitoring of TCAs. Comparing results obtained with this assay and HPLC, the reference method, the authors concluded that although the FPIA assay may be satisfactory for routine monitoring of desipramine, imipramine, nortriptyline, and trimipramine, this assay significantly overestimates the concentrations of amitriptyline, clomipramine, dothiepin, and doxepin. Moreover, there were significant discordance between values obtained by the HPLC reference method and FPIA, which may confuse physicians and cause errors in dose adjustments of TCAs. The authors concluded that they concur with the manufacturer's recommendation of using this assay only for the diagnosis of overdose (40).

Urine screening for the presence of TCAs using immunoassays are also available. Melanson et al compared serum TCAs using a reference HPLC method and urine TCAs qualitative results obtained by using Syva rapid Test and Biosite Triage method and observed that serum concentrations of amitriptyline, desipramine, doxepin, imipramine and nortriptyline ranging from sub-therapeutic to toxic triggered a positive response to both urine immunoassays for TCAs but neither immunoassay detected clomipramine even levels greater than therapeutic range. False-positive results were more common with Biosite assays if cyclobenzaprine was present in the urine. For virtually all positive urinary TCAs, it was not possible to identify whether the patient had sub-therapeutic, therapeutic or toxic serum TCAs level (41).

14.5.2. Interference of Phenothiazine and Metabolites

Schroeder et al evaluated the EMIT TCA assay in serum for an application in identifying overdose using 87 patients. The authors concluded that EMIT correctly identified 53 negative patients whose TCAs levels were less than 300 ng/mL, but of 34 remaining patients who showed TCAs concentrations higher than 300 ng/mL, only 22 patients had confirmed TCA levels over 300 ng/mL using the reference HPLC method. Phenothiazine and phenothiazine metabolites were present in the remaining 12 unconfirmed specimens. The authors concluded that the EMIT assay is useful to exclude the presence of TCAs overdose, but results over 300 ng/mL may not specifically confirm the presence of TCAs (42). Ryder and Glick reported a case in which a patient who ingested thioridazine and flurazepam tested positive for TCAs by immunoassays. Investigation showed that the false-positive TCA result was due to thioridazine, which was present in a therapeutic concentration of 125 ng/mL (43). It is important to note that even therapeutic concentrations of thioridazine (125 ng/mL) may cause interference. In another report, the authors observed false elevations of imipramine and desipramine in a HPLC method using a cyanopropyl column and the interference was caused by thioridazine (44). Wertz and Street also reported interference of thioridazine in a chromatographic method for the analysis of TCAs using a normal phase silica column. The authors tried to resolve the problem of coelution of imipramine and thioridazine by using a cyano column, but the attempt to circumvent coelution failed (45).

14.5.3. Interference of Carbamazepine and Its Epoxide Metabolite with Immunoassays for TCAs

Carbamazepine is metabolized to carbamazepine 10, 11-epoxide, an active metabolite. Both the parent drug and the epoxide metabolite interfere with immunoassays for TCAs due to structural similarities. Another structurally related drug, oxcarbazepine, also interferes with immunoassays for TCAs. The FPIA for TCAs for application on the AxSYM analyzer demonstrated significant cross-reactivities with both carbamazepine and its epoxide metabolite. In 30 patients receiving carbamazepine but no TCAs, the apparent TCAs ranged from 31.8 to 130 ng/mL (carbamazepine level ranged from 1.4 to 20.9 μg/mL), indicating that carbamazepine may falsely show the presence of TCAs in a patient never exposed to it. Carbamazepine demonstrated much higher cross-reactivity with the FPIA assay for TCAs compared with its active epoxide metabolite. When aliquots of a drug-free serum pool were supplemented with various concentrations of carbamazepine or its epoxide, we observed that the interference of carbamazepine is roughly threefold more than its epoxide metabolite based on observed apparent levels of TCAs. However, TCAs do not cross-react with the carbamazepine immunoassay on the AxSYM analyzer (Abbott Laboratories). Therefore, true concentration of TCAs can be calculated in the presence of carbamazepine by knowing the true carbamazepine concentration using a carbamazepine immunoassay. This is possible because the interference of epoxide in the FPIA assay for TCAs is low and can be ignored. Moreover, carbamazepine demonstrated a near linear response over a wide range of concentrations in the FPIA assay for TCAs (46).

$$\text{True TCAs concentration (ng/mL)} = \text{Observed TCAs concentration by FPIA} - \text{contribution of carbamazepine}$$

$$\text{Carbamazepine contribution (ng/mL)} = 4.87 \times \text{Carbamazepine in } \mu\text{g/mL} + 28.27$$

Alternatively,

$$\text{True TCAs concentration (nmol/L)} = \text{Observed TCAs concentration by FPIA} - \text{contribution of carbamazepine}$$

$$\text{Carbamazepine contribution (nmol/L)} = 4.12 \times \text{Carbamazepine in } \mu\text{mol/L} + 100.8$$

Case Study

A 16-year-old girl with a remote history of seizure and a tic disorder was comatose and admitted to the hospital. She was last seen awake at 11 P.M. the night before and complained about abdominal discomfort for which she was given Pepto-Bismol, ibuprofen, and St. John's wort. She was also receiving her regular dose of clonidine (0.1 mg) for her tic disorder. Her family denied that she was exposed to any illicit drug, alcohol, or any other medication. She was found unresponsive next morning when she was transferred to the hospital. On admission to the hospital, the child was found unresponsive with a Glasgow coma scale score of 7. However, there was no sign of trauma. Clonidine overdose was ruled out because the bottle contained the appropriate amount of remaining drug. However, the laboratory result indicated that the patient was positive for serum TCAs, although there was no evidence of QRS prolongation or axis deviation in a repeat electrocardiogram. On recommendation from the toxicology service, a serum carbamazepine level was ordered that was found to be 17.0 µg/mL (therapeutic range: 4–12 µg/mL), which was in the toxic range. Therefore, her false TCAs in serum were due to a carbamazepine overdose. By day 3 her carbamazepine level had dropped to 9 µg/mL and she was transferred to the medical floor and discharged subsequently with outpatient psychiatry follow-up (47).

Chattergoon et al reported that two patients with a history of ingestion of carbamazepine showed positive urinary screen for TCAs using an immunoassay. However, a confirmatory HLPC assay showed a negative result. The authors concluded that interference of carbamazepine with immunoassays for TCAs is a significant clinical problem, and a specific HPLC method must be used for the screening of TCAs in urine specimens (48). Saidinejad et al studied apparent TCAs levels in serum and urine specimens of 52 patients between 3 and 19 years of age who were prescribed either carbamazepine (33 patients) or oxcarbazepine (19 patients). The authors used FPIA TCAs assay for the determination of serum TCAs (Abbott Laboratories) levels using a cut-off of 50 ng/mL for a positive specimen and an enzyme-linked immunoassay ([EIA]; Microgenic, Fremont, CA) for TCAs for screening urine specimens (cut-off concentration: 150 ng/mL). Of 33 patients receiving carbamazepine, 13 (39.4%) tested positive for serum TCAs using the FPIA assay. However, none of the 19 patients receiving oxcarbazepine demonstrated a positive serum screen for TCAs. Of 52 patients, 43 patients were able to provide urine specimens, but none of these specimens tested positive for TCAs. Further analysis of results showed that positive screening for TCAs in serum using the FPIA assay invariably occurred at a serum carbamazepine concentration at and above 8 μg/mL because 12 of 13 patients (range: 8.0–11.6 μg/mL) with positive screen had levels about 8 μg/mL and one patient had a carbamazepine level below that suggested cut-off. Interestingly, using a linear regression model, the authors demonstrated that 4.2 ng/mL of TCAs is detected per microgram/milliliter of carbamazepine concentration. Oxcarbazepine showed a much lower cross-reactivity with the FPIA assays for TCAs in serum with only 0.7 ng/mL of TCAs detected per microgram of oxcarbazepine metabolite 10-monohydroxy carbamazepine (49).

14.5.4. Interference of Cyproheptadine with Immunoassays for TCAs

Cyproheptadine is an antihistamine used for treating various allergic symptoms. This drug is also effective in treating moderate to severe serotonin syndrome caused by an overdose of SSRIs and in carcinoid cancer where tumor cells produce excess serotonin. Wians et al reported a case of a 14-year-old girl who ingested approximately 120 mg of cyproheptadine. The patient tested positive for TCAs by EMIT assay (50). In vitro studies indicated that a cyproheptadine concentration of 400 ng/mL may cause false-positive TCAs results, although serum obtained from a volunteer who was given a 12-mg dose of cyproheptadine for 3 days showed no TCAs when tested by the same EMIT assay (50).

Case Study

A 5-year-old child was taken to the emergency department due to unusual behavior after awakening from a nap. The grandmother reported that she awoke from the nap crying, had slurred speech, and appeared to be warm. The child was in a good health before this episode, and the family denied any history of ingestion. In the household, the only medications available were the grandmother's nitroglycerine and over-the-counter cold medications. On examination, it was observed that the child was agitated and her toxicology screen showed positive for TCAs. The family was advised to bring all household medications, which cyproheptadine tablets, hydroxyzine syrup, carbinoxamine syrup, nitroglycerine tablets, and multivitamin tablets. Her aunt reported that the child was playing with some tablets on the floor and later those pills were identified as cyproheptadine using GC/mass spectrometric analysis (MS). When the child's serum was further analyzed by a reference HPLC method, no peak was observed for nordoxepin, doxepin, desipramine, imipramine, nortriptyline, and amitriptyline, but a peak was observed with the same retention time as the cyproheptadine standard, indicating that the false-positive serum TCAs level detected by the EMIT assay was indeed due to interference of the cyproheptadine metabolite because when a drug-free serum pool was supplemented with various amounts of cyproheptadine, no positive response was observed using the EMIT TCA assay (51).

14.5.5. Interference of Quetiapine with Immunoassays for TCAs

Quetiapine, an atypical antipsychotic used for treating schizophrenia, acute episodes of bipolar disorders, and also an augmentor for the treatment of depression cross-reacts with the plasma TCA immunoassay, and the false-positive response depends on the quetiapine concentration. When aliquots of a drug-free serum pool were supplemented with a quetiapine stock solution to achieve a concentration range of 1–540 ng/mL followed by analysis for an apparent tricyclic level using Abbott's FPIA assay for TCAs, Syva EMIT assay, and S TAD serum TCAs screen on the ACA-Star 300 autoanalyzer, the FPIA quantitative assay demonstrated apparent TCA levels starting at a quetiapine concentration of 5 ng/mL, and the 640 ng/mL spiked specimens produced apparent TCAs of 379 ng/mL and 385 ng/mL as measured in two laboratories using the same method. The two other assays that are qualitative produced positive TCAs at a level of 160 ng/mL of quetiapine (S TAD) and 320 ng/mL (Syva EMIT) (52).

Case Study

A 34-year-old man with a history of refractory-schizoaffective disorder and amphetamine dependence was admitted to the hospital for his depressive and psychotic symptoms. His only prescribed medication during admission was quetiapine 600 mg per day. On admission his urine toxicology screen showed positive for TCAs (at a level over the cut-off: 300 ng/mL) but negative for other illicit drugs including amphetamines, barbiturates, benzodiazepines, cocaine metabolite, opiates, methadone, and alcohol. The patient denied any use of TCAs but because that class could contribute to his present symptoms, further investigations were made. Because the patient was receiving quetiapine, which is structurally similar to TCAs, a 25-mg tablet of quetiapine was dissolved in water and diluted to concentrations between 1 and 10 μg/mL. The levels of 7 μg/mL and more yielded positive results in the TCAs screen, and the cross-reactivity of quetiapine with this homogeneous enzyme immunoassay for TCAs (Diagnostics Reagents Inc) on the Hitachi 911 analyzer was calculated to be 4.3%. Interestingly, another antipsychotic drug, olanzapine, showed only minimal cross-reactivity (0.06–0.7%) and did not interfere with the immunoassay used for TCAs (53).

Case Study

A 42-year-old woman with a diagnosis of schizoaffective disorder and borderline personality disorder was hospitalized due to the acute exacerbation of schizoaffective disorder along with psychosis, depression, and suicidal tendencies. Quetiapine was added to her medication regime (nortriptyline 25 mg four times a day, levothyroxine 0.1 mg/day, and lithium 300 mg three times a day). Her serum nortriptyline was 34 ng/mL on admission, but it was increased to 487 ng/mL several weeks later without any signs of acute toxicity from the nortriptyline overdose. This level was determined by the FPIA assay for TCAs (Abbott Laboratories) on the TDx analyzer. However, reanalysis of her serum level with a HPLC method in a reference laboratory showed a nortriptyline level of 216 ng/mL. Further analysis of her serum by a chromatographic method demonstrated the presence of quetiapine and its metabolites. The authors concluded that a falsely elevated serum nortriptyline level as determined by the immunoassay for TCAs was due to the interference of quetiapine (54).

Cerullo et al commented that five patients who initially tested negative for TCAs tested positive after being treated with topiramate or quetiapine. A quetiapine trough blood level was obtained as the same time as the urine toxicology screen of TCAs was performed, and the results paralleled positive urinary TCAs screen (when patients showed serum quetiapine levels between 5.1 and 85 ng/mL; urine TCAs was also positive). However, when two patients were noncompliant, the urine TCAs screen was also negative. The authors suggested that urine screen for TCAs may reflect adherence with quetiapine therapy (55).

14.5.6. Interference of Miscellaneous Other Drugs with Immunoassays for TCAs

Hydroxyzine and cetirizine are antihistamines that may interfere with the FPIA assay for TCAs for application on the Abbott AxSYM analyzer. However, neither hydroxyzine nor cetirizine should interfere after recommended low-dose therapy, and such interference is expected with ingestion of a higher dose or an overdose. When aliquots of a drug-free serum pool were supplemented with various amounts of hydroxyzine or cetirizine, no apparent TCAs were observed with hydroxyzine or cetirizine concentration up to 250 ng/mL (therapeutic level after low-dose use: 50–200 ng/mL). However, when an aliquot of the drug-free serum pool was supplemented to achieve a hydroxyzine concentration of 500 ng/mL, the apparent TCA was 21.9 ng/mL, and for the same concentration of cetirizine, the apparent TCAs level was 24.2 ng/mL. In the presence of 40 μg/mL of hydroxyzine, the apparent TCA level was 136.4 ng/mL (56). The expected in vivo concentration of hydroxyzine after a standard dose is less than 100 ng/mL, and toxicity is common at a serum hydroxyzine level of 1 μg/mL. Simons et al reported that the mean serum peak hydroxyzine level was 116.5 ng/mL in healthy volunteers when they ingested a single dose of 0.7 mg/kg of bodyweight (mean dose: 43.9 mg) (57). Johnson reported a case in which a 43-year-old woman who committed suicide with hydroxyzine had a blood level of 39 μg/mL (58). Magera et al published a case of hydroxyzine toxicity following accidental ingestion in a 13-month-old girl whose plasma hydroxyzine concentration reached 102.7 μg/mL after 8.5 hours. The infant recovered in 72 hours (59). Cetirizine is also used as a second-generation antihistamine agent, and a single oral dose of 5 mg in children led to a peak plasma level of 607 ng/mL (60). Cetirizine overdose was also reported in an 18-month-old boy whose serum cetirizine concentration was 2.4 μg/mL 14 hours after an overdose. The boy survived the episode (61).

However, if a patient is not taking any TCAs or other drugs that may cross-react with the FPIA assay for TCAs, it is possible to predict a serum level of hydroxyzine or cetirizine from apparent TCA level.

$$\log \text{TCAS observed (ng/mL)} = 0.4339[\log \text{Hydroxyzine ng/mL}] + 0.1562.$$

$$\text{Therefore } \log \text{Hydroxyzine (ng/mL)} = \frac{\log \text{TCAs observed (ng/mL)} - 0.1562}{0.4339}$$

Similarly, $\log \text{TCAS observed (ng/mL)} = 0.3771[\log \text{Cetirizine ng/mL}] + 0.3693$.

$$\text{Therefore } \log \text{Cetirizine (ng/mL)} = \frac{\log \text{TCAs observed (ng/mL)} - 0.3693}{0.3771}$$

Taking antilog, the concentration of hydroxyzine or cetirizine can be easily determined. Because there is no immunoassay for hydroxyzine or cetirizine, this is an alternative approach for rapidly but indirectly determining the presence of these antihistamines in serum in the case of an overdose (56).

Interference of diphenhydramine, an antihistamine in the EMIT assay for TCAs in serum, is interesting because unlike phenothiazine, diphenhydramine is an ethanolamine. A 21-year-old woman ingested 2 g of diphenhydramine. In vitro testing indicated that when drug-free serum was supplemented with up to 60 ng/mL of diphenhydramine, the response to the TCA screen was negative, and specimens containing 690 ng/mL and higher amounts of diphenhydramine were clearly positive. Usually the ingestion of 100 mg of diphenhydramine leads to a serum diphenhydramine level of approximately 112 ng/mL 2 hours after ingestion. Therefore, a positive screen for TCAs more likely results from a diphenhydramine overdose (62).

Case Study

A 16-year-old boy in a coma (Glasgow coma scale 3) with general seizure was admitted to the hospital. A blood specimen tested by the EMIT assay for TCAs (Behring, Cupertino, CA) was positive, but the same specimen tested negative using the FPIA assay for TCAs (Abbott Laboratories, Abbott Park, IL). A comprehensive drug screen using a HPLC method equipped with diode-array detection identified the presence of buflomedil, a peripheral vasodilator at a concentration of 28 μg/mL in the specimen. When aliquots of a drug-free serum pool were further supplemented with buflomedil concentration between 1 and 100 μg/mL, a positive screen for TCAs at a buflomedil concentration of 13 μg/mL and higher was obtained using the EMIT assay for TCAs. The FPIA assay for TCAs showed a positive response at a buflomedil concentration of 85 μg/mL or higher. Therapeutic plasma concentration of buflomedil is between 1 and 4 μg/mL, and cardiotoxicity is usually encountered at a concentration of 10 μg/mL. Therefore, interference of buflomedil in immunoassays for TCAs is expected in the case of overdose only (63).

Cyclobenzaprine, a muscle relaxant, and its major metabolite norcyclobenzaprine differ from amitriptyline and nortriptyline only by the presence of a double bond in the cycloheptane ring. Three patients developed enough serum levels of cyclobenzaprine and its metabolite due to rapid or long-term ingestion and caused false-positive TCAs using both the EMIT immunoassay and a HPLC protocol for the identification and quantification of TCAs. Cyclobenzaprine co-eluted with amitriptyline during chromatographic separation. However, GC/MS analysis after derivatization with trifluoroacetic anhydride can eliminate this interference (64).

14.6. WHY CHROMATOGRAPHIC METHODS SHOULD BE USED FOR THERAPEUTIC DRUG MONITORING OF TRICYCLIC ANTIDEPRESSANTS

As mentioned earlier, immunoassays for TCAs are suitable for screening patients for a suspected overdose with TCAs and should not be used for routine therapeutic drug monitoring. Various GC and LC methods for routine therapeutic drug monitoring of TCAs are available that are less subjected to interferences compared with various immunoassays. GC methods coupled with a flame ionization detector (FID), nitrogen phosphorus detector (NPD), or mass spectrometer (MS) are widely used for analysis of TCAs. Due to their lipophilic characters, these drugs can be extracted from serum or urine using liquid–liquid extraction, although numbers of solid-phase extraction procedures have also been reported. Columns are typically fused silica capillary columns with bonded nonpolar to intermediate polarity methyl silicone liquid phases (0–50% phenyl). Although sample derivatization is not necessary, some investigators recommended derivatization to improve peak quality and chromatographic separation. Trifluoroacetyl and heptafluorobutyryl are commonly used derivatives. However, these derivatives are not very stable. Way et al, using stable isotope dilution GC-MS, found that 4-carbethoxyhexafluorobutyryl chloride derivatives are more stable than trifluoroacetyl derivatives. The lower limit of quantitation was 25 ng/mL; the analytical measurement range was up to 1500 ng/mL. Although interferences from thioridazine, perphenazine, cyclobenzaprine, and norcyclobenzaprine have been reported in some HPLC assay for TCAs, this method is free from all such interferences (65).

LC, especially HPLC, is another widely used method for the analysis of TCAs. Problems posed in GC by polar secondary amines and hydroxy metabolites are easily overcome by HPLC. Ultraviolet (UV) detection is the most common method for quantitative analysis of TCAs using HPLC. The columns most commonly used for chromatographic separations are C18, C8, phenyl, and CN, and they permit simultaneous determination of tertiary and secondary amines. Common mobile phases are phosphate buffers with or without ion-pairing agents. HPLC methods with normal phase silica and aqueous base mobile phase with fluorescence or electrochemical detection methods are also available (66). A sensitive and specific HPLC combined with tandem mass spectrometry (HPLC/MS/MS) method was described for the rapid identification and quantitation of seven TCAs: amitriptyline, nortriptyline, doxepin, dosulepin, dibenzepin, opipramol, and melitracen. The method uses direct injection and online removal of proteins and other large biomolecules with a total analysis time of only 12 minutes (67). Titier et al described a LC combined with tandem mass spectrometric method for quantitation of amitriptyline, clomipramine, trimipramine, imipramine, doxepin, mianserin, maprotiline, dosulepine, amoxapine, and their active metabolites after a liquid–liquid extraction from the blood. Chromatographic separation was achieved using a reverse phase C-18 column using methylrisperidone as the internatal standard. The limit of quantitation was 2 ng/mL (68). Breaud et al described a fully automated method for the quantitation of TCAs (amitriptyline, desipramine, imipramine, and nortriptyline) in human serum using a direct online Cyclone P solid-phase extraction followed by chromatographic separation using a C-18 reverse phase column. Detection of peaks was achieved using electrospray ionization mass spectrometry with selected reaction monitoring. The limit of quantitation was 3 ng/mL (69). Bonke and Jensen reported a fast and sensitive LC combined with tandem mass spectrometry determination of nortriptyline and its optical isomers metabolites using a reverse phase C-18 column electrospray ionization mass spectrometry in the positive mode (70).

14.7. THERAPEUTIC DRUG MONITORING OF TETRACYCLIC AND NEWER ANTIDEPRESSANTS

After the discovery of TCAs and monoamine oxidase inhibitors, collectively called first-generation antidepressants, many other classes of antidepressants were developed later, for example, amoxapine and maprotiline, that affect the reuptake of monoamines similar to secondary amine TCAs. Trazodone is a weak inhibitor of serotonin reuptake but has little effect on norepinephrine uptake. Bupropion inhibits the reuptake of norepinephrine and dopamine. Venlafaxine and mirtazapine are other non-TCAs that inhibit both norepinephrine and serotonin. In general, these newer drugs have fewer toxic side effects compared with TCAs (71). Another class of antidepressants, the SSRIs, have become one of the most widely prescribed group of antidepressants in the United States due to their wider therapeutic ranges and lower side effects compared with TCAs. One side effect of SSRIs is hyponatremia, which may have harmful clinical ramifications. Elderly patients are especially at risk of developing hyponatremia from the use of SSRIs and if untreated could cause serious consequences (72). Although the risk of bleeding associated with SSRI therapy is low, routine monitoring of patients taking SSRIs should probably include questions about bruising and abnormal bleeding, especially in patients with blood dyscrasias and in elderly patients taking other medications that may affect platelet function. In addition, caution must be exercised in prescribing SSRIs to patients with preexisting bleeding risk as well as patients taking nonsteroidal anti-inflammatory drugs and other medications that may impair coagulation (73). Therefore antidepressants should be used in an evidence-based manner with frequent monitoring of side effects especially during initiation of therapy and dose adjustment (74). However with SSRIs, overdose, drug–drug, or drug–herb interactions may cause serotonin syndrome, which is a life-threatening syndrome requiring immediate medical intervention. Serotonin syndrome is caused by excessive serotonergic activity in the nervous system and is characterized by altered mental status, autonomic instability, and neuromuscular hyperactivity. Although this is more common in patients taking multiple serotonergic drugs such as SSRIs, overdose with a single agent may also cause serotonin syndrome. Diagnosis is made using Hunter Serotonin toxicity criteria. If serotonin syndrome is recognized and complications are managed appropriately (supportive care, withdrawing the offending agent, therapy with benzodiazepines, using cyproheptadine as antidotes, etc.), the prognosis is favorable (75).

Although most SSRIs do not require therapeutic drug monitoring, measuring the plasma level of a few SSRIs (citalopram, fluoxetine, fluvoxamine, paroxetine, and sertraline) may be clinically useful for certain patient populations

TABLE 14.3. Therapeutic Ranges of Some Newer Antidepressants

Antidepressant	Suggested Therapeutic Range
Amoxapine	200–400 ng/mL
Citalopram	40–100 ng/mL
Fluoxetine	300–1000 ng/mL
Fluvoxamine	20–400 ng/mL
Haloperidol	2–15 ng/mL
Maprotiline	150–300 ng/mL
Paroxetine	20–200 ng/mL
Sertraline	30–200 ng/mL
Trazodone	800–1600 ng/mL
Venlafaxine	250–500 ng/mL

and also in overdose situations. Katoh et al, based on a study of 12 depressed patients taking fluvoxamine, observed that based on the HAMD scores, there were seven responders and five nonresponders on day 28, and the plasm levels of responders (mean: 14.2 ng/mL) were significantly lower than the plasma levels of nonresponders (mean: 44.2 ng/mL). Based on the receiver operating characteristic curve analysis conducted on day 28, the upper concentration range of responders was 28.2 ng/mL because none of the responder patients showed a level above this cut-off. The authors concluded that HAMD scores and fluvoxamine plasma level could predict whether a patient is going to respond to fluvoxamine or not (76). The suggested therapeutic range of common SSRIs and non-TCAs are given in Table 14.3.

Case Study

A term infant (40 weeks 2 days) boy was born to a 29-year-old woman. Mother's past history was significant for depression and anxiety, and she was treated with paroxetine. The infant showed respiratory distress syndrome along with abnormalities in tone and was intubated. The child improved afterward and was first breastfed at approximately 50 hours after birth but still showed symptoms of paroxetine toxicity. Because the mother received paroxetine in her third trimester, the mother's blood, cord blood, and blood of the newborn were analyzed for paroxetine. The serum paroxetine concentration in the cord blood was high (368 nmol/L; adult therapeutic range of paroxetine: 12–155 mmol/L). In addition, the mother's milk also had high level of paroxetine (1128 nmol/L). This case illustrates the importance of paroxetine level in a newborn exposed to paroxetine in utero (77).

14.7.1. Chromatographic Methods for Monitoring Tetracyclic and Newer Antidepressants

There is no immunoassay for monitoring tetracyclic and newer antidepressants. Therefore, chromatographic techniques (a GC- or HPLC-based assay with different detectors) are used in clinical laboratories for monitoring these drugs in serum or plasma. In general, GC can be used for screening and quantitation of newer antidepressants. The methods involve either liquid–liquid or solid phase extraction. Single-step extractions are generally successful, but methods describing multiple extractions or back extractions have also been described. The method may be capable of analyzing a single drug or multiple drugs. A GC-MS method involving acid hydrolysis for the simultaneous determination of citalopram, fluoxetine, fluvoxamine, paroxetine, and sertraline in human urine samples has been reported (78). A capillary GC-MS method, using selected ion monitoring for the simultaneous determination five antidepressant drugs, fluoxetine, fluvoxamine, citalopram, sertraline, and paroxetine, was published (79).

Due to heat labiality, certain antidepressants such as trazodone and nefazodone cannot be analyzed by GC but can be analyzed by HPLC. In general, HPLC methods are more commonly reported for the analysis of non-TCAs in serum, plasma, or other body fluids. HPLC is a commonly used method for the determination of non-TCAs (80–83). A HPLC method for the simultaneous determination of citalopram, fluoxetine, paroxetine, and their metabolites has been described. The authors used solid phase extraction and reversed-phase HPLC with fluorescence and UV detection for analysis of these newer antidepressants. The limits of quantitation were 0.025 μg/L for citalopram and paroxetine, and 0.10 μg/L for fluoxetine and nor fluoxetine, respectively (82). Another reversed-phase HPLC method for the simultaneous determination of bicyclic, tricyclic, and tetracyclic, and their metabolites was also reported (83). Jain et al described a rapid and sensitive method for the determination of sertraline in human plasma using LC combined with tandem MS. The analytical measurement range of sertraline was 0.5–60 ng/mL (84). Paroxetine and its three major metabolites (M1, M2, and M3) can be analyzed in human plasma using HPLC combined with fluorescence detection (exciting at 294 nm and monitoring emission at 330 nm for paroxetine, exciting at 280 nm and measuring at 330 nm for both M1 and M2, and exciting at 268 nm and measuring emission at 290 nm for M3). Chromatographic separation was achieved by using a reverse phase C-18 column and a mobile phase composition of 66.7% aqueous phosphate buffer (pH 2.5) and 33.3% acetonitrile. The assay was linear for paroxetine concentration between 2.5 and 100 ng/mL with a detection limit of 1.2 ng/mL. The analytical measurement range for all metabolites was 5–100 ng/mL with a detection limit of 2 ng/mL (85). Recently, Unceta et al reviewed various chromatographic methods available for the therapeutic drug monitoring of citalopram, a highly selective serotonin reuptake inhibitor along with its metabolites (86). Chromatographic methods for the analysis of newer non-TCAs are mostly free from interferences.

14.8. CONCLUSIONS

Although there are many newer non-TCAs available for treating depression, older TCAs are still used. Both TCAs and newer SSRIs require an induction period of 2–4 before observable clinical effects of reduced depression. However, SSRIs and other non-TCAs have wider safety margins than TCAs and may not require routine therapeutic drug monitoring. Nevertheless, for certain patient populations, especially elderly patients and in overdose situations, the determination of serum or plasma levels of these newer antidepressants is useful. In contrast, for TCAs, routine therapeutic drug monitoring is required for all patients. Although immunoassays are available for measuring total TCA level (parent drug plus active metabolite if applicable), these assays cannot differentiate between the concentration of the parent drug and its active metabolite (for tertiary amines) and are only useful for assessing an overdose situation. Like newer antidepressants, only chromatographic methods (GC or HPLC) must be used for routine therapeutic drug monitoring of TCAs.

REFERENCES

1. Broquet K. Status of treatment of depression. South Med J 1999; 92: 846–856.
2. Kok RM, Heeren TJ, Nolen WA. Continuing treatment of depression in the elderly: a systematic review and meta-analysis of double blinded randomized controlled trials with antidepressants. Am J Geriatr Psychiatry 2011; 19: 249–255.
3. Zohar J, Westenberg HG. Anxiety disorders: a review of tricyclic antidepressants and selective serotonin reuptake inhibitors. Acta Psychiatr Scand Suppl 2000; 403: 39–49.
4. Feighner JP. Overview of antidepressants currently used to treat anxiety disorders. J Clin Psychiatry 1990; 60 (Suppl 22): 18–22.
5. Jann MW, Slade JH. Antidepressant agents for the treatment of chronic pain and depression. Pharmacotherapy 2007; 27: 1571–1587.
6. Jackson JL, Shimeall W, Sessums L, Dezee KJ et al. Tricyclic antidepressants and headaches: a systematic review and meta-analysis. BMJ 2010; 341: c5222.
7. Stassen HH, Angst J, Delini-Stula A. Delayed onset of antidepressant drugs? Survey of results of Zurich meta-analyzes. Pharmacopsychiatry 1996; 29: 87–96.
8. Frazer A. Pharmacology of antidepressants. J Clin Psychopharmacol 1997; 17 (Suppl 1): 2S–18S.
9. Reneric JP, Lucki I. Antidepressant behavioral effect by dual inhibition of monoamine reuptake in the rat forced swimming model. Psychopharmacology 1998; 136: 190–197.
10. Ziegler V, Biggs J, Wylie L et al. Doxepin kinetics. Clin Pharamcol Ther 1978; 23: 573–579.
11. Biggs J, Holland W, Sherman W. Steady state protriptyline levels in outpatient population. Am J Psychiatry 1975; 132: 960–962.
12. Spiker DG, Biggs JT. Tricyclic antidepressants. Prolonged plasma levels after overdose. JAMA 1976; 236: 1711.
13. Haddad LM. Managing tricyclic antidepressant overdose. Am Fam Physician 1992; 46: 153.
14. Thanacoody HK, Thomas SH. Tricyclic antidepressant poisoning: cardiovascular toxicity. Toxicol Rev 2005; 24: 205–214.

15. Clauser J. Tricyclic antidepressant poisoning. Cleve Clin J Med 2000; 67: 709–713.
16. Fujitani N, Koda Y, Okamura T, Hattori H et al. A fatal case of hyperthermia due to tricyclic antidepressant intoxication. Leg Med (Tokyo) 2000; 2: 152–155.
17. Chaturvedi AK, Hidding JT, Rao NG, Smith JC 2nd et al. Two tricyclic antidepressant poisonings: levels of amitriptyline, nortriptyline and desipramine in post-mortem biological samples. Forensic Sci Int 1987; 33: 93–101.
18. Koski A, Vuori E, Ojanpera I. Relation of postmortem blood alcohol and drug concentrations in fatal poisoning involving amitriptyline, propoxyphene and promazine. Hum Exp Toxicol 2005; 24: 389–396.
19. Margalho C, Barroso M, Gallardo E, Monsato P et al. Massive intoxication involving high concentration of amitriptyline. Hum Exp Toxicol 2007; 26: 667–670.
20. Brosen K. Some aspects of genetic polymorphism in the biotransformation of antidepressants. Therapie 2004; 59: 5–12.
21. Bertilsson L, SDahl ML, Tybring G. Pharmacogenetics of antidepressants: clinical aspects. Acta Psychiatr Scand Suppl 1997; 391: 14–21.
22. Kirchheiner J, Brøsen K, Dahl ML,Gram LF et al. CYP2D6 and CYP2C19 genotype-based dose recommendations for antidepressants: a first step towards subpopulation-specific dosages. Acta Psychiatr Scand 2001; 104: 173–192.
23. Steimer W, Müller B, Leucht S, Kissling W. Pharmacogenetics: a new diagnostic tool in the management of antidepressive drug therapy. Clin Chim Acta 2001; 308: 33–41.
24. Jiang ZP, Shu Y, Chan XP, Huang SL et al. The role of CYP2C19 in amitriptyline N-demethylation in Chinese subjects. Eur J Clin Pharamcol 2002; 58: 109–113.
25. Koshi A, Sistonen J, Ojanpera I, Gergov M et al. CYP2D6 and CYP2C19 genotypes and amitriptyline metabolite ratios in a series of medicolegal autopsies. Forensic Sci Int 2006; 158: 177–183.
26. Dalén P, Dahl ML, Bernal Ruiz ML, Nordin J et al. 10-Hydroxylation of nortriptyline in white persons with 0, 1, 2, 3, and 13 functional CYP2D6 genes. Clin Pharmacol Ther 1998; 63: 444–452.
27. Muller MJ, Dragicevic A, Fric M, Gaertner I et al. Therapeutic drug monitoring of tricyclic antidepressants: how does it work under clinical conditions? Pharmacopsychiatry 2003; 36: 98–104.
28. Preskprn SH. Tricyclic antidepressants: the whys and hows of therapeutic drug monitoring. J Clin Psychiatry 1989; 50 (Suppl 34–42); 43–46.
29. Nelson JC, Jatlow P, Quinlan DM, Bowers MB Jr. Desipramine plasma concentration and antidepressant response. Arch Gen Psychiatry 1982; 39: 1419–1422.
30. Stage KN, Brosen K. Insufficient effect of nortriptyline in a rapid metabolizer [in Danish]. Ugeskr Laeger 1993; 155: 1561–1563.
31. Petersen P, Brosen K. Severe nortriptyline poisoning in poor metabolizer of the sparteine type [in Danish]. Ugeskr Laeger 1991; 153: 443–444.
32. White N, Litovitz T, Clancy C. Suicidal antidepressant overdoses: a comparative analysis by antidepressant type. J Med Toxicol 2008; 4: 238–250.
33. Keers R, Aitchison KJ. Gender differences in antidepressant drug response. Int Rev Psychiatry 2010; 22: 485–500.

34. Ernst R, Williams L, Dalbey M, Collins C et al. Homogeneous enzyme immunoassay (EMIT) protocol for monitoring tricyclic antidepressants on the COBAS-BIO centrifugal analyzer. Ther Drug Monit 1987; 9: 85–90.
35. Pankey DC, Collins A, Jaklitsch A et al. Quantitative homogeneous enzyme immunoassays for amitriptyline, nortriptyline, imipramine, and desipramine. Clin Chem 1986; 32: 768–772.
36. Rao ML, Staberock U, Baumann P, Hiemke C et al. Monitoring tricyclic antidepressant concentrations in serum by fluorescence polarization immunoassay compared with gas chromatography and HPLC. Clin Chem 1994; 40: 929–933.
37. Schroeder TJ, Tasset JJ, Otten RJ, Hedges JR. Evaluation of Syva EMIT toxicology serum tricyclic antidepressant assay. J Anal Toxicol 1986; 10: 221–224.
38. Poklis A, Edinboro LE, Lee JS, Crooks CR. Evaluation of a colloidal metal immunoassay device for the detection of tricyclic antidepressants in urine. J Toxicol Clin Toxicol 1997; 35: 77–82.
39. Poklis A, Soghoian D, Crooks CR, Saddy JJ. Evaluation of the Abbott ADX total serum tricyclic immunoassay. J Toxicol Clin Toxicol 1990; 28: 235–248.
40. Hackett LP, Dusci LJ, Ilett KF. A comparison of high-performance liquid chromatography and fluorescence polarization immunoassay for therapeutic drug monitoring of tricyclic antidepressants. Ther Drug Monit 1998; 20: 30–34.
41. Melanson SE, Lewandrowski EL, Griggs DA, Flood JG. Interpreting tricyclic antidepressant measurements in urine in an emergency department: comparison of two qualitative point of care tricyclic antidepressant drug immunoassays with quantitative serum chromatographic analysis. J Anal Toxicol 2007; 31: 270–275.
42. Schroeder TJ, Tasset JJ, Otten EJ, Hedges JR. Evaluation of Syva EMIT toxicological serum tricyclic antidepressant assay. J Anal Toxicol 1986; 10: 221–224.
43. Ryder KW, Glick MR. The effect of thioridazine on the Automatic Clinical Analyzer serum tricyclic anti-depressant screen. Am J Clin Pathol 1986; 86: 248–249.
44. Maynard GL, Soni P. Thioridazine interferences with imipramine metabolism and measurement. Ther Drug Monit 1996; 18: 729–731.
45. Wertz PG, Street TL. Thioridazine interference in chromatographic and enzyme immunoassays for imipramine in serum. Clin Chem 1983; 29: 724–725.
46. Dasgupta A, McNeese C, Wells A. Interference of carbamazepine and carbamazepine 10, 11-epoxide in the fluorescence polarization immunoassay for tricyclic antidepressants: estimation of the true tricyclic antidepressant concentration in the presence of carbamazepine using a mathematical model. Am J Clin Pathol 2003; 121: 418–425.
47. Fleischman A, Chang V. Carbamazepine overdose recognized by a tricyclic antidepressant assay. Pediatrics 2001; 107: 176–178.
48. Chattergoon DS, Verjee Z, Anderson M, Johnson D et al. Carbamazepine interference with an immunoassay for tricyclic antidepressant. J Toxicol Clin Toxicol 1998; 36: 109–113.
49. Saidinejad M, Law T, Ewald MB. Interference of carbamazepine and oxcarbazepine with serum and urine screening for tricyclic antidepressants. Pediatrics 2007; 120: e504–509.
50. Wians FH Jr, Norton JT, Wirebaugh SR. False-positive serum tricyclic antidepressant screen with cyproheptadine. Clin Chem 1993; 39: 1355–1356.

51. Yuan CM, Spandorfer PR, Miller SL, Henretig FM et al. Evaluation of tricyclic antidepressant false positivity in a pediatric case of cyproheptadine (Periactin) overdose. Ther Drug Monit 2003; 25: 299–304.
52. Caravati EM, Juenke JM, Crouch BI, Anderson KT. Quetiapine cross-reactivity with plasma tricyclic antidepressant immunoassays. Ann Pharamcother 2005; 39: 1446–1449.
53. Sloan KL, Haver VM, Saxon AJ. Quetiapine and false-positive urine drug testing for tricyclic antidepressants [letter to the editor]. Am J Psychiatry 2000; 157: 148–149.
54. Schussler JM, Juenke JM, Schussler I. Quetiapine and falsely elevated nortriptyline level [letter to the editor]. Am J Psychiatry 2003; 160: 589.
55. Cerullo MA, Albertz AA, Bell JN, Anthenelli RM et al. Tricyclic antidepressant immunoassays may reflect quetiapine adherence [letter to the editor]. Am J Psychiatry 2008; 165: 919–920.
56. Dasgupta A, Wells A, Datta P. False positive serum tricyclic antidepressant concentrations using fluorescence polarization immunoassay due to the presence of hydroxyzine and cetirizine. Ther Drug Monit 2007; 29: 134–139.
57. Simons FE, Watson WT, Chen XY, Minuk GY, Simons KJ. The pharmacokinetics and pharmacodynamics of hydroxyzine in patients with primary biliary cirrhosis. J Clin Pharmacol 1989: 29: 809–815.
58. Johnson GR. A fatal case involving hydroxyzine. J Anal Toxicol 1982; 6: 69–70.
59. Magera BE, Betlach CJ, Sweatt AP, Derrick CW. Hydroxyzine intoxication in a 13 month old child. Pediatrics 1981; 67: 280–283.
60. Pariente-Khayat A, Rey E, Dubois MC, Vauzella-Kervroedan F et al. Pharmacokinetics of cetirizine in 2 to 6 year old children. Int J Clin Pharmacol Ther 1995; 33: 340–344.
61. Ridout SM, Tariq SM. Cetirizine overdose in a young child. J Allergy Clin Immunol 1997; 99: 860–861.
62. Sorisky A, Watson DC. Positive diphenhydramine interference in the EMIT-st assay for tricyclic antidepressants in serum. Clin Chem 1986; 32: 715.
63. Mura P, Kintz P, Robert R, Papet Y. Buflomedil is a potent interfering substance in immunoassay of tricyclic antidepressants [letter to the editor]. J Anal Toxicol 1998; 22: 254.
64. Wong EC, Koenig J, Turk J. Potential interference of cyclobenzaprine and norcyclobenzaprine with HPLC measurement of amitriptyline and nortriptyline: resolution by GC-MS analysis. J Anal Toxicol 1995; 19: 218–224.
65. Way BA, Stickle D, Mitchell ME, Koenig JW et al. Isotope dilution gas chromatographic-mass spectrometric measurement of tricyclic antidepressant drugs. Utility of the 4-carbethoxyhexafluorobutyryl derivatives of secondary amines. J Anal Toxicol 1998; 22: 374–382.
66. Wong SH. Measurement of antidepressants by liquid chromatography: a review of current methodology. Clin Chem 1988; 34: 848–855.
67. Kollroser M, Schober C. Simultaneous determination of seven tricyclic antidepressant drugs in human plasma by direct-injection HPLC-APCI-MS-MS with an ion trap detector. Ther Drug Monit 2002; 24: 537–544.

68. Titier K, Casting N, Le-Deodic M, Le-Bars D et al. Quantification of tricyclic antidepressants and monoamine oxidase inhibitors by high performance liquid chromatography-tandem mass spectrometry. J Anal Toxicol 2007; 31: 200–207.

69. Breaud AR, Harlan R, Di Bussolo JM, McMillin GA et al. A rapid and fully automated method for the quantitation of tricyclic antidepressants in serum using turbulent flow liquid chromatography-tandem mass spectrometry. Clin Chim Acta 2010; 411: 825–832.

70. Bonke G, Jensen BP. Fast and sensitive LC-MS/MS assay for quantification of nortriptyline and its active metabolite E and Z 10-hydroxynortriptytline in human plasma. Bioanalysis 2010; 2: 1553–1560.

71. Buckley NA, Faunce TA. "Atypical" antidepressants in overdose: clinical considerations with respect to safety. Drug Saf 2003; 26: 539–551.

72. Smith JM. Clinical implications of treating depressed older adults with SSRIs: possible risk of hyponatremia. J Gerontol Nurs 2010; 36: 22–27.

73. Reeves RR, Wise PM, Cox SK. SSRIs and the risk of abnormal bleeding. J Psychosoc Nurs Ment Health Serv 2007; 45: 15–21.

74. Murphy TK, Segarra A, Storch EA, Goodman WK. SSRI adverse events: how to monitor and manage. Int Rev Psychiatry 2008; 20: 203–208.

75. Ables AZ, Nagubilli R. Prevention, recognition and management of serotonin syndrome. Am Fam Physician 2010; 81: 1139–1142.

76. Katoh Y, Uchida S, Kawai M, Takei N et al. Onset of clinical effects and plasma concentration of fluvoxamine in Japanese population. Biol Pharm Bull 2010; 33: 1999–2002.

77. Knoppert DC, Mimkar R, Principi T, Yuen D. Paroxetine toxicity in a newborn after utero exposure: clinical symptoms with serum levels. Ther Drug Monit 2006; 28: 5–7.

78. Maurer HH, Bickeboeller-Friedrich J. Screening procedure for detection of antidepressants of the selective serotonin reuptake inhibitor type and their metabolites in urine as part of a modified systematic toxicological analysis procedure using gas chromatography-mass spectrometry. J Anal Toxicol 2000; 24: 340–347.

79. Berzas Nevado JJ, Villasenor Llerena MJ, Guiberteau Cabanillas C, Rodriguez Robledo V et al. Sensitive capillary GC-MS-SIM determination of selective serotonin reuptake inhibitors: reliability evaluation by validation and robustness study. J Sep Sci 2006; 29: 103–113.

80. Mandrioli R, Saracino MA, Ferrari S, Berardi D et al. HPLC analysis of the second-generation antidepressant sertraline and its main metabolite N-desmethylsertraline in human plasma. J Chromatogr B Analyt Technol Biomed Life Sci 2006; 836: 116–119.

81. Loboz KK, Gross AS, Ray J, McLachlan AJ. HPLC assay for bupropion and its major metabolites in human plasma. J Chromatogr B Analyt Technol Biomed Life Sci 2005; 823: 115–121.

82. Kristoffersen L, Bugge A, Lundanes E, Slordal L. Simultaneous determination of citalopram, fluoxetine, paroxetine and their metabolites in plasma and whole blood by high-performance liquid chromatography with ultraviolet and fluorescence detection. J Chromatogr B Biomed Sci Appl 1999; 734: 229–246.

83. Joron S, Robert H. Simultaneous determination of antidepressant drugs and metabolites by HPLC. Design and validation of a simple and reliable analytical procedure. Biomed Chromatogr 1994; 8: 158–164.
84. Jain DS, Sanyal M, Subbaiah G, Pande UC et al. Rapid and sensitive method for determination of sertraline in human plasma using liquid chromatography-tandem mass spectrometry (LC-MS/MS). J Chromatogr B Analyt Technol Biomed Life Sci 2005; 829: 69–74.
85. Mandrioli R, Mercolini L, Ferranti A, Furlanetto S et al. Determination of the antidepressant paroxetine and its three main metabolites in human plasma by liquid chromatography with fluorescence detection. Anal Chim Acta 2007; 591: 141–147.
86. Unceta N, Goicolea MA, Barrio RJ. Analytical procedures for the determination of the selective serotonin reuptake inhibitor antidepressant citalopram and its metabolites. Biomed Chromatogr 2011; 25: 238–257.

Chapter 15

Therapeutic Drug Monitoring of Immunosuppressants: Limitations of Immunoassays and the Need for Chromatographic Methods

15.1. INTRODUCTION

Immunosuppressive drugs nonspecifically diminish immune responses and are widely used to prevent organ rejection after transplantation. These agents are also used to treat autoimmune disease, allergic disorders, and several other diseases. The discovery that cyclosporine has immunosuppressive activity that specifically targets T lymphocytes was a major breakthrough in organ transplantation because it dramatically reduced acute rejection and improved

Resolving Erroneous Reports in Toxicology and Therapeutic Drug Monitoring: A Comprehensive Guide, First Edition. Amitava Dasgupta.

long-term graft and patient survival (1). The identification of other immunosuppressive drugs that modulate immune responses by additional molecular mechanisms provides a wide range of treatment options after solid organ transplant. Immunosuppressive agents used in organ transplants can be classified according to their mechanism of action. The classes are corticosteroids (prednisone, methylprednisolone, and dexamethasone), antimetabolite/proliferative agents (azathioprine, cyclophosphamide, mycophenolate mofetil, and mycophenolate sodium), calcineurin inhibitors (cyclosporine, tacrolimus), and mammalian target of rapamycin (mTOR) inhibitors (sirolimus, everolimus). In addition, polyclonal and monoclonal antibodies against T lymphocyte cell surface antigens are also used in combination with other immunosuppressive agents.

Therapeutic drug monitoring of certain immunosuppressants such as cyclosporine, tacrolimus, sirolimus, everolimus, and mycophenolic acid are needed due to their narrow therapeutic ranges, significant variability in blood concentrations between individual patients, gender differences in drug metabolism, and drug–nutrient as well as drug–drug interactions. For example, the rate of metabolism of mycophenolic acid is higher in male kidney recipients than female kidney recipients due to gender differences in the glucuronidation of mycophenolic acid (2). Age can also influence the rate of metabolism of various immunosuppressants. For example, pediatric transplant patients require higher dose per kilogram dosage than adults to maintain therapeutic cyclosporine blood concentration. In addition, poor absorption of cyclosporine in small children requires a considerable increase in dosage on a per kilogram body weight basis (3). Although both cyclosporine and tacrolimus can be used in children safely, cyclosporine should be selected when patients experience tacrolimus-induced side effects (4). Dosing by body surface area, age, and interval from transplantation are all correlated independently with mycophenolic acid level in serum, but higher doses to achieve desired levels are needed in younger patients (5).

The potential for drug interactions among the various classes of immunosuppressive agents is clinically significant, and therapeutic drug monitoring of immunosuppressants is essential to maintain an adequate level in blood to avoid organ rejection. Cyclosporine inhibits the transport of the mycophenolic acid metabolite, thus affecting the mycophenolic acid level in serum (6). The combination of cyclosporine and sirolimus or tacrolimus and sirolimus results in increased blood concentrations of sirolimus (6,7). Drug–herb interaction involving various immunosuppressants can cause a significant reduction in the blood concentration of cyclosporine and tacrolimus; for example, St. John's wort can reduce trough levels of both cyclosporine and tacrolimus significantly and may even cause treatment failure. However, St. John's wort does not interact with mycophenolic acid (see Chapter 16 for an in-depth discussion of this topic).

The calcineurin inhibitors cyclosporine and tacrolimus are available for both oral and intravenous administration. Tacrolimus is about 100 times more potent than cyclosporine and associated with better long-term graft survival. The calcineurin inhibitors block activation/proliferation of CD4+ and CD8+ T lymphocytes by inhibiting interleukin (IL)-2 production. Both cyclosporine and tacrolimus form complexes with specific cytoplasmic binding protein immunophilins. Cyclosporine binds cyclophilin, whereas tacrolimus binds the FK506-binding protein-12. The complexes block calcineurin activity and prevent dephosphorylation of the nuclear factor of activated T cells and subsequent translocation into the nucleus. This results in downregulated gene transcription for cytokines such as IL-2 (8,9). Cyclosporine and tacrolimus have at least 30 and 9 drug metabolites, respectively, and some of the metabolites have immunosuppressive activity. Metabolite cross-reactivity varies depending on the immunoassay and the transplant group studied. Nephrotoxicity is a major problem of cyclosporine therapy. Therefore, careful monitoring of whole blood cyclosporine level is crucial not only to avoid organ rejection but also to avoid nephrotoxicity. Nephrotoxicity with tacrolimus may be less of a problem than with cyclosporine, especially in renal transplantation. Nevertheless, nephrotoxicity may also be encountered with tacrolimus therapy.

The mTOR inhibitor sirolimus is available for oral administration. A structural analog of sirolimus, named everolimus, has improved pharmacokinetics and has greater oral bioavailability than sirolimus. Everolimus received U.S. Food and Drug Administration (FDA) clearance in 2010 for its therapeutic use as a prophylactic agent in organ transplant and also to treat advanced kidney cancer. Everolimus is a macrolide drug derived by chemical modification of rapamycin (sirolimus), a natural product derived from strains of *Streptomycin hygroscopicus.* Therefore, everolimus is structurally similar to sirolimus but contains a 2-hydroxyethyl chain substitution at position 40 on the sirolimus molecule (10,11). Similar to tacrolimus, sirolimus and everolimus bind to the intracellular immunophilin, FK506-binding protein-12. The complexes are highly specific inhibitors of mTOR, a cell cycle serine/threonine kinase involved in the protein kinase B signaling pathway (12). Both tacrolimus and everolimus thus exert immunosuppressant activity by blocking the proliferative signals of growth factors, thereby preventing cells from entering the S-phase, and the effects are not limited to IL-2–dependent proliferation of T cells because everolimus also inhibits growth factor–dependent proliferation in hematopoietic as well as nonhematopoietic cells (13,14). Therefore, the mechanism of everolimus is broader than calcineurin inhibitors such as cyclosporine, which acts by inhibiting T cell growth factors such as IL-2. Because of their complementary mechanisms of action, everolimus can be used along with cyclosporine for immunosuppression, and a lower dosage of cyclosporine can be used in combination with everolimus, thus reducing the dosage of cyclosporine and hence the toxicity of cyclosporine (15,16). Everolimus was developed to improve the pharmacokinetic profile of sirolimus, which has an elimination half-life of 60 hours compared with 18–35 hours for everolimus. Because of the shorter half-life of everolimus, it can be dosed twice daily compared with the once-daily dosing of sirolimus. Hence steady state can be achieved more quickly within 7 days compared with the 13 days required for sirolimus (17). However, like other immunosuppressants, whole blood therapeutic drug monitoring of everolimus is needed (18). Metabolite cross-reactivity can lead to significant positive assay bias for both tacrolimus and everolimus immunoassays.

Usually whole blood is used for therapeutic drug monitoring of cyclosporine, tacrolimus, sirolimus, and everolimus, whereas serum is the preferred specimen for therapeutic drug monitoring of mycophenolic acid. For all immunosuppressants, trough specimen is the preferred specimen for therapeutic drug monitoring, although for cyclosporine, there are benefits of C2 monitoring (monitoring whole blood cyclosporine concentration 2 hours after dosage). The therapeutic ranges of various immunosuppressants are summarized in a table in Chapter 1.

15.2. THERAPEUTIC DRUG MONITORING OF CYCLOSPORINE

Cyclosporine is a small cyclic polypeptide containing 11 amino acids (Fig. 15.1). This drug was isolated in 1970 from fungal cultures of *Tolypocladium inflatum Gams* and approved in the United States as an immunosuppressive drug to prolong organ and patient survival in kidney, liver, heart, and bone marrow transplantation. This drug is available for both oral and intravenous administration (Sandimmune). A microemulsion oral formulation of cyclosporine, called Neoral, exhibits more reproducible absorption characteristics and has largely replaced Sandimmune. Neoral is available in 25- and 100-mg capsules for oral administration, and as a 100 mg/mL sterile solution for intravenous administration. Several generic microemulsion formulations are also available. Cyclosporine forms complexes with a cytoplasmic binding protein immunophilin called cyclophilin that blocks calcineurin activity and prevents dephosphorylation of the nuclear factor of activated T cells (NF-AT), and subsequent translocation into the nucleus. The end result is downregulated cytokine gene transcription (IL-2 and others) that leads to a block in the activation/proliferation of CD4+ and CD8+ T lymphocytes (19).

Oral absorption of Sandimmune is low (5–30%) and highly variable, ranging from 4% to 89% in renal and liver transplant patients. Absorption of Neoral is more consistent, averaging approximately 40%. Peak blood concentrations typically occur between 1 and 3 hours and 2 and 6 hours following oral administration of Neoral and Sandimmune, respectively (20). The distribution of cyclosporine between plasma and erythrocytes is temperature and hematocrit dependent (21). Therefore, ethylene diamine tetra-acetic acid (EDTA)-anticoagulated whole blood is the recommended specimen type for therapeutic drug monitoring of cyclosporine. Cyclosporine is extensively metabolized by cytochrome P450 enzymes (CYP3A isoenzymes) located in the small intestine and liver. There is also a cellular transporter of immunosuppressive drugs, called P-glycoprotein, which influences cyclosporine metabolism by

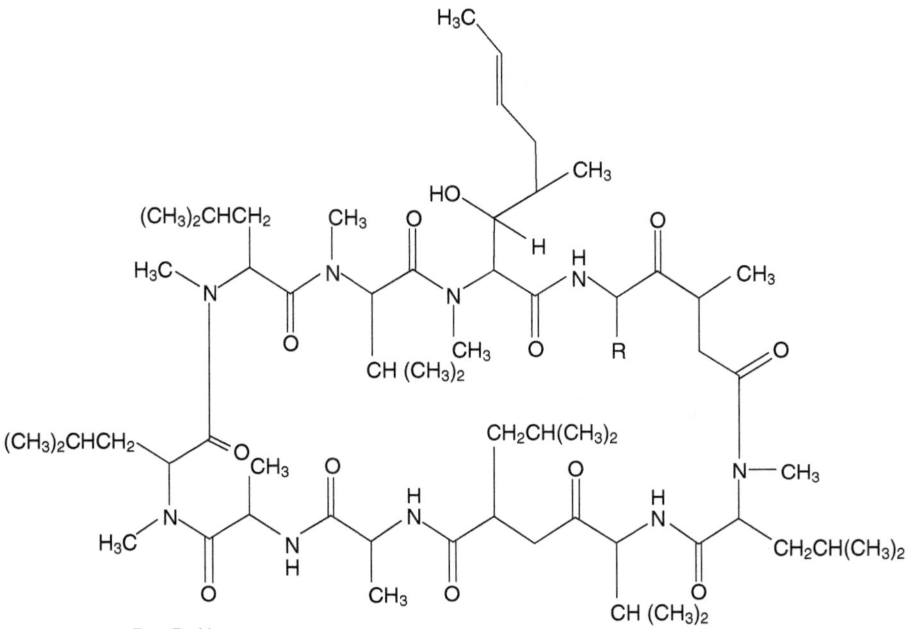

R = D-Ala

Cyclosporine

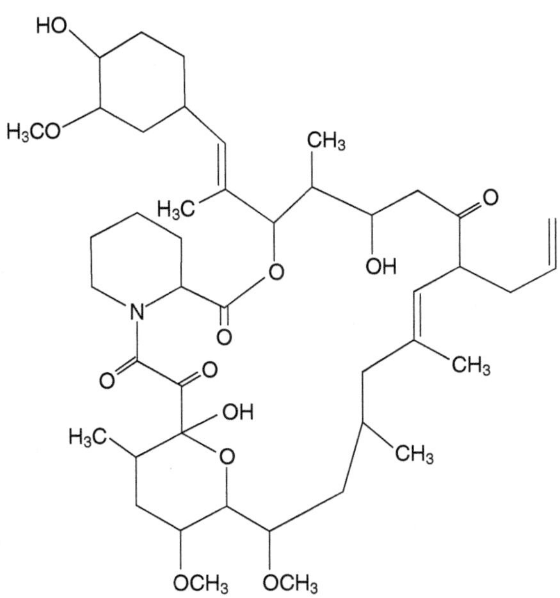

Tacrolimus

Figure 15.1. Chemical structure of cyclosporine, tacrolimus, sirolimus, and everolimus.

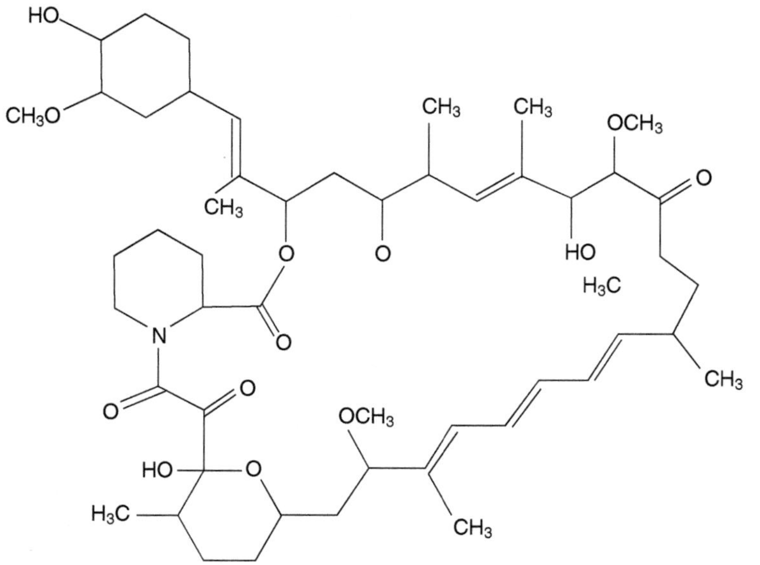

Sirolimus

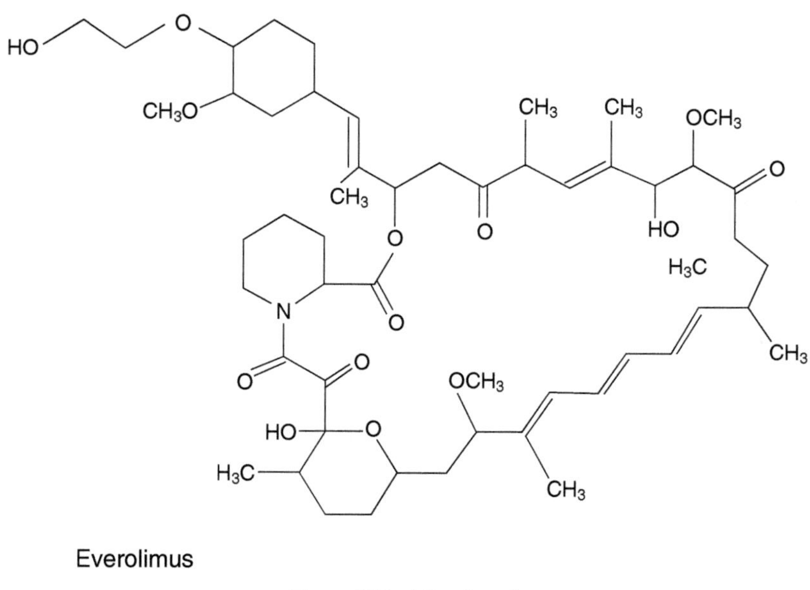

Everolimus

Figure 15.1. (*Continued*)

controlling bioavailability. This efflux pump probably contributes to the poor absorption rates observed after oral administration of cyclosporine. Cyclosporine is oxidized or N-demethylated to more than 30 metabolites. Most of the metabolites do not possess immunosuppressive activity and are not clinically significant. Two of the hydroxylated metabolites, AM1 and AM9, exhibit 10–20% of the immunosuppressive activity of the parent compound (22). The major route of cyclosporine elimination is biliary excretion into the feces. As expected, dosage adjustments are necessary in patients with hepatic dysfunction. Cyclosporine clearance also decreases in elderly patients (older the 65 years) because target C2 level can be achieved with lower cyclosporine dosage in elderly patients compared with younger patients. In addition, elderly patients also exhibit higher whole blood cyclosporine concentration located at the site of action (within T lymphocytes). Therefore, in elderly patients it might be safe to target lower whole blood concentration of cyclosporine than in younger patients (23).

The timing of specimen collection has always been right before administration of the next dose (trough level). For standardization purposes, the timing should be within 1 hour before the next dose. Two-hour postdosing cyclosporine concentrations (called C2 monitoring) have been shown to correlate better with total drug exposure and result in improved clinical outcomes (24). Therapeutic ranges for cyclosporine are often organ specific and can vary widely between transplant centers. In addition, target cyclosporine levels may be lower if cyclosporine is used in addition to another immunosuppressant. Trough whole blood cyclosporine levels following kidney transplants are typically between 150 and 250 ng/mL shortly after transplant and are tapered down to less than 150 mg/L during maintenance therapy. Recommended levels after liver and heart transplants are 250–350 ng/mL shortly after transplant and less than 150 mg/L during maintenance therapy. These target ranges were determined by high performance liquid chromatography (HPLC) and vary considerably when measured by immunoassay, depending on the amount of metabolite cross-reactivity. Min et al demonstrated that up to 1 month after renal transplantation, the cyclosporine therapeutic response threshold was 182 ng/mL and the nephrotoxicity threshold was 204 ng/mL. Between 1 month and 3 months after transplantation, the respective therapeutic and toxic threshold for cyclosporine was 175 ng/mL and 189 ng/mL, respectively. However, between 3 and 12 months after transplantation, the therapeutic and toxic cyclosporine threshold became 135 ng/mL and 204 ng/mL, respectively (25). Therefore, maintaining cyclosporine level less than 150 ng/mL in stable transplant patients is justified. However, for C2 monitoring, target concentrations vary between 600 and 1700 ng/mL depending on the type of graft and the time after transplantation (26).

Genetic polymorphism also plays a role in cyclosporine pharmacokinetics. Therefore, genetic testing prior to transplantation has some adjustment to predict cyclosporine dosage. Interindividual differences in the activity and expression of the metabolizing enzymes CYP3A4 and CYP3A5 and multidrug efflux pump P-glycoprotein contribute considerably to cyclosporine pharmacokinetics. Genetic polymorphism of CYP3A5 may be responsible in part for large interindividual variability of cyclosporine pharmacokinetics especially during the early phase of renal transplant, and patients with a CYP3A5*3 variant genotype require a lower dose of cyclosporine to reach target therapeutic level than patients with the CYP3A5*1 allele (27).

15.3. THERAPEUTIC DRUG MONITORING OF TACROLIMUS

Tacrolimus (also known as FK-506) is a macrolide antibiotic that was originally isolated from the fungus *Streptomyces tsukubaensis*. In the United States, tacrolimus (brand name Prograf) was approved for use in liver transplantation in 1994 and in kidney transplantation in 1997. It is approximately 100 times more potent than cyclosporine and superior to cyclosporine in preventing organ rejection (28). Tacrolimus is available as 0.5-, 1-, and 5-mg capsules for oral administration, and as a 5 mg/mL sterile solution for intravenous administration. Tacrolimus forms complexes with the binding protein immunophilin called FK506-binding protein-12 in the cytoplasm of lymphocytes. The complexes block calcineurin activity and prevent dephosphorylation of the NF-AT and subsequent translocation into the nucleus. The end result is downregulated cytokine gene transcription (IL-2 and others) that leads to a block in the activation/proliferation of CD4+ and CD8+ T lymphocytes (29). Similar to cyclosporine, oral absorption of tacrolimus from the gut is poor and highly variable, averaging 25% (30). Peak blood concentrations occur within 0.5–6 hours.

Tacrolimus is metabolized by cytochrome P450 isoenzymes (CYP3A4/5) located in the small intestine and liver. Metabolism of tacrolimus occurs by demethylation, hydroxylation, and oxidative reactions. At least nine metabolites have been identified based on in vitro studies, and all with the exception of 31-*O*-demethyl tacrolimus (M-II) have very little immunosuppressive activity (31). M-II has been shown in vitro to have the same immunosuppressive activity as the parent compound. Metabolites represent 10–20% of whole blood tacrolimus concentrations. Tacrolimus is eliminated primarily by biliary excretion into the feces. Patients with hepatic dysfunction require dosage adjustments. Very little tacrolimus is found in the urine, and blood concentrations are not altered in renal dysfunction. Like cyclosporine, single nucleotide polymorphism of the gene governing CYP3A5 activity plays a role in the pharmacokinetics of tacrolimus. Carriers of CYP3A5 wild-type allele have a higher CYP3A5 expression compared with those patients who are homozygous for nonexpression of CYP3A5 activity (CYP3A5*3). Therefore, patients with CYP3A5*3 allele would exhibit significantly lower tacrolimus clearance than the carrier of the wild-type gene and would require a lower dosage of tacrolimus than patients with normal activity of CYP3A5 (32).

Tacrolimus is primarily bound to albumin, α_1-acid glycoprotein, and lipoproteins in the plasma. However, most tacrolimus is found within erythrocytes. The distribution of tacrolimus between plasma and erythrocytes is temperature dependent and varies with changes in the hematocrit. Therefore, EDTA-anticoagulated whole blood is recommended for the therapeutic drug monitoring of tacrolimus. Trough blood tacrolimus concentrations are almost exclusively used for routine monitoring and are believed to be a good indicator of total drug exposure (33). Initial therapeutic range of tacrolimus was suggested to be 10–25 ng/mL, but levels of approximately 20 ng/mL were frequently associated with toxicity. The therapeutic range of tacrolimus as 5–15 ng/mL has been widely accepted, but lower limits have not been defined clearly (34).

15.4. THERAPEUTIC DRUG MONITORING OF SIROLIMUS

Sirolimus (also known as rapamycin) is a lipophilic macrocyclic lactone derived from *Streptomyces hygroscopicus* in the early 1970s and was approved by the FDA in 1999 for use with cyclosporine to reduce the incidence of acute rejection in renal transplantation. Sirolimus readily crosses lymphocyte plasma membranes and binds to the intracellular immunophilin, FK506-binding protein-12. In contrast to tacrolimus, sirolimus-immunophilin complexes do not inhibit calcineurin activity. Instead, the complexes are highly specific inhibitors of the mTOR, a cell cycle serine/threonine kinase involved in the protein kinase B signaling pathway. This leads to suppressed cytokine-induced T lymphocyte proliferation, with a block in progression from the G1 to S phase of the cell cycle. The mTOR inhibitors work synergistically with the calcineurin inhibitors to produce a profound immunosuppressive effect on T lymphocytes (34). Sirolimus (brand name Rapamune) is available as 1- and 2-mg tablets and as a 1 mg/mL solution for oral administration. Its long half-life of approximately 60 hours allows once-daily dosing. Sirolimus is rapidly absorbed from the gastrointestinal tract, and peak blood concentrations can be observed 1–2 hours after an oral dose. Oral bioavailability is low, ranging from 5% to 15% and is considerably reduced when administered within 4 hours or concomitantly with cyclosporine. There is a considerable interpatient variability in total drug exposure that can vary by as much as 50%. Sirolimus is primarily bound within erythrocytes (95%), with approximately 3% and 1% partitioning into the plasma and lymphocytes/granulocytes, respectively (35). Almost all of the plasma sirolimus is bound to proteins, with albumin, α-1-glycoprotein, and lipoproteins the major binding proteins. The incidence of adverse effects of sirolimus is dose related and includes metabolic, hematological, and dermatological effects. Metabolic side effects include hypercholesterolemia, hyper- and hypokalemia, hypophosphatemia, hyperlipidemia, and abnormal liver function tests. Anemia can be problematic, with decreases in leukocyte, erythrocytes, and platelet counts the most common. Skin rashes, acne, and mouth ulcers are also observed in patients switched to sirolimus. As with other immunosuppressive drugs, there is an increased risk of infection and an association with lymphoma development. Interstitial pneumonitis is also associated with sirolimus therapy.

Sirolimus is metabolized in the intestine and liver by cytochrome P450 enzymes (CYP3A). The multidrug efflux pump P-glycoprotein in the gastrointestinal tract also controls metabolism by regulating bioavailability. Sirolimus is hydroxylated and demethylated to more than seven metabolites with the hydroxyl forms the most abundant. Metabolites represent approximately 55% of whole blood sirolimus levels (36). The biological activity of metabolites has not been fully investigated due to difficulties associated with their isolation. However, preliminary studies indicate that the immunosuppressive activity of metabolites is less than 30% of that observed for the parent compound. Sirolimus is eliminated primarily by biliary and fecal pathways, with small quantities appearing in urine. As with the calcineurin inhibitors, dosage adjustments are needed in patients with hepatic dysfunction. EDTA anticoagulated whole blood is the recommended specimen matrix (37). This is because almost all of the sirolimus (about 95%) is concentrated in erythrocytes, and plasma levels are too low for most analytical methods. Therapeutic monitoring of sirolimus is critical because the administered dose is a poor predictor of total drug exposure due to individual patient variables. A sirolimus therapeutic range of 4–12 ng/mL has been suggested when used in combination with a calcineurin inhibitor such as cyclosporine. However, if cyclosporine therapy is discontinued, a target trough concentration of 12–20 ng/mL has been recommended (38). Wong suggested a trough therapeutic range of 5–15 ng/mL for sirolimus (39).

15.5. THERAPEUTIC DRUG MONITORING OF EVEROLIMUS

Everolimus (also known as SZD RAD) is a structural analog of sirolimus containing an additional 2-hydroxyethyl group at position 40 of the sirolimus molecule. Everolimus (40-*O*-[2-hydroxy] ethylrampamycin) was formulated to have improved pharmacokinetic characteristics compared with sirolimus. The hydroxyethyl chain increases the polarity of the molecule, resulting in greater oral bioavailability. Everolimus is a potent antiproliferative agent with a mechanism of action similar to sirolimus. It blocks cell cycle progression into the S-phase. In limited clinical trials, everolimus was shown to be less nephrotoxic than cyclosporine and tacrolimus and may reduce the incidence of allograft vasculopathy in heart transplantation (40,41). Oral bioavailability of everolimus is approximately 16%, which is an improvement over the low bioavailability of sirolimus. Everolimus has a shorter elimination half-life (mean of 25 hours) compared with sirolimus. This requires twice-daily dosing. Intrapatient total drug exposure for everolimus is approximately 40% and can be considerably more variable between patients (more than 85%) (40). At therapeutic concentrations, more than 75% of everolimus is found within erythrocytes. Therefore, whole blood specimens are used for everolimus quantitation. Approximately 75% of the everolimus found within the plasma is bound to protein. Similar to sirolimus, everolimus is metabolized in the intestine and liver by cytochrome P450 enzymes. At least 20 metabolites have been identified, with mono-hydroxyl, di-hydroxyl, demethylated, and an open ring form the major metabolites (42). For therapeutic drug monitoring of everolimus, a trough concentration of least 3 ng/mL must be reached in the first week post-transplant to prevent graft rejection (43). Therapeutic range of everolimus is usually considered as 3–8 ng/mL (44). Similar to other immunosuppressive agents, the incidence of adverse effects of everolimus is related to dose. Hyperlipidemia is of major concern, with increases in serum triglycerides and cholesterol observed, especially when everolimus is combined with cyclosporine and corticosteroids. Leukopenia and thrombocytopenia are also observed, and pancytopenia occurs rarely. There is also an increased risk of infectious episodes such as pneumonia, pharyngitis, sinusitis, and multiple herpetic oral lesions.

15.6. THERAPEUTIC DRUG MONITORING OF MYCOPHENOLIC ACID

Mycophenolic acid (MPA) is a fermentation product of *Penicillium* species that has antibacterial, antifungal, and immunosuppressive potential. However, due to poor bioavailability, mycophenolic acid is administered as mycophenolate mofetil (brand name CellCept), the 2-morpholinoethyl ester of mycophenolic acid. Both oral and intravenous preparations are commercially available. This drug received FDA approval for use as an immunosuppressant with corticosteroids and cyclosporine to prevent organ rejection in 1995. Mycophenolate mofetil is available in 250-mg capsules, 500-mg tablets, as a powder to make an oral suspension at 200 mg/mL, and as a sterile powder (500 mg) for intravenous administration. Mycophenolic acid is a potent noncompetitive inhibitor of inosine monophosphate dehydrogenase (IMPDH) enzymatic activity and thus selectively inhibits lymphocyte proliferation (45). More recently, mycophenolic acid sodium became available for clinical use. Figure 15.2 shows the chemical structures of mycophenolic acid, mycophenolic acid mofetil, and mycophenolic acid sodium.

Mycophenolate mofetil is rapidly and completely absorbed, and it is quickly deesterified in the blood and tissues into the pharmacologically active mycophenolic acid. The half-life of mycophenolate mofetil during intravenous administration is less than 2 minutes. Following an oral dose of mycophenolate mofetil, MPA reaches a maximum concentration within 1 hour (46). Almost all the drug (more than 99%) can be found in the plasma compartment. For this reason, serum or plasma is used for MPA quantitation instead of whole blood, which is the preferred specimen for therapeutic drug monitoring of

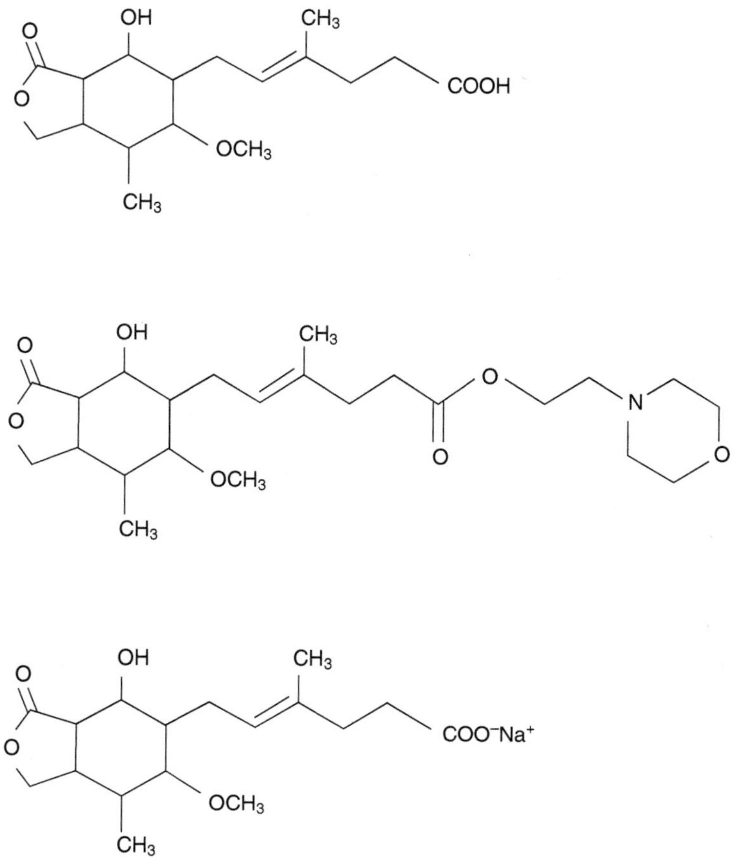

Figure 15.2. Chemical structures of (top) mycophenolic acid, (middle) mycophenolic acid mofetil, and (bottom) mycophenolate sodium.

other immunosuppressants. Mycophenolic acid has an elimination half-life of 8–18 hours (mycophenolate mofetil) and is conjugated with glucuronic acid in the liver to the primary inactive metabolite, 7-O-glucuronide mycophenolic acid (47). Another metabolite produced in small quantities is acyl glucuronide, an active metabolite that may contribute to the adverse gastrointestinal effects of mycophenolic acid. Mycophenolic acid is strongly bound to serum albumin with typical concentrations of free mycophenolic acid ranging from 1.25% to 2.5% of the total concentration. Free mycophenolic acid (pharmacologically active fraction) concentrations are increased in hypoalbuminemia, hyperbilirubinemia, and uremia (48). The suggested therapeutic range of mycophenolic acid (trough serum or plasma level) is 1–3.5 μg/mL (49). Adverse effects from mycophenolate mofetil include diarrhea, nausea, vomiting, and abdominal pain. Marrow suppression and anemia can also occur. An increased risk of cytomegalovirus, candida, and herpes simplex infections has also been reported.

15.7. IMMUNOASSAYS VERSUS CHROMATOGRAPHY FOR THERAPEUTIC DRUG MONITORING OF IMMUNOSUPPRESSANTS: LIMITATION OF IMMUNOASSAYS

For therapeutic drug monitoring of immunosuppressants, several immunoassays are available commercially (Table 15.1). Antibody conjugate magnetic immunoassays (ACMIA), marketed by Siemens Diagnostics (formerly Dade Behring), includes CSA Flex (cyclosporine), CSA-E Flex reagent cartridges (cyclosporine extended range), and TAC-R Flex reagent cartridges (tacrolimus) for application on the Dimension systems. The ACMIA assay is the only assay that does not require any specimen pretreatment and relies on online mixing and ultrasonic lysing of whole blood followed by exposure to β-galactosidase-antibody conjugate, removal of free conjugate using analyte-coated magnetic particles, and detection via a spectrometric β-galactosidase

TABLE 15.1. Commercially Available Immunoassays for Various Immunosuppressants from Major Diagnostic Companies (FDA-Approved Assays)

Immunosuppressant	Immunoassay	Analytical Range
Cyclosporine	ACMIA Cyclosporine Flex	25–500 ng/mL
	ACMIA Cyclosporine Extended Range Flex	350–2000 ng/mL
	CEDIA Cyclosporine Plus	25–400 ng/mL
	CEDIA Plus High Range	450–2000 ng/mL
	CMIA Cyclosporine	30–1500 ng/mL
	EMIT 2000 Cyclosporine	25–500 ng/mL
	EMIT 2000 Cyclosporine Specific Assay	350–2000 ng/mL
	ADVIA Centaur Cyclosporine immunoassay	30–1500 ng/mL
Tacrolimus	ACMIA TAC-R Flex	1.2–30 ng/mL
	CEDIA Tacrolimus assay	2.0–30 ng/mL
	CMIA Tacrolimus assay	2.0–30 ng/mL
	MEIA Tacrolimus assay	3.0–30 ng/mL
	EMIT 2000 Tacrolimus assay	2.0–30 ng/mL
Sirolimus	CEDIA Sirolimus assay	5.0–30 ng/mL
	CMIA Sirolimus assay	2.0–30 ng/mL
	MEIA Sirolimus assay	2.5–30 ng/mL
Everolimus	QMS Everolimus assay	1.5–20 ng/mL
Mycophenolic acid	CEDIA Mycophenolic acid	0.3–10 µg/mL
	EMIT 2000 Mycophenolic acid	0.1–15 µg/mL

ACMIA, Antibody conjugated magnetic immunoassay from Dade Behring, now Siemens Diagnostics; CEDIA; Cloned enzyme donor immunoassay developed by Microgenics Corporation, now Thermo Fischer; CMIA, chemiluminescent microparticle immunoassay developed by Abbott Laboratories; EMIT, enzyme multiplied immunoassay developed by Syva and now in Siemens Diagnostics product line. Roche Diagnostics also marketed the EMIT assay.

reaction. Cloned enzyme donor immunoassay (CEDIA) products are available commercially from Microgenics (now part of Thermo Fisher) as the CEDIA Cyclosporine Plus Assay (also available are reagents for a high-range assay), CEDIA Mycophenolic Acid Assay, CEDIA Sirolimus Assay, and CEDIA Tacrolimus Assay. Analyses can be performed on a variety of instrument systems including the Hitachi 917, Beckman SYNCHRON, Olympus, and other automated analyzers. The principle of the CEDIA technology is to lyse whole blood, followed by exposure to enzyme donor (analyte-inactive β-galactosidase fragment conjugate) and anti-analyte antibody, then bind remaining free enzyme donor to enzyme acceptor (second inactive β-galactosidase fragment) to produce an active enzyme that catalyzes a spectrometric reaction.

Chemiluminescent microparticle immunoassay (CMIA) is a new product line from Abbott Diagnostics. Reagent kits are available for cyclosporine, sirolimus, and tacrolimus for analysis on the Abbott Diagnostics Architect *i* system. The principle of CMIA technology is to lyse whole blood with subsequent exposure to anti-analyte–coated paramagnetic particles, followed by addition of acridinium-labeled analyte conjugate, and then pretrigger and trigger solutions to initiate a chemiluminescent reaction. Enzyme-multiplied immunoassay technique (EMIT) reagents are available commercially from Siemens as the Dade Behring Syva product line. EMIT 2000 reagent kits are available for cyclosporine, mycophenolic acid, and tacrolimus on Viva E or V-Twin Drug Testing Systems. Roche Diagnostics offers an EMIT assay for cyclosporine. EMIT technology uses a whole blood pretreatment followed by exposure to analyte-glucose-6-phosphate dehydrogenase conjugate and specific antibody; the remaining free enzyme conjugate forms nicotinamide adenine dinucleotide (NADH) in a spectrometric reaction. Enzyme-mediated spectrometric assay is an antibody-independent technology recently commercialized by Roche Diagnostics for mycophenolic acid (Total MPA). This product is approved for use on COBAS INTEGRA systems. The principle of the test mimics inhibition by mycophenolic acid of the nicotinamide adenine dinucleotide (NAD)-dependent enzymatic reaction of inosine monophosphate dehydrogenase II (IMPDH) with inosine monophosphate (IMP) but uses a mutant enzyme. NADH product formation is proportional to absorbance at 340 nm. For everolimus only, the quantitative microsphere system (QMS) everolimus assay (Thermo Fischer) is approved by the FDA for use in the United States. A fluorescence polarization immunoassay for everolimus is also available.

In addition to immunoassays, HPLC with ultraviolet detection (HPLC-UV) is another well-developed technology with many published procedures. Ultraviolet light at a wavelength selective for analyte and internal standard is used for the detection and concentration of an eluting compound proportional to the amount of light absorbed. However, tacrolimus cannot be quantified by HPLC-UV because it lacks a significant chromophore. Liquid chromatography with mass spectrometric detection or tandem mass spectrometric detection (LC/MS and LC/MS/MS) technology has been applied to therapeutic drug monitoring of all of the immunosuppressive drugs and is considered the gold standard. In principle, analyte must be extracted from the biological matrix before application onto a short column where water-soluble substances are washed away; then the compounds of interest are eluted (sometimes selectively) from the column using a methanol- or acetonitrile-based mobile phase. Using user-defined instrument specifications, column eluate is vaporized and compounds are ionized in the mass spectrometer source where specific ions are directed into the mass spectrometer. Because only ions with specific mass-to-charge ratios (e.g., analyte and internal standard) are permitted to pass though the quadrupoles to the photomultiplier for detection, this technology is highly selective. However, electrospray ionization commonly used for immunosuppressive drug analysis is prone to matrix-associated problems.

Although immunoassays can be adapted easily on an automated analyzer for the rapid therapeutic drug monitoring of immunosuppressants, all immunoassays suffer from metabolite cross-reactivities of corresponding drug metabolite, causing a significant positive bias in value determined by an immunoassay compared with a value determined by a specific chromatographic technique. The disadvantage of liquid chromatography combined with mass spectrometric method is the initial high cost of the instrument and availability of highly skilled scientific staff for operation. However, after initial investment, cost per specimen is substantially low compared with immunoassays. However, advantages are high sensitivity, high specificity, small sample requirement, minimal sample preparation, and rapid throughput. In addition, all immunosuppressants can be measured simultaneously (50).

15.7.1. Limitations of Cyclosporine and Tacrolimus Immunoassays

Bias between immunoassays and chromatography-based methods in the analysis of various immunosuppressants has been well documented. The mean cyclosporine concentrations have been found to be approximately 12%, 13%, 17%, 22%, and 40% higher than chromatographic method compared with measurements by various immunoassays; Dimension ACMIA, Syva EMIT, CEDIA PLUS, FPIA (fluorescence polarization immunoassay on the TDx, and FPIA on the AxSYM, respectively) (51–55). The FPIA assay for application on the TDx analyzer (Abbott Laboratories, Abbott Park, IL) is no longer commercially available. More recently, Abbott Laboratories has marketed a chemiluminescent microparticle immunoassay (CMIA) for application on the Architect analyzer. The assay is accomplished by incubating an extracted whole blood sample (whole blood pretreatment with organic reagents to precipitate protein and extracting the drug) with magnetic microparticle coated with a very low amount of anticyclosporine antibody. After a wash step, the microparticles were incubated with a chemiluminescent cyclosporine tracer following a second wash step and measuring the chemiluminescent signal (56).

Wallemacq et al reported findings from the multicenter evaluation of Abbott Architect cyclosporine assay using seven clinical laboratories. Values obtained by the immunoassay were compared with corresponding values obtained by LC/MS/MS. The authors observed minimal cross-reactivity of cyclosporine metabolites in the cyclosporine Architect immunoassay. The AM1 and AM9, the two major cyclosporine metabolites, exhibited −2.5 to 0.2% and 0.8–2.2% cross-reactivity, respectively, with the cyclosporine immunoassay. Comparison testing with Roche Integra assay (Integra 800; Roche Diagnostics) showed 2.4% cross-reactivity for AM1 metabolite and 10.7% cross-reactivity with the AM9 metabolite. The Architect immunoassay showed an average bias of 31 ng/mL of cyclosporine compared with the LC/MS/MS method. The authors concluded that the cross-reactivity of cyclosporine metabolites was significantly reduced in the Architect cyclosporine immunoassay (57). Soldin et al evaluated performance evaluation of a new ADVIA Centaur cyclosporine immunoassay that requires a single-step extraction and observed excellent correlation between cyclosporine values obtained by the LC/MS/MS assay and ADVIA Centaur cyclosporine assay. The regression analysis using LC/MS/MS values as the reference method and ADVIA Centaur cyclosporine value as the method to be compared, the authors observed the following regression equation:

$$y = 0.94\,x + 19$$

The authors concluded that the new ADVIA centaur assay compares favorably with the LC/MS/MS assay (58). However, falsely elevated blood cyclosporine levels due to the presence of endogenous antibody was reported to the ACMIA cyclosporine assay run on the Dimension RXL analyzer (Siemens Diagnostics). De Jonge et al reported a falsely elevated cyclosporine level of 492 ng/mL in a 77-year-old patient. However, using LC/MS, the cyclosporine level was undetectable. In addition, the Architect cyclosporine assay also yielded a value lower than the detection limit. Treating a specimen with polyethylene glycol and remeasuring cyclosporine in the supernatant by the same ACMIA assay showed no detected level of cyclosporine, confirming that the interfering substance was a protein, most likely an endogenous antibody (59).

Case Report

A 23-month-old pediatric patient underwent allogeneic hematopoietic stem cell transplantation from cord blood obtained from an unrelated donor for familial hemophagocytic lymphohistiocytosis with a homozygous 284–285 mutation in the perforin gene where both parents were heterozygous for this molecular defect. The patient received oral cyclosporine before transplant and then to prevent graft-versus host disease. Blood cyclosporine levels were routinely tested by ACMIA assay. Two months after transplantation a trend toward increasing cyclosporine blood concentration was observed. Cyclosporine treatment was discontinued on day 58 and restarted on day 68 (5 mg/kg per day). However, cyclosporine concentration increased from 222 ng/mL on day 58 to 312 ng/mL on day 60. The highest value observed was 850 ng/mL on day 67, and then after restarting of cyclosporine on day 68, the observed value was 1034 ng/mL. The cyclosporine value was increased to 1146 ng/mL on day 72. In contrast, using the EMIT cyclosporine assay, all cyclosporine values were below 40 ng/mL, the detection limit of the assay between days 65 and 68 (no values available before day 65). On day 72, the corresponding value obtained by EMIT was 64 ng/mL, which was significantly lower than the 1146 ng/mL value observed by the ACMIA assay. When five of the blood samples that showed high cyclosporine levels using the ACMIA assay were purified by Protein A-Sepharose chromatography to remove endogenous immunoglobulin G, the cyclosporine values reassayed by the same ACMIA assay showed values less than 25 ng/mL, the detection limit of the assay. Therefore, falsely high cyclosporine values as determined by the ACMIA assay were due to interference by an endogenous antibody in the assay (60).

Although HPLC-based assays are considered the reference method for the analysis of tacrolimus in whole blood, immunoassays are routinely used in clinical laboratories due to their speed and ease of operation. Several studies have reported false-positive tacrolimus concentrations in patients with low hematocrit values and high imprecision at tacrolimus values less than 9 ng/mL with the microparticle enzyme immunoassay (MEIA) tacrolimus assay for application on the AxSYM platform (Abbott Laboratories). However, the EMIT assay was not affected. In one study when specimens were divided into three groups (group A, hematocrit less than 25%; group B, hematocrit 25–35%; and group C, hematocrit more than 35%), the difference between methods (MEIA versus EMIT 2000) increases as hematocrit decreases. Moreover, false-positive results were reported in 63% of specimens with MEIA where patients did not receive any tacrolimus, but only 2.2% of specimens showed false-positive tacrolimus value using EMIT. The median difference between tacrolimus results were 2.2 ng/mL using MEIA assay compared with corresponding values using the EMIT assay in patients with hematocrit equal to or less than 25%. In contrast, no difference was observed in tacrolimus values measured by the MEIA and the EMIT assay in patients with hematocrit over 35%. The false-positive values ranged from 0.0 to 3.7 ng/mL using MEIA in patients who never received tacrolimus (61). Immunoassays for tacrolimus are also affected by the cross-reactivity from tacrolimus metabolites. Westley et al evaluated CEDIA tacrolimus assay by measuring values obtained by the CEDIA assay, LC/MS/MS, and the MEIA assay. The authors observed a 33.1% bias with the CEDIA assay and 20.1% bias with the MEIA assay in tacrolimus values compared with the tacrolimus value measured by the LC/MS/MS method in renal transplant recipients (62). The new Architect tacrolimus assay

(CMIA tacrolimus assay) is free from interferences of low hematocrit that affect MEIA assay manufactured by the same Abbott Laboratories. The average bias of tacrolimus values determined by the Architect tacrolimus assay and specific LC/MS/MS assay evaluated at two sites was 0.51 ng/mL and 1.63 ng/mL, respectively. The authors concluded that the semiautomated Architect tacrolimus assay is an alternative approach of therapeutic drug monitoring for tacrolimus in laboratories not equipped with LC/MS/MS (63). Bazin et al also evaluated the CMIA architect tacrolimus assay and observed an average bias of 20% between values determined by the CMIA assay and LC/MS/MS as evidenced by this regression equation:

$$\text{CMIA} = 1.2\ \text{LC/MS/MS} + 0.14 \quad (64)$$

Like cyclosporine, the ACMIA tacrolimus assay is also affected by rheumatoid factors and endogenous heterophilic antibodies. In one study, the authors analyzed blood samples from patients who never received tacrolimus but had various concentrations of rheumatoid factors in their blood. No positive tacrolimus value was observed (above 2.3 ng/mL, the detection limit of the assay) in patients with rheumatoid factor concentrations less than 20 U/mL. However, two patients out of 50 analyzed showed false-positive tacrolimus values, and the concentration range of rheumatoid factors in these patients varied from 110 to 2650 U/mL. However, when two specimens that showed positive tacrolimus values were treated with an immunoglobin-blocking agent, the tacrolimus values became nondetected when reanalyzed with the same ACMIA assay, indicating that the rheumatoid factor may interfere with the ACMIA tacrolimus assay (65). Altinier et al also described the interference of heterophilic antibody in the ACMIA tacrolimus assay. A sample of a patient showed tacrolimus values in the range of 49 to 12.5 ng/mL even after interruption of the treatment. The authors confirmed that the elevated tacrolimus levels were due to the presence of heterophilic antibody by treating samples with heterophilic blocking tubes and protein G resin that removed such interference (66). In another report, the authors observed a high tacrolimus value (79.7 ng/mL) in a liver transplant recipient using the ACMIA tacrolimus assay despite discontinuation of tacrolimus therapy. The authors identified β-galactosidase antibodies as the cause of interference because in this assay anti-tacrolimus antibody conjugated to β-galactosidase is used (67). Rostaing et al observed a falsely elevated tacrolimus level of 24 ng/mL using the ACMIA tacrolimus assay, but no detected level of tacrolimus using LC/MS/MS as well as the EMIT tacrolimus assay. The authors identified positive anti–double-stranded DNA autoantibodies as the cause of interference in the ACMIA assay (68).

Case Study

A 59-year-old man underwent a kidney transplant and was treated with tacrolimus and corticosteroids. In the posttransplant period he experienced recurrent bacterial and fungal infections and was treated with antibiotics. For the first 3 weeks after transplant, the patient's tacrolimus whole blood concentrations were consistent with dosage and were below 12 ng/mL. Twenty-five days after transplant, his tacrolimus level was measured by the ACMIA tacrolimus assay and Dimension analyzer and was found to be highly elevated to 21.5 ng/mL. The whole blood tacrolimus levels were monitored continuously, but values, although reduced, were discordant with the discontinuation of tacrolimus therapy. The tacrolimus values measured by the ACMIA assay after discontinuation of tacrolimus were all above 10 ng/mL, whereas corresponding tacrolimus values measured by the microparticle enzyme immunoassay (MEIA assay, Abbott Laboratories) were below 2 ng/mL, indicating interference in the tacrolimus measurement using the ACMIA assay. The authors diluted whole blood samples showing high tacrolimus concentrations with zero calibrator and then remeasured tacrolimus values using the ACMIA assay and observed nonlinearity, thus further confirming the interference in the ACMIA assay. However, washed erythrocytes showed significantly lower tacrolimus values as measured by the ACMIA assay as compared with the corresponding whole blood specimens. The authors concluded that the interference with the ACMIA tacrolimus assay was method specific and due to a factor present in the plasma because plasma tacrolimus values as measured by the ACMIA assay was higher than whole blood values. The authors also suggested that if the tacrolimus value measured by the ACMIA assay does not match the clinical picture, tacrolimus must be measured by an alternative method before any clinical intervention (69).

15.7.2. Limitations of Sirolimus, Everolimus, and Mycophenolic Acid Immunoassays

Colantonio et al compared the suitability of CEDIA and MEIA sirolimus immunoassay for therapeutic drug monitoring of sirolimus. The limit of detection was 1.1 ng/mL, and the limit of quantitation was found to be 1.5 ng/mL for the MEIA assay. In contrast, the limit of detection was much higher with the CEDIA assay (4.8 ng/mL), and the authors determined the limit of quantitation as 5.2 ng/mL. Comparison of sirolimus values obtained by these assays and a specific LC/MS analytical method for measuring sirolimus concentration showed a positive bias of an average 0.9 ng/mL with the MEIA assay and a larger positive bias of 2.1 ng/mL with the CEDIA sirolimus assay. The authors concluded that although the MEIA assay is suitable for therapeutic drug monitoring of sirolimus, the CEDIA assay based on a higher limit of quantitation that falls within the therapeutic range is not suitable for the therapeutic drug monitoring of sirolimus (70). Morris et al compared the MEIA sirolimus assay with the values obtained by using a specific LC/MS/MS assay and observed a mean bias of 49.2% (range: –2.4% to 128%) in tacrolimus values obtained by the MEIA assay with the corresponding values obtained by the LC/MS/MS assay. The authors did not investigate the cause of large and variable bias between the MEIA assay and the specific LC/MS/MS assay, but they speculated that sirolimus metabolite cross-reactivity with the MEIA assay could be a substantive contributing factor. The authors concluded that the MEIA sirolimus assay may be an adjunct to sirolimus dosage individualization in transplant recipients and that physicians must be aware of substantial and variable bias when interpreting sirolimus blood concentration measured by the MEIA assay. For selected patients with difficult clinical issue a specific chromatographic method must be used for therapeutic drug monitoring of sirolimus (71). Schmidt et al evaluated the new CMIA sirolimus assay for application on the Architect analyzer (Abbott Laboratories) and concluded that the assay only cross-reacts with sirolimus metabolite F4 and F5, but hematocrit has no effect on the assay. In a multisite clinical trial, the authors observed an average of 14%, 25%, and 39% mean bias with the CMIA assay and three different LC/MS/MS methods for determination of sirolimus in three different sites that evaluated the CMIA assay. The authors concluded that although the CMIA assay correlated well with LC/MS/MS method, it shows a positive bias in sirolimus values compared with values determined by more specific LC/MS/MS assays (72).

The quantitative microparticle system (QMS) everolimus assay received FDA approval for clinical use in 2011, and the assay is linear between 1.5 ng/mL and 20 ng/mL, covering the entire therapeutic range of everolimus. The limit of quantitation for this assay is 1.3 ng/mL. According to one report, this assay is not affected by 70 commonly used drugs, but structurally similar sirolimus exhibited an average of 46% cross-reactivity. In addition, the average bias everolimus values determined by the QMS everolimus assay and corresponding values obtained by a specific LC/MS/MS method was 11% based on the comparison of 90 specimens obtained from patients receiving everolimus. The authors concluded that the QMS everolimus assay showed adequate sensitivity and specificity and can be used for routine therapeutic drug monitoring of everolimus (73). Sallustio et al observed an average bias of over 30% in everolimus concentration as determined by the Seradyn fluorescence polarization immunoassay (FPIA) and a specific LC/MS/MS for everolimus and concluded that further investigation is needed before this assay can be used for the routine therapeutic monitoring of everolimus (74).

Mycophenolic acid is monitored in serum or plasma. Hosotsubo et al studied the analytical performance of the EMIT mycophenolic acid immunoassay and observed no interference from major metabolite mycophenolic acid glucuronide. In addition, the EMIT assay also correlated well with HPLC combined with an ultraviolet detection method for the determination of mycophenolic acid (75). In another study the authors investigated the analytical performance of the Roche total mycophenolic acid assay for application on Cobas Integra and the Cobas 6000 analyzer by comparing these methods with a specific LC/MS/MS method for the determination of mycophenolic acid in specimens obtained from liver transplant recipients. The authors did not observe any significant bias between values obtained by Roche total mycophenolic acid immunoassay and the values determined by LC/MS/MS. The Passing-Bablok regression analysis showed the Cobas Integra (mg/L) = 1.02 × LC/MS/MS (mg/L) − 0.50 and Cobas 6000 (mg/L) = 0.98 × LC/MS/MS (mg/L) − 0.47. Due to excellent correlation with LC/MS/MS values, the authors concluded that the Roche immunoassay is suitable for therapeutic drug monitoring of mycophenolic acid (76). However, Westley et al observed significant bias in mycophenolic acid determined by another immunoassay (CEDIA assay on Hitachi 911 analyzer) and values obtained by a chromatographic method (HPLC combined with ultraviolet detection [HPLC-UV]). The regression analysis of patients samples who were transplant recipients gave this equation: CEDIA (mg/L) = 1.18 HPLC/UV (mg/L) + 0.45 indicating an average positive bias of 18% with the CEDIA immunoassay results compared with HPLC/UV. The package insert indicated 192% cross-reactivity with the mycophenolic acid glucuronide metabolite. The authors concluded that CEDIA (Thermo Fischer, formerly Microgenics) mycophenolic acid run on the Hitachi 911 analyzer overestimated plasma mycophenolic acid values that are influenced by the transplant type (77). Shipkova et al observed a mean positive bias of 15% with samples obtained from patients after heart transplantation, but the mean bias was 41.7% and 52.3%, respectively, in specimens obtained from kidney transplant recipients and liver transplant recipients, respectively, using the CEDIA mycophenolic acid immunoassay compared with values obtained by the more specific HPLC-UV method (78). Major limitations of immunoassays available for monitoring various immunosuppressants are listed in Table 15.2.

15.8. CHROMATOGRAPHIC METHODS FOR THE THERAPEUTIC DRUG MONITORING OF IMMUNOSUPPRESSANTS

Although HPLC-UV can be used for the determination of cyclosporine and mycophenolic acid, tacrolimus cannot be determined by HPLC-UV due to the lack of absorption peak at the ultraviolet range. However, currently only mycophenolic acid is determined by HPLC-UV assay; cyclosporine, tacrolimus, sirolimus, and everolimus are measured using specific LC/MS or LC/MS/MS methods. In the early 1980s, cyclosporine became the first of the targeted immunosuppressants approved for clinical use, and HPLC-UV methods were developed for the monitoring of cyclosporine. For cyclosporine, under isocratic HPLC conditions, Gaussian chromatographic peak shape and a reasonable retention time could be achieved using C8- or C18-based column material but only with an elevated column temperature (70–80°C).

Compared with the more complex HPLC-UV techniques required to determine cyclosporine, sirolimus, and everolimus, HPLC-UV procedures for mycophenolic acid are relatively straightforward. A relatively interference-free

TABLE 15.2. Major Interferences in Commercially Available Immunoassays (FDA Approved) Used for Therapeutic Drug Monitoring of Immunosuppressants

Immunosuppressant/Immunoassay	Comments on Major Interferences
Cyclosporine	
ACMIA Cyclosporine (all assays)	Only cyclosporine assay affected by endogenous antibodies
CEDIA Cyclosporine (all assays)	Average 17% positive bias compared with chromatographic method
EMIT 2000 Cyclosporine (all assays)	Average 17% positive bias compared with chromatographic method
ADVIA Centaur Cyclosporine	No significant interference reported in this relatively new assay
Tacrolimus	
ACMIA TAC-R Flex	Only tacrolimus assay affected by endogenous antibodies
CEDIA Tacrolimus assay	Average 33.1% bias compared with chromatographic methods
CMIA Tacrolimus assay	Average 20% bias compared with chromatographic methods
MEIA Tacrolimus assay	False-positive results with low hematocrit (<25%) Average 20.1% bias compared with chromatographic methods
EMIT 2000 Tacrolimus assay	Not affected by hematocrit by variable cross-reactivity with tacrolimus metabolites
Sirolimus	
CEDIA Sirolimus assay	Limit of quantitation according to one report is 5.2 ng/mL, which overlaps with therapeutic range. This is a major limitation according to the report (reference 70)
CMIA Sirolimus assay	Average bias ranged from 14% to 39% compared with chromatographic methods
MEIA Sirolimus assay	Average 49.2% bias compared with chromatographic methods
Everolimus	
QMS Everolimus assay	Average 11% positive bias compared with chromatographic methods
Mycophenolic Acid	
CEDIA Mycophenolic acid	Significant cross-reactivity with mycophenolic acid glucuronide metabolite, positive bias ranging from 15% to 52.3% compared with chromatographic methods
EMIT 2000 Mycophenolic acid	No significant bias reported, no cross-reactivity with mycophenolic acid metabolite

ACMIA, antibody conjugated magnetic immunoassay from Dade Behring, now Siemens Diagnostics; CEDIA, cloned enzyme donor immunoassay developed by Microgenics Corporation, now Thermo Fischer; CMIA, chemiluminescent microparticle immunoassay developed by Abbott Laboratories; Chromatographic liquid chromatography combined with ultraviolet detection, mass spectrometry, or tandem mass spectrometry; EMIT, enzyme multiplied immunoassay developed by Syva and now in Siemens Diagnostics product line. Roche Diagnostics also marketed the EMIT assay.

measurement of mycophenolic acid is possible because sample cleanup begins with plasma (or serum) rather than whole blood, and therapeutic trough levels are in the μg/mL rather than the ng/mL range. In addition, mycophenolic acid exhibits two relatively unique and one more common ultraviolet absorbance maxima (216, 251, and 305 nm) from which detection wavelength can be selected (79). As an example, changing detection at 254 from 214 nm has been shown to reduce base line noise by 87%, but peak height was reduced by only 15% (80). With regard to analysis time, determination of mycophenolic acid alone allows for a "quick" isocratic chromatographic method (less than 10 min), whereas when both parent drug and metabolite(s) are measured simultaneously, a time-consuming isocratic or gradient mobile phase program must be applied that may require up to 30 minutes. One novel way to circumvent this problem is to use β-glucuronidase on a portion of the sample to hydrolyze the glucuronide metabolites to mycophenolic acid and then calculate the difference between the two mycophenolic acid measurements to quantitate the amount of metabolites (81). Moreover, automation of sample preparation has replaced tedious solid phase extraction, thus saving analysis time (82).

Although there is significant interest in measurement of sirolimus, there have been only a few procedures published for sirolimus by HPLC-UV because such techniques lack the sensitivity easily achieved by LC/MS/MS. The first description of an HPLC-UV procedure for whole blood everolimus was an adaptation of a procedure developed for sirolimus (83). A few other published procedures describe the quantitation of everolimus using HPLC-UV. These procedures begin with steps similar to those commonly applied to sample preparation for HPLC-MS analysis; that is, protein precipitation using a strong organic solvent in the presence of zinc metal ions. However, to remove interfering lipophilic substances, it was necessary to apply the resulting supernatant to solid phase extraction or basify the supernatant and extract into the same or another solvent and finally to wash with hexane or heptane. Chromatography accomplished on C-8 or C-18–based columns using elevated column temperature (55–60°C) improved peak shape and reduced run time to less than 15 minutes per sample. For these reasons, LC/MS/MS methods are favored by therapeutic drug monitoring of everolimus.

The capability of LC-MS technology to sort through the unintended interfering compounds in biological matrices and to detect compounds of interest selectively and sensitively is far greater than any other technology applied to the quantification of immunosuppressive drugs. The compounds of interest must be ionizable for detection using commonly used mass spectrometric techniques such as electrospray ionization and atmospheric pressure chemical ionization. Mobile phase additives such as acetic and formic acids facilitate ionization and conditions that promote labile ammoniated ions, that is, supplementation of the mobile phase with ammonium acetate or ammonium formate, as compared with the more stable species preferred in many of the LC/MS procedures developed for the determination of the immunosuppressive drugs.

To achieve the most accurate measurements, the performance of the internal standard must parallel that of the analyte as closely as possible, including chromatographic retention time and ionization efficiency. Therefore, deuterated standards are preferred. Quantifier mass transitions are monitored by selection of only ions with specific mass-to-charge ratios (i.e., ammoniated adducts of the intact analyte and internal standard), followed by fragmentation of these precursor or parent ions into one or more product or daughter ions and then selective detection of only the most abundant product ions. For confirmation, use of the ratio of the quantifier mass transition and qualifier mass transition response (fragmentation of the precursor ion to a second product ion) for each compound of interest is recommended. Matrix effects can be minimized in a variety of ways including optimization of sample preparation, manipulation of chromatography, compensation through use of a suitable internal standard, or, if possible, change to a different mass spectrometer interface. Zinc sulfate with either acetonitrile or methanol is commonly used in LC/MS procedures for the determination of immunosuppressive drugs. Zinc sulfate (10% in a 1:1 ratio with 0.5 N sodium hydroxide) in a 4:1 ratio with human plasma removes more than 99.9% of proteins; acetonitrile removes 92–95%, and methanol removes 91–92% of proteins. However, sulfate causes severe ion suppression and may pollute the mass spectrometer interface. Therefore, for each sample extract injected into the HPLC system, an aqueous mobile phase must be flushed through the HPLC column and to waste through use of a diverter valve before the compounds of interest are eluted into the ionization chamber of the mass spectrometer. Liquid–liquid extraction can produce clean extracts, but recovery of the compounds of interest may be low or variable. Mixed-mode (reverse phase and ion-exchange) solid phase extraction produces the cleanest extracts but adds a multiple-step procedure to the sample treatment. The choice of a mobile phase organic solvent, usually acetonitrile or methanol, can also influence ionization efficiency.

Using LC/MS or LC/MS/MS, not only individual immunosuppressant but also the simultaneous determination of several immunosuppressants is possible using one method. There are many published methods for the analysis of various immunosuppressants using LC/MS or LC/MS/MS. Ivanova et al described a LC/MS/MS method for the simultaneous determination of sirolimus and everolimus using a C-18 column held at 60°C after online solid phase extraction. The authors achieved complete analysis by 2.6 minutes using a gradient elution profile. The intra- and interassay CVs were less than 7% and detection limit as 2 ng/mL (84). Koster et al described a LC/MS/MS protocol for the analysis of cyclosporine, tacrolimus, sirolimus, and everolimus in whole blood with an analysis time of 2.6 minutes. The authors used a C-18 column held at 60°C while sirolimus and everolimus were measured separately from cyclosporine and tacrolimus (85). Meinitzer et al described a protocol for the simultaneous analysis of cyclosporine, tacrolimus, sirolimus, and everolimus using LC/MS/MS. The authors used solid phase online extraction requiring only 1 minute, and protein precipitation was achieved using zinc sulfate and methanol. The analytical measurement range of cyclosporine was 10–1500 ng/mL for cyclosporine, 1.0–44 ng/mL for tacrolimus, 1.0–48 ng/mL for sirolimus, and 1.2–48 ng/mL for everolimus. The authors used deuterated internal standards for cyclosporine and everolimus; ascomycin was used as the internal standard for the analysis of tacrolimus and sirolimus (86). Figurski et al described a LC/MS/MS method for the simultaneous quantification of both total and free fraction of mycophenolic acid, mycophenolic acid glucuronide, mycophenolic acid acyl glucuronide while total drug concentrations were determined in plasma and free drug or metabolite concentrations were determined in protein free ultrafiltrates. The authors used carboxy-butoxy ether of mycophenolic acid as the internal standard. Specimens (plasma or ultrafiltrate) were extracted with acetonitrile/formic acid (99.5:0.5 by volume), and the composition of the mobile phase was acetonitrile/water/formic acid (10:89.5:0.5 by volume). The analytical column used was a C-18 reverse phase column (87).

15.9. CONCLUSIONS

Therapeutic drug monitoring of immunosuppressants is essential for proper patient management. Currently cyclosporine, tacrolimus, sirolimus, everolimus, and mycophenolic acid are used as major immunosuppressant drugs. Although whole blood is the preferred specimen for therapeutic drug monitoring of cyclosporine, tacrolimus, sirolimus, and everolimus, plasma or serum is the preferred specimen for therapeutic drug monitoring of mycophenolic acid. Immunoassays are commercially available for routine therapeutic drug monitoring of cyclosporine, tacrolimus, sirolimus, everolimus, and mycophenolic acid, but immunoassays are subjected to interferences mostly from drug metabolites. In addition, ACMIA tacrolimus and cyclosporine immunoassays (Siemens Diagnostics, Deerfield, IL) are affected by endogenous antibodies. Therefore, chromatographic methods, especially chromatography combined with mass spectrometry or chromatography combined with tandem mass spectrometry, are the preferred analytical methods for therapeutic drug monitoring of immunosuppressants, although therapeutic drug monitoring of mycophenolic acid can be adequately conducted using chromatography combined with an ultraviolet detector.

REFERENCES

1. Lindholm A, Albrechtsen D, Tufveson G, Karlberg I et al. A randomized trial of cyclosporine and prednisolone versus cyclosporine, azathioprine, and prednisolone in primary cadaveric renal transplantation. Transplantation 1992; 54: 624–631.
2. Morissette P, Albert C, Busque S, St-Louis G et al. In vivo higher glucuronidation of mycophenolic acid in male than female recipients of a cadaveric kidney allograft

and under immunosuppressant therapy with mycophenolic acid mofetil. Ther Drug Monit 2001; 23: 520–525.

3. Frauca E, Diaz MC, de le Vega A, Hierro L et al. Cyclosporine monitoring in the early post-transplant period in pediatric liver transplant recipients. Pediatr Transplant 2007; 11: 530–535.
4. Filler G. Calcineurin inhibitor in pediatric renal transplant recipients. Pediatr Drugs 2007; 8: 165–174.
5. Dipchand AL, Peitra B, McCrindle BW, Rosenbrook-Bicknell HL et al. Mycophenolic acid levels in pediatric heart transplant recipients receiving mycophenolate mofetil. J Heart Lung Transplant 2001; 20: 1035–1043.
6. Filler G, Lepage N, Delisle B, Mai I. Effect of cyclosporine on mycophenolic acid area under the concentration-curve in pediatric kidney transplant recipients. Ther Drug Monit 2001; 23: 514–519.
7. Undre NA. Pharmacokinetics of tacrolimus-based combination therapies. Nephrol Dial Transplant 2003; 18 (Suppl 1): i12–i15.
8. First MR. Tacrolimus based immunosuppression. J Nephrol 2004; 17: 25–31.
9. Schreiber SL, Crabtree GR. The mechanism of action of cyclosporin A and FK-506. Immunol Today 1992; 13: 136–142.
10. Augustine JJ, Hricik DE. Experience with everolimus. Transplant Proc 2004; 36 (Suppl 2S): 500WS–503WS.
11. Kirchner GI, Meier-Wiedenbach I, Manns MP. Clinical pharmacokinetics of everolimus. Clin Pharmacokinet 2004; 43: 83–95.
12. Abraham RT, Wiederrecht GJ. Immunopharmacology of rapamycin. Ann Rev Immunol 1996; 14: 483–510.
13. Schuler W, Sedrani R, Cottens S, Haberlin B et al. SDZ RAD a new rapamycin derivative: pharmacological properties in vitro and in vivo. Transplantation 1997; 64: 36–42.
14. Kahan BD, Kaplan B, Lorber M, Winkler M et al. RAD in de novo renal transplantation: comparison of three dosages on the incidence and severity of acute rejection. Transplantation 2001; 71: 1400–1406.
15. Neumayer HH, Paradis K, Korn A, Jean C et al. Entry into human study with the novel immunosuppressant SDA RAD in stable renal transplant recipients. Br J Clin Pharmacol 1999; 48: 694–703.
16. Deters M, Kirchner G, Resch K, Kaever V. Simultaneous quantification of sirolimus, everolimus, tacrolimus and cyclosporine by liquid chromatography-mass spectrometry. Clin Chem Lab Med 2002; 40: 285–292.
17. Kovarik JM, Kahan BD, Kaplan B, Lober M et al. Longitudinal assessment of everolimus in de novo renal transplant recipients over the first post-transplant year: pharmacokinetics, exposure response relationships and influence on cyclosporine. Clin Pharamcol Ther 2001; 69: 48–56.
18. Mabasa VH, Ensom MH. The role of therapeutic drug monitoring of everolimus in solid organ transplantation. Ther Drug Monit 2005; 27: 666–676.
19. Borel JF, Feurer C, Gubler HU, Stahelin H. Biological effects on cyclosporin A: a new antilymphocytic agent. Agents and Actions 1976; 6: 468–475.

20. Kovarik JM, Mueller EA, van Bree JB, Fluckinger SS et al. Cyclosporine pharmacokinetics and variability from a microemulsion formulation-a multicenter investigation in kidney transplant patients. Transplantation 1994; 58: 658–663.
21. Frauca E, Diaz MC, de le Vega A, Hierro L et al. Cyclosporine monitoring in the early post-transplant period in pediatric liver transplant recipients. Pediatr Transplant 2007; 11: 530–535.
22. Chrintians U, Sewing KF. Cyclosporin metabolism in transplant patients. Pharmacol Ther 1993; 57: 291–345.
23. Falck P, Asberg A, Byberg KT, Bremer S et al. Reduced elimination of cyclosporine A in elderly (>65 years) kidney transplant recipients. Transplantation 2008; 86: 1379–1383.
24. Dipchand AL, Peitra B, McCrindle BW, Rosenbrook-Bicknell HL et al. Mycophenolic acid levels in pediatric heart transplant recipients receiving mycophenolate mofetil. J Heart Lung Transplant 2001; 20: 1035–1043.
25. Min DI, Perry PJ, Chen HY, Hunsicker LG. Cyclosporine trough concentrations in pediatric allograft rejection and renal toxicity up to 12 months after renal transplantation. Pharmacotherapy 1998; 18: 282–287.
26. Sukhavasharin NH, Praditpornsilpa K, Avihingsanon Y, Kuoatawintu P et al. Study of cyclosporine level at 2 hours after administration in preoperative kidney transplant recipients for prediction of postoperative optimal cyclosporine dose. J Med Assoc Thai 2006; 89 (Suppl 2): S15–S20.
27. Hu YF, Qui W, Liu ZQ, Zhu LJ et al. Effects of genetic polymorphism of CYP3A4, CYP3A5 and MDR1 on cyclosporine pharmacokinetics after renal transplantation. Clin Exp Pharamacol Physiol 2006; 33: 1093–1098.
28. First MR. Tacrolimus based immunosuppression. J Nephrol 2004; 17: 25–31.
29. Jorgensen KA, Koefoed-Nielsen PB, Karamperis N. Calcineurin phosphatase activity and immunosuppression. A review on the role of calcineurin phosphatase activity and the immunosuppressive effect of cyclosporin A and tacrolimus. Scand J Immunol 2003; 57: 93–98.
30. Venkataramanan R, Swaminathan A, Prasad T, Jain A et al. Clinical pharmacokinetics of tacrolimus. Clin Pharmacokinet 1995; 29: 404–430.
31. Kelly P, Kahan BD. Review: metabolism of immunosuppressant drugs. Current Drug Metab 2002; 3: 275–287.
32. Coto E, Tavira B. Pharmacogenetics of calcineurin inhibitors in renal transplantation. Transplantation 2009; 88 (Suppl): S62–S67.
33. Holt DW. Therapeutic drug monitoring of immunosuppressive drugs in kidney transplantation. Curr Opin Nephrol Hypertens 2002; 11: 657–663.
34. McMaster P, Mirza DF, Ismail T, Vennarecci G et al. Therapeutic drug monitoring of tacrolimus in clinical transplantation. Ther Drug Monit 1995; 17: 602–605.
35. Kimball PM, Derman RK, Van Buren CT, Lewis RM et al. Cyclosporine and rapamycin affect protein kinase C induction of intracellular activation signal, activator of DNA replication. Transplantation 1993; 55: 1128–1132.
36. Yatscoff R, LeGatt D, Keenan R, Chackowsky P. Blood distribution of rapamycin. Transplantation 1993; 56: 1202–1206.
37. Gallant-Haidner HL, Trepanier DJ, Freitag DG, Yatscoff RW. Pharmacokinetics and metabolism of sirolimus. Ther Drug Monit 2000; 22: 31–35.

38. Stenton SB, Partovi N, Ensom MH. Sirolimus: the evidence for clinical pharmacokinetic monitoring. Clin Pharmacokinetic 2005; 44: 769–786.
39. Wong SH. Therapeutic drug monitoring for immunosuppressants. Clin Chim Acta 2001; 313: 241–253.
40. Kirchner GI, Meier-Wiedenbach I, Manns MP. Clinical pharmacokinetics of everolimus. Clin Pharmacokinet 2004; 43: 83–95.
41. Eisen HJ, Tuzcu EM, Dorent R, Kobashigawa J et al; the RAD B253 Study Group. Everolimus for the prevention of allograft rejection and vasculopathy in cardiac-transplant recipients. N Engl J Med 2003; 349: 847–858.
42. Kirchner GI, Winkler M, Mueller L, Vidal C et al. Pharmacokinetics of SDZ RAD and cyclosporin including their metabolites in seven kidney graft patients after the first dose of SDZ RAD. Br J Clin Pharmacol 2000; 50: 449–454.
43. Romagnoli J, Citterio F, Favi E, Salerno MP et al. Higher incidence of acute rejection in renal transplant recipients with low everolimus exposure. Transplant Proc 2007; 39: 1823–1826.
44. Mabasa VH, Ensom MH. The role of therapeutic drug monitoring of everolimus in solid organ transplantation. Ther Drug Monit 2005; 27: 666–676.
45. Wu JC. Mycophenolate mofetil: molecular mechanisms of action. Perspex Drug Discov Design 1994; 2: 185–204.
46. Bullingham RE, Nicholls AJ, Kamm BR. Clinical pharmacokinetics of mycophenolate mofetil. Clin Pharmacokinet 1998; 34: 429–455.
47. Korecka M, Nikolic D, van Breemen RB, Shaw LM. Inhibition of inosine monophosphate dehydrogenase by mycophenolic acid glucuronide is attributable to the presence of trace quantities of mycophenolic acid. Clin Chem 1999; 45: 1047–1050.
48. Kaplan B, Meier-Kriesche HU, Friedman G, Mulgaonkar S et al. The effect of renal insufficiency on mycophenolic acid protein binding. J Clin Pharmacol 1999; 39: 715–720.
49. Tredger JM, Brown NW, Adams J, Gonde CE et al. Monitoring mycophenolate in liver transplant recipients: toward a therapeutic range. Liver Transpl 2004; 10: 492–502.
50. Taylor PJ. Therapeutic drug monitoring of immunosuppressant drugs by high performance liquid chromatography-mass spectrometry. Ther Drug Monit 2004; 26: 215–219.
51. Steimer W. Performance and specificity of monoclonal immunoassays for cyclosporine monitoring: how specific is specific? Clin Chem 1999; 45: 371–381.
52. Schutz E, Svinarov D, Shipkova M, Niedmann PD et al. Cyclosporin whole blood immunoassays (AxSYM, CEDIA, and Emit): a critical overview of performance characteristics and comparison with HPLC. Clin Chem 1998; 44: 2158–2164.
53. Hamwi A, Veitl M, Manner G, Ruzicka K et al. Evaluation of four automated methods for determination of whole blood cyclosporine concentrations. Am J Clin Pathol 1999; 112: 358–365.
54. Terrell AR, Daly TM, Hock KG, Kilgore DC et al. Evaluation of a no-pretreatment cyclosporin A assay on the Dade Behring Dimension RxL clinical analyzer. Clin Chem 2002; 48: 1059–1065.

55. Butch AW, Fukuchi AM. Analytical performance of the CEDIA® cyclosporine PLUS whole blood immunoassay. J Anal Toxicol 2004; 28: 204–210.
56. Brate EM, Finley DM, Grote J, Holets-McCormack S et al. Development of an Abbott Architect cyclosporine immunoassay without metabolite cross-reactivity. Clin Biochem 2010; 43: 1152–1157.
57. Wallemacq P, Maine GT, Berg K, Rosiere T et al. Multisite analytical evaluation of Abbott Architect cyclosporine assay. Ther Drug Monit 2010; 32: 145–151.
58. Soldin SJ, Hardy RW, Wians FH, Balko JA et al. Performance evaluation of the new ADVIA Centaur system cyclosporine assay (single-step extraction). Clin Chim Acta 2010; 411: 806–811.
59. De Jonge H, Geerts I, Declercq P, de Loor H et al. Apparent elevation of cyclosporine whole blood concentration in a renal allograft recipient. Ther Drug Monit 2010; 32: 529–531.
60. Bartoli A, Molinaro M, Visal L. Falsely elevated whole blood cyclosporine concentrations measured by an immunoassay with automated pretreatment [letter to the editor]. Ther Drug Monit 2010; 32: 791–792.
61. Armedariz Y, Garcia S, Lopez R, Pou L et al. Hematocrit influences immunoassay performance for the measurement of tacrolimus in whole blood. Ther Drug Monit 2005; 27: 766–769.
62. Westley IS, Taylor PJ, Salm P, Morris RG. Cloned enzyme donor immunoassay tacrolimus assay compared with high-performance liquid chromatography-tandem mass spectrometry in liver and renal transplant recipients. Ther Drug Monit 2007; 29: 584–591.
63. Wallemacq P, Goffinet JS, O'Morchoe S, Rosiere T et al. Multi-site analytical evaluation of the Abbott Architect tacrolimus assay. Ther Drug Monit 2009; 31: 198–204.
64. Bazin C, Guinedor A, Barau C, Gozalo C et al. Evaluation of the Architect tacrolimus assay in kidney, liver and heart transplant recipients. J Pharm Biomed Appl 2010; 53: 997–1002.
65. Barcelo-Martin B, Marquet P, Ferrer JM, Castanyer Puig B et al. Rheumatoid factor interference in a tacrolimus immunoassay. Ther Drug Monit 2009; 31: 743–745.
66. Altinier S, Varagnolo M, Zaninotto M, Boccagni P et al. Heterophilic antibody interference in a non-endogenous molecule assay: an apparent elevation in the tacrolimus concentartion. Clin Chim Acta 2009; 402: 193–195.
67. Knorr JP, Grewal KS, Balasubramanian M, Zaki R et al. Falsely elevated tacrolimus levels caused by immunoassay interference secondary to beta-galactosidase antibodies in an infected liver transplant recipient. Pharmacotherapy 2010; 30: 954.
68. Rostaing L, Cointault O, Marquet P, Josse AG et al. Falsely elevated whole blood tacrolimus concentrations in a kidney transplant patient: potential hazards. Transpl Int 2010; 23: 227–230.
69. D'Alessandro M, Mariani P, Mennini G, Severi D et al. Falsely elevated tacrolimus concentrations measures using the ACMIA method due to circulating endogenous antibodies in a kidney transplant recipient. Clin Chim Acta 2011; 412: 245–248.
70. Colantonio DA, Borden KK, Clarke W. Comparison of the CEDIA and MEIA assays for the measurement of sirolimus in organ transplant recipients. Clin Biochem 2007; 40: 680–687.

71. Morris RG, Salm P, Taylor PJ, Wicks FA. Comparison of the reintroduced MEIA assay with HPLC-MS/MS for the determination of whole blood sirolimus from transplant recipients. Ther Drug Monit 2006; 28: 164–168.

72. Schmidt RW, Lotz J, Schweigert R, Lackner K et al. Multi-site analytical evaluation of a chemiluminescent magnetic microparticle immunoassay (CMIA) for sirolimus on the Abbott ARCHITECT analyzer. Clin Biochem 2009; 42: 1543–1548.

73. Dasgupta A, Davis B, Chow L. Evaluation of QMS everolimus assay using Hitachi 917 analyzer: comparison with liquid chromatography/mass spectrometry. Ther Drug Monit 2011; 33: 149–154.

74. Sallustio BC, Noll BD, Morris RG. Comparison of blood sirolimus, tacrolimus and everolimus concentrations measured by LC-MS/MS, HPLC-UV and immunoassay methods. Clin Biochem 2011; 44: 231–236.

75. Hosotsubo H, Takahara S, Imamura R, Kyakuno M et al. Analytical validation of the enzyme multiplied immunoassay technique for the determination of mycophenolic acid in plasma from renal transplant recipients compared with a high performance liquid chromatographic assay. Ther Drug Monit 2001; 23: 669–674.

76. Decavele AS, Favoreel N, Heyden FV, Verstraete AG. Performance of the Roche total mycophenolic acid assay on the Cobas Integra 400, Cobas 6000 and comparison to LC-MS/MS in liver transplant patients. Clin Chem Lab Med 2011; 49: 1159–1165.

77. Westley IS, Ray JE, Morris RG. CEDIA mycophenolic acid assay compared with HPLC-UV in specimens from transplant recipients. Ther Drug Monit 2006; 28: 632–536.

78. Shipkova M, Schutz E, Besenthal I, Fraunberger P et al. Investigation of the cross-reactivity of mycophenolic acid glucuronide metabolites and of mycophenolic acid mofetil in the Cedia MPA assay. Ther Drug Monit 2010; 32: 79–85.

79. Shipkova M, Armstrong VW, Wieland E, Niedman PD et al. Identification of glucoside and carboxyl-linked glucuronide conjugates of mycophenolic acid in plasma of transplant recipients treated with mycophenolic acid. Br J Pharmacol 1999; 126: 1075–1082.

80. Patel CG, Akhlaghi F. High-performance liquid chromatography method for the determination of mycophenolic acid and its acyl and phenol glucuronide metabolites in human plasma. Ther Drug Monit 2006; 28: 116–122.

81. Seebacher G, Weigel G, Wolner E, Grimm M et al. A simple HPLC method for monitoring mycophenolic acid and its glucuronidated metabolite in transplant recipients. Clin Chem Lab Med 1999; 37: 509–515.

82. Daurel-Receveur M, Titier K, Picard S, Ducint D et al. Fully automated analytical method for mycophenolic acid quantification in human plasma using on-line solid phase extraction and high performance liquid chromatography with diode array detection. Ther Drug Monit 2006; 28: 505–511.

83. French DC, Saltzgueber M, Hicks DR, Cowper AL et al. HPLC assay with ultraviolet detection for therapeutic drug monitoring of sirolimus. Clin Chem 2001; 47: 1316–1319.

84. Ivanova M, Srtusi C, Polo C, Zaninotto M et al. High throughput LC-MS/MS method for monitoring sirolimus and everolimus in the routine clinical laboratory. Clin Chem Lab Med 2011; 49: 1151–1158.

85. Koster RA, Dijkers EX, Uges DR. Robust high-throughput LC/MS/MS method for therapeutic drug monitoring of cyclosporine, tacrolimus, everolimus and sirolimus in whole blood. Ther Drug Monit 2009; 31: 116–125.
86. Meinitzer A, Gartner G, Pilz S, Stettin M. Ultra fast liquid chromatography-tandem mass spectrometric routine method for simultaneous determination of cyclosporine A, tacrolimus, sirolimus and everolimus in whole blood using deuterated internal standards for cyclosporine and everolimus. Ther Drug Monit 2010; 32: 61–66.
87. Figurski MJ, Korecka M, Fields L, Waligorska T et al. High performance liquid chromatography-mass spectroscopy/mass spectroscopy method for simultaneous determination of total or free fraction of mycophenolic acid and its glucuronide metabolites. Ther Drug Monit 2009; 31: 717–726.

Chapter 16

Effect of Drug–Herb Interactions on Therapeutic Drug Monitoring

16.1. INTRODUCTION

Herbal remedies are consumed by a significant percentage of the general population in the United States, and in many developing countries 50–70% of the general population relies on native medicines that are mostly herbal preparations. In the United States, herbal remedies are sold according to the 1994 Dietary Supplement Health and Education Act and classified as food supplements. Manufacturers of herbal remedies are not allowed by law to claim any medical benefit from using their products, but at the same time they are not under the surveillance of the U.S. Food and Drug Administration (FDA). Jiang commented that the Dietary Supplement Act was passed due to intense lobbying by the supplement industry, and although some manufacturers of such supplements produce a "mimic drug" rather than a "dietary supplement," the FDA cannot do anything except wait for another dietary supplement to produce serious adverse effects and then publish a warning for the public (1).

The sales of herbal supplements are also poorly regulated in most developed countries. However, in Germany, herbal monographs called the German Commission E monographs are prepared by an interdisciplinary committee

Resolving Erroneous Reports in Toxicology and Therapeutic Drug Monitoring: A Comprehensive Guide, First Edition. Amitava Dasgupta.

using historical information; chemical, pharmacological, clinical, and toxicological study findings; case reports; epidemiological data; and unpublished manufacturers' data. If an herb has an approved monograph, it can be marketed. Currently, there is a push for harmonization of the European market for herbal supplements. The European Directive 2004/24/EC released in 2004 by the European Parliament and also by the Council of Europe now provides the basis of regulation of herbal supplements in the European market. This directive outlines that herbal medicines released in the market require authorization by the national regulatory authorities of each European country and that these products must be safe. The safety of a supplement is established based on published scientific literature, and when safety is not sufficient, it is communicated to consumers. In Europe, there will be two kinds of herbal supplements in the future: herbal supplements with well-established safety and efficacy, and traditional herbal supplements that do not have a recognized level of efficacy but are relatively safe (2). The Australian government created a Complementary Medicine Evaluation Committee in 1997 to address regulatory issues regarding herbal remedies. In Canada, the federal government implemented a policy on January 1, 2004, to regulate natural health products; naturopaths, traditional Chinese medicine practitioners, homeopaths, and Western herbalists are concerned that this policy will eventually affect their access to the products they need to practice effectively (3).

In the United States, the sale of herbal remedies skyrocketed from $200 million in 1988 to more than $3.3 billion in 1997. Within the European community, the sale of herbal remedies is also widespread with an estimated annual sale of US $7 billion in 2001 (4). Sale of herbal supplements worldwide was estimated to be $15.7 billion in 2000, and in 2003 it increased to an estimated $18.8 billion (5). Cavaliere et al reported that in 2009, the sale of herbal supplements in the United States alone was estimated to be $5.03 billion (6). The popularity of using herbal supplements is steadily increasing among the general population in the United States. It is estimated that roughly 20,000 herbal products are available, and in one survey; approximately 1 of 5 adults have used an herbal supplement within the past year. The 10 most commonly used herbal supplements are echinacea, ginseng, ginkgo biloba, garlic, St. John's wort, peppermint, ginger, soy, chamomile, and kava (7). Another study estimated that 28.9% of U.S. adults used at least one form of complementary and alternative therapy in the past year; use of such therapy was most prevalent among women between 35 and 54 years of age. The overall use of such therapies was higher in white non-Hispanics (30.8%) and black non-Hispanics (24.1%), and the use of complementary and alternative therapies was higher among people who have health insurance (8).

16.2. WHY DRUG–HERB INTERACTIONS ARE IMPORTANT

Individuals with an illness tend to use herbal supplements more than healthy people. For example, the use of complementary and alternative medicines is common among patients infected with the human immunodeficiency virus (HIV). According to one nationwide survey, 42% of participants admitted using complementary and alternative medicines, but among patients infected with HIV, this percentage was approximately 80%. The complementary and alternative therapies used by these patients include megavitamins, herbal supplements, lifestyle and exercise, diet, counseling, and prayer (9). Use of complementary and alternative medicines is also common among cancer patients and cancer survivors. In one report, 66% women with breast cancers used at least one alternative therapy during the previous 12 months of the survey, and most of these patients did not inform their physicians regarding their use of alternative therapies. The reason for using such therapies was the belief that such therapy could prevent cancer recurrences and would also improve the quality of life (10). Interestingly, health care providers also widely use herbal supplements. In one survey of 1249 health care professionals, 51% reported the use of herbal supplements, and the rates of use were highest among physicians' assistants and nurse practitioners (63%), clinical nurses (59%), and students (52%). According to this survey, 48% of physicians, 40% of dietitians and 37% of pharmacists also used herbal supplements (11). In addition, the use of herbal supplements is common in older adults, especially those with health issues. In one survey, an overall 62.9% of respondents used one or more complementary and alternative medicine modalities with herbal supplements the most common modality (12). Individuals with various health conditions who are likely to use herbal remedies are listed in Table 16.1.

Because people with health conditions, especially with chronic conditions, are more likely to use herbal supplements than younger and healthier adults, drug–herb interaction is an important public health and safety issue. Many relatively safe herbal supplements such as the herbal antidepressant St. John's wort, garlic, ginseng, and so on, can react with many Western drugs, causing either treatment failure or drug toxicity. In addition, most patients consider herbal supplements as safe and natural, and they do not report to their physicians regarding their use of herbal supplements. Mehta et al reported that overall only 33% of herbal and dietary supplement users reported their use of herbal supplements to their conventional health care providers. In addition, Hispanic and Asian American adults are very much less likely than non-Hispanic white Americans to disclose herbal and dietary supplement use to

TABLE 16.1. People with Health Conditions Who Use More Complementary and Alternative Medicines Than the General Population

HIV Infection	Cancer	Diabetes
Arthritis	Headache	Chronic pain
Anxiety	Depression	Stress management
Fatigue	Dementia	Women's issues*
Transplant recipients	Life-threatening or terminal Illness	

*Women's issues include problem with menstrual cycle, urinary tract infection, infertility, nausea and vomiting during pregnancy and menopause.

their doctors. Having less than a high school education and not having insurance were also associated with being less likely to disclose the use of herbal or dietary supplements use to conventional health care providers. The authors concluded that the low rate of reporting the use of herbal supplements by patients to health care professionals raises concerns regarding the concurrent use of these supplements and conventional health care (13). Graham et al reported that the use of herbal and dietary supplements is common among adult ambulatory care patients, and the most frequently used herbal supplements are echinacea, ginkgo biloba, glucosamine, omega-3 fatty acids, garlic, St. John's wort, and ginseng. The authors commented that the risk of interactions between these herbal supplements and Western medications are worrisome (14). Sood et al based on a survey of 1818 patients commented that the five most common herbal supplements with a potential for interaction with Western drugs accounted for 68% of potential clinically significant drug–herb interactions. These herbal supplements include St. John's wort, garlic, valerian, kava-kava, and ginkgo biloba. In addition, the four most common classes of Western medications (antithrombotics, sedatives, antidepressants, and antidiabetics) accounted for 94% of the potentially clinically significant drug–herb interactions (15).

Many drug–herb interactions lead to unexpected levels of a therapeutic drug during a routine outpatient therapeutic drug monitoring. The most commonly encountered herbal supplement involved in a clinically significant drug–herb interaction is St. John's wort. In the experience of this author, when a clinician calls with an unexpectedly low drug level of cyclosporine where no other explanation is found, often the undisclosed use of St. John's wort by the patient is a probable cause. A drug–herb interaction can be either pharmacokinetic or pharmacodynamic. Monitoring the serum, plasma, or whole blood concentration of a therapeutic drug is capable of identifying drug–herb interactions that are only pharmacokinetic. Certain patient populations are more susceptible to adverse drug–herb interaction (see Table 16.2).

16.3. DRUG–HERB INTERACTIONS INVOLVING ST. JOHN'S WORT

St. John's wort is a popular herbal antidepressant and one of the most commonly used herbal supplements by the general population. The commonly sold St. John's wort preparation in the United States is a dried alcoholic extract of or alcohol/water extract of hypericum, a perennial aromatic shrub with bright yellow flowers that blooms from June to September. The flowers are believed to be most abundant and brightest around June 24, the day traditionally believed to be the birthday of John the Baptist. Therefore, the name St. John's wort became popular for this herbal product. The German Commission E monograph indicates that St. John's wort can be used in the supportive treatment of anxiety and depression. It is also used as a sedative and anti-inflammatory agent.

TABLE 16.2. Patients Vulnerable to Drug–Herb Interactions

Patient Group	Comments
Elderly	Many elderly patients take multiple drugs and may have reduced renal or liver function. Therefore, they are susceptible to drug–herb interactions.
Transplant recipients	St. John's wort and other drugs may reduce the efficacy of cyclosporine and tacrolimus and may cause organ rejection.
AIDS patients	AIDS patient take multiple drugs every day, and these drugs undergo various interactions with herbal supplements. Treatment failure may result.
Warfarin therapy	Warfarin interacts with many herbal supplements causing under- or overcoagulation. See Chapter 9 for more details.
Depressed patients	Many antidepressants including selective serotonin reuptake inhibitor (SSRIs) may interact pharmacodynamically with St. John's wort, kava, and valerian causing serious adverse reactions.
Diabetic patients	Several herbs may cause hypoglycemia or render antidiabetic medicines less effective.
Pregnant women/Nursing mother	Avoid taking herbal supplements because some components may cross the placenta and have adverse effects on the fetus or may be secreted in milk.
Infants and young children	Avoid giving infants and young children any herbal supplements because they may be more susceptible to toxicity than adults.

Many chemicals have been isolated from St. John's wort including hypericin, pseudohypericin, quercetin, isoquercitrin, rutin, amentoflavone, hyperforin, other flavonoids, and xanthone. However, hypericin, hyperforin, and 1,3,5,7-tetrahydroxyxanthone are unique to St. John's wort and are active components that have antidepressant properties (16). In a meta-analysis of randomized clinical trials using 1757 patients, hypericum extracts were found to be significantly superior to placebo (55.1% response to St. John's wort versus 22.3% response to placebo). St. John's wort was found to be similar in efficacy as standard antidepressants such as amitriptyline or imipramine. More interestingly, St. John's wort showed fewer adverse effects. Woelk reported a study comparing the effect of hypericum extract and imipramine in treating 324 psychiatric outpatients with mild to moderate depression. The author concluded that hypericum is theoretically equivalent to imipramine but better tolerated by patients (17). Moreover, patients with anxiety associated with depression may derive more benefit from treatment with hypericum than with imipramine (18).

Although the adverse effects of St. John's wort appear to be less than prescription antidepressants, side effects and toxicity have been reported. St. John's wort may induce photosensitivity, so fair skinned persons should be cautious about exposure to bright sunlight. Photosensitivity may also be present as neuropathy, possibly due to the demyelination of cutaneous axons by photo-activated hypericins. After taking St. John's wort for 4 weeks, a 35-year-old woman complained about stinging pain on sun-exposed areas. The neuropathy improved 2 months after she discontinued the product (19). There are a few case reports describing episodes of hypomania (irritability, disinhibition, agitation, anger, insomnia, and difficulty in concentrating) after using St. John's wort. In one report, the authors described two cases of hypomania occurring 6 weeks and after 3 months of use of St. John's wort (20). Other adverse effects reported with St. John's wort include gastrointestinal irritations, headache, allergic reactions, tiredness, and restlessness (21).

The active components of St. John's wort, hypericin and hyperforin, are responsible for the pharmacokinetic interactions between many Western medications and St. John's wort. Mannel commented that the drugs most significantly affected and contraindicated for concomitant use with St. John's wort are metabolized via both cytochrome P450 mixed function enzymes and the P-glycoprotein pathway. These drugs include HIV protease inhibitors, HIV nonnucleoside reverse transcriptase inhibitors (metabolized by cytochrome P450 mixed function enzymes), immunosuppressants such as cyclosporine and tacrolimus, antineoplastic agents, for example, irinotecan and imatinib mesylate, and other drugs. The combination of St. John's wort with serotonergic agents and other antidepressants should be restricted due to potential pharmacodynamic interactions (22). Di et al also reported significant interactions of St. John's wort with many classes of medications including anticancer agents (e.g., imatinib and irinotecan), anti-HIV agents (e.g., indinavir, lamivudine, and nevirapine), anti-inflammatory agents (e.g., ibuprofen and fexofenadine), antimicrobial agents (e.g., erythromycin and voriconazole), cardioactive drugs (e.g., digoxin, ivabradine, warfarin, verapamil, nifedipine, and talinolol), central nervous system (CNS) agents (e.g., amitriptyline, buspirone, phenytoin, methadone, midazolam, alprazolam, and sertraline), antidiabetic agents (e.g., tolbutamide and gliclazide), immunosuppressants (e.g., cyclosporine and tacrolimus), oral contraceptives, proton pump inhibitor (omeprazole), stains (e.g., atorvastatin and pravastatin), and the antiasthma agent theophylline (23).

16.3.1. Mechanism of Interaction Between St. John's Wort and Various Drugs

The main component of the cytochrome P450 mixed function oxidase family of enzymes is CYP3A4, which is expressed mainly in the liver and small intestine and responsible for metabolism of more than 50% of the prescription medications currently available in the United States (24). CYP3A4 has the broadest substrate specificity among all the enzymes of the cytochrome P450 family and is involved in the metabolism of medications (e.g., carbamazepine, cyclosporine, tacrolimus, and HIV protease inhibitors) as well as endogenous compounds such as bile acids, steroid hormones (e.g., estrogens, including synthetic drugs such as ethinyl estradiol), and fat-soluble vitamins (especially calcitriol and other vitamin D compounds) (25). Other clinically important enzymes in the cytochrome P450 family of enzymes include CYP1A2, CYP2C9, and CYP2D6. CYP1A2 catalyzes the metabolism of a number of medications including amitriptyline, caffeine, clozapine (antipsychotic), cyclobenzaprine (muscle relaxant), and fluvoxamine (antidepressant). CYP2C9 is best known as the main enzyme involved in the metabolism of warfarin but also catalyzes the metabolism of ibuprofen, sulfonylurea hypoglycemics, and losartan. CYP2D6 is responsible for the metabolism of approximately 25% of the prescription medications on the market in the United States including β-adrenergic receptor blockers, many antidepressants, and tamoxifen (26).

The efflux transporter P-glycoprotein is expressed in hepatocytes, intestinal epithelium, and capillary endothelial cells comprising the blood-brain barrier. Similar to the CYP3A enzymes, P-glycoprotein has broad substrate specificity and thus plays an important role in the clearance of many clinically important drugs. Currently, an extensive body of evidence, including both clinical pharmacology studies in humans and in vitro models, demonstrates the induction of CYP3A and P-glycoprotein by St. John's wort as major mechanisms of interaction between St. John's wort and various drugs. In the primary human hepatocyte model, St. John's wort increases CYP3A4 mRNA levels, protein expression, and overall enzyme activity. However, St. John's wort does not alter the activity of CYP2D6, and there are conflicting reports on the effect of St. John's wort on CYP1A2 and CYP2C9 expression (27,28).

The molecular mechanism by which drugs induce the liver enzymes responsible for the metabolism of drugs is through a nuclear hormone receptor (NHR) known as pregnane X receptor (also called steroid and xenobiotic receptor). In 2000, two research groups independently reported that St. John's wort strongly activates the human Pregnane receptor X, particularly due to the presence of the active ingredient hyperforin, thus causing the induction of the cytochrome P450 mixed function oxidase family of enzymes, particularly CYP3A4 (29,30). In addition, the role of hyperforin in CYP3A induction by St. John's wort was further demonstrated by the fact that enzyme induction depended on hyperforin content, with St. John's wort containing low hyperforin usually not causing enzyme induction (31). The chemical structure of hyperforin is given in Figure 16.1.

Hypericin, another active component of St. John's wort, induces P-glycoprotein, the gene product of the multidrug resistant gene 1 (MDR1/ABCB1). St. John's wort increased MDR1 messenger ribonucleic acid (mRNA) as well as the P-glycoprotein level in duodenal mucosa. Individuals with combined MDR1 genotype comprising 1236 > T, 2677 > G/T/A, and 3435 > T polymorphism usually show lower intestinal MDR1 mRNA level, but St. John's wort is capable of inducing P-glycoprotein even in these subjects as assessed by talinolol disposition (32).

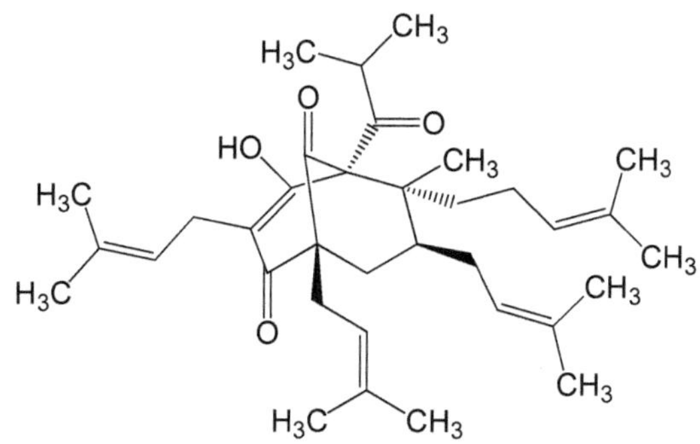

Figure 16.1. Chemical structure of hyperforin.

16.3.2. Interaction of St. John's Wort with Immunosuppressants

St. John's wort significantly reduces the efficacy of the immunosuppressants cyclosporine and tacrolimus, thus increasing the risk of organ rejection in stable transplant recipients after self-medication with St. John's wort. It is advisable that transplant recipients should refrain from taking any herbal supplement to avoid any potential drug–herb interaction. Both cyclosporine and tacrolimus are metabolized by CYP3A4, and St. John's wort induces the activity of this enzyme, thus significantly increasing the metabolism of both cyclosporine and tacrolimus. Barone et al reported two cases in which renal transplant recipients experienced subtherapeutic concentrations of cyclosporine after they started self-medication with St. John's wort. One patient developed acute graft rejection due to low cyclosporine concentration. Upon termination of use of St. John's wort, both patients' cyclosporine concentrations returned to therapeutic levels (33). Ruschitzka et al reported two cases of acute heart transplant rejections due to the interaction of St. John's wort with cyclosporine (34). Mai et al reported that the hyperforin content of St. John's wort determines the magnitude of interaction between the herb and cyclosporine. Patients who received low hyperforin containing St. John's wort showed minimal changes in pharmacokinetic parameters and needed no dose adjustment. In contrast, the patients who received high amounts of hyperforin containing St. John's wort needed dose increases within 3 days to maintain their trough therapeutic concentration of cyclosporine (35).

Case Study

A 57-year-old male kidney transplant recipient who was stable on cyclosporine (125–150 mg/day) and prednisolone (5 mg/day) and showed a therapeutic cyclosporine trough level (100–130 ng/mL) over the past 2years, suddenly had a drop in cyclosporine blood level to 70 ng/mL despite the daily cyclosporine dose increased to 250 mg/day from 150 mg/day. The patient admitted taking an herbal tea mixture for depression (Greek remedy) that contained St. John's wort and also controlled his seasonal depression. He was advised to discontinue the herbal tea, and 5 days later his cyclosporine trough level increased from 70 ng/mL to 170 ng/mL (250 mg of cyclosporine per day). Subsequently, his dose was reduced to 175 mg/day, and his trough cyclosporine level again was within therapeutic range of 130 ng/mL (36).

Case Study

A 55-year-old woman had diabetes mellitus beginning in childhood, and a kidney transplant was performed in 1985. Her trough cyclosporin level had been within the therapeutic range since then. In 1995 she experienced medium grade depression and self-medicated with St. John's wort (300 mg three times daily). Her mean cyclosporine level before taking St. John's wort was 210.0 ng/mL, but after taking St. John's wort it dropped to a mean value of 81.1 ng/mL. Her cyclosporine dosage was increased by 8.2 mg/kg bodyweight daily to maintain therapeutic cyclosporine level. In April 2000, the cyclosporine and St. John's wort interaction was suspected, and the patient was advised to stop taking St. John's wort. Her cyclosporine level again returned to a therapeutic level. This case report also showed the unnecessary increased cost of her medication when she was taking St. John's wort to make her stable with cyclosporine (37).

St. John's wort also interacts with tacrolimus, another immunosuppressant, by significantly reducing its efficacy due to a lower whole blood level. Significant reductions in the area under the curve (AUC) for tacrolimus were observed in 10 stable renal transplant patients receiving St. John's wort. The maximum concentration of tacrolimus was also reduced from a mean value of 29.0 ng/ml to 22.4 ng/ml following coadministration of St. John's wort (38).

Case Study

A 65-year-old patient who received a renal transplant in November 1998 had a trough whole blood level tacrolimus concentration between 6 and 10 ng/mL, which was within the therapeutic range. The patient experienced depression in July 2000 and started self-medication with St. John's wort (600 mg/day) and in August 2000 showed an unexpected low tacrolimus concentration of 1.6 ng/mL. Interestingly, the serum creatinine was also decreased to 0.8 mg/dL from an initial value of between 1.6 and 1.7 mg/dL. When the patient stopped taking St. John's wort, the tacrolimus level returned to the previous range of 6–10 ng/mL. After 1 month, the creatinine value was also gradually increased to 1.3 mg/dL. Because the patient showed no rejection episode, the new tacrolimus target level was set to 4–6 ng/mL by adjusting the tacrolimus dosage (39).

Mai et al studied interaction of St. John's wort with tacrolimus and mycophenolic acid using 10 stable renal transplant patients. Coadministration of St. John's wort significantly reduced the AUC as well as both the peak and trough blood concentrations of tacrolimus. To achieve sufficient immunosuppression, tacrolimus doses were increased in all patients (median: 4.5–8.0 mg/day). The tacrolimus trough levels after corrected for dose decreased from a median value of 10.8 ng/mL (pre–St John's wort) to 3.8 ng/mL 2 weeks after the initiation of use of St. John's wort. Two weeks after the discontinuation of St. John's wort treatment, trough concentrations were increased again to 7.6 ng/mL, and patients were adjusted again to their previous doses approximately 4 weeks after the end of the study. Interestingly, the pharmacokinetic parameters of mycophenolic acid, another immunosuppressant, were not affected by the coadministration of St. John's wort (40).

16.3.3. Interaction of St. John's Wort with Anti-HIV Agents

St. John's wort interacts with many anti-HIV drugs, which are primarily metabolized by CYP3A4, and reduces their efficacy. As mentioned earlier, patients with acquired immunodeficiency syndrome (AIDS) tend to use herbal supplements more than the general population, but it may turn out to be a dangerous practice because of potential drug–herb interactions. Patients receiving highly active antiretroviral therapy (HAART) must refrain from using any herbal supplements. Interactions of many protease inhibitors with St. John's wort have been reported in the literature. In one report, St John's wort reduced the AUC of the HIV-1 protease inhibitor indinavir by a mean of 57% when subjects received 300 mg of St. John's wort three times a day for 14 days. The mean peak concentration (Cmax) decreased from 12.3 ng/mL to 8.9 ng/mL in healthy volunteers taking both indinavir and St. John's wort. Reduction in indinavir concentrations of these magnitudes are clinically significant and could lead to treatment failure (41). Busti et al reported that the efficacy of atazanavir therapy can be reduced due to the simultaneous use of St. John's wort (42). Coadministration of lopinavir/ritonavir with St. John's wort is contraindicated due to the substantial reduction in lopinavir plasma concentrations (43). De Matt et al reported an increased oral clearance of nevirapine in five patients due to the concomitant use of St. John's wort (44).

16.3.4. Interaction of St. John's Wort with Warfarin

St. John's wort reduces the efficacy of warfarin due to the reduced blood level. In addition, many other commonly used herbal supplements affect warfarin's disposition (either by increasing or decreasing its efficacy). Patients receiving warfarin must refrain from using any herbal supplements due to the tight international normalization ratio (INR) needed for successful warfarin therapy. This important topic is discussed in detail in Chapter 9.

16.3.5. Interaction of St. John's Wort with Cardioactive Drugs

Digoxin is not metabolized by the liver cytochrome P450 mixed function oxidase family of enzyme, but P-glycoprotein plays an important role in the disposition of digoxin. The interaction between St. John's wort and digoxin is clinically significant. Johne et al reported that the use of 10 days of St. John's wort an average reduced the peak concentration by 33% and the trough concentration by 26%. The mean peak digoxin concentration was 1.9 ng/mL in the placebo group and 1.4 ng/mL in the group taking St. John's wort. (45). The dosage and preparation of St. John's wort also affect the pharmacokinetics of digoxin. Mueller et al reported that low daily dose of hyperforin containing St. John's wort does not affect the pharmacokinetics of digoxin. In contrast, comedication with a high dose of a hyperforin-rich extract resulted in a 24.8 % decrease in AUC from time zero to 24 hours. A 37% reduction in the peak plasma concentration was also observed (46).

Nifedipine and verapamil are calcium channel blockers metabolized by CYP3A4. Therefore, as expected, St. John's wort interacts with both these agents in a clinically significant manner. The induction of first-pass metabolism of verapamil by CYP3A4 in the gut also contributes to the lower bioavailability of verapamil if concurrently taken with St. John's wort (47).

Talinolol, a β-1-adrenergic blocker is used in treating hypertension and arrhythmia. In one report, the authors investigated the effects of long-term use of St. John's wort on oral and intravenous pharmacokinetics of talinolol in relation to intestinal P-glycoprotein expression. In a controlled randomized study involving nine subjects, the pharmacokinetics or oral (50 mg) and intervenous (30 mg) talinolol were determined before and after 12 days of use of St. John's wort (900 mg daily). St. John's wort reduced the oral talinolol bioavailability by 25%. In addition, a 93% increase in the clearance of talinol after oral delivery was also observed. Duodenal biopsies taken from the subjects revealed that St. John's wort increased the MDR1 messenger ribonucleic acid level as well as the P-glycoprotein level in the duodenal mucosa (32). Ivabradine is a new class of antianginal drug that is extensively metabolized by CYP3A4. In one study, the authors demonstrated that administration of St. John's wort significantly reduced the AUC of ivabradine (144 versus 44 ng hour/mL) as well as the peak plasma level (33 versus 15 ng/mL), indicating that the use of St. John's wort may significantly reduce the efficacy of ivabradine (48).

16.3.6. Interaction of St. John's Wort with Anticancer Drugs

Imatinib is used in the treatment of Philadelphia chromosome–positive chronic myeloid leukemia and gastrointestinal stromal tumors. Clearance of imatinib mesylate is increased due to the administration of St. John's wort. In one study involving 10 healthy volunteers, a 2-week treatment with St. John's wort significantly reduced maximum plasma concentration by 29% and AUC by 32%. The half-life of the drug was reduced by 21%. Therefore, efficacy of this drug could be compromised if a patient also takes St. John's wort (49). St. John's wort also showed a significant interaction with irinotecan, another anticancer drug. One study involving the ingestion by five patients of St. John's wort (900 mg/day) for 18 days resulted in an average 42% reduction in the concentration of SN-39, the active metabolite of irinotecan. This reduction also caused a reduced efficacy of irinotecan as evidenced by decreased myelosuppression (50).

16.3.7. Interaction of St. John's Wort with Benzodiazepines and Antidepressants

Alprazolam and midazolam are metabolized by CYP3A4. Although short-term ingestion of St. John's wort (900 mg/day for 1–3 days) does not alter the pharmacokinetics of alprazolam and midazolam in healthy volunteers, long-term ingestion (900 mg/day for 2 weeks) significantly increased the oral clearance of midazolam and decreased oral bioavailability by 39.3% (51). Quazepam is a short-acting benzodiazepine that is very effective in treating insomnia. This drug is metabolized by both CYP3A4 and CYP2C19, and, as expected, St. John's wort significantly reduces its efficacy by reducing both the peak plasma concentration and AUC based on a study using 13 healthy subjects. All subjects took 15 mg of quazepam after treatment with St. John's wort (900 mg/day) for 14 days or placebo. The study was performed using a randomized placebo-controlled crossover design with an interval of 4 weeks between treatments. However, St. John's wort did not affect the pharmacodynamic profile of quazepam (52).

St. John's wort reduces the efficacy of tricyclic antidepressants by increasing their metabolism. The demethylation of amitriptyline to nortriptyline is catalyzed by CYP3A4 and CYP2C19; further metabolism of nortriptyline through hydroxylation at position 10 is mediated by CYP3A4 and CYP2D6. The concomitant intake of St. John's wort for at least 2 weeks in 12 depressed patients decreased the AUC between zero and 12 hours of amitriptyline by 22% and nortriptyline by 41%. The AUC of all hydroxylated metabolites except 10-E-hydroxynortriptyline were also decreased. The mean peak concentration of amitriptyline was reduced from 69.8 ng/mL to 54.1 ng/mL when patients also took St. John's wort (53).

St. John's wort interacts pharmacodynamically with several selective serotonin reuptake inhibitors (SSRIs) and may potentially cause serotonin syndrome, a medical emergency. St. John's wort interacts with paroxetine, sertraline, venlafaxine, and nefazodone resulting in symptoms of the serotonin syndrome (47). Buspirone, an anxiolytic agent, is a 5-hydroxytryptamine receptor agonist. Dannawi reported a case of possible serotonin syndrome due to the combination of buspirone and St. John's wort in a 27-year-old woman with generalized anxiety disorder (54). A 28-year-old man who developed depression following bilateral orchidectomy for cryptorchidism was treated with sertraline. He was advised to discontinue St. John's wort but he continued taking it against medical advice. He subsequently developed a manic episode (55).

Case Study

A 42-year-old woman was referred for cognitive remediation 17 months after a motor vehicle accident in which she was injured with a cerebral concussion. She also exhibited symptoms of depression including excessive self-criticism, hypersomnia, and weight gain from excessive eating. She was taking fluoxetine (20 mg twice a day) and buspirone (15 mg twice a day). Two months after beginning cognitive remediation; her buspirone dose was increased to 20 mg twice a day due to her persistent anxiety. Over the next few weeks she appeared increasingly anxious during her sessions, and she revealed she was also taking herbal supplements including ginkgo biloba and St. John's wort for the past few weeks along with her prescribed medications, fluoxetine and buspirone. Her symptoms of hypomania were related to the interaction of St. John's wort with her medications. She was advised to stop herbal symptoms, and following discontinuation of herbal supplements her symptoms of hypomania also resolved (56).

16.3.8. Interaction of St. John's Wort with Agents That Act on the Respiratory System

St. John's wort induces CYP1A2, which is responsible for the metabolism of theophylline. Therefore, St. John's wort may potentially interact with theophylline increasing its clearance and thus reducing its efficacy. An in vitro study with intestinal LS 180 cells indicated that St. John's wort increased the expression of CYP1A2 in a concentration-dependent manner and this induction may be responsible for reduced plasma theophylline concentrations upon coadministration of St. John's wort (57). However, Morimoto et al found no significant interaction between St. John's wort and theophylline in a study conducted with healthy volunteers who consumed St. John's wort for 15 days (300 mg three times a day) and on day 14 received a single oral dose of 400 mg of theophylline (58).

Case Study

A 42-year-old woman taking an unusually high dose of theophylline (1600 mg/day) indicated that prior to her recent hospitalization she was stabilized on a twice-daily theophylline dose of 300 mg. That dosage also produced a theophylline serum concentration of 9.2 μg/mL, which was within the therapeutic range of theophylline. The patient revealed during an interview that she had been taking St. John's wort (0.3% hypericin)

300 mg/day for the last 2 months when her dosage was increased. She decided to discontinue St. John's wort, and 7 days later, her theophylline level increased to 19.6 μg/mL, which was toxic. At that time her dosage was lowered again (59).

16.3.9. Interaction of St. John's Wort with Antihyperlipidemic Agents

St. John's wort increases the clearance of statins that are metabolized via CYP3A4. Sugimoto et al reported interactions of St. John's wort with simvastatin and pravastatin. In a double-blind crossover study using 16 healthy male volunteers, the authors demonstrated that use of St. John's wort (900 mg/day) for 14 days decreased the peak serum concentration of simvastatin hydroxyl acid, the active metabolite of simvastatin from an average of 2.3 ng/mL in the placebo group to 1.1 ng/ml in the group taking St. John's wort. The AUC was also decreased in the group of volunteers taking St. John's wort compared with the placebo group. Simvastatin is extensively metabolized by CYP3A4 in the intestinal wall and liver, and St. John's wort induces this enzyme, but it did not influence the plasma pravastatin concentration because it is not metabolized by CYP3A4 (60). Andren et al studied the interaction between St. John's wort and atorvastatin, another lipid-lowering drug using 16 patients with hypercholesterolemia. Patients receiving atorvastatin for at least 3 months also took one tablet containing St. John's wort (300 mg) twice daily or a placebo multivitamin tablet for 4 weeks. A clinically significant decrease in the efficacy of the atorvastatin serum level due to the interaction with St. John's wort was revealed by a significant increase in total cholesterol concentrations (197 mg/dL in the St. John's–treated group versus 184.5 in the placebo group) as well as low density lipoprotein cholesterol concentration (102.7 mg/dL in the St. John's wort–treated group versus 90.3 mg/dL in the placebo group). No significant change in triglyceride or high density cholesterol concentration was observed between patients receiving St. John's wort versus placebo (61). St. John's wort also reduces the efficacy of rosuvastatin, another statin (62).

16.3.10. Interaction of St. John's Wort with Antiepileptic Agents

Interestingly, St. John's wort does not interact with carbamazepine, which is also metabolized by CYP3A4. Burstein et al reported that the intake of St. John's wort (900 mg/day) for 2 weeks did not alter the pharmacokinetics of the antiepileptic drug carbamazepine. This lack of interaction may be due to the inducing property of carbamazepine itself on cytochrome P450 mixed function enzymes, and therefore further induction by St. John's wort may not occur (63). The antiepileptic agent mephenytoin is metabolized by CYP2C19. In one study, the authors observed an increased urinary excretion of 4′-hydroxymephenytoin, the major metabolite of mephenytoin in CYP2C19 wild-genotype subjects due to the consumption of St. John's wort, but no effect was observed in CYP2C19-poor metabolizers (64).

16.3.11. Interaction of St. John's Wort with Hypoglycemic Agents

The hypoglycemic agent gliclazide is metabolized by CYP2C9. In one study, the authors used 21 healthy subjects to study the possible interaction between St. John's wort and gliclazide. Each subject received 80 mg of gliclazide either alone or during 15 days of treatment with St. John's wort. St. John's wort significantly altered gliclazide pharmacokinetics in 17 subjects, but 4 subjects did not show any significant interaction between St. John's wort and gliclazide. The authors concluded that treatment with St. John's wort increases the apparent clearance of gliclazide that was independent of the CYP2C9 genotype but had no significant effect on glucose or insulin levels (65).

16.3.12. Interaction of St. John's Wort with Oral Contraceptives/ Hormonal Agents

Oral contraceptives are divided into two types; progestogen only and combined estrogen and progestogen. Most oral contraceptives are metabolized by CYP3A4. 17-Ethynylestradiol is a major component of many oral contraceptive preparations. This drug is also used in hormonal replacement therapy in postmenopausal women. 17-Ethynylestradiol is metabolized through hydroxylation in position 2 by CYP3A4 (66). St. John's wort has a significant interaction with oral contraceptives due to its inductive effect on CYP3A4. Murphy et al studied the interaction between St. John's wort and oral contraceptives by investigating the pharmacokinetics of norethindrone and ethinyl estradiol using 16 healthy women. Treatment with St. John's wort (300 mg three times a day for 28 days) resulted in a 13–15 % reduction in dose exposure from oral contraceptives. Breakthrough bleeding increased in the treatment cycle as did evidence of follicle growth and probable ovulation. The authors concluded that St. John's wort increased the metabolism of norethindrone and ethinyl estradiol and thus interfered with contraceptive effectiveness (67). Reports of women getting pregnant while using oral contraceptives and St. John's wort have been reported to the German, Swedish, and British authorities. Both intermenstrual bleeding and reduced efficacy of oral contraceptives due to interaction with St. John's wort is probably related to reduced plasma concentrations of oral contraceptives due to the increased metabolism secondary to induction of CYP3A4 by St. John's wort (47).

Case Study

A 36-year-old woman with depression and hypercholesterolemia presented to a gynecology practice with an unexpected pregnancy that was confirmed by ultrasound. The patient was using an oral contraceptive (ethinyl estradiol/ dienoestrol) regularly for the last year. She was treated with fluvastatin for the past 2 years and also received antidepressants (SSRIs and tricyclic antidepressants) after a suicide attempt. She was self-medicating with the herbal antidepressant St. John's wort (1700 mg/day) for approximately 3 months prior to conception. She elected an abortion, but her unwanted pregnancy was due to the interaction between St. John's wort and the oral contraceptive (68).

Tibolone, an analog of progestin, influences the synthesis and metabolism of endogenous progesterone and androgen. Therefore, this drug is used to treat menopausal symptoms. A patient who was receiving tibolone therapy for 2 years developed a mixed-type liver injury with prolonged cholestasis after receiving St. John's wort for 10 weeks (69).

16.3.13. Interaction of St. John's Wort with Antibiotics/Antifungals

Erythromycin is metabolized by CYP3A4. Because St. John's wort induces CYP3A4, interaction between erythromycin and St. John's wort is plausible (47). Voriconazole is a triazole antifungal agent used for the treatment of severe fungal infections. In a controlled open-label study using 16 healthy men subjects who received a single oral dose of 400 mg of voriconazole, short-term use of St. John's wort (initial 10 hours of first day treatment of St. John's wort; 300 mg thrice daily), a 22% increase in the AUC of voriconazole was observed, but prolonged treatment with St. John's wort for 15 days significantly reduced the AUC by 59% compared with the control. The extensive metabolism of voriconazole is catalyzed primarily by CYP2C19 and CYP3A4 and to a lesser extent by CYP2C9. The authors concluded that coadministration of St. John's wort leads to a short term but clinically irrelevant increase followed by a prolonged extensive reduction in voriconazole exposure. Therefore, St. John's wort might put CYP2C19 wild-type individuals at higher risk for potential voriconazole treatment failure (70).

16.3.14. Interaction of Various Other Drugs with St. John's Wort

A reduced plasma level of methadone was observed in the presence of St. John's wort. Long- term treatment with St. John's wort (900 mg/day) for a median period of 31 days (range: 14–47 days) decreased the trough concentrations of methadone by an average of 47% in four patients. Two patients experienced withdrawal symptoms due to reduced plasma levels of methadone (71). St. John's wort also induces both CYP3A4 catalyzed sulfoxidation and 2C19-dependent hydroxylation of omeprazole. In a study involving 12 healthy adult men who received St. John's wort (900 mg/day) for 14 days or a placebo and then ingested a single dose of omeprazole (20 mg) orally, significant decreases in peak plasma concentrations of omeprazole were observed in volunteers taking St. John's wort, indicating significant interactions between St. John's wort and omeprazole (72). Loperamide is an antidiarrheal medication that is available over the counter. A brief episode of acute delirium possibly induced by a combination of St. John's wort, valeriana, and loperamide was reported; the mechanism of such interaction was unclear (47). Common drug–herb interactions involving St. John's wort are listed in Table 16.3.

16.4. INTERACTION OF VARIOUS DRUGS WITH GINSENGS

Ginseng, a widely used herbal product in China, other Asian countries, and also in the United States is promoted as a tonic and also as a reliever of stress. Ginseng may also be effective in the treatment of mild hyperglycemia. The Chinese ginseng that grows in Manchuria is *Panax ginseng,* whereas the ginseng that grows in North America is *Panax quinquefolius*. The common preparation of ginseng is ginseng root. Ginseng is one of the most popular herbal supplements used by the general population. In one survey among patients with chronic diseases who visit outpatient clinics, ginseng was the second most popular supplement used by these patients after vitamin supplements (73). Lee et al studied the interaction between Panax ginseng and alcohol in 14 healthy male volunteers using each subject as their own control. At 40 minutes after the last drink, the blood alcohol in the test group receiving ginseng extract along with alcohol was about 35% lower than the control value (74). A study using mice indicated that decreased plasma concentrations of alcohol in the presence of ginseng may be due to a delay in gastric emptying (75). Interaction between ginseng and phenelzine, a monoamine oxidase inhibitor, has been reported. The interaction may be related to the psychoactive effect of ginseng (76). Interaction between ginseng and the anticancer agent imatinib may cause hepatotoxicity. Interaction of ginseng and other herbal supplements with various drugs are listed in Table 16.4.

Case Study

A 26-year-old man with chronic myelogenous leukemia had taken a 400-mg imatinib tablet daily for 7 years with no apparent complications. He presented to the clinic with right upper quadrant pain. His laboratory test results indicated abnormal liver function tests with elevated levels of alanine aminotransferase (1069 U/L), aspartate aminotransferase (481 U/L), alkaline phosphatase (124 U/L), and bilirubin (1.4 mg/dL). His albumin concentration was normal (4.0 g/dL), and his INR was1.08. Liver biopsy demonstrated acute lobular hepatitis indicating a drug-induced etiology. A tentative diagnosis of imatinib-induced hepatotoxicity was made. The patient admitted daily ingestion of Panax ginseng via energy drinks for the past 3 months before presenting to the clinic with his complaint of right upper quadrant pain. Imatinib-associated hepatotoxicity usually occurs within 1–2 years after initiation of therapy, with the median time to hepatotoxicity 100 days. Although ginseng does not induce hepatotoxicity, it inhibits CYP3A4, the primary enzyme responsible for the metabolism of imatinib. The authors proposed that the hepatotoxicity experienced by the patient was related to the interaction between ginseng and imatinib (77).

TABLE 16.3. Important Drug–Herb Interactions Involving St. John's Wort

Class of Drug	Name of the Drug	Mechanism/Comments
Immunosuppressants	Cyclosporine, tacrolimus	Reduced efficacy due to CYP3A4 Induction that results in increased clearance of immunosuppressants
Protease inhibitors	Atazanavir, lopinavir, indinavir	Reduced efficacy due to increased CYP3A4 mediated clearance
Anticoagulant	Warfarin	Reduced efficacy due to increased CYP2C9-induced clearance of warfarin
Cardioactive	Digoxin	Increased clearance of digoxin due to modulation of P-glycoprotein pump by St. John's wort
	Verapamil	Reduced efficacy due to CYP3A4 induction in gut (increased first-pass metabolism) and liver that results in increased clearance of verapamil
	Nifedipine	Reduced efficacy due to CYP3A4-induced increased clearance
Benzodiazepines	Alprazolam, midazolam, quazepam	Reduced efficacy due to CYP3A4-induced increased clearance
Anticancer	Irinotecan, imatinib	Reduced efficacy due to CYP3A4-induced increased clearance
Tricyclic antidepressant	Amitriptyline	Reduced efficacy due to CYP3A4, CYP2C19-induced increased clearance
Cholesterol-lowering drug	Simvastatin	Reduced efficacy due to CYP3A4-induced increased clearance
	Atovarstatin	Reduced efficacy due to CYP3A4-induced increased clearance
Synthetic opioid	Methadone	Reduced efficacy due to CYP3A4-induced increased clearance
Oral contraceptives	Norethindrone, ethinyl estradiol	Contraceptive failure due to increased clearance
Proton pump inhibitor	Omeprazole	Reduced efficacy because St. John's wort induces both CYP3A4-catalyzed sulfoxidation and CYP2C19-dependent hydroxylation of omeprazole thus reducing plasma level
Antidiabetic	Gliclazide	Increased apparent clearance
Respiratory agent	Theophylline	Reduced efficacy due to increased metabolism due to induction of CYP2E1 and CYP3A4
Antibiotic	Erythromycin	Reduced efficacy due to increased metabolism due to induction of CYP3A4
SSRI	Fluoxetine, sertraline, paroxetine	Possible serotonin syndrome due to pharmacodynamic interaction with St. John's wort

TABLE 16.4. Miscellaneous Drug–Herb Interactions

Herbal Supplement	Interacting Drugs	Effect
Ginkgo biloba	Aspirin, clopidogrel,	Bleeding
	Dipyridamole, ticlopidine	Bleeding
	Trazodone	Coma
	Phenytoin, valproic acid	Reduced effects
Garlic	Saquinavir	Reduced effect
	Ibuprofen	Bleeding
	Chlorpropamide	Hypoglycemia
Feverfew	Iron	Reduced absorption
	Cyclosporine, azathioprine tacrolimus	Reduced immunosuppression
Echinacea	Ketoconazole, methotrexate	Liver toxicity
	Simvastatin, lansoprazole	Increased effect
Flaxseed oil	Aspirin	Increased bleeding
Kava	Alprazolam, benzodiazepines	Coma
Milk thistle		
Valerian	Midazolam, barbiturates	Increased sedation
Licorice	Atenolol	Hypotension
Saw Palmetto	Aspirin	Bleeding
Bromelain	Naproxen	Increased bleeding

16.5. INTERACTION OF VARIOUS DRUGS WITH GINKGO BILOBA

Ginkgo biloba is prepared from the dried leaves of the ginkgo tree by organic extraction (acetone/water) followed by drying the extract and standardization. Most commercial preparations of ginkgo biloba contain 40 mg of this extract and are sold as a dietary supplement for improving the blood flow in the brain and peripheral circulation. It is also used to sharpen mental focus and to improve diabetes-related circulatory disorders. However Fransen et al commented that scientific research failed to substantiate three presumed health benefits of ginkgo biloba that includes improvement of blood circulation, improvement of symptoms of old age, and improvement of memory (78). Ginkgo leaf contains kaempterol-3-rhamnoglucoside, ginkgetin, isoginketin, and bilobetin. In addition, several glycosides have also been isolated (ginkgolide A and B). A number of compounds found in ginkgo extracts, especially ginkgolide B, are potent antagonists of platelet activity.

Ginkgo biloba exerts an inductive effect on CYP3A4, CYP2C9, CYP2C19, and CYP1A2 activity. In addition, ginkgo biloba can inhibit P-glycoprotein, thus reducing the efficacy of certain drugs. Yin et al investigated the interaction of ginkgo biloba with omeprazole using 18 healthy subjects. All subjects received a single dose of omeprazole (40 mg) at baseline and after 12 days of treatment with ginkgo biloba (140 mg twice daily). Plasma concentrations of omeprazole and omeprazole sulfone were significantly decreased, and the concentration of 5-hydroxyomeprazole was significantly increased following the treatment with ginkgo biloba. The authors concluded that ginkgo biloba can induce omeprazole hydroxylation in a CYP2C19 genotype-dependent manner and concurrently reduce the renal clearance of 5-hydroxyomeprazole (79). Yang et al studied the bioavailability of cyclosporine in the presence of ginkgo and onion in rats. Cyclosporine was administered both orally and intravenously with or without ginkgo or onion in a crossover design. The oral administration of ginkgo significantly decreased the Cmax by 62% and also reduced the AUC by 51 %. The average maximum serum concentration of cyclosporine in the control group was 169.4 ng/mL and in the group receiving ginkgo was 65.2 ng/mL (80). Granger reported the cases of two patients who were stable with valproic acid but developed seizures within 2 weeks of using ginkgo products. After discontinuation of ginkgo, both patients were again seizure free without any increases in the dose of valproic acid (81). Ginkgo biloba reduces the hypotensive action of nicardipine and nifedipine by inducing CYP3A4 while also reducing the hypotensive effect of talinolol by inhibiting P-glycoprotein. In addition, ginkgo biloba also reduces the bioavailability of cyclosporine, midazolam, propranolol, theophylline, and omeprazole based on clinical or animal studies. In contrast, ginkgo biloba increases the efficacy of the glucose-lowering drug tolbutamide (82).

Case Study

A 55-year-old man taking antiepileptic drugs for several years had a fatal breakthrough seizure with no evidence his noncompliance with his anticonvulsant medications. The postmortem femoral blood concentrations of both phenytoin (2.5 μg/mL) and valproic acid (26 μg/mL) were subtherapeutic. Interestingly, his phenytoin serum concentrations were within therapeutic range in the preceding 6 months (range: 9.6–21.2 μg/mL), and the last phenytoin value prior to his death was 13.9 μg/mL. The patient was taking a variety of herbal supplements, but ginkgo was a main component. Phenytoin is primarily metabolized by CYP2C9 and secondarily by CYP2C19; valproic acid metabolism is also modulated by CYP29 and CYP2C19. Ginkgo biloba induces CYP2C19 activity and thus may be responsible for the subtherapeutic levels of anticonvulsant medications in this patient (83).

16.6. INTERACTION OF GARLIC SUPPLEMENTS WITH DRUGS

Garlic (*Allium sativum*) is widely used as both as a spice and also as an herbal supplement promoted for lowering cholesterol and blood pressure. Garlic is rich in the sulfur-containing compounds allicin and alliin, but use of garlic as a spice is not likely to cause any significant interaction with a drug because the concentrations of active ingredients are relatively low. In contrast, garlic dietary supplements sold in health food stores and herbal stores contain much higher levels of these compounds, and taking garlic supplements may cause clinically significant drug interactions. Garlic supplements can inhibit platelet aggregation and thus interact pharmacodynamically with warfarin (this important drug interaction with garlic is discussed in Chapter 9).

Piscitelli et al studied the effect of garlic on the pharmacokinetics of saquinavir using 10 healthy volunteers. In the presence of garlic, the mean saquinavir AUC during the 8-hour dosing interval decreased by 51% and the trough serum concentration 8 hours after dosing was reduced by 54%. After a 10-day washout period, the AUC and trough serum concentrations returned to 60–70% of the baseline values. The altered pharmacokinetics of saquinavir was related to the decreased bioavailability of saquinavir, which may be due to the modulation of the P-glycoprotein pump in the gut mucosa by garlic (84).

16.7. INTERACTION OF KAVA AND VALERIAN WITH DRUGS

Kava is an herbal sedative prepared from a South Pacific plant (*Piper methysticum*). The sedative as well as the anxiolytic activity of kava is related to the modulation of γ-hydroxybutyric acid (GABA) receptor activity via alteration of the lipid membrane structure and sodium channel function, monoamine oxidase inhibition, and noradrenaline and dopamine reuptake inhibition. In traditional practice, a kava drink is prepared by mixing the fresh or dried root with cold water or coconut milk. Kava is available from a variety of manufacturers in the U.S. market, but such preparations usually use alcohol/water as an extraction solvent for kava roots. However, the use of the water extract of kava is safer than the commercial preparation that uses alcohol along with water for extraction of kava roots (85). The neurological effects of kava are attributed to a group of substituted dihydropyrones called kava lactones. Heavy consumption of kava has been associated with increased concentrations of γ-glutamyltransferase, suggesting potential hepatotoxicity. In January 2003, kava extracts were banned in the entire European Union, Canada, and also in the United States (the FDA strongly cautioned against using kava). There are at least 11 cases of serious hepatic failure and 4 deaths directly linked to kava extract consumption, and there are also 23 reports indirectly linking kava with hepatotoxicity (86).

Several kava lactones are a potent inhibitor of several enzymes of the cytochrome P450 system (CYP1A2, CYP2C19, CYP2C9, CYP2D6, CYP3A4, and CYP4A9/11). Therefore there is a potential of drug interaction with kava, especially for drugs metabolized by the cytochrome P450 system, but actual systemic studies with human subjects to demonstrate such pharmacokinetic drug interactions are very limited. A study involving six healthy human volunteers who consumed the traditional aqueous extract of kava indicated that the caffeine metabolic ratio increased twofold from 0.3 with consumption of kava to 0.6 at 30 days after the subjects stopped using kava. The later value corresponds to metabolic ratios in healthy subjects. The authors concluded that kava drinking inhibits CYP1A2 (87). However, pharmacodynamic interaction between kava and CNS depressants such as benzodiazepines, alcohol, and barbiturates are well documented in the literature.

Case Study

A 54-year-old man was hospitalized in a lethargic and disoriented state. During admission, he was taking alprazolam, cimetidine, and terazosin. His laboratory test results and vital signs were within acceptable limits and blood alcohol test was negative. However, his urine drug screen was positive for benzodiazepines. He was more alert after several hours and admitted taking a natural tranquilizer containing kava 3 days prior to his hospital admission. The authors suggested that both kavalactones and alprazolam have an additive effect because both act on the same GABA receptors. Moreover, kavalactones are a potent inhibitor of CYP3A4, which metabolizes alprazolam (88).

Valerian is a perennial herb that grows in North America, Europe, and western Asia. The crude valerian root, rhizome, or stolon is dried and used as it is or as an extract as a sedative. Commercially, valerian is available as a capsule, liquid extract, or tea. More than 40 compounds have been isolated from valerian root, but valepotriates are probably responsible for the sedative activity. Valerenic acid, another component of valerian, may also have a pentobarbital-like pharmacological action. Leathwood et al conducted a study with 128 volunteers and concluded that valerian significantly improved subjective sleep quality in habitually poor or irregular sleepers (89). Studies have indicated that valerian is generally safe, but the long-term safety of valerian has not been reported. Side effects are mild such as dizziness, upset stomach, headache, and sleepiness in the morning.

Valerian has a potential to act pharmacodynamically with benzodiazepines, barbiturates, and other antidepressants. Carrasco et al reported a case in which a patient taking lorazepam self-medicated himself with valerian and passiflora incarnate (passionflower). He experienced handshaking, dizziness, throbbing, and muscular fatigue 32 hours before presenting to the hospital. His medical history revealed a generalized anxiety disorder and a history of medicinal plant consumption but he did not have Parkinson's disease or Wilson's disease. The authors concluded that his symptoms were due to the additive effect of valerian and passionflower with the benzodiazepine lorazepam because valerian and passionflower may increase the inhibitory activity of benzodiazepines binding to GABA receptors (90). An adverse interaction between haloperidol and valerian is also likely (91).

16.8. MISCELLANEOUS DRUG–HERB INTERACTIONS

Echinacea is a popular herbal supplement used for boosting the immune system and preventing colds and other upper respiratory tract infections. Echinacea should be avoided if taking immunosuppressant drugs such as cyclosporine. In addition, echinacea may be toxic to the liver and should not be used with acetaminophen, other hepatotoxic drugs, and the anticancer drug methotrexate. Another immunomodulatory herbal supplement known as astragalus should be avoided when taking cyclosporine, azathioprine, and methotrexate. Commonly used herbal remedies for menopause are black cohosh, red clover, and dong quai. These herbal remedies contain hormones of plant origin (phytoestrogen) and may interfere with hormone replacement therapy.

Licorice is commonly taken to alleviate the symptoms of ulcers. Licorice tends to increase blood pressure and should be avoided by anyone being treated for high blood pressure. Licorice may also increase loss of potassium from the body and may interfere with treatment using diuretics (hydrochlorothiazide and related drugs). Yasue et al described the case of a 93-year-old woman who suffered from severe hypokalemia (1.3 mmol/L) and developed paralysis in all extremities associated with metabolic alkalosis, hypercapnia, and a high level of creatinine kinase (up to 9280 U/L), myoglobin, and myoglobinuria due to rhabdomyolysis. She had been taking licorice for last 7 years. After discontinuation of licorice and treatment with spironolactone, her symptoms improved in 2 weeks (92). Lin et al described a case of an Asian man who presented to the emergency department with muscular weakness that progressed to paralysis. His major biochemical abnormalities were hypokalemia (1.8 mmol/L) and metabolic alkalosis related to his use of tea flavored with 100 g of natural licorice daily for 3 years (93). These cases illustrate how clinical laboratory tests can aid in the diagnosis of toxicity related to the use of herbal supplements. Another herbal supplement, goldenseal, also interacts with antihypertensive medicines and should be avoided by anyone taking medication to control high blood pressure. Hawthorn is taken by people with heart disease, and this herbal supplement (although relatively safe) should not be taken with any cardioactive drug such as digoxin. Similarly, many Chinese herbs are indicated for treating heart conditions, but such herbs also interact with cardioactive drugs. Guggul reduces the efficacy of β-blockers and calcium channel blockers used in treating various conditions including heart disease. Goldenseal may increase stomach acid secretion and should be avoided by anyone who suffers from acidity or acid reflux and takes acid suppression medication such as omeprazole. Ipriflavone is a synthetic derivative derived from soy that may have bone protective effect. However, ipriflavone can increase the effects of theophylline, warfarin, and diabetic medications, thus causing toxicity (94,95).

16.9. ROLE OF THE LABORATORY IN IDENTIFYING DRUG–HERB INTERACTIONS

Because many patients do not report their use of herbal supplements to their clinicians, the clinical laboratory may play an important role in helping clinicians identify a potential drug–herb interaction. For example, if the drug level observed during a routine therapeutic drug monitoring is significantly lower than the previous measurement and if noncompliance can be ruled out, it is a strong indication of a potential drug–herb interaction. The most probable cause is use of St. John's wort and on discontinuation of St. John's wort, the drug level usually returns to pre–herbal supplement use level within 2 weeks. Similarly, observing a usual INR during routine monitoring of a patient taking warfarin is also indicative of a potential interaction between warfarin and an herbal supplement (96).

Many drugs not routinely monitored also interact with herbal supplements. These herb–drug interactions are more difficult to detect, but other laboratory tests may help identify such interactions. For example, an elevated cholesterol level in a patient taking a statin that controlled his or her cholesterol level earlier may be indicative of the lower efficacy of the statin drug due to lower serum level secondary to a drug interaction. Similarly, hypoglycemia in a patient receiving a hypoglycemic agent may be related to a drug–herb interaction. Chromium, a mineral supplement, is known to increase the hypoglycemic effect of a drug. Similarly, interaction with ginseng may modulate the hypoglycemic effect of an antidiabetic medication. Lower glucose in a patient with no episode of previous hypoglycemia may be indicative of such an interaction (96).

In addition, many herbal supplements may cause liver toxicity with kava the most common. Abnormal liver function tests may be the first indication of such an adverse effect (97). Clinical laboratory tests are helpful in identifying such toxicity (see Chapter 10 for more details).

16.10. CONCLUSIONS

The popular use of herbal remedies among the general population is on the rise, and such practice also increases the risks of drug–herb interactions. Although some drug–herb interactions may be benign, other drug herb interactions may have a devastating effect on patient management, for example, treatment failure in a patients infected with HIV due to self-medication of St. John's wort. It is important for laboratory personnel to indicate to the ordering clinician that an unusually low or high drug level during a routine therapeutic drug monitoring to a patient who previously showed a drug level within therapeutic range may be due to a drug–herb interaction. In this case the clinician can ask the patient regarding his or her use of any herbal supplement that may have cause the unexpected test result during a routine therapeutic drug monitoring. Fortunately, in most cases after discontinuing the herbal remedy, the drug level returns to the original preherbal use level. In addition, transplant recipients, patients receiving HAART for AIDS treatment as well as patients receiving warfarin or any related anticoagulants must refrain from using any herbal remedy to avoid any potential life-threatening drug–herb interaction. Moreover, an abnormal laboratory test result in a otherwise healthy individual during a routine checkup may be an early indication of toxicity or organ damage caused by an herbal supplement. An abnormal liver function test due to the use of kava, chaparral, or germander, or hypokalemia due to the use of licorice are common examples of such abnormal test results due to use of herbal supplements.

REFERENCES

1. Jiang T. Re-thinking dietary supplement laws and regulations 14 years after the dietary supplement health and education act implementation. Int J Food Sci Nutr 2009; 60: 293–301.
2. Calapai G. European legislation on herbal medicines: a look into the future. Drug Saf 2008; 31: 428–431.
3. Moss K, Boon H, Ballantyne P, Kachan N. New Canadian natural health product regulations: a qualitative study on how CAM practitioners perceive they will be impacted. BMC Complement Altern Med 2006; 10: 6:18.
4. Mahady GB. Global harmonization of herbal health claims. J Nutr 2001; 131: 1120S–1123S.
5. Kelly JP, Kaufman DW, Kelley K, Rosenberg L et al. Recent trends in use of herbal and other natural products. Arch Intern Med 2005; 165: 281–286.
6. Cavaliere C, Rea P, Lynch MA, Blumenthal M. Herbal supplement sales rise in all channels in 2009. HearbalGram 2010; 86: 62–65.
7. Bent S. Herbal medicine in the United States: review of efficacy, safety and regulation. J Gen Intern Med 2008; 23: 854–859.
8. Ni H, Simile C, Hardy AM. Utilization of complementary and alternative medicine by United States adults. Med Care 2002; 40: 353–358.
9. Duggan J, Peterson WS, Schutz M, Khunder S et al. Use of complementary and alternative therapies in HIV-infected patients. AIDS Patient Care STDs 2001; 15: 159–167.
10. Henderson JW, Donatelle RJ. Complementary and alternative medicine use by women after completion of allopathic treatment for breast cancer. Altern Ther Health Med 2004; 10: 52–57.
11. Gardiner P, Legedza A, Woods C, Phillips RS et al. Herb use among health care professionals enrolled in an online curriculum on herbs and dietary supplements. J Herb Pharmacother 2006; 6: 51–64.
12. Cheung CK, Wyman JF, Halcon LL. Use of complementary and alternative therapies in community dwelling older adults. J Altern Complement Med 2007; 13: 997–1006.
13. Mehta DH, Gardiner PM, Phillips RS, McCarthy EP. Herbal and dietary supplement disclosure to health care providers by individuals with chronic condition. J Altern Complement Med 2008; 14: 1263–1269.
14. Graham RE, Gandhi TK, Borus J, Seger AC et al. Risk of concurrent use of prescription drugs with herbal and dietary supplements in ambulatory care. In: Henriksen K, Battles JB, Keyes MA, Grady ML eds. *Technology and Medication Safety.*

Rockville, MD: Agency for Healthcare Research and Quality, 2008. *Advances in Patient Safety: New Directions and Approaches*; vol. 4.

15. Sood A, Sood R, Brinker FJ, Mann R et al. Potential for interactions between dietary supplements and prescription medications. Am J Med 2008; 121: 207–211.
16. Raffa R. Screen of receptor and uptake site activity of hypericin components of St John's Wort reveal σ receptor binding. Life Sci 1998; 62: PL 265–270.
17. Woelk H. Comparison of St John's wort and imipramine for treating depression: randomized controlled trial. Br Med J 2000; 321: 536–539.
18. Shelton RC, Keller MB, Gelenberg A et al. Effectiveness of St. John's wort in major depression: a randomized controlled trial. JAMA 2001; 285: 1978–1986.
19. O'Breasail AM, Argouarch A. Hypomania and St. John's wort. Can J Psychiatry 1998; 43: 746–747.
20. Awang DVC. Herbal medicine: St. John's wort. Can Pharm J 1991; 124: 33–35.
21. Gordon JB. SSRIs and St. John's wort: possible toxicity. Am Fam Physician 1998; 57: 950–953.
22. Mannel M. Drug interactions with St. John's wort: mechanisms and clinical implications. Drug Saf 2004; 27: 773–797.
23. Di YM, Li CG, Xue CC, Zhou SF. Clinical drugs that interact with St. John's wort and implications in drug development. Curr Pharm Des 2008; 14: 1723–1742.
24. Li AP, Kaminski DL, Rasmussen A: Substrates of human hepatic cytochrome P450 3A4. Toxicology 1995, 104: 1–8.
25. Wrighton SA, Schuetz EG, Thummel KE, Shen DD et al. The human CYP3A subfamily: practical considerations. Drug Metab Rev 2000, 32: 339–361.
26. Rettie AE, Jones JP: Clinical and toxicological relevance of CYP2C9: drug-drug interactions and pharmacogenetics. Annu Rev Pharmacol Toxicol 2005, 45: 477–494.
27. Komoroski BJ, Zhang S, Cai H, Hutzler JM et al. Induction and inhibition of cytochromes P450 by the St. John's wort constituent hyperforin in human hepatocyte cultures. Drug Metab Dispos 2004; 32: 512–518.
28. Wang Z, Gorski JC, Hamman MA, Huang SM et al. The effects of St John's wort (*Hypericum perforatum*) on human cytochrome P450 activity. Clin Pharmacol Ther 2001, 70: 317–326.
29. Moore LB, Goodwin B, Jones SA, Wisely GB et al. St. John's wort induces hepatic drug metabolism through activation of the pregnane X receptor. Proc Natl Acad Sci U S A 2000, 97: 7500–7502.
30. Wentworth JM, Agostini M, Love J, Schwabe JW et al. St John's wort, a herbal antidepressant, activates the steroid X receptor. J Endocrinol 2000, 166: R11–R16.
31. Godel-Armbrust U, Metzger A, Kroll U, Kelber O et al. Variability in PXR mediated induction of CYP3A4 by commercial preparations and dry extracts of St. John's wort. Naunyn Schmiedebergs Arch Pharmacol 2007; 375: 377–382.
32. Schwarz UI, Hanso H, Oertel R, Miehlke S et al. Induction of intestinal P-glycoprotein by St. John's wort reduces the oral bioavailability of talinolol. Clin Pharmacol Ther 2007; 81: 669–678.
33. Barone GW, Gurley BJ, Ketel BL, Abul-Ezz SR. Herbal supplements; a potential for drug interactions in transplant recipients. Transplantation 2001; 71: 239–241.

34. Ruschitzka F, Meier PJ, Turina M, Luscher TF et al. Acute heart transplant rejection due to Saint John's wort. Lancet 2000; 355: 548–549.

35. Mai I, Bauer S, Perloff ES, Johne A, et al. Hyperforin content determines the magnitude of the St. John's wort-cyclosporine drug interaction. Clin Pharmacol 2004; 76: 330–340.

36. Alscher DM, Klotz U. Drug interaction of herbal tea containing St. John's wort with cyclosporine [letter]. Transpl Int 2003; 16: 543–544.

37. Beer AM, Ostermann T. St. John's wort: interaction with cyclosporine increased risk of rejection for the kidney transplant and raises daily cost of medication [in German]. Med Klin (Munich) 2001; 96: 480–483.

38. Hebert MF, Park JM, Chen YL, Akhtar S et al. Effects of St John's wort (*Hypericum perforatum*) on tacrolimus pharmacokinetics in healthy volunteers. Clin Pharmacol 2004; 44: 89–94.

39. Bolley R, Zulke C, Kammerl M, Fischereder M et al. Tacrolimus induced nephrotoxicity unmasked by induction of CYP3A4 system with St. John's wort [letter]. Transplantation 2002; 73: 1009.

40. Mai I, Stormer E, Bauer S, Kruger H et al. Impact of St. John's wort treatment on the pharmacokinetics of tacrolimus and mycophenolic acid in renal transplant patients. Nephrol Dial Transplant 2003; 18: 819–822.

41. Piscitelli SC, Burstein AH, Chaitt D, Alfaro RM et al. Indinavir concentrations and St. John's wort. Lancet 2000; 355: 547–548.

42. Busti AJ, Hall RJ, Margolis DM. Atazanavir for the treatment of human immunodeficiency virus infection. Pharmacotherapy 2004; 24: 1732–1747.

43. Cvetkovic RS, Goa KL. Lopinavir/ritonavir: a review of its use in the management of HIV infection. Drugs 2003; 63: 769–802.

44. De Matt MM, Hoetelmans RM, Math RA, van Gorp EC et al. Drug interaction between St. John's wort and nevirapine. AIDS 2001; 15: 420–421.

45. Johne A, Brockmoller J, Bauer S, Maurer A et al. Pharmacokinetic interaction of digoxin with an herbal extract from St John's wort (*Hypericum perforatum*). Clin Pharmacol Ther 1999; 66: 338–345.

46. Mueller SC, Uehleke B, Woehling H, Petzsch M et al. Effect of St. John's wort dose and preparation on the pharmacokinetics of digoxin. Clin Pharmacol Ther 2004; 75: 546–557.

47. Borrelli F, Izzo AA. Herb-drug interactions with St. John's wort (*Hypericum perforatum*): an update on clinical observations. AAPS J 2009; 11: 710–727.

48. Portoles A, Terleira A, Calvo A, Martinez I et al. Hypericum perforatum on ivabradine pharmacokinetics in healthy volunteers: an open labeled pharmacokinetic interaction trial. J Clin Pharmacol Ther 2006; 46: 1188–1194.

49. Smith P. The influence of St. John's wort on the pharmacokinetics and protein binding of imatinib mesylate. Pharmacotherapy 2004; 24: 1508–1514.

50. Mathijssen RH, Verweij J, de Bruijn P, Loos WJ et al. Effects of St. John's wort on irinotecan metabolism. J Natl Cancer Inst 2002; 94: 1247–1249.

51. Wang Z, Gorski JC, Hamman MA, Huang SM, et al. The effects of St. John's wort (*Hypericum perforatum*) on human cytochrome P450 activity. Clin Pharmacol Ther 2001; 70: 317–326.

52. Kawaguchi A, Ohmori M, Tsuruoka S, Nishiki K et al. Drug interaction between St. John's wort and quazepam. Br J Clin Pharmacol 2004; 58: 403–410.
53. Johne A, Schmider J, Brockmoller J, Stadelman AM, et al. Decreased plasma levels of amitriptyline and its metabolites on comedication with an extract from St. John's wort (*Hypericum perforatum*). J Clin Psychopharmacol 2002; 22: 46–54.
54. Dannawi M. Possible serotonin syndrome after combination of buspirone and St. John's wort. J Psychopharmacol 2002; 16: 401.
55. Barbenel DM, Yusufi B, O'Shea D, Bench CJ. Mania in a patient receiving testosterone replacement postorchiectomy taking St. John's wort and sertraline. J Psychopharmacol 2000; 14: 84–86.
56. Spinelli M, Eaton LA. Hypomania induced by herbal and pharmaceutical psychotropic medicines following mild traumatic brain injury. Brain Inj 2002; 16: 359–367.
57. Kayaker CS, Edenton ND, Dowling TC. Effect of St. John's wort extract on intestinal expression of cytochrome P4501A2: studies with LS180 cells. J Post Grad Med 2002; 48: 97–100.
58. Morimoto T, Tonegawa T, Satsuma K, Obtain Y, et al. Effect of St. John's wort on the pharmacokinetics of theophylline in healthy volunteers. J Clin Pharmacol 2004; 44: 95–101.
59. Nobel A, Schneider BJ, Kroll DJ. Potential metabolic interaction between St. John's Wort and theophylline. Ann Pharmacother 1999; 33: 502.
60. Sugimoto K, Ohmori M, Tsuruoka S, Nishiki K et al. Different effect of St. John's wort on the pharmacokinetics of simvastatin and pravastatin. Clin Pharmacol Ther 2001; 70: 518–524.
61. Andren L, Andreessen A, Egerses R. Interaction between a commercially available St. John's wort produce (Movana) and atorvastatin in patients with hypercholesterolemia. Euro J Clin Pharmacol 2007; 63: 913–916.
62. Gordon RY, Becker DJ, Rader DJ. Reduced efficacy of rosuvastatin by St. John's wort. Am J Med 2009; 122: e1–2.
63. Burstein AH, Horton RL, Dunn T, Alfaro RM, et al. Lack of effect of St. John's wort on carbamazepine pharmacokinetics in healthy volunteers. Clin Pharmacol Ther 2000; 68: 605–612.
64. Wang LS, Zhu B, Bad El-AtYa AM, Zhou G et al. The influence of St. John's wort on CYP2C19 activity with respect to genotype. J Clin Pharmacol 2004; 44: 577–581.
65. Xu H, Williams KM, Lieu WS, Murray M et al. Effects of St. John's wort and CYP2C9 genotype on the pharmacokinetics and pharmacodynamics of gliclazide. Br J Pharmacal 2008; 153: 1579–1586.
66. Guengerich FP. Oxidation of 17-ethynylestradiaol by human liver cytochrome P 450. Mol Pharmacol 1998; 33: 500–508.
67. Murphy PA, Kern SE, Stanczyk FZ, Westhoff CL. Interaction of St. John's wort with oral contraceptives: effects on the pharmacokinetics of norethindrone and ethinyl estradiol, ovarian activity and breakthrough bleeding. Contraception 2005; 71: 402–408.
68. Schwartz UI, Buschel B, Kirch W. Unwanted pregnancy on self-medication with St. John's wort despite hormonal contraception. Br J Clin Pharmacol 2003; 55: 112–113.

69. Etogo-Asse F, Boemer F, Sempoux C, Geubel A. Acute hepatis with prolonged cholestasis and disappearance of interlobular bile ducts following tibolone and *Hypericum perforatum* (St. John's wort): a case of drug interaction? Acta Gastroenterol Belg 2008; 71: 36–38.
70. Rengelshausen J, Banfield M, Riedel KD, Burhenne J et al. Opposite effects of short term and long term St. John's wort intake on voriconazole pharmacokinetics. Clin Pharmacol Ther 2005; 78: 25–33.
71. Eich-Hochli D, Oppliger R, Golay KP, Baumann P, Eap CB. Methadone maintenance treatment and St. John's wort—a case study. Pharmacopsychiatry 2003; 36: 35–37.
72. Wang LS, Zhou G, Zhu B, Wu J et al. St. John's wort induces both cytochrome P450 3A4 catalyzed sulfoxidation and CYP2C19 dependent hydroxylation of omeprazole. Clin Pharmacol Ther 2004; 75: 191–197.
73. Hasan SS, Ahmad AI, Bukhari NI, Loon WC. Use of complementary and alternative medicine among patients with chronic diseases at outpatient clinics. Complement Ther Clin Prac 2009; 15: 152–157.
74. Lee FC, Ko JH, Park JK, Lee JS. Effects of Panax ginseng on blood alcohol clearance in man. Clin Exp Pharmacol Physiol 1987; 14: 543–546.
75. Koo MW. Effects of ginseng on ethanol induced sedation in mice. Life Sci 1999; 64: 153–160.
76. Jones BD, Runikis AM. Interaction of ginseng with phenelzine. J Clin Psychopharmacol 1987; 3: 201–202.
77. Bilgi N, Bell K, Ananthakrishnan AN, Atallah E. Imatinib and Panax ginseng: a potential interaction resulting in liver toxicity. Ann Pharmacother 2010; 44: 926–928.
78. Fransen HP, Pelgrom SM, Stewart-Knox N, de Kaste D et al. Assessment of health claim, content and safety of herbal supplements containing ginkgo biloba. Food Nutr Res 2010; 54: doi 10:340/fnr.v54io.5221.
79. Yin OQ, Tomlinson B, Waye MM, Chow AH et al. Pharmacogenetics and herb-drug interactions: experience with ginkgo. Pharmacogenetics 2004; 14: 841–850.
80. Yang CY, Chao PD, Hou YC, Tsai SY et al. Marked decrease of cyclosporine bioavailability caused by coadministration of ginkgo and onion. Food Chem Toxicol 2006; 44: 1572–1578.
81. Granger AS. Ginkgo biloba precipitating epileptic seizures. Age Ageing 2001; 30: 523–525.
82. Abad MJ, Bedoya LM, Bermejo P. An update on drug interactions with herbal medicine ginkgo biloba. Curr Drug Metab 2010; 11: 171–181.
83. Kupiec T, Raj V. Fetal seizures due to potential herb-drug interactions with ginkgo biloba. J Anal Toxicol 2005; 29: 755–758.
84. Piscitelli SC, Brustein AH, Welden N, Gallicano KD et al. The effect of garlic supplements on the pharmacokinetics of saquinavir. Clin Infect Dis 2002; 34: 234–238.
85. Clouatre DL. Kava kava: examining new reports of toxicity. Toxicol Lett 2004; 150: 85–96.
86. Sarris J, LaPorte E, Schweitzer I. Kava: a comprehensive review of efficacy, safety and psychopharmacology. Aust NZ J Psychiatry 2011; 45: 27–35.

87. Russmann S, Lauterburg BH, Barguil Y, Choblet E et al. Traditional aqueous kava extracts inhibit P450 1A2 in humans: protective effect against environmental carcinogens? [letter]. Clin Pharmacol Ther 2005; 77: 453–454.

88. Almeida JC, Grimsley EW. Coma from the health food store: interaction between kava and alprazolam. Ann Intern Med 1996; 125: 940–941.

89. Leathwood PD, Chauffard F, Heck F et al. Aqueous extract of valerian root (*Valeriana officinalis L*) improves sleep quality in man. Pharmacol Biochem Behav 1982; 17: 65–71.

90. Carrasco MC, Vallejo JR, Pardo-Santayana M, Pearl D et al. Interactions of valerian officinalis L and passiflora incarnata L in a patient treated with lorazepam. Phyto Res 2009; 23: 1795–1796.

91. Dalla-Corte CL, Fachinetto R, Colle D, Pereira RP et al. Potentially adverse interactions between haloperidol and valerian. Food Chem Toxicol 2008; 46: 2369–2375.

92. Yasue H, Itoh T, Mizuno Y, Harada E. Severe hypokalemia, rhabdomyolysis, muscle paralysis, and respiratory impairment in a hypertensive patient taking herbal medicine containing licorice. Intern Med 2007; 46: 575–578.

93. Lin SH, Yang SS, Chau T, Halperin ML. An unusual cause of hypokalemic paralysis: chronic licorice ingestion. Am J Med 2003; 325: 153–156.

94. Colatlo C. Herbal interactions on absorption of drugs: mechanisms of action and clinical risk assessment. Pharmacol Res 2010; 62: 207–227.

95. Skalli S, Soulaymani R. Drug interactions with herbal medicines. Ther Drug Monit 2007; 29: 679–686.

96. Dasgupta A, Bernard DW. Complementary and alternative medicines: effects on clinical laboratory tests. Arch Pathol Lab Med 2006; 130: 521–528.

97. Moulds RF, Malani J. Kava: herbal panacea or liver poison? Med J Aus 2003; 178: 451–453.

Chapter 17

Pharmacogenomics and the Toxicology Laboratory

17.1. INTRODUCTION

Therapeutic drug monitoring and pharmacogenomics testing are both clinical laboratory-based methods for supporting pharmacotherapy. The goal of traditional therapeutic drug monitoring is the individualization of a dosage of a drug achieved by a trial-and-error method using the serum drug level as a guide after administration of a standard dosage to the patient. Pharmacogenomics deals with the effects of the genetic variations of a patient and his or her response to a particular drug. This can be achieved by correlating gene expression or single nucleotide polymorphism (SNP) with a drug's efficacy or toxicity. The goal of pharmacogenomics testing is to determine a patient's genotype before the initiation of drug therapy to develop a rational means of selecting a drug: to optimize dosing to get the maximum benefit from the drug and minimum toxicity. The first pharmacogenomics discovery occurred more than 50 years ago when it was demonstrated that patients with a genetic polymorphism leading to glucose-6-phosphate dehydrogenase deficiency developed hemolysis after treatment with primaquine (1). With the completion of the Human Genome Project and the availability of pharmacogenomics tests, currently there are more than 121 drugs where pharmacogenomics may help with patient management; drug labels also contain the pharmacogenomics information. Polymorphism of CYP2D6 and its effect on the pharmacokinetics

Resolving Erroneous Reports in Toxicology and Therapeutic Drug Monitoring: A Comprehensive Guide, First Edition. Amitava Dasgupta.

of various drugs was the most common information (2). The human genome consists of approximately 3 billion base pairs, and there are genetic variations among individuals. These variations include SNPs, base insertions, copy number variations, and variable numbers of tandem repeats that alter the functions of various proteins that interact with a drug (drug-metabolizing enzyme, drug receptors, etc.). Understanding such genetic variations can help a physician tailor pharmacotherapy based on the need of an individual patient. Although the approach of traditional therapeutic drug monitoring is also used to personalize the dosage for a particular patient, this approach cannot determine before the initiation of therapy whether a person will respond to a particular drug and how the patient would metabolize that drug. Pharmacogenomics methods can help identify before the therapy whether a person would be a poor or fast metabolizer and for certain drugs whether a patient is likely to respond or not. It is a superior approach to traditional therapeutic drug monitoring.

The words *pharmacogenetics* and *pharmacogenomics* are often used interchangeably. In a stricter sense, pharmacogenetics refers to the study of inherent differences in drug metabolism and the response due to genetic variations among different people, whereas pharmacogenomics refers to the general study of all of the different genes that determine the overall response of an individual to a particular drug. Therefore, pharmacogenomics is the whole genome application and includes genes that affect a drug's metabolism and disposition.

The pharmacological response of a particular drug given in a selected dosing regimen depends on many factors, including patient compliance, bioavailability, drug clearance, drug protein binding as well as many others such as if medication is taken with food, if the patient is a smoker, and so on. The genetic makeup of an individual is also very important in determining whether a person will respond to a particular drug and how well the drug, if effective, can exert its pharmacological action. Pharmacotherapy in many ways could be ineffective because as many as 30–60% of patients may be nonresponders to a particular drug (3). In addition, serious adverse drug reactions lead to 6–7% of all hospitalizations; an estimated 2,216,000 patients were admitted to hospitals in 1994 in the United States due to serious drug toxicity including 106,000 deaths (4). Therapeutic drug monitoring is helpful to avoid adverse drug reactions. For example, Slaughter and Cappelletty reported that the therapeutic drug monitoring of aminoglycoside can significantly reduce the incidences of nephrotoxicity from therapy with aminoglycosides. The authors also concluded that from a pharmacoeconomics point of view, therapeutic drug monitoring of aminoglycosides is beneficial because therapeutic drug monitoring for an individual patient during therapy cost approximately $301, whereas each case of nephrotoxicity costs the hospital $4583 (5). In another report, the authors concluded that therapeutic drug monitoring of β-lactam antibiotics in critically ill patients twice a week and dosage adjustments based on blood level (antibiotics were monitored using a high performance liquid chromatographic method) resulted in the prevention of drug toxicity and a positive outcome in 87.3% of the patients studied (6). However, therapeutic drug monitoring is based on the hypothesis that the concentration of a drug in blood reflects the concentration of the drug in the target site better than the given dosage, and there is a correlation between blood concentration of the drug and clinical efficacy as well as toxicity (when drug level is higher than therapeutic range). In the 1970s and 1980s, therapeutic drug monitoring was the only approach available to individualize a patient's drug regime, but therapeutic drug monitoring is a long process to adjust dosage because information whether a patient is a poor metabolizer of a drug or not can only be obtained after prescribing the drug and then measuring drug levels for a few days. In the 1990s, traditional therapeutic drug monitoring was supplemented by pharmacogenetic methods for phenotyping and then later improved by genotyping methods. Although the practice of therapeutic drug monitoring involves approximately 25 commonly monitored drugs and another 30–40 less frequently monitored drugs, today pharmacogenomics principles hold promise to reduce adverse reactions from certain drugs dramatically and improve patient safety by genetically determining interindividual differences in response to a particular drug. However, not all drugs require prior pharmacogenomics testing. At present, successful pharmacogenomics testing has improved patient care in the area of warfarin therapy, pain management with opioids, the treatment of depression with tricyclic antidepressants and selective serotonin reuptake inhibitors (SSRIs), cancer patients receiving certain antineoplastic agents, transplant recipients receiving specific immunosuppressants as well as patients receiving β-blockers and type I antiarrhythmic agents.

17.2. METABOLISM AND THE PHARMACOLOGICAL ACTION OF A DRUG

When a drug is administered orally, it undergoes several steps in the body that eventually determine the concentration of that drug in serum or whole blood. These steps include liberation, the release of the drug from the tablet, capsule, or extended-release formulation; absorption, which is the movement of drug from the site of administration (for drugs taken orally) to blood circulation; distribution, the movement of the drug from the blood circulation to tissues; protein binding; and metabolism. The cytochrome P450 enzyme system is the major drug-metabolizing agent of the body. Metabolism may follow linear or nonlinear kinetics and may be zero order, first order, or may follow more complicated kinetics. Drug metabolites may be active, further complicating the pharmacokinetic parameters. Finally, a drug is excreted from the body through urine or feces.

Liberation of a drug after oral administration depends on the formulation of the dosage. Immediate-release formulation releases the drugs at once from the dosage form when administered, and the same drug may also be available in a sustained-release formulation. Local and systematic adverse effects of a drug can also be reduced by the use of controlled-release delivery systems (7). Enteric-coded formulations resist gastric acid degradation and deliver drugs into the distal small intestine and proximal colon. An enteric-coded formulation of mycophenolic acid mofetil, a prodrug of the immunosuppressant mycophenolic acid, is commercially available (8).

Absorption of a drug depends on the route of administration. Generally, oral administration is the route of choice, but under certain circumstances (nausea, vomiting, and convulsion), the rectal route may present a practical alternative for delivering anticonvulsants, nonnarcotic and narcotic analgesics, theophylline, and antibacterial and antiemetic agents. When a drug enters the blood circulation, it is distributed throughout the body into various tissues, and the pharmacokinetic parameter is called the volume of distribution. The amount of a drug at a target receptor site, where it exerts its pharmacological activity, is usually a very small fraction of the total amount of the drug in the body because most of the drug is distributed in tissue and blood. Protein binding of a drug also limits its movement into tissues. Drugs usually undergo chemical transformation (metabolism) before elimination. Drug metabolism may occur in any tissue including the blood. For example, plasma butylcholinesterase metabolizes drugs such as succinylcholine. The role of metabolism is to convert lipophilic nonpolar molecules to water-soluble polar compounds that are excreted in urine. Many drugs are metabolized in the liver in two phases by various enzymes, but cytochrome P450 mixed function oxidase is the major liver enzyme responsible for the metabolism of most drugs. Enzymes responsible for most phase 1 and phase 2 reactions are listed in Table 17.1. See Chapter 1 for a more in-depth discussion of this topic.

A drug may also undergo extensive metabolism before fully entering the blood circulation. This process is called first-pass metabolism. Drugs may also be excreted via other routes, such as biliary excretion. The factors that determine elimination of a drug through the biliary tract include chemical structure, polarity, and molecular weight as well as active transport sites within the liver cell membranes for that particular drug. A drug excreted in bile may also be reabsorbed from the gastrointestinal tract, or a drug conjugate may be hydrolyzed by the bacteria of the gut, liberating the original drug, which can return into the blood circulation. Enterohepatic circulation may prolong the effects of a drug. Cholestatic disease states, in which the flow of normal bile flow is diminished, reduce the bile clearance of a drug and may cause drug toxicity (9).

There is a great deal of variability at the DNA level between individuals that governs many characteristics of the person including his or her ability to respond to a particular drug therapy. SNPs accounts for more than 90% of the genetic variations in the human genome. The rest of the genetic variations include insertions and deletions, tandem repeats, and microsatellites. With the completion of the human genome project, there is an initial estimate of approximately 1.42 million polymorphisms in the human genome (10). The

TABLE 17.1. Major Drug-Metabolizing Enzymes

Reaction	Phase	Name of Enzyme	Genetic Polymorphism/ Significance in Pharmacogenomics Tests
Oxidation	Phase I	Cytochrome P450	Yes/Significant
		Alcohol dehydrogenase	Yes/Nonsignificant
		Aldehyde dehydrogenase	Yes/Nonsignificant
		Monoamine oxidase	Yes/Significant
Reduction	Phase I	Various reductases	Yes/May be significant for some type of reductases
Hydrolysis	Phase I	Butylcholinesterase	Yes/Significant
		Epoxide hydrolase	Yes/Nonsignificant
		Amidases	Yes/Nonsignificant
Glucuronidation	Phase II	Glucuronosyltransferase	Yes/Significant
Acetylation	Phase II	N-Acetyltransferase	Yes/Significant
Methylation	Phase II	Methyltransferase	Yes/Significant
Amino acid conjugation	Phase II	Glutathione transferase	Yes/Nonsignificant
Sulfation	Phase II	Sulfotransferase	Yes/Nonsignificant

science of pharmacogenomics focuses on genetic variability in drug response due to molecular alterations such as SNP, gene deletion, or gene duplication at the level of drug-metabolizing enzymes, drug-transporting proteins as well as drug targets/receptors. However, most of the pharmacogenomics studies so far have focused on the impact of genetic variations on the expression and function of drug-metabolizing enzymes and transporter proteins. The effect of genetic polymorphism of drug metabolism butylcholinesterase on the metabolism of neuromuscular blocking agent succinylcholine and mivacurium used during general anesthesia has been well documented. More than 30 genetic variations of butylcholinesterase (also known as serum cholinesterases) have been described, and some of the carriers (who are homozygous for atypical or silent variants) are known to have low activity for hydrolysis of succinylcholine and mivacurium, resulting in the prolonged action of these drugs causing severe toxicity (11). Although traditional tests determining the butylcholinesterase phenotype by measuring enzymatic activities with a variety of inhibitors, most noticeably dibucaine (dibucaine number), these approaches are inadequate for identifying all variants of this enzyme and leave many cases of prolonged response to succinylcholine unexplained. The application of molecular genetics has resulted in the identification of previously unrecognized variants (12).

Based on the response to a drug, individuals can be classified as poor or extensive metabolizers. Molecular genetic testing can characterize an enzyme's gene to demonstrate which alleles (genetic polymorphisms) are present and how such alleles may affect enzymatic activity. Some of these alleles may be associated with the loss or reduction of gene function. Alleles are denoted by an asterisk (*), and different alleles are given a different number; *1 usually means a normally functioning gene, hence a normally functioning enzyme. Usually individuals who have two normal genes are extensive metabolizers (EMs), and these individuals are normal metabolizers of drugs. In contrast, individuals with two nonfunctional genes are poor metabolizers (PMs) because they metabolize a drug very slowly compared with a normal individual (EM). Ultrarapid metabolizers (UMs) may have multiple copies of active genes, and these individuals metabolize a particular drug so fast that the drug may not have any pharmacological effect. Intermediate metabolizers (IMs) may have one active allele and one nonactive allele for the same gene. In general, these individuals have enzymatic activity that is intermediate between PMs and EMs, but particular activity may vary between a particular gene, ethnic group, or drug. These different types of metabolizers are more evident in individuals with the CYP2C9, CYP2C19, and CYP2D6 genotype, whereas the polymorphism of CYP2D6 affects a vast majority of drugs.

17.2.1. Polymorphism in Cytochrome P450

The cytochrome P450 proteins (CYPs) comprise a large group of heme-containing mono-oxygenase proteins that localize to the endoplasmic reticulum and mitochondrial membrane. Nicotinamide adenine dinucleotide phosphate (NADPH) is a required cofactor for CYP-mediated biotransformation, and oxygen serves as a substrate. The CYP superfamily is found in many organisms, with more than 7700 known members across all species studied. At present, 57 functional genes and 58 pseudogenes, with members 1, 2, and 3, are known to encode CYP isoforms; of these, at least 15 are associated with xenobiotic metabolism. CYP isoenzymes are named according to sequence homology: amino acid sequence similarity more than 40% assigns the numeric family (e.g., CYP1, CYP2); more than 55% similarity determines the subfamily letter (e.g., CYP2C, CYP2D); isoforms with more than 97% similarities are given an additional number (e.g., CYP2C9, CYP2C19) to distinguish them. The major CYP isoforms responsible for metabolism of drugs include CYP1A2, CYP2B6, CYP2C9, CYP2C19, CYP2D6, CYP2E1, and CYP3A4/CYP3A5. However, CYP3A4 is the predominant isoform of the CYP family (almost 30%), usually responsible for the metabolism of approximately 37% of all drugs followed by CYP2C9 (17% of drugs), CYP2D6 (15% of drugs), CYP2C19 (10% of drugs), CYP1A2 (9% of drugs), CYP2C8 (6% of drugs), and CYP2B6 (4% of drugs). Polymorphisms in the CYP family greatly influence the metabolism of many therapeutic drugs. CYP2D6, 2C19, and 2C9 polymorphisms account for the most frequent variations in phase I metabolism of drugs, and many drugs are metabolized by these enzymes. Approximately 5–14% of whites, 0–5% of Africans, and 0–1% of Asians lack CYP2D6 activity, and these individuals are known as PMs because they carry two defective alleles (13). There are more than 50 mutations and 70 alleles of CYP2D6, with many of these leading to the PM phenotype. For example, catalytically inactive CYP2D6*4 is present in 21% of whites but in only 1% of Asians. In contrast, CYP2D6*10, which has reduced catalytic activity compared with CYP2D6*1, is expressed in 50–70% of Asians but in only 2% of whites (14). For patients who are CYP2D6 poor metabolizers, a dosage reduction is recommended for clomipramine, flecainide, haloperidol, zuclopenthixol, doxepin, nortriptyline, imipramine, propafenone, and metoprolol (15).

Another polymorphically expressed human CYP enzyme is CYP2C19. Thera are eight major alleles of CYP2C19, and among then CYP2C19*2 and CYP2C19*3 are catalytically inactive. CYP2C8*2 is expressed at greater frequency in Asians (25%) than whites; CYP2C19*3 is found in approximately 8% of Asians but less than 1% in whites. Individuals who have inactive CYP2C19 are highly sensitive to drugs such as the proton pump inhibitors omeprazole, propranolol, and amitriptyline. Patients with CYP2C19*2 and CYP2C19*3 are poor metabolizers and have higher plasma concentrations of omeprazole and a greater pharmacological response (16). An ultrarapid genotype of CYP2C19*17*17 has also been recently reported.

CYP2C9 is another clinically significant enzyme that demonstrates six allelic variations with a potentially functional impact on the efficacy and adverse effects of drugs that are mainly eliminated by this enzyme. Studies into the CYP2C9 polymorphism have demonstrated the importance of the CYP2C9*2 and *3 alleles. These alleles are expressed at a greater frequency in whites (7–10%) compared with Asians (less than 3%) causing impaired metabolism of a variety of CYP2C9 substrates including warfarin, phenytoin, tolbutamide, and diclofenac. Extensive polymorphism also occurs in other CYP genes, such as CYP1A1, 2A6, 2A13, 2C8, 3A4, and 3A5. Because several of these CYPs (e.g., CYP1A1 and 1A2) play a role in the bioactivation of many procarcinogens, polymorphisms of these enzymes may contribute to the variable susceptibility to carcinogenesis. The distribution of the common variant alleles of CYP genes varies among different ethnic populations (13). Other polymorphically expressed drug-metabolizing enzymes are N-acetyltransferase (NAT 1 and NAT 2) and thiopurine-S-methyltransferase (TPMT). The slow acetylator phenotype of the NAT1/2 polymorphism results in isoniazid-induced peripheral neuropathy and sulfonamide-induced hypersensitivity reaction; TPMT catalyzes the inactivation of various anticancer and anti-inflammatory drugs (14).

17.2.2. Polymorphism of Transporter Proteins and Receptors

Most drug responses are determined by the interplay of several gene products that influence pharmacokinetics and pharmacodynamics, that is, drug-metabolizing enzymes, drug transporters, and drug targets. With the complete sequencing of the human genome, it has been estimated that there are approximately 500–1200 genes code for drug transporters, and today the best characterized drug transporter is the multidrug-resistant transporter P-glycoprotein/MDR1, the gene product of multiple drug-resistant protein (MDR1). MDR1 is a glycosylated membrane protein of 170 kDa of the ATP-binding cassette superfamily expressed mainly in the intestines, liver, kidneys, and brain. A large number of structurally unrelated drugs are substrates for MDR1, and their intestinal absorption, hepatobiliary secretion, renal secretion, and brain transport are regulated by MDR1. A silent SNP, C3435T in exon 26, was found to be associated with the duodenal expression of MDR1, and thereby the plasma concentration of digoxin after oral administration. Sakaeda et al demonstrated that the serum concentration of digoxin after a single dose of oral administration was lower in subjects harboring the mutant allele C3435T at exon 26 of the MDR1 gene (17). Although a total of 28 SNPs have been found at 27 positions on the MDR1 gene, clinical investigations on the association of MDR1 genotypes with the expression and function of MDR1 in tissues, and with pharmacokinetics and pharmacodynamics, have mainly focused on C3435T. C3435T is also reported to be a risk factor for a certain class of diseases including the inflammatory bowel diseases, Parkinson's disease, and renal epithelial tumor, and this also might be explained by the effects on MDR1 expression and function (18).

Compared with drug-metabolizing enzymes, much less is known about the genetic polymorphisms of drug targets and receptors, but molecular research has revealed that many of the genes that encode drug targets demonstrate genetic polymorphism. These variations can result in an altered sensitivity of the receptor to a drug affecting its efficacy as well as toxicity. For example, the β-2 adrenoreceptor that is encoded in the ADRB2 gene demonstrates clinically significant genetic variations in drug targets. The variable number of polymorphisms of the serotonin transporter gene are associated with a variable response to some antidepressants (19). Polymorphism of the gene-encoding enzyme vitamin K epoxide reductase is also of significant clinical importance (20).

17.2.3. Polymorphism of Thiopurine Methyl Transferase

The enzyme thiopurine methyl transferase (TPMT) converts thiopurine drugs azathioprine, thioguanine, and 6-mercaptopurine through a series of steps to the toxic thioguanine nucleotide, which is pharmacologically active as well as to other less toxic metabolites. Theoretically, only 5% of the toxic thioguanine metabolite is needed for therapeutic action. If TRMT is deficient either completely as a deficient homozygous or partially as a heterozygous, the patient is at a greater risk of developing white cell depression including neutropenia. One in 300 whites are deficient homozygous and require a much lower dosage of the thiopurine drug (50–90% reduction in dosage) to avoid toxicity (21). Selection of an alternative drug is also advised in these patients (PMs and IMs).

17.3. GLUCOSE-6-PHOSPHATE DEHYDROGENASE DEFICIENCY AND MEDICATIONS

Glucose-6-phosphate dehydrogenase (G6PD)is an enzyme that helps produce energy from glucose and also protects red blood cells from oxidation. If there

TABLE 17.2. Drugs to Avoid in Patients with Glucose-6-Phosphate Dehydrogenase Deficiency

Acetanilid	Aspirin	Chloroquine
Dapsone	Dimercaptosuccinic acid	Furazolidone
Isobutyl nitrite	Methylene blue	Nalidixic acid
Niridazole	Nitrofurantoin	Pamaquine
Phenacetin	Phenazopyridine	Primaquine
Probenecid	Mepacrine	Sulfacetamide
Sulfamethoxazole	Sulfapyridine	Thiazolsulfone
Urate oxidase	Vitamin K analogs	

is a deficiency of the enzyme, red blood cells are more prone to oxidative damage and may break down, releasing hemoglobin and causing hemolysis. Deficiency of G6PD is a very common X-linked genetic disorder. A number of different variants may offer protection against malaria infection; however, these individuals cannot eat fava beans. Male individuals with certain deficient alleles are susceptible to neonatal jaundice and acute hemolytic anemia usually during an infection and after the treatment with certain drugs. Altogether 160 different mutations have been described, most mutations cause red cell enzyme deficiency by decreasing enzyme stability (22). It is further estimated that 400 million people worldwide carry a mutation in the G6PD gene that causes deficiency of the enzyme. Among the white population, the prevalence of G6PD deficiency varies from less than 1 in 1000 among northern Europeans to 50% of males among Kurdish Jews. G6PD deficiency may also be observed among Asians but is rare in the Japanese population. G6PD deficiency is also common among Africans. Drug-induced hemolysis is considered the most common adverse effect of G6PD deficiency. Youngster et al recently compiled a list of drugs that should be avoided in patients with G6PD deficiency (23). Drugs that are not recommended for patients with G6PD deficiency are listed in Table 17.2. G6PD deficiency was one of the first genetic disorders described in the medical literature.

Although related to certain drug-related hemolysis, G6PD testing is ordered primarily for patients who have symptoms of anemia or jaundice where the common causes for anemia or jaundice have been ruled out. The most common enzymatic activity of G6PD is tested, and if the activity is less than normal the patient may be deficient in G6PD activity. The G6PD activity can be easily measured in clinical laboratories using automated analyzers. Glucose 6-phosphate is converted into 6-phosphogluconate in the presence of G6PD, and in this process the cofactor nicotinamide adenine dinucleotide phosphate (NADP) is converted to NAPDH. NADP has no absorption at 340 nm, whereas NADPH absorbs at 340 nm. Therefore, the absorbance change is proportional to G6PD activity. A similar reaction where erythrocyte 6-phosphogluconate dehydrogenase may generate NADPH from NADP is prevented by maleimide.

Although laboratory tests measuring G6PD deficiency are common, there is also a fluorescent spot test and molecular methods for testing G6PD deficiency. Molecular methods, although capable of identifying particular variants, are less frequently ordered. Wang et al reported that a G6PD level less than 8.5 U/g of hemoglobin is a good predictor of severe neonatal hyperbilirubinemia (24). Nafa et al reported based on molecular analysis (single-strand conformation polymorphism analysis) that 92% of all abnormalities were due to five variants, and the two most common variants are G6PD A-variant (46%), and G6PD Mediterranean (23%) (the area also associated with favism) (25). For molecular testing, polymerase chain reaction (PCR) and related methods are useful.

17.4. ROLE OF PHARMACOGENOMICS IN PHARMACOTHERAPY

Although traditional therapeutic drug monitoring is successful in discerning the pharmacotherapy of certain drugs under a variety of altered pathophysiologies as well as identifying pharmacokinetic interactions between drugs, drugs and herbs, and between drugs and components of grapefruit juice, it also has certain limitations. Genetic polymorphism of drug-metabolizing enzymes, transporters, and target receptors can only be determined by pharmacogenomics testing. The biggest advantage of genotyping is to determine what initial dosage must be used for a particular patient based on the genetic polymorphism of drug-metabolizing enzymes prior to the initiation of therapy, thus preventing an adverse drug event or treatment failure. Because genotyping involves the analysis of a person's DNA directly, it is not influenced by concurrent drug administration, alteration in hormonal level, and altered pathophysiology. Phenotyping the polymorphism of a drug-metabolizing enzyme is the indirect analysis of genetic variation by examining an individual's capacity to metabolize a particular drug. This can be achieved by administrating a probe drug and the measurement of drug metabolites, thus classifying an individual as a PM, IM, EM, or UM. Therapeutic drug monitoring can also be used for phenotyping an individual and personalized dose adjustment, but major limitations of phenotyping include potential adverse effects from drug administration and the effects of other administered drugs, disease state, and alteration in hormonal level in phenotyping of the drug. The ultimate goal of pharmacogenomics testing is to determine the safest and most effective dosage of the best drug for a particular patient and to predict a patient's response to a given drug prior to its use. Therefore, pharmacogenomics testing is a proactive approach to personalized therapy (26). Traditional therapeutic drug monitoring is a reactive approach where dosage adjustments are made based on the serum level of the drug after administration to a patient. In addition, therapeutic drug monitoring can provide predictive value of only one drug and dosage adjustment based on serum concentrations of the particular drug. In contrast, pharmacogenomics testing to identify a patient's genotype can provide predictive value for multiple drugs (for example, CYP2D6 polymorphism will affect all drugs metabolized via this isoform in the patient) and also provide a mechanistic explanation for why a patient may require a higher or lower dosage of the drug or a different drug altogether. However, from an economic point of view, therapeutic drug-monitoring tests are inexpensive, whereas pharmacogenomics tests is currently significantly more expensive.

Although therapeutic drug monitoring is the most common test for phenotyping a type of metabolizer for the individualization of drug dosage, other biochemical tests are often used for the purpose of phenotyping. The most common biochemical test is the determination of the enzymatic activity of butylcholinesterase (dibucaine number) before the use of succinylcholine as a muscle relaxant during anesthesia. However, genotyping using pharmacogenomics is superior to the biochemical phenotype approach. Drugs that may benefit from pharmacogenomics testing during therapy are listed in Table 17.3.

TABLE 17.3. Examples of Some Drugs Where Pharmacogenomics Testing Has Clinical Significance

Drug	Polymorphism	Comments
Anticoagulants		
Warfarin	CYP2C19/ VKORC1	In poor metabolizer with "A" haplotype VKORC1 increases risk of bleeding; need much lower dose
Antineoplastic Drugs		
Irinotecan	UTG1A1	Accumulation of active metabolite SN-38 is homozygous for UTG1A1828 causing toxicity
6-Marcaptopurine, thiopurine, azathioprine	TPMT	Neutropenia in poor metabolizer
tamoxifen	CYP2D6	Poor metabolizers have lower blood level of active metabolite and are unresponsive
Antidepressants		
Amitriptyline	CYP2D6	Poor metabolizers have more adverse effects
Doxepin	CYP2D6	Poor metabolizers have more adverse effects
Nortriptyline	CYP2D6	Poor metabolizers have higher levels of drug
Paroxetine	CYP2D6	Poor metabolizers have more adverse effects; ultrarapid metabolizers are nonresponsive
Sertraline	CYP2C19	Poor metabolizers have more adverse effects; ultrarapid metabolizers are nonresponsive
Cardioactive		
Clopidogrel	CYP2C19	Poor metabolizers have low levels of active metabolite in blood and show poor response
Flecainide	CYP2D6	Reduce dosage in poor metabolizers
Opiate Analgesics		
Codeine	CYP2D6	Poor metabolizers unable to convert codeine to morphine and do not get pain relief
Tramadol	CYP2D6	Select alternative drug for poor metabolizers
Other Agents		
Omeprazole	CYP2C19	Ultrarapid metabolizers have treatment failure

17.4.1. Pharmacogenomics and Warfarin Therapy

Usually warfarin (Coumadin) therapy is monitored pharmacodynamically using the international normalization ratio. Warfarin, the most widely used oral anticoagulant, is a synthetic compound available as a racemic mixture of 50% R-warfarin and 50% S-warfarin. The considerable variability in the warfarin dose–response relationship among individuals is explained mainly by the genetic polymorphism of CYP2C9, the major enzyme that metabolizes warfarin and the target enzyme (VKORC1) (27). Finkelman et al recently found that pharmacogenomics algorithms are better predictors of warfarin dosage than empirical dosing, clinical algorithms, or warfarin labels (28). See Chapter 9 for a more in-depth discussion of this topic.

17.4.2. Pharmacogenomics and Certain Anticancer Drugs

Cancer chemotherapy is characterized by wide variations of efficacy and toxicity among different patients due to the wide interindividual variability in pharmacokinetics. In addition, most anticancer agents also have narrow therapeutic windows. Today, pharmacogenomics testing is very useful in the pharmacotherapy of several anticancer drugs because the candidate gene approach has provided important clues for pharmacogenomics-based personalized chemotherapy with 6-mercaptopurine (6-MP), solely metabolized by thiopurine S-methyltransferase (TPMT), and irinotecan, mainly detoxified by UDP-glucuronosyltransferase 1A1 (UGT1A1). The reduced activity of TPMT caused by polymorphisms in the TPMT gene and decreased activity of UGT1A1 caused by UGT1A1*28 are related to the severe toxic effects of 6-MP and irinotecan, respectively. In response to these findings, the U.S. Food and Drug Administration (FDA) has supported clinical pharmacogenetic testing by revising the package inserts for these anticancer drugs (29). Studies have shown that genotype-phenotype correlates in breast cancer patients and the polymorphism of the pregnane X receptor (PXR) have effects on the efficacy of anticancer drugs, for example, breast cancer patients with PXR*1B demonstrate a low clearance of doxorubicin (30). Tamoxifen, used as a therapy for estrogen receptor–positive breast cancer, is a prodrug that is converted into the active metabolite endoxifen (4-hydroxy-N-desmethyl-tamoxifen) mostly by CYP2D6. PMs have lower levels of endoxifen in their blood and are poor responders to the therapy. In addition, drugs that inhibit the activity of CYP2D6 can have an impact on the outcome of therapy with tamoxifen (31). Endoxifen blood level can be measured to assess antitumor activity, and it has been shown that PMs and IMs have lower endoxifen blood levels than EMs (32). However, the CYP2D6 genotype could be useful in deciding on the nonuse of tamoxifen because there are alternative treatments.

Although pharmacogenomics testing has some application in cancer chemotherapy, traditional therapeutic drug monitoring is also useful for certain anticancer drugs. Methotrexate is a commonly monitored therapeutic drug in clinical laboratories. Anticancer drugs demonstrate wide interindividual pharmacokinetic variability. In addition, sometimes toxicity better correlates with serum drug level than dosage. Although the pharmacokinetic behavior of many anticancer drugs such as cyclophosphamide, cisplatin, methotrexate, cytarabine, 5-fluorouracil, doxorubicin, daunorubicin, bleomycin, vincristine, and vinblastine in humans has been described, the complexity of drug metabolism and the lack of established therapeutic ranges for many anticancer drugs had limited applications for therapeutic drug monitoring of anticancer drugs in the past (33). However, the development of chromatography combined with high sensitivity detection techniques including mass spectrometry has contributed to a wider application of therapeutic drug monitoring of anticancer drugs at present with published protocols for monitoring busulfan, treosulfan, cyclophosphamide, ifosfamide, 5-fluorouracil, cytarabine, cytosine arabinoside, tamoxifen, and a variety of other drugs (34).

17.4.3. Pharmacogenomics and Opioid Therapy

Drug-metabolizing enzymes represent major associations between an individual's response to narcotic analgesics (opioids) and genetic profile, and candidate genes, such as those encoding (opioid) receptors, transporters, and other molecules, also play important roles in the pharmacotherapy of pain management (35). In general, most of the therapeutic drugs used in pain management, for example, codeine, dihydrocodeine, fentanyl, hydrocodone, methadone, morphine, oxycodone, tramadol, and tricyclic antidepressants, are metabolized by polymorphic CYP450 enzymes such as CYP2D6, CYP3A4, and/or uridine diphosphate glucuronosyltransferase 2B7 (UGT2B7). The wide range of genetic polymorphisms of *CYP2D6* leads to four distinct groups of metabolizers including UMs containing multiple copies of the *CYP2D6* gene, EMs with a single wild-type copy of the *CYP2D6* gene, IMs showing decreased enzymatic activity, and PMs with almost no detectable activity. Differences in drug metabolism due to the polymorphism of these genes can lead to toxicity or therapeutic failure by altering the relationship between the dose and steady state blood concentration of the pharmacologically active drug. In addition, genetic polymorphisms in drug targets can have a greater influence on the efficacy and toxicity of medications compared with other clinical variables. Although most of the population are EMs of *CYP2D6*, 5–10% of whites and 1–4% of most other ethnic groups have decreased *CYP2D6* activity and risk toxic effects if they receive the routine clinical dose of a drug inactivated by this enzyme, for example codeine. PMs in general tend to demonstrate the highest drug levels in serum followed by IMs and EMs. A total of 80% of patients reporting adverse drug reactions had impaired CYP2D6 metabolism. There is also a correlation between oxycodone steady state serum concentration and pain relief. The authors suggested that genotyping following therapeutic drug monitoring concentrations are beneficial in pain management (36). Madadi et al reported a fatal hydrocodone overdose in a child who had a reduced capability of metabolizing hydrocodone to hydromorphone by CYP2D6 (CYP2D6*2A/*41); a high postmortem level of hydrocodone was in the range that may have caused the fatality (37).

Case Study

A healthy 2-year-old boy weighing 13 kg with a history of sleep apnea underwent an elective adenotonsillectomy. The surgery was uneventful and the boy received 10 mg meperidine and 12.5 mg dimenhydrinate intramuscularly. The family was sent home with instructions to give the boy 10–12.5 mg codeine and 120 mg acetaminophen syrup orally every 4–6 hours as needed for pain. On the second day after surgery, fever and wheezing developed, and on the following day the boy died. A postmortem investigation showed 700 ng/mL of codeine and 32 ng/mL of morphine in his femoral blood as determined by gas chromatography/mass spectrometry. The genetic analysis revealed that the boy was a UM with duplication of the CYP2D6 allele. The prescribed dosage of codeine was within the therapeutic range, but the boy metabolized codeine faster than a normal individual resulting in a much higher morphine level than expected. Usually morphine concentration exceeding 20 ng/mL causes respiratory depression in small children. However, the boy also had other complicating factors that may have contributed to his death including bronchopneumonia, which may have added to his hypoxia (38).

Case Study

A 66-year-old man taking nifedipine was scheduled for a surgery for recurring renal carcinoma. He underwent an uneventful surgery 4 years ago. In the recovery room, he complained of severe pain and received two boluses of tramadol 50 mg administered within 35 minutes. The patient was eventually discharged from the recovery room. At 10.5 hours, the patient was unresponsive, but later he regained consciousness after receiving naloxone. He eventually recovered fully. The genetic analysis of the patient revealed that he was a UM with CYP2D6 gene duplication that contributed to a higher than expected concentration of O-desmethyltramadol, the active metabolite of tramadol causing severe respiratory depression in the patient (39).

17.4.4. Pharmacogenomics and Immunosuppressants

Organ transplant recipients receive immunosuppressants to prevent organ rejection. At present all immunosuppressants including cyclosporine, tacrolimus, sirolimus, mycophenolic acid, and recently FDA-approved (in 2010) everolimus require routine therapeutic drug monitoring. Although therapeutic drug monitoring of cyclosporine, tacrolimus, sirolimus, and everolimus are conducted in whole blood, another immunosuppressant, mycophenolic acid, is monitored in serum or plasma. Recent reports indicate that pharmacogenomics testing can complement the practice of therapeutic drug monitoring of immunosuppressants in certain patient populations. Immunosuppressant drugs have a narrow therapeutic window with wide variations in blood concentrations among individuals after a standard loading dose, and it may take 5–7 days to reach the target blood concentration of a desired immunosuppressant using traditional therapeutic drug monitoring. This timeline is maybe too long to achieve enough concentration of the immunosuppressant in all patients to prevent graft rejection without exposing certain patients to excessive toxicity over the critical 2- to 3-day period after transplant. Polymorphisms of drug-metabolizing enzyme and transporter proteins have ben associated with altered metabolism and/or response of an immunosuppressant. Pharmacogenomics testing can be used to identify the polymorphism of CYP3A5 to predict the optimal initial dosage of tacrolimus (40). Tavira et al, based on a study of 200 patients who received kidney transplants and were managed on tacrolimus as the primary immunosuppressant, concluded that the CYP3A5 genotype was the strongest predictor of tacrolimus dosage. Patients who were CYP3A5*3*3 (CYP3A5 nonexpressor) received a significantly higher dosage of tacrolimus compared with other patients (41). Jacobson et al also commented that although CYP3A4*3 is the top variant associated with lower trough tacrolimus concentration despite higher dosages, other variants also play an important role in determining tacrolimus pharmacokinetics (42).

Case Study

A 5-year-old white boy underwent a renal transplant, and for 9 years afterward the boy had normal renal function. At age 14 the boy presented with acute viral gastroenteritis and was hospitalized. One week later he developed acute renal failure, and his tacrolimus blood level was 27.2 ng/mL, which was toxic. His tacrolimus dosage was reduced to a third of the original dosage. Despite the reduced dosage and controlling his hypertension with amlodipine, his renal function deteriorated requiring hemodialysis, and tacrolimus trough concentration was increased to a very toxic level of 38.8 ng/mL. No other explanation for renal failure was found except for the high tacrolimus level. Tacrolimus was discontinued, and the tacrolimus level returned to therapeutic range within 7 days. Hemodialysis was discontinued after 20 days when the boy recovered fully. The patient was found to have a CYP3A5 deficiency with the nonexpressor CYP3A5*3*3 genotype and apparently normal CYP3A4 activity. The authors speculated that the high initial tacrolimus level in the boy was due to diarrhea and the calcium channel blocker amlodipine; the genetic factor also contributed to his clinical condition (43).

17.4.5. Pharmacogenomics and Cardioactive Drugs

Therapeutic drug monitoring of cardioactive drugs such as digoxin, digitoxin, lidocaine, procainamide and its metabolite N-acetylprocainamide, and quinidine is essential due to the narrow therapeutic ranges of these drugs, and many hospital laboratories routinely offer these tests. Digoxin is transported by P-glycoprotein, and patients with polymorphisms in exon 26 (C345T) of the MDR-1 gene appear to have a higher plasma level of digoxin (15). Clopidogrel decreases the morbidity and mortality associated with severe cardiovascular diseases, but clopidogrel is a prodrug that needs to be transformed into its active metabolite through the cytochrome P450 family of enzymes. Polymorphisms in CYP2C19 and to a lesser extent the adenosine 5′ triphosphate binding cassette gene, ABCB1, contribute to the variability of the responsiveness of this drug. Patients who possess at least one variant CYP2C19 allele (CYPC219*2*3) have an impaired response to clopidogrel due to the decreased formation of the pharmacologically active metabolite (44). Evidence also demonstrates that genetic factors interact with numerous heart failure medications including β-blockers, vasodilators, and angiotensin-converting enzyme inhibitors (45).

17.4.6. Pharmacogenomics and Antiretroviral Drugs

The observed interindividual variation in antiretroviral pharmacokinetics results in a wide range of drug exposures from the fixed-dose regimes to different patients. To circumvent that problem, therapeutic drug monitoring of certain antiretrovirals, especially protease inhibitors, to individualize dose regimens of antiretrovirals has been strongly recommended in patients with acquired immunodeficiency syndrome. There is evidence for the clinical benefits of therapeutic drug monitoring of nelfinavir, indinavir, ritonavir, amprenavir, saquinavir, lopinavir, efavirenz, delavirdine, zidovudine, and nevirapine (46). Rendon et al reported that of 151 requests from 137 patients for therapeutic monitoring of antiretroviral drugs, 59% of the requests were due to evaluating drug toxicity, 39% for assessing virological failure, and only 2% to investigate drug–drug interactions. Nonnucleoside reverse transcriptase inhibitor (NNRTI) levels were more often requested because of toxicity, whereas protease inhibitors levels were more often requested due to suspected virological failure. Elevated drug levels were confirmed in 36% of patients with suspected drug toxicity, and subtherapeutic levels of antiretrovirals were found in 37% of patients with virological failure. The authors concluded that therapeutic drug monitoring is useful in identifying toxic levels of NNRTIs and subtherapeutic concentrations of protease inhibitors (47). In addition to therapeutic drug monitoring, pharmacogenomics testing can be clinically useful for certain antiretroviral agents. The strong association of the abacavir hypersensitivity reaction with HLA-B*5701 permits testing patients for the allele and if it is present then avoiding the drug. Persons with the allele CYP2B6*6 gene present higher efavirenz area under the curve and have an increased risk of neuropsychological toxicity from the drug (48). There is also an association between the UGT1A1 allele and atazanavir-associated hyperbilirubinemia as well as a correlation between HLA class II allele HLA-DRB*101 and nevirapine-associated hypersensitivity (49).

17.4.7. Pharmacogenomics of Psychoactive Drugs

Therapeutic drug monitoring of the tricyclic antidepressants amitriptyline, clomipramine, and imipramine is standard practice in patients receiving these medications. In addition, there is evidence of the benefits of therapeutic drug monitoring of new-generation antidepressants such as citalopram, escitalopram, fluoxetine, fluvoxamine, paroxetine, sertraline, venlafaxine, mianserin, mirtazapine, bupropion, and milnacipran. The objectives of therapeutic drug monitoring for old- and new-generation antidepressants differs because older antidepressants have narrow therapeutic windows with higher risks of severe drug interactions, whereas the newer antidepressants have wide therapeutic ranges but an unclear plasma concentration–effect relationship. Therefore, the purpose of therapeutic drug monitoring for the new-generation antidepressants is useful in monitoring patient compliance and special patient groups such as the elderly, patients with liver and kidney impairment, patients with poor metabolism by CYP 450, and comedication with inhibitors and inducers of those enzymes (50). At present, psychotropic medications for the treatment of mental illnesses, including antidepressants, mood stabilizers, and antipsychotics, are clinically suboptimal because they are effective in only a subset of patients or produce partial responses, and they are often associated with significant side effects that discourage adherence (51). There are a number of pharmacogenetic studies on antidepressant drugs demonstrating the effects of metabolizing enzymes of the cytochrome P450 families and genes within the monoaminergic system on the response of individual patients to various psychoactive drugs. There is compelling evidence for an effect of CYP2D6 polymorphisms on antidepressant drug plasma levels and of a serotonin transporter promoter polymorphism on the clinical response to a specific class of antidepressants, especially SSRIs. There is also evidence that polymorphisms in genes regulating the hypothalamus-pituitary adrenal axis have an important impact on the response to antidepressants (52). Tricyclic antidepressant amitriptyline is metabolized by CYP2C19 to active metabolite nortriptyline, and then CYP2D6 is needed for deactivation of nortriptyline. Adverse drug reactions tend to be associated with nortriptyline concentrations, and CYP2D6-poor metabolizers are more likely to suffer from adverse effects due to the buildup of nortriptyline concentration. Smith and Curry recently reported the case of a comatose woman who intentionally overdosed with amitriptyline and demonstrated a rising total tricyclic antidepressant concentration over the 6 days after admission. The level started declining on day 7, but the patient did not achieve normal mental status until day 12. Genotyping showed the patient to be homozygous for the CYP2D6*4 allele, the most common cause of CYP2D6 enzymatic deficiency among whites. The authors commented that patients taking tricyclic antidepressants who are homozygous for CYP2D6*4 demonstrate over three times the concentration time curve and prolonged half-lives of tricyclic antidepressants (53).

Paroxetine, a SSRI metabolized by CYP2D6, is also an inhibitor of CYP2D6. PMs are at a higher risk of adverse effects from paroxetine (54). Charlier et al commented that 7–10% of the white population are PMs; about 5% are UMs for CYP2D6-mediated drug metabolism. Based on a study involving 49 patients, the authors found a significant correlation between the plasma concentrations of fluoxetine and paroxetine based on the genetic polymorphism of CYP2D6. In one patient who demonstrated CYP2D6 gene duplication (a UM), the plasma paroxetine level was extremely low (55). Laine et al reported a case of third-trimester paroxetine exposure with an adverse birth outcome. The symptoms of the newborn included severe tremor and rigidity as well as loose stools during the first 4 days of life. Plasma paroxetine concentration in the newborn was low at birth, and the infant was genotyped to be a poor metabolizer of CYP2D6, the enzyme responsible for the metabolism of paroxetine. The authors concluded that even a low paroxetine level may be associated with perinatal complications in infants exposed to paroxetine during late pregnancy, and the poor metabolizer genotype of CYP2D6 may be at a risk factor for these complications (56).

Case Study

A 39-year-old Japanese man had been treated with paroxetine (20 mg/day) for the last 18 months. He presented with rhabdomyolysis, renal failure, fulminant liver failure, cardiac conduction disturbance, and disseminated intravascular coagulation. In addition, he exhibited classical symptoms of serotonin syndrome. Other than paroxetine, the patient was also receiving digoxin, disopyramide, ticlopidine, and omeprazole. All medications were discontinued, and 24 hours after continuous venovenous hemofiltration, diuresis resumed and his renal as well as liver functions improved rapidly. The patient recovered after 1 week. He had a CYP2D6*1/*5, a heterozygous of an inactivated CYP2D6 allele that metabolizes paroxetine. This genetic polymorphism accounted for his high serum level of paroxetine (70 ng/mL) because the patient was an IM (57).

17.4.8. Miscellaneous Other Issues

Recent developments in understanding severe immunological adverse reactions following therapy with certain drugs indicate that the HLA-B type is a biomarker to strongly predict the onset of a serious skin rash such as the Stevens-Johnson syndrome. HLA-B57 and HLA-B58 are major histocompatability class I allotypes that are predictive of clinically important immune phenotypes. It has been demonstrated that HLA-B*1502, which is more abundant among Asians, is associated with severe skin rash including the Stevens-Johnson syndrome following treatment with carbamazepine. The FDA has released a warning suggesting HLA-B testing in Asians before carbamazepine therapy. Although strong correlation between carbamazepine-induced Stevens-Johnson syndrome has been found among Han Chinese patients, HLA-B*1511, a member of HLA-B75 found among Japanese, is also a risk factor for such adverse reactions from carbamazepine therapy (58). HLA-B*5701 is strongly associated with hypersensitivity to the anti-HIV drug abacavir. HLA-B*5801 is associated with hypersensitivity to allopurinol (59).

17.5. METHODOLOGIES OF PHARMACOGENOMICS

A variety of methods have been used to identify DNA polymorphisms affecting genes involved in the enzymes responsible for drug metabolism or drug target receptor genes. These techniques either focused on analyses of selected number of genes (examples: classic Sanger sequencing, pyrosequencing, real-time PCR, and melting curve analysis) or a highly multiplexed analysis of a large number of genes (various microarray, microchip, and microbead techniques). There are two different ways a pharmacogenomics test can be introduced in the toxicology laboratory. In the first and most convenient way, the FDA approves a test kit for pharmacogenomics testing. The first test for which the FDA has granted market approval using a DNA microarray is the AmpliChip CYP450 (Roche) that genotypes cytochrome P450 (CYP2D6 and CYP2C19). The test was approved to be used with a scanner (Affymetrix Genetic Chip Microarray Instrument) and can predict phenotypes by testing for 27 CYP2D6 alleles. The assay uses the patient's blood or a specimen from a buccal swab and consists of five steps including PCR, amplification of the gene, fragmentation of PCR products, hybridization, and staining on the AmpliChip DNA microarray, scanning the chip by matching the sample DNA with a probe of known DNA molecules arranged in a chip-like microarray, and finally data analysis (60). Some currently available pharmacogenomics tests in addition to DNA microarray are a test that detects variations in the UGT1A1 gene, which produces the enzyme UDP-glucuronosyltransferase, and another test that detect genetic variants of the CYP2C9 and vitamin K epoxide reductase (VKORC1) enzymes. However, compared with therapeutic drug monitoring, pharmacogenomics testing is very expensive; for example, the AmpliChip CYP450 test costs $600 or more. In the second mechanism, an individual laboratory develops a pharmacogenomics test (home-brew test). Although these tests do not require formal FDA approval, the quality of such tests is closely monitored by agencies (eg, the Centers for Medicare and Medicaid Services) that certify clinical laboratories following the rules of the Clinical Laboratory Improvement Act (CLIA) of 1988. Under CLIA rules, home-brew pharmacogenomics tests are considered highly complex tests and must follow the strict guidelines qualifications of the laboratory director and testing personnel. A few reference laboratories offer pharmacogenomics testing.

TABLE 17.4. FDA-Approved Pharmacogenomics Tests

Pharmacogenomics test	Gene Tested	Company
AmpliChip	CYP2C19, CYP2ED6	Roche, Indianapolis, IN
Invander UGT1A1 Molecular Assay	UGT1A1	Third Wave/Hologic, Madison, WI
Verigene warfarin, nucleic acid test	CYP2C9, VKORC1*	Nanosphere, Northbrook, IL
Infinity CYP2C9-VKORC1 Multiplex assay	CYP2C9, VKORCI*	AutoGenomics, Carlsbad, CA
Paragon Dx Rapid Genotype Assay	CYP2C9, VKORC1*	Paragon Dx, Morrisville, NC
e-Sensor warfarin sensitivity	CYP2C9, VKORC1*	Osmetech Molecular Diagnosis, Pasadena, CA
eQ-PCR-LC warfarin genotype	CYP2C9, VKORC1*	TrimGen, Sparks, MD
Xtag CYP2D6	CYP2D6	Luminex, Austin, TX
Infinity CYP2C19 Assay	VYP2C19	AutoGenomics, Carlsbad, CA

* Pharmacogenomics test intended for warfarin.

Currently, there are only nine FDA-approved tests for pharmacogenomic testing (Table 17.4). Other than the AmpliChip test system, the recently approved Infinity CYP2C19 assay (Infinity CYP2C19 assay, AutoGenomics) also uses microarray technology. The assay can identify gene products especially with *2, *3, and *17 alleles. AutoGenomics also has another pharmacogenomics assay (Infinity CYP2C9 and VKORCI for warfarin) assay approved by the FDA, and the xTAG CYP2D6 assay is also FDA approved. The xTAG assay (Luminex Corporation, Austin, TX) takes advantage of the universal tag/anti-tag system consisting of 24-mer oligonucleotides with isothermal melting temperatures that have been designed to minimize nonspecific hybridization to genomic DNA and between each other, allowing high-level multiplexing to occur in a single tube. After multiplex PCR, allele-specific primer extension generates single-stranded DNA containing the allele-specific tag in the 5′ region and incorporated biotinylated nucleotides. The biotin moieties allow for a streptavidin-conjugated fluorophore (typically phycoerythrin) to bind to the tagged DNA; the 5′-tag sequence allows the DNA to be captured by 5.6-micron polystyrene beads containing the corresponding anti-tag probes. The beads are then read by the xMAP analyzer, which separates and identifies each bead by flow cytometry coupled with laser-induced fluorescence detection. Therefore, each allele, bound by a specific bead, can be identified and quantified, and up to 100 alleles or 50 biallelic variants can be analyzed simultaneously. The eQ-PCR LC warfarin assay (TrimGen) is also FDA approved and is an in vitro diagnostic test for genotyping SNPs in the CYP2C9 (CYP2C9*2, CYP2C9*3) and VKORCI gene (1639G>A). The test uses the melting curve method to identify the SNP.

17.6. CONCLUSIONS

Traditional therapeutic drug monitoring is useful for the individualization of pharmacotherapy for certain drugs with narrow therapeutic range, but this

TABLE 17.5. Drugs That Carry Pharmacogenomics Biomarkers in FDA-Approved Drug Labels

Abacavir	Aripiprazole	Arsenic trioxide
Atomoxetine	Azathioprine	Boceprevir
Busulfan	Capecitabine	Carbamazepine
Carvedilol	Celecoxib	Cetuximab
Cevimeline	Chloroquine	Clopidogrel
Clozapine	Codeine	Dapsone
Dasatinib	Quinidine	Diazepam
Doxepin	Drospirenone and ethinyl estradiol	
Erlotinib	Esomeprazole	Fluorouracil
Fluoxetine	Fulvestrant	Gefitinib
Imatinib	Irinotecan	Isosorbide and hydralazine
Lapatinib	Lenalidomide	Mercaptopurine
Metoprolol	Nelfinavir	Nilotinib
Panitumumab	Propafenone	Propranolol
Protriptyline	Quinidine	Rabeprazole
Rasburicase	Rifampin	Tamoxifen
Telaprevir	Terbinafine	Tetrabenazine
Thioguanine	Thioridazine	Timolol
Tiotropium	Tolterodine	Tositumomab
Tramadol	trastuzumab	Valproic acid
Venlafaxine	Voriconazole	Warfarin

traditional approach has many limitations. Pharmacogenomics testing holds the promise to predict a drug's response or if a patient is likely to respond to a particular drug before initiation of testing, and this genotype approach certainly has many advantages over the traditional phenotype approach of therapeutic drug monitoring. Currently, FDA-approved modifications in more than 30 drug labels include pharmacogenomics information; some drug labels also carry black box warning. Drugs that carry FDA labels regarding pharmacogenomics biomarkers are listed in Table 17.5. The Royal Dutch Association for Advancement of Pharmacy established a pharmacogenomics working group that has developed a list of 53 drugs for which pharmacogenomics testing may be valuable (15). Most hospital-based clinical laboratories at present do not offer pharmacogenomics testing, but with the rapid development of new tests and automation, pharmacogenomics tests will gain a more prominent role in therapeutic drug-monitoring practice.

REFERENCES

1. Beutler E. Drug induced hemolytic anemia. Pharmacol Rev 1969; 21: 73–103.
2. Frueh FW, Amur S, Mummaneni P, Epstein RS et al. Pharamacogenomics biomarkers information in drug labels approved by the United States Food and Drug Administartion: prevaslance and related drug use. Pharmacotherapy 2008; 28: 992–998.

3. Spear BB, Health-Chiozzi M, Huff J. Clinical application of pharmacogenetics. Trends Mol Med 2001; 7: 201–204.
4. Lazarou J, Pomeranz BH, Corey PN. Incidence of adverse drug reactions in hospitalized patients: a meta-analysis of prospective studies. JAMA 1998; 279: 1200–1205.
5. Slaughter RL, Cappelletty DM. Economic impact of aminoglycoside toxicty and its prevention through therapeutic drug monitoring. 1998; 14: 385–394.
6. Roberts JA, Ulldemolins M, Roberts MS, McWhinney B et al. Therapeutic drug monitoring of beta-lactams in crtitically ill patients: proof of concept. Int J Antimicrob Agents 2010; 36: 332–339.
7. Florence AT, Jani PU. Novel oral drug formulations: their potential in modulating adverse effects. Drug Saf 1994; 10: 233–266.
8. Tedesco-Silva H, Bastien MC, Choi L, Felipe C et al. Mycophenolic acid metabolite profile in renal transplant patients receiving enteric coated mycophenolate sodium or mycophenolate mofetil. Transplant Proc 2005; 37: 852–855.
9. Rollins DE, Klaassen CD. Biliary excretion of drugs in man. Clin Pharmacokinet 1979; 4: 368–379.
10. Sachidanaadam R, Weissman D, Schmidt SC, Kakol JM et al. A map of human genome sequence variation containing 1.42 million single nucleotide polymorphisms. Nature 2001; 409: 928–933.
11. Gatke MR, Bundgaard JR, Viby-Mogensen J. Two novel mutations in the BCHE gene in patients with prolonged duration of action of mivacurium or succinylcholine during anesthesia. Pharamcogenet Genomics 2007; 17: 995–999.
12. Pantuck EJ. Plasma cholinesterase: gene and variations. Anesth Analg 1993; 77: 380–386.
13. Zhou SF. Liu JP, Chowbay B. Polymorphism of human cytochrome P 450 enzymes and its clinical impact. Drug Metab Rev 2009; 41: 89–295.
14. Ensom MHH, Chang T, Patel P. Pharmacogenomics: the therapeutic drug monitoring of the future? Clin Pharmacokinetic 2010; 40: 783–802.
15. Swen JJ, Nijenhuis M, de Boer A, Grandia L et al. Pharmacogenomics from bench to byte: an updated guideline. Clin Pharmacol Ther 2011; 89: 662–673.
16. Jin SK, Kang TS, Eom SO, Kim JI et al. CYP2C19 haplotypes in Koreans as a marker of enzyme activity with omeprazole. J Clin Pharm 2009; 34: 437–446.
17. Sakaeda T, Nakamura T, Horinouchi M, Kakumoto M et al. MDR1 genotype related pharmacokinetics of digoxin after oral administration in healthy Japanese subjects. Pharm Res 2001; 18: 1400–1404.
18. Sakaeda T, Nakamura T, Okumura K. Pharmacogenetics of drug transporters and its impact on the pharmacotherapy. Curr Top Med Chem 2004; 4: 1385–1398.
19. Zhou SF, Di YM, Chan E, Du YM et al. Clinical pharmacogenetics and potential application in personalized medicine. Curr Drug Metab 2008; 9: 738–784.
20. Daly AK. Pharmacogenetics and human genetic polymorphisms. Biochem J 2010; 429: 435–449.
21. Gardiner SJ, Gearry RB, Begg EJ, Zhang M et al. Thiopurine dose in intermediate and normal metabolizers of thiopurine methyltransferase may differ three fold. Clin Gastroenterol Hepatol 2008; 6: 654–660.

22. Mason PJ, Bautista JM, Gilsanz F. G6PD deficiency: the genotype-phenotype association. Blood Rev 2007; 21: 267–283.
23. Youngster I, Arcavi L, Schechmaster R, Akayzen Y et al. Medications and glucose-6-phosphate dehydrogenase deficiency: an evidenced based review. Drug Saf 2010; 33: 713–726.
24. Wang FL, Boo NY, Ainoon O, Wong MK. Comparison of detection of glucose-6-phosphate dehydrogenase deficiency using fluorescent spot test, enzyme assay and molecular method for prediction of severe neonatal hyperbilirubinemia. Singapore Med J 2009; 50: 62–67.
25. Nafa K, Reghis A, Osmani N, Baghli L et al. At least five polymorphic mutants account for the prevalence of glucose-6-phosphate dehydrogenase deficiency in Algeria. Hum Genet 1994; 94: 513–517.
26. Marsh S, McLeod HL. Pharmacogenomics: from bedside to clinical practice. Hum Mol Gen 2006; 15: R89-R93.
27. Lurie Y, Loebstein P, Kurnik D, Almog S et al. Warfarin and vitamin K intake in the era of pharmacogenomics. Br J Clin Pharamcol 2010; 70: 164–170.
28. Finkelman BS, Gage BF, Johnson JA, Brensinger CM et al. Genetic warfarin dosing: tables versus algorithms. J Am Coll Cardiol 2011; 57: 6122–618.
29. Fujita K, Sasaki Y. Pharmacogenomics in drug-metabolizing enzymes catalyzing anticancer drugs for personalized cancer chemotherapy. Curr Drug Metab 2007; 8: 554–562.
30. Sandanaraj E, Lal S, Selvarajan V, Ooi LL et al. PXR pharmacogenetics: association of haplotypes and ABCB! Messenger RNA expression and doxorubicin clearance in Asian breast cancer patients. Clin Cancer Res 2008; 14: 7116–7126.
31. Ingle JN. Pharmacogenomics of tamoxifen and aromatase inhibitors. Cancer 2008; 112: (3 Suppl): 695–699.
32. Goetz MP, Kamal A, Ames NM. Tamoxifen pharmacogenomics: the role of CYP2D6 as a predictor of drug response. Clin Pharamcol Ther 2008; 83: 160–166.
33. Balis FM, Holcenberg JS, Bleyer WA. Clinical pharmacokinetics of commonly used anticancer drugs. Clin Pharmacokinet 1983; 8: 202–232.
34. Dasgupta A. Therapeutic drug monitoring in cancer patients: application of chromatographic techniques. In: Dasgupta A, ed. *Advances in Chromatographic Techniques for Therapeutic Drug Monitoring*. Boca Raton, FL: CRC Press, 2009:299–322.
35. Stamer UM, Zhang L, Stuber F. Personalized therapy in pain management: where do we stand? Pharmacogenomics 2010; 11: 843–864.
36. Jannetto PJ, Bratanow NC. Utilization of pharmacogenomics and therapeutic drug monitoring for opioid pain management. Pharmacogenomics 2009; 10: 1157–1167.
37. Madadi P, Hildebrandt D, Gong IY, Schwarz UI et al. Fatal hydrocodone overdose in a child: pharmacogenetics and drug interaction. Pediatrics 2010; 126: e986–989.
38. Ciszkowski C, Madadi P, Phillips MS, Lauwers AE et al. Codeine, ultrarapid-metabolism genotype and postoperative death. N Eng J Med 2009; 361: 827–828.
39. Stamer UM, Stuber F, Muders T, Musshoff F. Respiratory depression with tramadol in a patient with renal impairment and CYP2D6 gene duplication. Anesth Analg 2008; 107: 926–929.

40. Warne N, MacPhee IA. Current progress in pharmacogenomics and individualized immunosuppressive drug dosing in organ transplantation. Curr Opin Mol Ther 2010; 12: 270–283.
41. Tavira B, Garcia EC, Diaz-Corte C, Ortega F et al. Pharmacogenomics of tacrolimus after renal transplantation: analysis of polymorphism in genes encoding 16 drug metabolizing enzymes. Clin Chem Lab Med 2011; 49: 825–833.
42. Jacobson PA, Oetting WS, Brearley AM, Leduc R et al. Novel polymorphism associated with tacrolimus trough concentrations: results from a multicenter kidney transplant consortium. Transplantation 2011; 91: 300–308.
43. Leroy S, Isapof A, Fargue S, Fakhoury M et al. Tacrolimus nephrotoxicity: beware of the association of diarrhea, drug interaction and pharmacogenomics. Pediatr Nephrol 2010; 25: 965–969.
44. Momary KM, Dorsch MP, Bates ER. Genetic causes of clopidogrel nonresponsiveness: which ones really count? Pharmacotherapy 2010; 30: 265–274.
45. Lowes BD, Buttrick PM. Genetic determinants of drug response in heart failure. Curr Cardiol Rep 2008; 10: 176–181.
46. Slish JC, Catanazaro LM, Ma Q, Okusanya OO et al. Update on the pharmacokinetic aspects of antiretroviral agents: implications in therapeutic drug monitoring. Curr Pharm Des 2006; 12: 1129–1145.
47. Rendon A, Nunez M, Jimenez-Nacher I, Gonzalez de Requena D et al. Clinical benefit of interventions driven by therapeutic drug monitoring. HIV Med 2005; 6: 360–365.
48. Roca B. Pharmacogenomics of antiretrovirals. Recent Pat Antiinfect Drug Discov 2008; 3: 132–135.
49. Tozzi V. Pharmacogenomics of antiretrovirals. Antiviral Res 2010; 85: 190–200.
50. Wille SM, Cooreman SG, Neels HM, Lambert WE. Relevant issues in the monitoring and toxicology of antidepressants. Crit Rev Clin Lab Sci 2008; 45: 25–89.
51. Zandi PP, Judy JT. The promise and reality of pharmacogenetics in psychiatry. Clin Lab Med 2010; 30: 931–974.
52. Binder EB, Holsboer F. Pharmacogenomics and antidepressant drugs. Ann Med 2006; 38: 82–94.
53. Smith JC, Curry SC. Prolonged toxicity after amitriptyline overdose in a patient deficient in CYP2D6 activity. J Med Toxicol 2011; 7: 220–223.
54. Sheffield LJ, Ohillimore HE. Clinical use of pharmacogenomics tests in 2009. Clin Biochem Rev 2009; 30: 55–65.
55. Charlier C, Broly F, Lhermitte M, Pinto E et al. Polymorphism in the CYP2D6 gene: association with plasma concentrations of fluoxetine and paroxetine. Ther Drug Monit 2003; 25: 738–742.
56. Laine K, Kytola J, Bertilsson L. Severe adverse effects in a newborn with two defective CYP2D6 alleles after exposure to paroxetine during late pregnancy. Ther Drug Monit 2004; 26: 685–687.
57. Sato A, Okura Y, Minagawa S, Ohno Y et al. Life threatening serotonin syndrome in a patient with chronic heart failure and CYP2D6*1/*5. Mayo Clin Proc 2004; 79: 1444–1448.

58. Kaniwa N, Saito Y, Aihara M, Matsunaga K et al. HLA-B*1511 is a risk factor for carbamazepine induced Stevens-Johnson syndrome and toxic epidermal necrolysis in Japanese patients. Epilepsia 2010; 51: 2461–2465.
59. Kostenko L, Kjer-Nielsen L, Nicholson I, Hudson F et al. Rapid screening for the detection of HLA-B57 and HLA-B58 in prevention of drug hypersensitivity. Tissue Antigens 2011; 78: 11–20.
60. de Leon J. Amplichip CYP450 test: personalized medicine has arrived in psychiatry. Expert Rev Mol Diag 2006; 6: 277–286.

Chapter 18

Approaches for Eliminating Interference/ Discordant Specimens in Therapeutic Drug Monitoring and Drugs of Abuse Testing

18.1. INTRODUCTION

Erroneous results in therapeutic drug monitoring and drugs of abuse testing may occur due to errors in the preanalytical stage, analytical stage, and post-analytical stage. The major cause of erroneous results in the analytical phase is due to interference from an endogenous factor or exogenous factor in immunoassays used in toxicology laboratories for therapeutic drug monitoring and drug of abuse testing. The matrix effect is more a problem when a serum or plasma specimen is analyzed compared with urine specimens. Sources of erroneous results in therapeutic drug monitoring were discussed in detail in Chapter 11 including problems of collecting specimens in serum separator gel tube for certain drugs, as well as the effect of the presence of high bilirubin, high lipids, and excessive hemolysis on various aspects of therapeutic drug

Resolving Erroneous Reports in Toxicology and Therapeutic Drug Monitoring: A Comprehensive Guide, First Edition. Amitava Dasgupta.

monitoring, False-positive test results in drugs of abuse testing using urine specimens was discussed in detail in Chapter 3. In addition, the adulteration of urine specimen submitted for workplace drug testing may invalidate urine drug screen by immunoassays. This topic was addressed in Chapter 2 along with guidelines on how to identify such adulterated urine at the preanalytical stage. Throughout the book, various aspects of interferences and other causes of erroneous or unexpected test results in therapeutic drug monitoring and drugs of abuse testing were discussed in detail. For example, unexpected test results in therapeutic drug monitoring due to drug–herb interaction was addressed in Chapter 16. This chapter summarizes some practical tips for eliminating interferences when possible as well as practical methods of resolving discordant specimens.

18.2. IDENTIFYING DISCORDANT SPECIMEN/INTERFERENCE

Discordant results may cause clinical misdiagnosis and mistreatment, and it is very important for a laboratory professional to identify such specimens. For example, the presence of heterophilic antibody may cause a false-positive test result in a serum human chorionic gonadotropin (hCG) test, or an unexpected digoxin test may result in a patient overdosed with digoxin being treated with an antidigoxin antibody (Digibind). Discordant specimen can be identified when a laboratory test result does not fit with the clinical condition of a patient and a clinician calls the laboratory for a consultation or from other test results for the same patient that are a part of the test panel. Reanalysis of the same specimen using an alternative method can also identify discordance when the values do not match. A discordant result is defined when the difference between the two methods exhibits an absolute difference in value and/or a percentage difference (to cover discordances over low and high ranges, respectively). Usually the absolute difference in two values is determined by the imprecision of the two assays, and the percentage difference is based on the clinical accuracy needed at the cut-off concentration or reference range. In general, a difference of more than 20% in values is considered clinically significant and the specimen is identified as discordant. However, as stated earlier, an absolute difference is analyte dependent; for example, a 0.5 ng/mL difference in value in a digoxin measurement is considered a discordant value, but a difference of even 0.5 μg/mL in a phenytoin value may not be considered as discordant, although more than a 20% difference in respective values for both digoxin and phenytoin should be considered significant.

18.3. INVESTIGATING A DISCORDANT SPECIMEN

To investigate further a potential discordant specimen or a suspected case of the presence of an interfering substance in the specimen, several approaches can be taken. One of the first things to check is the quality control data for that day for the specific analyte to ensure that it is within the acceptable limits for the day where the specimen in question was analyzed as well as 1 or 2 days before the date of analysis. If all quality control data are within acceptable limits, a random error can be suspected during the analysis of that particular specimen such as short sampling by the probe of the analyzer or the presence of a micro-clot. Although most automated analyzers are equipped with a clot detection mechanism, this type of error can still occur albeit very infrequently. Also it is important to examine the raw data generated by the instrument to ensure that the result was not flagged for hemolysis or high bilirubin or any other cause. If such causes can be ruled out, repeating the assay with the same specimen and same analyzer can rule out a random error in the result. If the repeat result is similar to the original result, interference in the assay can be suspected. The best way to resolve a false-positive or a false-negative result in a test due to interference is to retest the specimen using a different assay. However, a different assay method may not available in the laboratory, and in that case specimen can be sent to a reference laboratory for retesting using a more specific method. Talking to the clinician and investigating a patient's record is very useful in resolving discordant specimens. Good coordination between the clinical laboratory and medical staff of the hospital is essential to avoid errors in laboratory results, a major patient safety issue. Plebani commented that the pre- and postanalytical phase of clinical laboratory testing are more prone to errors than the analytical phase. However, most errors are identified outside the laboratory. In the interest of patient safety, any direct or negative consequence related to a laboratory test must be addressed regardless of whether the error was due to a laboratory professional (e.g., calibration/testing error) or nonlaboratory personnel (e.g., inappropriate test request, error in patient identification, or blood collection process). In addition, grading laboratory errors on the basis of their seriousness may help to identify priorities for quality improvement and encourage a focus on corrective or preventive action (1). Physicians need to communicate openly following medical errors. In one report, the authors invited 260 practicing pathologists and 81 laboratory medical directors to take a survey to better understand their attitudes regarding laboratory errors. Most of the respondents (95%) stated that they were involved in resolving laboratory errors, and there was a near unanimous belief that errors should be disclosed to the hospital, colleagues, and patients; however, pathologists expressed their discomfort with their communication skills regarding error disclosure. The authors commented that improving error reporting systems and developing robust disclosure training could help prevent future errors, thus improving patient safety and trust (2). Golemboski commented that other industries and certain health care specialities have used a variety of methods to improve the quality and safety of their services. Such approaches such as industrywide standardized collection and reporting errors, standardization of practice through checklists, application of electronic health records, and simulator-based interdisciplinary training have improved outcomes in aviation, anesthesiology, and surgery. Although laboratory professionals are focused on eliminating errors in the analytical phase of operation only, performance of the pre- and postanalytical phases in the clinical laboratory may also be improved through application of these approaches that are already implemented successfully in other industries for minimizing errors (3).

18.4. CLINICAL INVESTIGATION TO RESOLVE DISCORDANT SPECIMEN/INTERFERENCE

Consulting with the attending physician and a critical reading of the patient record is very helpful in identifying errors in the clinical laboratory. When the clinical picture of a patient does not correlate with a laboratory test result or a particular test result in a panel of test does not match with other results, it may indicate a laboratory error. Lack of correlation between the laboratory and the clinical condition of patient may occur from many aspects, summarized in Table 18.1.

18.4.1. Overdosed Patient But Negative Toxicology Report

Frequently a clinician calls the laboratory and questions a negative toxicology result where clinical symptoms clearly indicate a drug overdose. This may be due to the fact that the particular drug was not detected by the toxicology panel. For example, if an opiate immunoassay is used and an oxycodone immunoassay is not a part of the toxicology panel, an oxycodone overdose could be missed because oxycodone exhibits poor cross-reactivity with most opiate immunoassays (see Chapter 2). Moreover, if a patient has abused a designer drug, the toxicology screen would be negative (see details in Chapter 5). A particular caution must be exercised for γ-hydroxybutyric acid (GHB) because this drug is not tested routinely in the toxicology laboratory but is used in date rape because it is tasteless and odorless and can be easily mixed with a drink. If a woman appeared overdosed with drug and the toxicology test is negative but physical examination indicated sexual activity but no sign of struggle, GHB analysis in serum or urine must be conducted. Many reference laboratories offer this test, and the specimen can be shipped under chain of custody so that positive results can be used in a court of law for prosecuting the assaulter. Another date rape drug, Rohypnol (flunitrazepam), is also hard to detect using the benzodiazepine immunoassay due to its low concentration in urine. Du Mont et al studied drug-facilitated sexual assaults using urine drug testing as well as testing of oral, vaginal, or rectal swabs for the unexpected presence of male DNA. Positive swabs for unexpected DNA were determined by whether the person reported having consensual sex in the previous week of assault. The authors studied 184 individuals. Urine specimens were positive for drugs in 44.9% cases, alcohol only in 1.9% cases, and both drug and alcohol in 18.0%

TABLE 18.1. Examples of Probable Causes for a Lack of Correlation Between Laboratory Test Result and Clinical Condition

Clinical Condition of a Patient	Laboratory Test	Comment
Overdosed patient	Negative toxicology	It is possible that the drug is not detected by the toxicology panel (e.g., designer drug, GHB)
Patient denies drug abuse	Positive toxicology	In winter months many false-positive test results for amphetamine and phencyclidine occur due to use of over-the-counter cold and cough medication
Patient denies alcohol use	Positive alcohol/ Alcohol biomarker	False-positive alcohol may be due to high lactate and LDH if enzymatic method used or carryover in alcohol done by GC
Patient denies using a medication	Positive result	It may be due to interference (e.g., false-positive TCA due to carbamazepine or contamination of herb with a Western drug)
Patient is compliant with drug	Subtherapeutic drug level	Probably due to drug–drug or drug–herb interaction (e.g., St. John's wort can reduce drug concentration by increasing clearance). It may also be due to negative interference
No change in drug dosage	Higher/lower than expected level of the drug	Probably due to drug–drug or drug–herb interaction (e.g., quinidine increases digoxin level). It may also be due to positive/negative interference
Patient is experiencing drug toxicity	Total drug level is within therapeutic range	Probably due to increased free drug level. Measure free drug, especially free phenytoin
A medication is discontinued	Positive result	May be due to interference
No symptom of drug toxicity	High drug level	May be specimen collected from a central line or due to interfering substance
Drug level does not match clinical condition	High or low drug level	May be due to interference or in rare cases interference from paraproteins or heterophilic antibodies

GHB, γ-hydroxybutyric acid; LDH, lactate dehydrogenase; TCA, tricyclic antidepressant.

cases. The drugs found on toxicology testing were unexpected in 64.4% of the cases, and positive drugs found included cannabinoids (40.2%), cocaine (32.2%), amphetamines (13.8%), 3,4-methylenedioxymethamophetamine (9.2%), ketamine (2.3%), and GHB (1.1%). Male DNA was unexpected in 30 of 64 cases where it was found. The authors concluded that the presence of unexpected drugs and male DNA was common with a suspicious drug-facilitated sexual assault, and such cases most typically involved date rape drugs such as ketamine and GHB (4).

Case Study

A 24-year-old woman had been studying abroad about 1 year and often complained about a general unhealthy feeling in the morning. At the end of the semester she returned to Italy and received home video clips shot by a mobile phone camera where she was having sex with a young man she met when she was studying abroad. Toxicological analysis of her hair (20 cm long) was performed using gas chromatography/mass spectrometry (GC/MS) that included a 2- to 3-cm segmentation of the entire length of hair specimen collected. Morphine and GHB were detected in the hair segment related to the period of time she was abroad. A higher amount of GHB was also found in the period associated with the possible drug-facilitated sexual assault of the young woman (5).

18.4.2. Patient Denied Drug Abuse But Positive Toxicology Result

Sometimes a patient denies drug abuse but the toxicology screen is positive for one or more drugs. It is common to observe high numbers of false-positive amphetamine/methamphetamine drug screens, especially in the winter months due to the presence of pseudoephedrine and related drugs in many over-the-counter cold medications. A positive phencyclidine screen can also be observed due to the presence of dextromethorphan in urine that is common after the use of over-the-counter cough medications. However, such false-positive test results can be easily circumvented by doing a GC/MS confirmation (see Chapter 3 for more details).

18.4.3. Patient Denied Alcohol Use But Positive Alcohol/ Biomarker Test

Sometimes an alcohol test is positive where the patient did not drink alcohol. This may happen in a critically ill patient with lactic acidosis or a trauma patient with high lactate and lactate dehydrogenase (LDH) that may interfere with an enzymatic method for alcohol determination. False-positive breath alcohol test result may occur due to different reasons (see Chapter 7). Even a false-positive alcohol or ethylene glycol result may be observed in alcohol determination using gas chromatography due to a carryover problem where the patient specimen was analyzed after analyzing a specimen containing a high amount of alcohol. Running a solvent blank can easily circumvent this problem. It is recommended to run a solvent blank after analyzing a specimen containing a high amount of alcohol or other volatiles such as ethylene glycol. Interference due to carryover may also occur in automated analyzers, but such an incidence in modern automated analyzers is minimal due to the adaptation of various approaches including using disposable parts that come in contact with the specimen, probe washing, and using disposable reaction cells because carryover can be eliminated by using a disposable probe or by washing the probe between analyses. For example, automated analyzer Vista 1500 (Siemens Diagnostics) uses automated probe washing between specimens and disposable reaction cells to eliminate the carryover problem. The Elecsys analyzer (Roche Diagnostics) uses a disposable probe to eliminate carryover. Armbuster et al reported that carryover between specimens is less than 0.1 ppm in Architect ci8200 analyzer (6).

In addition to sample carryover that most commonly occurs in tests with a wide reportable range, for example, β-hCG, creatinine kinase, or cocaine metabolite benzoylecgonine and can be easily detected by observing a high analyte value in a specimen after analyzing another high analyte value specimen, reagent carryover may affect a batch of tests and be more difficult to identify. In one report, the author observed multiple patient specimens with cholesterol values less than 100 mg/dL using Cobas c501 and Cobas 6000 analyzers (Roche Diagnostics). Such a problem can also arise in toxicology testing. The instrument was functioning properly and the room temperature in the laboratory was within acceptable limits. The possibility of fibrin clots or micro-clots was also ruled out. The author suspected reagent carryover and determined that apparently a component of the creatinine kinase (CK) reagent was interfering with the cholesterol test procedure. The CK reagent contained N-acetylcysteine that was carried over to the cholesterol test, and this component destroyed the hydrogen peroxide generated by oxidation of cholesterol using cholesterol oxidase in the cholesterol assay. Because hydrogen peroxide then reacts with 4-aminophenazone and phenol to generate a red dye that is measured at 505 nm, destruction of hydrogen peroxide by N-acetylcysteine caused the false-negative result in the cholesterol assay. However, the authors eliminated reagent carryover by implementing an additional probe wash (7).

Case Study

A 85-year-old woman presented to the emergency department with a history of unconscious collapse, and she tested 3 of 15 on the Glasgow coma scale. A blood alcohol test was performed using an alcohol enzymatic assay. The value was 200 mg/dL and the alcohol value was 400 mg/dL in a different specimen after 1 hour when she was transferred to the intensive care unit. However, 8 hours later, a third blood specimen showed alcohol level of 100 mg/dL. The family insisted that she consumed no alcohol. Fortunately, the patient improved and was eventually discharged from the hospital on day 18. The family was deeply offended by the assumption that alcohol was the cause of her collapse and requested that all such references be struck from the record. On day 11, three blood specimens showing positive alcohol test results were retrieved from the laboratory refrigerator and were sent to a different laboratory for reanalysis using GC, and no alcohol was found in any specimen. The LDH of the patient was mildly elevated (294 U/L), and the authors commented that it was unlikely that the mildly elevated LDH caused the interference. The lactate level was not determined. The cause of interference in the alcohol assay was not clear (8). Therefore, it is important to use an alternative method for alcohol analysis when the laboratory result does not agree with the clinical picture.

False-positive alcohol biomarkers test results have been reported. Wolff et al reported positive carbohydrate-deficient transferrin (an alcohol biomarker) results in a chronic hemodialysis patient using a particle-enhanced immunonephelometry assay that was assay specific. However, reanalysis of the specimen using capillary electrophoresis as well as chromatography resolved such discordance (9). During pregnancy the relative disialo, pentasialo, and hexasilo-transferrin levels increase gradually, whereas concentrations of trisialo and tetrasialo-transferrin are reduced. Such an effect is most pronounced in the third trimester and may cause false-positive interpretation in the % disialo-transferrin of %Carbohydrate deficient transferrin test result because heavy drinking also increases these carbohydrate-deficient transferrins (10). Arndt reported false-positive ethyl glucuronide (a marker for alcohol abuse) concentration in urine using the DRI ethyl glucuronide immunoassay (Thermo Fischer Scientific) on a Hitachi 912 analyzer in patients receiving chloral hydrate, most likely due to the cross-reactivity of trichloroethyl glucuronide, an important chloral hydrate metabolite (11).

Determination of ethylene glycol in serum spectrophotometrically may be inaccurate in the presence of elevated serum lactate or ketoacids, for example, β-hydroxybutyric acid that is structurally related to ethylene glycol. In one report, the authors observed a false-positive ethylene glycol level of 33.5 mg/dL using spectrometry, but no ethylene glycol was detected by the GC method. The patient had ketoacidosis (pH: 7.09; urine ketone: 8 mmol/L) (12).

18.4.4. Patient Denied Using a Medication

When a patient denies using a medication but that particular drug is detected, either interference or adulteration of an herbal supplement with that particular drug must be suspected. False-positive tricyclic antidepressants due to the presence of carbamazepine and other drugs has been reported (see Chapter 14 for more details). In Chapter 13 interferences in immunoassays used for the therapeutic drug monitoring of carbamazepine and other classical anticonvulsants was discussed. Herbal remedies are frequently adulterated with analgesics, hypoglycemic agents, psychotropic drugs, steroids and corticosteroids, anticonvulsants, and other Western drugs.. Huang et al collected samples from eight major hospitals in Taiwan. Of 2609 samples, 618 (23.7%) samples were adulterated with Western drugs. More than half of the adulterated samples contained two or more drugs. Of 618 adulterated samples, approximately a quarter of the samples was adulterated with acetaminophen (167 samples [27%]) and/or indomethacin (152 samples [25%]). In addition, a number of herbal remedies contained other analgesics including phenylbutazone, oxy-phenylbutazone, diclofenac, ibuprofen, ketoprofen, mefenamic acid, piroxicam, and salicylamide. Diazepam was found in 20 herbal supplements, and a significant number of samples were adulterated with prednisolone, betamethasone, dexamethasone, prednisone, cortisone, and hydrocortisone (13). Phenylbutazone and oxy-phenylbutazone are banned by the U.S. Food and Drug Administration. Therefore, if a patient denied any drug use but tested positive for the drug, analysis of the herbal supplement the patient is taking is useful to determine if a contaminated herbal supplement is the source of that drug. Usually herbal supplements prepared in the United States are free from contamination with Western drugs.

Case Study

A 33-year-old patient with an 8-year history of epilepsy was followed up regularly, and she was managed with valproate, carbamazepine, and phenobarbital but was not receiving phenytoin. One month before admission to the hospital, she was consuming proprietary Chinese medicines, but package inserts indicated that these medicines contained only pure Chinese medicines and are effective in controlling epilepsy. On the second day of her hospital stay, the serum phenytoin was 48.5 μg/mL despite the fact she was never prescribed phenytoin. The Chinese medicine capsules were submitted to the toxicology laboratory for analysis using chromatography. The Chinese medicine Hue Dian Shen Ying Jiao Nang Yi Hao contained 41 mg phenytoin in each capsule, explaining the presence of phenytoin in her serum. The other two Chinese medicines the patient took were contaminated with carbamazepine (14).

18.4.5. Patient Is Compliant with Drug But Subtherapeutic Drug Level Reported

When a patient is compliant and demonstrated a therapeutic range of the drug previously but certainly shows subtherapeutic level of a drug, a drug–drug interaction or drug–herb interaction could be suspected. Another explanation is negative interference in drug measurement due to the presence of an interfering substance. For a patient complaint with warfarin and suddenly showing an international normalized ratio (INR) outside the normal range while previous INR values were within the reference range, a warfarin–drug or warfarin–herb interaction is probably the cause. Review of the patient chart with the clinician can easily identify the drug–drug interaction as the cause of a subtherapeutic drug level, but identifying a drug–herb interaction is more difficult because most patients do not disclose their use of herbal supplements to their clinicians. A subtherapeutic drug level may be due to the use of St. John's wort, and such a low drug level may cause treatment failure. This important topic is discussed in Chapter 16. Asking a patient regarding use of any herbal supplement is essential to identify such clinically important drug–herb interactions. This is especially critical if the patient is taking any immunosuppressant, warfarin, or highly active antiretroviral therapy because the consequence of treatment failure is clinically very significant, and such interactions may also lead to fatality, for example, severe organ rejection may occur in a patient taking cyclosporine or tacrolimus and then self-medicated with St. John's wort.

18.4.6. No Change in Drug Dosage But Higher or Lower Than Expected Drug Level

Sometimes clinicians question the validity of a therapeutic drug monitoring test result because the drug level is substantially higher or lower than previous values but there was no change in drug dosage. After checking the quality control values and ensuring that all values are within expected limits, the specimen must be reanalyzed to ensure that no random error occurred during drug testing. If drug level upon reanalysis agrees with previous data, any random laboratory error can be ruled out. Drug–drug interaction can increase the level of a drug. For example, if quinidine is introduced in the treatment regime, the digoxin dose must be reduced. Otherwise the serum digoxin concentration would increase. See Chapter 12 for more discussion on this topic. In addition, interference may cause a higher than expected or lower than expected drug level using immunoassays for therapeutic drug monitoring. Issues of interferences in therapeutic drug monitoring was addressed in multiple chapters.

In general, reanalysis of the specimen using a chromatographic technique usually eliminates such interference. Alternatively, reanalysis of the specimen using a different commercially available assay for the same drug is also useful to identify such interferences. A common cause of a subtherapeutic drug level in a patient who showed a therapeutic drug level before is noncompliance. Another common cause of a high drug level is the analysis of the specimen using a wrong collection time. If a specimen is collected mistakenly after administration of a drug, it represents a peak sample not a trough sample, which is the specimen of choice for most therapeutic drugs (except aminoglycosides and vancomycin where both peak and trough levels may be required).

18.4.7. Patient Is Experiencing Drug Toxicity But Drug Level Is Within Therapeutic Range

This is a frequent observation with therapeutic drug monitoring of classical anticonvulsants where patients may experience phenytoin or valproic acid toxicity but total phenytoin or total valproic acid concentrations are within the therapeutic range. Phenytoin, carbamazepine, and valproic acid are strongly bound to serum proteins, especially albumin, and it is the only free level that is pharmacologically active. Although free level can be predicted from the total drug level, under certain pathological conditions the free drug level may increase disproportionately without altering the total drug concentration. Certain strongly protein-bound drugs can also displace phenytoin, carbamazepine, or valproic acid causing an elevated free drug level without altering the total drug level. Monitoring the free drug level is essential in these patients. In addition, the accumulation of carbamazepine 10,11-epoxide in a renally compromised patient taking carbamazepine may cause drug toxicity while total carbamazepine concentration determined using an immunoassay has low cross-reactivity with the epoxide metabolite. In this case, directly measuring the epoxide metabolite using a chromatography technique is recommended (see Chapter 13). Another possibility is negative interference in the drug level measurement. In this case reanalysis of the specimen using an alternative method can resolve the issue.

18.4.8. Medication Discontinued But Patient Showed Drug Level in Serum

Most likely cause of this observation is interference, probably from a heterophilic antibody in the serum digoxin measurement or whole blood cyclosporine and tacrolimus measurement using the Antibody Conjugate Magnetic Immunoassays (marketed by the Siemens Diagnostics; formerly Dade Behring for application on the Dimension analyzer). Interference of the heterophilic antibody in serum digoxin measurement was discussed in Chapter 12; interference of endogenous antibodies in certain assays for cyclosporine and tacrolimus was discussed in Chapter 15. Also see the section, "Drug Level Does Not Match Clinical Condition" in this chapter for more discussion of interference caused by the heterophilic antibody.

18.4.9. No Symptom of Drug Toxicity But High Drug Level Reported

The most plausible cause is collection of the specimen from a contaminated central line (see Chapter 11). In another scenario, concentration of a therapeutic drug is significantly above the therapeutic level, but the patient is experiencing no drug toxicity, probably due to a mislabelled specimen or interference including interference from other drug or herbal supplements. A number of herbal supplements may interfere with serum digoxin measurement using immunoassays and falsely increase the serum digoxin level, but the patient may not show any symptom of digoxin toxicity (see Chapter 12 for a more detailed discussion).

18.4.10. Drug Level Does Not Match Clinical Condition

This may occur from interferences because interfering substances most likely cause positive interference but they may also cause negative interference. As discussed earlier, due to the presence of the heterophilic antibody, a high amount of digoxin, cyclosporine, or tacrolimus can be detected even after discontinuation of the drug. Another source of interference is paraprotein, which may cause significant negative interference with certain immunoassays used for therapeutic drug monitoring. However, paraproteins and heterophilic antibodies are large molecules and are absent in protein-free ultrafiltrate or urine, thus not causing any interference in drug testing using oral fluids or urine specimens even using immunoassays.

Although interference of the heterophilic antibody in therapeutic drug monitoring is limited to digoxin, cyclosporine, and tacrolimus by certain assays, interference of the heterophilic antibody in other clinical laboratory tests, especially big molecules, has clinical significance. Therefore, a brief discussion on this topic is presented here. Heterophilic antibodies are endogenous human antibodies that may interact with assay antibodies (raised in various animals) causing false-positive or false-negative results. The heterophilic antibodies are polyclonal and heterogeneous, consisting of three types:

1. Heterophilic antibodies interacting poorly and nonspecifically with the assay antibodies.
2. Anti-animal antibodies that interact strongly and specifically with the antibody of a specific or a number of assays.
3. A therapeutic antibody that interferes with an assay.

The heterophilic antibody may arise in a patient in response to exposure to certain animals or animal products (handling an animal, therapy) although many of the immunoglobulin clones in normal human serum may display anti-animal antibody properties. Among the human anti-animal antibodies (HAAAs), the most common occurrence is of human anti-mouse antibody (HAMA) because of the wide use of murine monoclonal antibody products in therapy or imaging. The human anti-animal antibodies can belong to immunoglobulin (Ig)G, IgA, IgM, and rarely, the IgE class. In addition, rheumatoid factors (IgG) that may be present in 10% of patients in high amounts may interfere with immunoassays. Heterophilic antibody and anti-animal antibody interference are often grouped together as heterophilic antibody interference. Such interference has been mostly found with immunometric sandwich assays used for big molecules such as hCG but less often with competition assays used for therapeutic drug monitoring and drugs of abuse testing. Heterophilic antibodies due to large molecular weight are absent in urine the and measuring of urinary β-hCG eliminates such effects (15). In one study the author investigated the effect of heterophilic antibodies on immunoassay results by surveying results of 74 analytes (21 hormones, 18 tumor markers, 8 therapeutic drugs, 5 cardiac markers, 4 proteins, 2 vitamins, and 16 miscellaneous other tests) in 10 donors (having rheumatoid factors in their sera) from 66 laboratories in 7 countries. The author observed that overall approximately 8.7% of the 3445 results were considered false positive, and 21% of all erroneous results (1.8% of all results) were potentially misleading. However, some of these false-positive results could be resolved by treating sera with heterophilic blocking agents, but approximately half of those misleading results could not be resolved by using blocking agents. The authors further reported that although assays such as plasma β-hCG, α-fetoprotein, myoglobin, CA 19–9, estradiol, and erythropoietin were affected, no therapeutic drug monitoring assay was affected due to the presence of the heterophilic antibody in the specimen (16).

Plasma specimens that have been refrigerated for prolonged periods or have undergone freeze-thaw cycles demonstrate protein interference from fibrins that may precipitate under such conditions. These fibrin clots may block sample probe of an autoanalyzer, generating incorrect results. Such samples should be centrifuged to remove any precipitates prior to analysis. However, most modern autoanalyzers include clot detection and alert systems to flag results suspected of interference from clots. Paraproteins circulate as the result of multiple myeloma or similar diseases. Paraproteins are known to interfere with many types of clinical chemistry assays including immunoassays. For example, in one report the authors observed significant interference of paraprotein in total bilirubin, direct bilirubin, and high-density lipoprotein cholesterol using the Roche Modular and Olympus AU2700 analyzer (17).

Case Study

A 77-year-old man admitted to the hospital after cerebrovascular accident was noted to have IgMκ monoclonal component of 5 g/L and was diagnosed with monoclonal gammopathy. On a later occasion he was admitted to the accident and emergency department after he ingested 100 tablets of acetaminophen 18 hours prior to reporting to hospital. The antidote of acetaminophen N-acetylcysteine was administered immediately before getting any laboratory test results. Over succeeding days, the alanine aminotransferase of the patient peaked to 2515 U/L; his monoclonal component was 7 g/L, and his serum acetaminophen concentration determined using a commercial enzyme assay kit for application on the Cobas Fara II analyzer was 5.3 mg/dL, but the blank reading was very high. Suspecting interference, acetaminophen concentration was measured using high performance liquid chromatography (HPLC) and the observed value was 8.6 mg/dL. Diluting the specimen and then reanalyzing by the same enzymatic assay showed a nonlinear response, thus confirming the suspicion of interference. The presence of paraprotein in the sample was also indicated by the formation of flocculant precipitate when a drop of serum was added to water (the Sia water test) (18).

Case Study

A 73-year-old female patient was admitted to the hospital and treated with intravenous phenytoin on day 3, but the phenytoin level was undetectable using the particle-enhanced turbidimetric inhibition immunoassay (PETINIA) phenytoin assay analyzed using the Dimension analyzer. The patient was also receiving valproic acid on day 4 of her hospital stay, and on day 5 her phenytoin concentration was less than 0.4 μg/mL using the PETINIA assay, but the valproic acid concentration of 32.6 μg/mL was measured by the PETINIA valproic acid assay using the same serum specimen. Despite additional intravenous injections of 175 mg and later 250 mg of phenytoin on day 5, her serum phenytoin was still undetectable on day 6 using the PETINIA assay. On day 8 the serum phenytoin determined by the PETINIA assay was hardly detectable (1 μg/mL), but the same specimen showed a serum concentration of 17.2 μg/mL of phenytoin, using HPLC. The serum samples from day 5 and 6 were subsequently evaluated, and observed values were 11.0 and 13.5 μg/mL, respectively. Analysis of these specimens using a different immunoassay (fluorescence polarization immunoassay) showed phenytoin concentrations of 15.6 μg/mL (day 8), 10.1 μg/mL (day 5), and 12.6 μg/mL (day 6). Serum protein electrophoresis of this patient revealed a monoclonal immunoglobulin (IGMλ). The authors concluded that immunoglobulins may cause negative interference in serum phenytoin measurement using the PETINIA assay because after protein precipitation of the specimen the phenytoin concentration as determined by the PETINIA assay was 18.9 μg/mL on day 11 and the concentration of phenytoin as determined by the HPLC method was 19.8 μg/mL. Unfortunately, the patient died on hospital day 16 from a bilateral cerebrovascular insult (19).

18.5. RESOLVING ANALYTICAL INTERFERENCES

Resolving a discordant specimen due to interferences is an important function of laboratory professionals in order to run an error-free laboratory. There are four general approaches to resolve analytical interferences:

1. Serial diluting a specimen. In the absence of interference, the response should be linear, whereas the response should be nonlinear if an interfering substance is present in the specimen.
2. Monitoring the drug or analyte in protein-free ultrafiltrate.
3. Using an alternative method to reanalyze the specimen.
4. Treating the specimen with heterophilic antibody blocking agent.

Serial dilution is the most straightforward approach to confirm the presence of an interfering substance in the specimen. If no interfering substance is

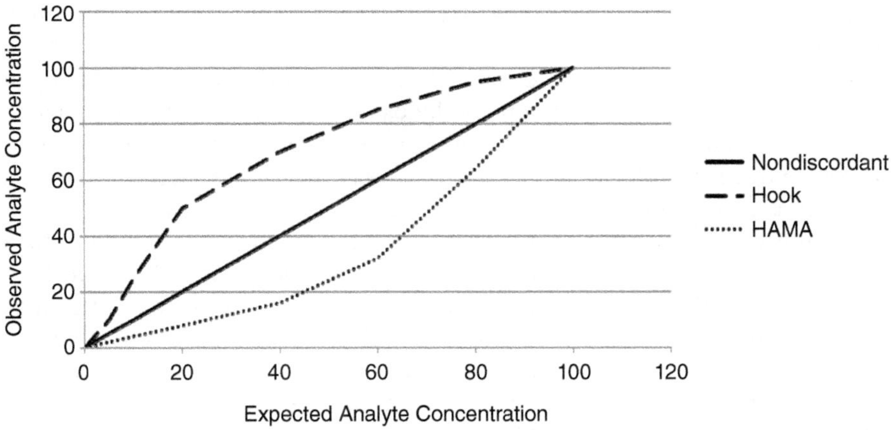

Figure 18.1. How serial dilution can be used to identify the hook effect of the human anti-mouse antibody (HAMA) effect.

present, the response to dilution should be a linear response (Fig. 18.1). However, if any interfering substance is present, the response should be non-linear. The expected pattern of dilution in the presence of the heterophilic antibody in the specimen is also shown in Figure 18.1; however, other interfering substances if present may cause a similar response. The hook effect is uncommon in immunoassays for therapeutic drug monitoring and drugs of abuse testing using immunoassays. It is caused by excess antigen and usually cause false-negative test results. However, the hook effect can occur in assays for tumor markers such as prostate-specific antigen, CA 125, as well as in assays for C-reactive protein, ferritin, and prolactin. The false-negative result in a urine β-hCG assay due to the hook effect has been well cited in the medical literature. Pang et al reported a case in which a false-negative urine β-hCG test resulted in exclusion of pregnancy as a diagnosis, but the patient had high concentrations of β-hCG in her urine specimen due to a molar pregnancy (20). In another report, the urine pregnancy test (as β-hCG) was negative despite the clinical symptoms of the patient that suggested a molar pregnancy that included abdominal pain and vaginal bleeding. The authors determined the cause of false-negative β-hCG was the hook effect because high amounts of β-hCG were present in her urine (21). Again, serial dilution is useful in identifying the hook effect as shown in Figure 18.1.

Many interfering substances are absent in the protein-free ultrafiltrate, and monitoring the free drug concentration can eliminate some interferences. For example, if a patient is overdosed with digoxin and being treated with Digibind, only free digoxin should be monitored. In addition, certain interferences in digoxin immunoassays can be eliminated by monitoring free digoxin (see Chapter 12). Moreover, interferences from heterophilic antibody and paraprotein can also be eliminated by monitoring drug concentration in the ultrafiltrate as mentioned earlier in this chapter.

Another approach to overcome interference is reanalyzing the specimen using another method, preferably a chromatographic method. For example, interferences of metabolites of various immunosuppressants in respective immunoassays can be easily eliminated by using chromatographic methods. In general, chromatographic methods for therapeutic drug monitoring are relatively free from the interferences of drug metabolites, other interfering drugs, and various endogenous factors such as high bilirubin or hemolysis. Moreover, during the extraction process, proteins are removed from the analyte; thus interference from paraproteins and heterophilic antibodies are absent in chromatographic methods. However, reanalysis of a specimen by another immunoassay may also eliminate interference. Heterophilic antibody blocking agents are commercially available, and treatment of a specimen with a heterophilic antibody blocking agent is an alternative to the use of protein-free ultrafiltrate in eliminating the interference of the heterophilic antibody in an immunoassay. Most of the heterophilic antibody blocker agents are mouse immunoglobulin, polyclonal, polymerized, or monoclonal of a particular subclass. The heterophilic antibody can also be removed by treating the specimen with Protein G, mouse IgG, or polyethylene glycol.

18.6. CONCLUSIONS

Recognizing discordant specimens and eliminating interference is a major challenge in the clinical laboratory. In this chapter various approaches to eliminate interference in therapeutic drug monitoring and drugs of abuse/alcohol testing were addressed. If a toxicology report does not match the clinical picture of the patient, it is important to communicate with the attending physician, nursing staff, pharmacologist, or other health care professionals to investigate this serious patient safety issue.

REFERENCES

1. Plebani M. The detection and prevention of errors in laboratory medicine. Ann Clin Biochem 2010; 47: 101–110.
2. Dintzis SM, Stetsenko GY, Sitlani CM, Gronowski AM et al. Communicating pathology and laboratory errors: anatomic and laboratory medical director's attitude and experience. Am J Clin Pathol 2011; 135: 760–765.
3. Golemboski K. Improving patient safety: lessons from other disciplines. Clin Lab Sci 2011; 24: 114–119.
4. Du Mont J, Macdonald S, Rotbard N, Bainbridge D et al. Drug facilitated sexual assault in Ontario, Canada: toxicological and DNA findings. J Forensic Leg Med 2010; 17: 333–338.
5. Rossi R, Lancia M, Gambelunghe C, Oliva A et a;. Identification of GHB and morphine in hair in a case of drug-facilitated sexual assault. Forensic Sci Int 2009; 186: e9–e11.

6. Armbruster DA, Alexander DB. Sample to sample carryover: a source of analytical laboratory error and its relevance to integrated clinical chemistry/immunoassay system. Clin Chim Acta 2006; 373: 37–43.
7. Kyle PB. Beware of carryover in modern chemistry analyzers. Clin Chem Lab Med 2010; 48: 519–521.
8. Jones TE. False positive ethanol concentrations leading to clinical confusion on Christmas Day. Clin Biochem 2011; 44: 1355–1357.
9. Wolff F, Mesquite M, Corazza F, Demulder A et al. False positive carbohydrate-deficient transferrin results in chronic hemodialysis patients related to the analytical methodology. Clin Biochem 2010; 43: 1148–1151.
10. Kenan N, Larsson A, Axelsson O, Helander A. Changes in transferrin glycosylation during pregnancy may lead to false-positive carbohydrate deficient transferrin (CDT) results testing for riskful alcohol consumption. Clin Chim Acta 2011; 412: 129–133.
11. Arndt T, Gierten B, Gussregen B, Werle A et al. False positive ethyl glucuronide immunoassay screening associated with chloral hydrate medication as confirmed by LC/MS/MS and self medication. Forensic Sci Int 2009; 184: e27–e29.
12. Boissier F, Weiss N, Faisy C. False positive ethylene glycol determination by spectrophotometry in the presence of severe lactic acidosis and ketosis [letter to the editor]. Ann Emerg Med 2010; 56: 75–76.
13. Huang WF, Wen KC, Hsiao ML. Adulteration by synthetic therapeutic substances of traditional Chinese medicines in Taiwan. J Clin Pharmacol 1997; 37: 344–350.
14. Lau KK, Lai CK, Chan AW. Phenytoin poisoning after using Chinese proprietary medicines. Hum Exp Toxicol 2000; 19: 385–386.
15. ACOG Committee on Gynecologic Practice. Avoiding inappropriate clinical decisions based on false positive human chorionic gonadotropin test results. Int J Gynaecol Obstet 2003; 80: 231–233.
16. Marks V. False-positive immunoassay results: a multicenter survey of erroneous immunoassay results from assays of 74 analytes in 10 donors from 66 laboratories in seven countries. Clin Chem 2002; 48: 2008–2016.
17. Yang Y, Howanitz PJ, Howanitz JH, Gorfajin H et al. Paraproteins are a common cause of interferences with automated chemistry methods. Arch Pathol Lab Med 2008; 132: 217–223.
18. Hullin DA. An IgM paraprotein causing a falsely low result in an enzymatic assay for acetaminophen. Clin Chem 1999; 45: 155–156.
19. Brauchili YB, Scholer A, Schwietert M, Krahenbuhl S. Undetectable phenytoin serum levels by an automated particle-enhanced turbidimetric inhibition immunoassay in a patient with monoclonal IGMλ. Clin Chimica Acta 2008; 389: 174–176.
20. Pang YP, Rajesh H, Tan LK. Molar pregnancy with false negative urine hCG: the hook effect. Singapore Med J 2010; 51: e58–e61.
21. Hunter CL, Ladde J. Molar pregnancy with false negative β-hCG urine in the emergency department. West J Emerg Med 2011; 12: 213–215.

Index

Page numbers in *italics* represent figures; those in **bold**, tables.

Resolving Erroneous Reports in Toxicology and Therapeutic Drug Monitoring: A Comprehensive Guide, First Edition. Amitava Dasgupta.

CPSIA information can be obtained
at www.ICGtesting.com
Printed in the USA
BVHW041021201219
567226BV00013B/74/P

9 781118 149652